Bio-Nano Interface

Manoranjan Arakha •
Arun Kumar Pradhan • Suman Jha
Editors

Bio-Nano Interface

Applications in Food, Healthcare and Sustainability

Springer

Editors
Manoranjan Arakha
Centre for Biotechnology
Siksha 'O' Anusandhan
(Deemed to be University)
Bhubaneswar, Odisha, India

Arun Kumar Pradhan
Centre for Biotechnology
Siksha 'O' Anusandhan
(Deemed to be University)
Bhubaneswar, Odisha, India

Suman Jha
Department of Life Science
National Institute of Technology Rourkela
Rourkela, Odisha, India

ISBN 978-981-16-2515-2 ISBN 978-981-16-2516-9 (eBook)
https://doi.org/10.1007/978-981-16-2516-9

This Springer imprint is published by the registered company Springer Nature Singapore Pte Ltd.
The registered company address is: 152 Beach Road, #21-01/04 Gateway East, Singapore 189721, Singapore

Preface

Nanoparticles (NPs), for their advanced physicochemical properties, have revolutionized different fields of biological sciences, like biomedical and pharmaceutical sciences. However, inside the biological milieu, NPs interact with different biological moieties to lower their surface free energy content and adopt stable shape, size, and surface functionalities. The interaction gives rise to nano-biomolecular complexes in the biological milieu, with dynamic biomolecular layer (corona) as per interaction affinity and concentration of the biomolecules present in the milieu. The interaction pattern at the interface to form complexes predominantly determines the overall fate of interacting biomolecules and NPs inside the biological system. Additionally, considering the multifaceted role of biomolecule and NP elements, use of nanoparticle-mediated approaches in biological system is like double-edged sword. Hence, understanding the interaction pattern at nano-bio interface is a key challenge to reduce the negative impact of NPs' use in biological sciences.

In view of the above discussion, the overarching aim of this book is to discuss about the interactions of NPs with different biomolecules in different biological environments and the consequences of this interaction on different aspects, like floura and fauna of the niche, cell proliferations, etc. Hence, the book, initially, discusses about some novel eco-friendly methods for NPs synthesis, followed by discussions on the interaction of NPs with biomolecules. The later chapters of the book discuss about the potential impacts of NPs in different fields of biological sciences like agriculture, food, healthcare, environment, etc.

The book, in total, comprises 19 chapters, written by experts working in the respective aspects of nanotechnology. For instance, Chaps. 1 and 2 describe synthetic strategies for development of different types of nanoparticles. Additionally, Chap. 1 also illustrates the importance of metal nanoparticles that have tremendously improved as functional nanoprobes to detect adulterants and contaminants for food safety management. Chapter 3 discusses the application of nanoparticle as a novel antibiotic with their antibacterial mechanisms, which could be potential solutions against different infectious diseases caused due to various microorganisms. Chapters 4–6 deal with the applications of nanotechnology in food science. Chapter 5 discusses in detail the applications of nanotechnology towards current use in food industries as food formulations, processing, and storage, whereas Chap. 6 focuses on nanoparticle-mediated approaches in the packaging of crops to increase its shelf life.

It also discusses different nanomaterials to safeguard the food from microorganisms and other pathogens, especially during storage time. Chapter 7 intends to discuss the longer use of nanoparticles to control plant diseases, either as nanoparticle alone or acting as protestants or acting as nanocarriers for insecticides, pesticides, and fungicides. Chapter 8 deals with various nanosystems and their applications in cancer therapy. Chapters 9–13 focus on nanotechnology impact in healthcare, where Chap. 9 describes phytoplankton-mediated nanoparticles formulation for cancer therapy and prediction of an optimal diagnostic therapy for cancer patients, which remained a major challenge for the last few decades. Currently, ovary cancer has become an excruciating factor with major cause of morbidity. Hence, Chap. 10 discerns the potential implications of nanotechnology and its application in several dimensions of ovary cancers. Chapter 11 evaluates different nanoparticle-based theranostic applications in protein amyloidogenesis and cancer, both are incurable human diseases. Chapter 12 deliberates the role of nanotechnology in premature detection of AD and explores targeting outgrowth in diagnosis and treatment. Chapter 13 concentrates on the formulation of different nanocurcumin particles and their implications in tuberculosis treatment. Chapters 14–17 discuss the impact of nanotechnology on the flora and fauna, where Chap. 14 focuses on the application of nanobiosensor in health care sector. Chapter 15 deals with bioactive nanoparticles as a next-generation smart nanomaterial for pollution abatement and ecological sustainability. Chapter 16 deals with smart nanomaterials for bio-imaging applications. Chapter 17 deals with effects of nanomaterials on the biology of naturally available scavengers like earthworm. Chapter 18 discusses the production of bioethanol from agricultural wastes with the help of nanotechnology. The closing chapter, Chap. 19, discusses current progress in the use of nanoparticles in biofuel production. Thus, this edited book is a contribution of experts from nanobiotechnology, microbiology, cancer biology, pharmaceutical science to help newcomers and the professionals in the field to potential applications of nanotechnology. We are very much thankful to all the contributing authors for timely cooperation and support, without which this book would not have taken its final shape. We are also thankful to the Publishing editors Dr. Madhurima Kahali, Dr. Akash Chakraborty, and Production editor Mr. Lenold Esithor for their critical reviews and positive remarks to bring positive developments to the book.

Bhubaneswar, Odisha, India — Manoranjan Arakha
Bhubaneswar, Odisha, India — Arun Kumar Pradhan
Rourkela, Odisha, India — Suman Jha

Contents

About the Editors

Manoranjan Arakha is an Assistant Professor at the Centre for Biotechnology, Siksha 'O' Anusandhan (deemed to be university), Bhubaneswar, Odisha, India. He received his doctorate in life science from the National Institute of Technology, Rourkela, India, in 2017, and subsequently pursued postdoctoral research in bio-nano interface at University College Dublin, Ireland. His research interests include green synthesis of different metal nanoparticles and moderate their physico-chemical parameters for antimicrobial and cytotoxic properties. For antimicrobial, he is mainly interested in nanoformulation effective against biofilm forming and antimicrobial peptide-resistant bacteria.

Arun Kumar Pradhan is an Assistant Professor in Centre for Biotechnology, School of Pharmaceutical Sciences, Siksha 'O' Anusandhan (deemed to be university), Bhubaneswar, Odisha, India. He received his PhD degree in 2016 from the Department of Environment and Sustainability, CSIR-Institute of Minerals and Materials Technology, Bhubaneswar, India. His area of interest includes microbial biosurfactants and their various biotechnological applications like effect of biosurfactant on cancer cells, study of biofilm inhibitory capacity of biosurfactants, nanoparticle approach (liposome) of heavy metal detection, etc.

Currently, he is working on green synthesis of various nanoparticles and clinical applications of biosurfactants to cure different disorders.

Suman Jha is an Associate Professor in the Department of Life Science, National Institute of Technology, Rourkela, Odisha, India. He received his PhD from the Department of Physical Chemistry-I, Technical University of Dortmund, Germany in 2009. He completed his post-doctoral research in protein biophysics from the Department of Molecular and Cell Biology, University of Connecticut, USA (2010–2011). Currently, he is investigating different nanoformulations for biological applications like in TIIDM, Parkinson's, and cancer. His research interests involve studying nanoparticle–biomolecular interfacial interactions, protein–metal ions, and protein–protein interactions.

Impact of Isotropic and Anisotropic Plasmonic Metal Nanoparticles on Healthcare and Food Safety Management

1

Nabojit Das and Rayavarapu Raja Gopal

Abstract

Nanoscale materials in the size range of 1–100 nm have advanced tremendously impacting a variety of consumer applications. Of the various nanomaterials, colloidal metal nanoparticles are widely used in biomedicine and biotechnology owing to their interesting optical and chemical properties. Metal (gold/silver/platinum) nanoparticle's size, shape, and surface chemistry govern the fate of nanoparticles for end-user applications. The precise control on the growth of metal nanoparticles governs their morphology (isotropic/anisotropic) that determines their plasmonic properties. The potential of metal nanoparticles has enabled futuristic technical capabilities that will enable them as chemical sensors, nano-carriers, and specific-targeting molecular probes. In this chapter, we have described detailed synthetic strategies for developing tunable isotropic and anisotropic metal nanoparticles. The absorption bands of metal nanoparticles are governed by their shape and size which determine their role in health care and food safety management technologies. The chapter illustrates the importance of metal nanoparticles that have tremendously improved as functional nanoprobes to detect adulterants and contaminants in food safety management. The chapter elaborately discusses the toxicity responses of metal nanoparticles where safety of nanomaterials is warranted for versatile health care applications. In summary, the safe-by-design approach of nanomaterials is paramount for enabling translational applications related to health care and food safety management.

N. Das · R. R. Gopal (✉)
Nanomaterial Toxicology Laboratory, Nanomaterial Toxicology Group, CSIR-Indian Institute of Toxicology Research (CSIR-IITR), Lucknow, India

Academy of Scientific and Innovative Research (AcSIR), Ghaziabad, India
e-mail: rajagopal@iitr.res.in

M. Arakha et al. (eds.), *Bio-Nano Interface*,
https://doi.org/10.1007/978-981-16-2516-9_1

Keywords

Metal nanoparticles · Food · Healthcare · Anisotropy

1.1 Introduction

Plasmonic metal nanoparticles have fascinated researchers across the world over few decades due to its unique physico-chemical properties and their direct implication in biological applications. The unique properties are attributed to higher surface area to volume ratio of the nanoparticles wherein the size range is less than 100 nm at least in one dimension. Plasmonic metal nanoparticles of gold, silver, platinum, and palladium exhibit surface plasmon resonance (SPR) due to oscillation of electrons over the nanoparticle surface upon incidence of electromagnetic spectrum (Fong and Yung 2013). Local surface plasmons get excited when irradiated with incident light to plasmonic metal nanostructures having size smaller than the wavelength of the incident light. Platinum and palladium nanoparticles exhibit absorption spectrum in ultraviolet (UV) region, whereas silver and gold nanoparticles exhibit absorption spectrum in visible as well as near infra-red (NIR) region. SPR is highly influenced by the shape and size of plasmonic nanoparticles determining absorption spectrum exhibited by the nanoparticles (Mustafa et al. 2010). According to Mie's theory, isotropic (spherical) plasmonic nanoparticles exhibit a single plasmon peak and the absorption spectrum is dependent on factors such as diameter of the nanoparticle, composition, and dielectric environment (Paramasivam et al. 2017). Whereas anisotropy refers to the shapes other than spherical such as rod-shaped, branched, triangle shaped, etc. Anisotropic plasmonic nanoparticles exhibit plasmon bands in both transverse and longitudinal directions which is attributed to difference in surface energies and strains on the crystal facets (Reguera et al. 2017). Hence, anisotropy results in multiple plasmon bands as a function of number of symmetric planes present in the nanoparticle. The resultant shape achieved by nanoparticles is determined by growth kinetics of nuclei during synthesis that can either be thermodynamically controlled or kinetically controlled. Thermodynamically controlled synthesis regime yields isotropic/spherical nanoparticles, whereas kinetically controlled regime leads to anisotropic growth of nanocrystals giving different shapes to the synthesized nanoparticles (Wang et al. 2015). During the controlled kinetic growth in an anisotropic nanoparticle, a subtle change in the reaction parameters can bring substantial difference in the surface free energy at different crystal facets leading to formation of anisotropic nanoparticles. Therefore, surface chemistry of nanoparticles which is highly influenced by the capping agents used during synthesis has a predominant role in determining shape and size of the nanoparticles. Anisotropic plasmonic nanoparticles are known for their enhanced optical absorption in near infra-red (NIR) region, enhanced photothermal capacity, greater extinction cross sections, and biocompatibility (Kim et al. 2019a; Kumar et al. 2020).

With the emergence of interdisciplinary sciences, the research on nanomaterials and biomedicine has ever grown from last few decades. Noble metal nanoparticles offer numerous advantages in biomedical research due to their unique property, that is, localized surface plasmon resonance (LSPR). Noble metal nanoparticles, particularly, are being utilized in bioimaging, nanovehicles, and as contrast agents that are desirable for targeted ligands/drugs delivery. In recent times, surface-enhanced Raman scattering (SERS) technique has gained interest due to its high sensitivity. It is a label-free and specific technique which is highly sensitive due to non-interference of water molecules with the Raman scattering signal wherein biomolecules are Raman-active. Noble metal nanoparticles although being a promising candidate in nanomedicine have limitations such as controlled nanoparticle growth, stability, and cytotoxicity which is highly governed by several parameters during synthesis such as temperature, stoichiometric volumes of the reagents, and capping/stabilizing agents. Among all the parameters, capping agents highly determine the surface chemistry of the synthesized nanoparticles which concurrently determine the safety of the nanoparticles when exposed to cells (Kumar et al. 2020). They also act as template, hence providing desired anisotropy to the nanoparticles in the presence of weak reducing agent.

Apart from biomedical applications, the unique properties of noble metal nanoparticles are also being utilized in agriculture and food safety. Food safety has been a continuous concern for the consumers, manufacturers, and producers in the society. Food contamination is caused by pathogenic bacteria and other microbes or sometimes due to use of excessive fertilizers and pesticides resulting in ill health of the human population. Sometimes for the greed of more profit, people are often indulged in unethical use of compounds and this is known as food adulteration. Food intoxication refers to the ingestion of toxins present in food due to bacterial or microbial growth. The toxins are the byproducts of the microbes which remain even after the elimination of the microbes and is the primary reason for causing food infection. Food borne infectious diseases commonly caused by pathogenic bacteria is the world's primary cause of premature death and overall third after cardiovascular diseases (Ali et al. 2014). Conventional methods for identification of bacteria or the toxic compounds take 3–4 days for results and 7–10 days for confirmation. Therefore, rapid, selective, and specific tests are a necessity for ensuring food safety to the blooming world population. Nanotechnology is one of the tools that has an immense potential to mitigate the uprising concerns regarding food safety due to its low cost, rapid, and specificity towards detection of food contaminants.

1.2 Synthetic Strategies for Metal Nanoparticles

Two broadly classified approaches are mainly adopted for nanoparticle synthesis: top down and bottom up. Top-down approach involves progressive conversion of the bulk dimension of material into nano-domain through physical or chemical means. The latter involves synthesis of nano-sized materials through reacting constituent atoms or molecules in a pertinent synthetic condition. Bottom-up approach is considered to be the most favorable method for nanoparticle synthesis since the synthetic environment provides control over nanoparticle growth giving it desired size and shape. Conversely, one has no control over size and shape of the nanoparticles yielded through top-down approach.

Plasmonic metal nanoparticles of both isotropic and anisotropic shapes are synthesized via methods such as electrochemical, photo-chemical, sonochemical, and wet-chemical methods (Herrera et al. 2013). Among all, wet chemical method is widely adopted because of the ability to generate monodisperse nanoparticles with high yield. Wet chemical synthesis includes different methods such as hot colloidal chemistry, seed-mediated synthesis, polyol synthesis, template assisted synthesis, green synthesis, etc. that has been adopted for synthesizing plasmonic metal nanoparticles. Hot colloidal approach allows manipulation of optical and electrical properties of the nanoparticles in a defined way with greater precision. This method generally employs three components: (1) precursor-corresponding metal salt, (2) capping agent preferably a surfactant containing a hydrophobic end and hydrophilic head containing a functional group such as oleic acid or oleylamine, and (3) a boiling solvent. A high temperature required for decomposition and solubilization of the reactants, as well as for nanoparticles crystallization is provided by the boiling solvent. Seed-mediated synthesis is a popular and powerful tool for synthesizing nanoparticles of anisotropic shapes. The vast popularity lies in the fact that it separates nucleation and growth process during synthesis allowing a great extent over nanoparticle growth which influences the properties of nanoparticles determining their performance when employed for an application (Liyu Chen et al. 2016). It is a two-step process where a metal salt is reduced by using a strong reducing agent resulting in formation of extremely small nanoparticles of size 2–3 nm that are known as seed. The seeds are then allowed to grow slowly in a growth solution containing a surfactant acting as a template/capping agent, ions as directing agent, and a weak reducing agent. The binding of capping agent would block specific crystal facets or reduce its surface energies significantly, thus resulting in confined growth in a particular direction. For example: halides, such as cetyltrimethylammonium bromide (CTAB), cetyltrimethylammonium chloride (CTAC), cetyltrimethylammonium iodide are considered as critical components determining the morphology of various nanoparticles. Beside halides, a variety of reagents (used during synthesis) such as the concentration of gold salt, ascorbic acid, and silver nitrate can be tailored to obtain nanospheres, pentagonal twinned nanorods, and trisoctahedral gold nanoparticles (Lohse et al. 2014). The amount of seed used into the growth solution also determines the desired shape and size of the nanoparticles. For instance, small amount of seed results into longer gold nanorods and vice versa (Burrows et al. 2017). Polyols are organic compound containing multiple –OH groups. Due to their polar nature and good solubility property, they are being used as a solvent for the precursor metal salts. Polyols at elevated temperature also act as reducing agent making them suitable for one-pot synthesis of nanoparticles. Most of the chemical methods discussed involve chemicals that are toxic when it comes to biomedical, food, and agriculture applications. Therefore, researchers are moving towards the use of plant material extracts to synthesize nanoparticles. The secondary metabolites present in plants act as both reducing and capping agents during synthesis. However, the qualities of nanoparticles synthesized are inferior in terms to monodispersity and it is not possible to control the growth of the nanoparticle, thus unable to provide desired size and shape. The only benefit of such nanoparticles is its non-toxic nature. Therefore, an amalgamation of green synthesis as well as other synthetic approaches is required for synthesizing non-toxic nanoparticles with controllable growth over size and morphology.

1.3 Physico-Chemical Properties of Nanoparticles and their Impact on Biological Milieu

The most widely used plasmonic nanoparticles for biological applications are gold and silver nanoparticles because of their ability to exhibit LSPR phenomenon due to unique plasmonic and electronic properties. However, the shape, size, surface chemistry, and functionalization of these nanoparticles have a great impact on biological effect in vitro and in vivo (Albanese et al. 2012; Kim et al. 2019b) (Fig. 1.1). Therefore, systematic understanding of shape-dependent effects of nanoparticles is essential for elucidating their biological implications. The cellular uptake of nanoparticles have attracted a greater deal of attention, one of the main reasons for being used as nanovehicles for drug delivery, bioimaging, theranostics, and nanotherapeutics (Kumar et al. 2016) as shown in Fig. 1.1. Smaller sized nanoparticles are found to be more cytotoxic as compared to nanoparticles of greater size. This is due to easy internalization of smaller nanoparticles into cells and other cellular compartments. The nanoparticle mediated cytotoxicity is dependent on several factors such as colloidal stability, particle uptake, cell surface, and other physiochemical factors (Nel et al. 2009). Shape can highly determine the ingestion of nanoparticles by the cells. High uptake rate and ratio of nanoparticles by cells can cause disruption of normal cellular processes leading to adverse cytotoxic effects.

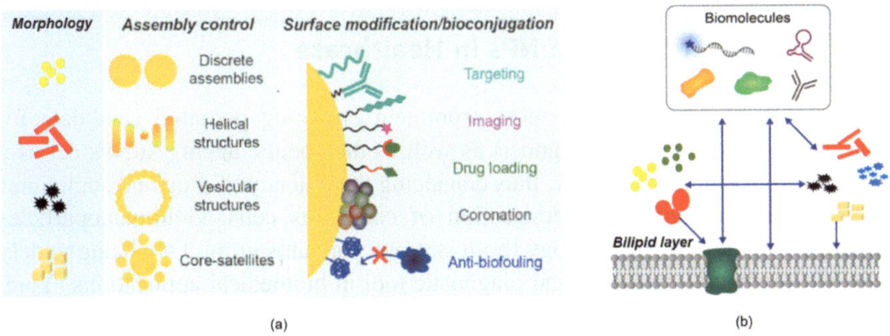

Fig. 1.1 (**a**) Effects of plasmonic nanoparticles on biological system: particle–cell interface, biocompatibility, and controllability. (**b**) Controlled morphology, surface modification of plasmonic nanoparticles determining various biological applications. Adapted with permission from reference Kumar et al. (2016) and Kim et al. (2019b), respectively. Copyright (2016 and 2019) American Chemical Society (ACS)

This induces excessive endocytosis ultimately leading to oxidative stress mediated mitochondrial damage (Manshian et al. 2017). In a study, cytotoxicity of silver nanoparticles with high aspect ratio was observed in A549 cells; silver nanowires exhibited shape-dependent toxicity when compared to nanoparticles of spherical shape (Stoehr et al. 2011). In case of anisotropic nanoparticles, the order from high to low cellular uptake efficiency is stars, rods, and triangles (Xie et al. 2017). It is also reported that the degree of accumulation and clearance of nanoparticles in target organs is highly influenced by size and surface charge (Schleh et al. 2012). Apart from exhibiting multiple plasmon bands by the anisotropic nanoparticles, they also show other superior characteristics over conventional isotropic (spherical) nanoparticles. The characteristics are high cellular internalization rate, high catalytic activity, longer blood-circulation time, and prominent tumor penetration. For instance, Huang et al. reported that anisotropic nanoparticles of larger aspect ratios are uptaken in greater amount and internalized rapidly by A375 human melanoma cells (Zhao et al. 2017). Uptake of nanoparticles in HeLa and mesenchymal stem cells (MSC) showed negative correlation between uptake rate and aspect ratio, i.e. non-spherical/anisotropic nanoparticles were found to possess less cellular uptake when compared to isotropic/spherical nanoparticles. This phenomenon observed is due to greater average radius of curvature in case of anisotropic nanoparticles. A similar study reported for the cellular uptake of anisotropic nanoparticles depends on shape and orientation of the nanoparticles, elucidating systematic understanding of the process of membrane wrapping (Dasgupta et al. 2014). Shape of the nanoparticles highly modulates their cellular uptake and undergoes various endocytosis pathways. A submarine mode of endocytosis is favored for nanorods with high aspect ratios and round tips, whereas rocket mode of endocytosis is preferred for nanorods with small aspect ratios and flat tips (Yang et al. 2019). Therefore, tuning the morphology along with cytotoxicity assessment, biocompatibility, and cellular interaction, it is important for plasmonic nanoparticles being specifically used in healthcare and food safety management applications.

1.4 Applications of AuNPs in Healthcare

AuNPs are considered as the most prominent emerging potential candidate for cancer biology in terms of diagnosis as well as therapeutic agent. AuNPs act as a cargo for targeted drug delivery, thus enhancing the potency of drug and sometimes also used in heat mediated destruction of cancerous cells. Gold nanoparticles depending upon their morphology (both isotropic and anisotropic) are being widely used as therapeutic, imaging, and diagnostic tool in biomedical applications (Yang et al. 2015; Mieszawska et al. 2013) as shown in Fig. 1.2. Anisotropic gold nanostructures which exhibit SPR at NIR region are extensively used as contrast agents in bio-imaging for locating tumors in the body tissue. However, spherical AuNPs which absorb light in the visible region are less likely to be used as contrast agent as compared to anisotropic AuNPs. This is because the biological tissue and water hardly absorb any radiation at NIR region, resulting in enhanced scattering

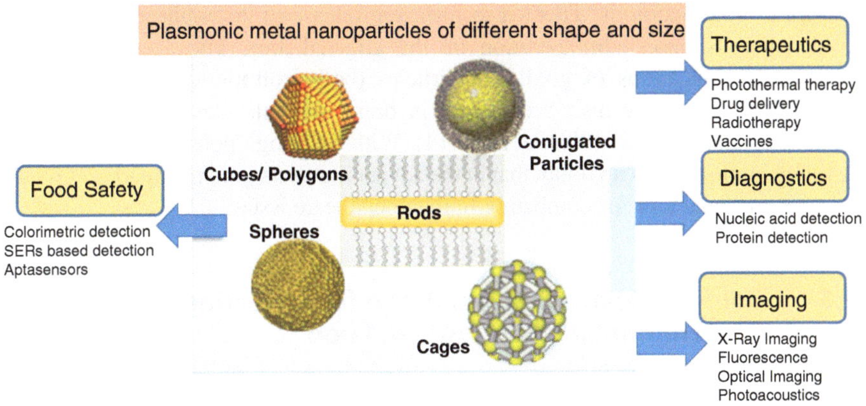

Fig. 1.2 Applications of isotropic and anisotropic plasmonic metal nanoparticles in healthcare and food safety. Adapted with permission from reference Mieszawska et al. (2013). Copyright (2013) American Chemical Society (ACS)

signal by the anisotropic AuNPs. Such anisotropic nanoparticles are further conjugated with tumor specific antibody for diagnosis and drug delivery with high accuracy (Paramasivam et al. 2017). In addition to diagnosis of tumor, AuNPs are also being widely used for the treatment of cancer. The most commonly used treatment using anisotropic AuNPs is carried out via photothermal therapy where the nanoparticles exhibiting SPR at NIR region are administered following the exposure of radiation. This results in heat mediated destruction of cancerous cells. Apart from the targeted localization, the AuNPs are more specific towards cancerous cells as compared to the normal cells (Huang and El-Sayed 2011). Anisotropic AuNPs have shown to possess greater half-life in circulatory system resulting in better adsorption onto tumor cells. This phenomenon is called the non-directional enhanced permeability and retention effect (EPR) (Kalyane et al. 2019). Gold nanoparticles due to their unique optical properties exhibiting different colors have laid the working principle to various biological sensors for detection of numerous biological molecules, entities and compounds. The designing of detectors using AuNPs mostly depends on colorimetric sensing, electrochemical sensing, fluorescence-based sensing, surface-enhanced Raman scattering (SERS) based sensing, enzymatic biosensors SPR-based sensors, etc. Canadian scientists Zagorovsky and Chan designed a test kit based on DNAzymes and AuNPs that can be used to detect infectious disease causing bacteria, viruses, and parasites (Zagorovsky and Chan 2013). It involved colorimetric detection where 13 nm sized AuNPs aggregated and the color turned from ruby red to blue-violet upon detection of the pathogen.

Gold nanoparticles are not only limited to diagnosis and treatment of a disease but they also have significant applications in critical conditions such as blood clotting and wound healing. The gold nanoparticles are conjugated with DNA thrombin inhibitor by suspending the gold nanorods in human serum protein and DNA rich medium leading to the attachment of greater amount of DNA molecules onto the surface of gold nanorods. This led to sufficient release of thrombin inhibitors from gold nanorods upon administration on the affected area. The aura of medical/ healthcare applications of gold nanoparticles (both isotropic and anisotropic) is multivariate. They are also being used in brain implants, designing of artificial skin, heart diseases, etc. (Shah et al. 2014). With such high potential and extensive research input, there is no doubt that in the advancing years gold nanoparticles would become a crucial tool for combating human healthcare issues.

1.5 Gold Nanoparticles as a Probe for Detecting Contaminants/Adulterants in Food

Apart from healthcare applications, AuNPs are also being used as sensitive probes for rapid detection of food contaminants. The visible color change of the AuNPs when interacted with certain compound present in food contaminants indicates successful detection. Currently available validation techniques for detecting food contamination uses complex machinery or cumbersome procedures such as high-performance liquid chromatography (HPLC), enzyme linked immunosorbent assay (ELISA), liquid chromatography/mass spectrometry (LC/MS), surface plasmon resonance (SPR), and electrochemical detection (Duncan 2011). Conversely the ease of nanoparticle synthesis and simple colorimetric method involves no complex procedure making them a suitable candidate for the detection of food contaminants. Among the most common and hazardous toxin that contaminates food grains is aflatoxin B1 (AFB1), produced by *Aspergillus flavus* and *Aspergillus parasiticus* which are most potent carcinogenic and genotoxic among the aflatoxins. Rapid and reproducible trace analysis of AFB1 have been achieved successfully using magnetic bead (as probe) and silica-encapsulated gold nanoparticles through SERS-based immuno-analytical detection (Ko et al. 2015). Gold nanoparticle coated starch magnetic beads upon functionalization with antibody were found to be more effective. 4 X-gold binding peptide-tagged Streptococcal proteins G (4GS) is a bifunctional linker protein used for antibody immobilization. The gold nanoparticle was responsible for the colloidal stability of the system which later was used for SERS-based detection system for *E. coli* in food (You et al. 2020) as shown in Fig. 1.3. Microbial contamination is not the only concern regarding food safety, the residual pesticides or fertilizers on crops/fruits/vegetables also possess serious threat to human health. Phosmet and thiabendazole are fungicides and insecticides, respectively, that are applied extensively in apple cultivation. These adversely affect human health and environment. Gold nanoparticles were used to detect trace amounts of phosmet and thiabendazole using a simple technique, UV-Vis spectrophotometry which was further validated through Raman spectrometry. The shift in absorption spectra indicated aggregation of the gold nanoparticles in the presence of the compounds (fungicide/insecticide) which was evident from the color change. Further the Raman spectra of the individual compounds were matched with the spectra of test samples containing gold nanoparticles validating the results obtained through UV-Vis spectrophotometry. To meet the demand of current population,

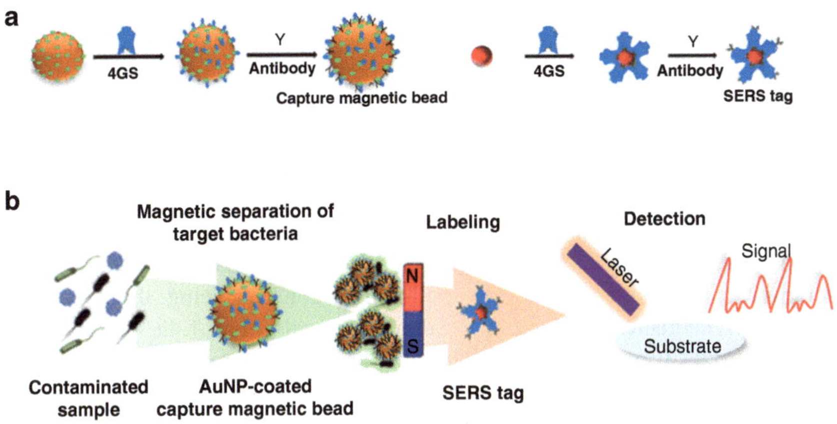

Fig. 1.3 (**a**) Preparation of antibody tagged AuNP-coated magnetic bead and SERS tag. (**b**) Gold nanoparticles coated magnetic beads using SERS-based detection system for detection of bacterial contamination (*E.coli*). Adapted with permission from reference You et al. (2020). Copyright (2020) American Chemical Society (ACS)

farmers are indulged in illegal use of veterinary drug to increase feed conversion efficiency. These drug residues finally get accumulated in milk, meat, fish, and eggs posing not only threat to human exposure, but also aids increased human exposure to antibiotics (Landers et al. 2012). In the recent years, gold nanoparticles based assays are being used with the benefit of target-selectivity and on-site analysis. For example, clenbuterol, a kind of adrenal neuronal stimulant which accelerates the rate of transformation and decomposition of fat resulting in acute poisoning. The plasmonic absorption of the gold nanoparticles is used to detect clenbuterol with limit of detection (LOD) of 50 nM (Kang et al. 2016). Melamine is one of the hazardous organic compounds that is being used unethically to increase protein content in milk due to its high content of nitrogen (66% by mass). A series of gold nanoparticles based assay have been established since the outbreak of renal failure in dogs and cats in 2007 and crisis of tainted milk in 2008 (Puschner and Reimschuessel 2011; Delman and Yang 2012). Among the entire gold nanoparticle based assay, citrate capped gold nanoparticles used in detection of melamine are the simplest one because of easy synthesis and on-site rapid colorimetric detection. The mechanism behind the detection is that melamine interacts with gold nanoparticles via ligand exchange with the negatively charged citrate ions present as capping (Chi et al. 2010).

1.6 Application of Silver Nanoparticles in Healthcare

Although silver nanoparticles are well known for their extensive use in industrial applications such as surface coatings, batteries, conductors, etc. but healthcare applications have attracted the use of silver nanoparticles due to its excellent plasmonic properties. Even if the economic status of utilization of silver nanoparticles is not realized, however the versatility over wide range of bacterial infections as an antimicrobial agent is well appreciated. The antibacterial effect of silver nanoparticles is due to the release of silver ions from the particles upon dissolution. The silver ions being positively charged interact electrostatically with the negatively charged cell membrane of the bacteria resulting in disruption, hence killing the bacteria (Li et al. 2010). Silver nanoparticles likely disrupt the growth signaling pathway in bacterial cells by modulating the tyrosine phosphorylation of certain proteins that are essential for cell viability (Mubarak Ali et al. 2011). Moreover, silver nanoparticles are well-known for enhancing the efficacy of antibiotics synergistically (Deng et al. 2016) as shown in Fig. 1.4a. Several studies have employed silver nanoparticles subjecting to various biomedical properties such as anti-biofilm, anti-cancer, anti-larval, and insecticidal properties (Martinez-Gutierrez et al. 2013; Lin et al. 2014; Rajakumar and Rahuman 2011; Moorthi et al. 2015).

Biofilm is considered as the prime mode of life for microorganisms with densely packed communities of microbial cells forming polymeric matrices over biotic or abiotic surfaces (Taheri et al. 2015) as shown in Fig. 1.4b. Biofilms formed by pathogenic microbes cause severe clinical infections adding complexities to a disease. The use of silver nanoparticles against biofilm formation is well documented. Silver nanoparticles have attracted substantial attention as a potent replacement for antibiotics. Biologically inspired synthesis approach of silver nanoparticles and their capability to immobilize plasma polymer interlayer functionalization for generating antibacterial coatings was also reported (Abbasi et al. 2016). The anti-biofilm formation property of biosynthesized silver nanoparticles was studied on clinically prominent bacteria *Klebsiella pneumoniae* (Shahwany et al. 2016). The electrostatic interaction between silver ions and bacterial cell membrane resulted in cellular disruption, thus inhibiting biofilm formation. The nanoparticles due to their high surface area to volume ratio can penetrate deeply into the mature biofilms making it suitable candidate for destructing biofilms. Silver nanoparticles are also being projected as a potential candidate in the arena of cancer biology. In the present time, several anti-cancer drugs are available for therapeutic options and for managing mortality due to cancer, although use of these drugs shows significant side effects and systemic toxicity. Interestingly, plasmonic nanoparticles created a platform which is attracting towards deciphering the avenues regarding cancer therapy and diagnosis due to their unique properties and target specificity. The anti-inflammatory properties of biosynthesized AgNPs with potential anti-cancer activity are tabulated below.

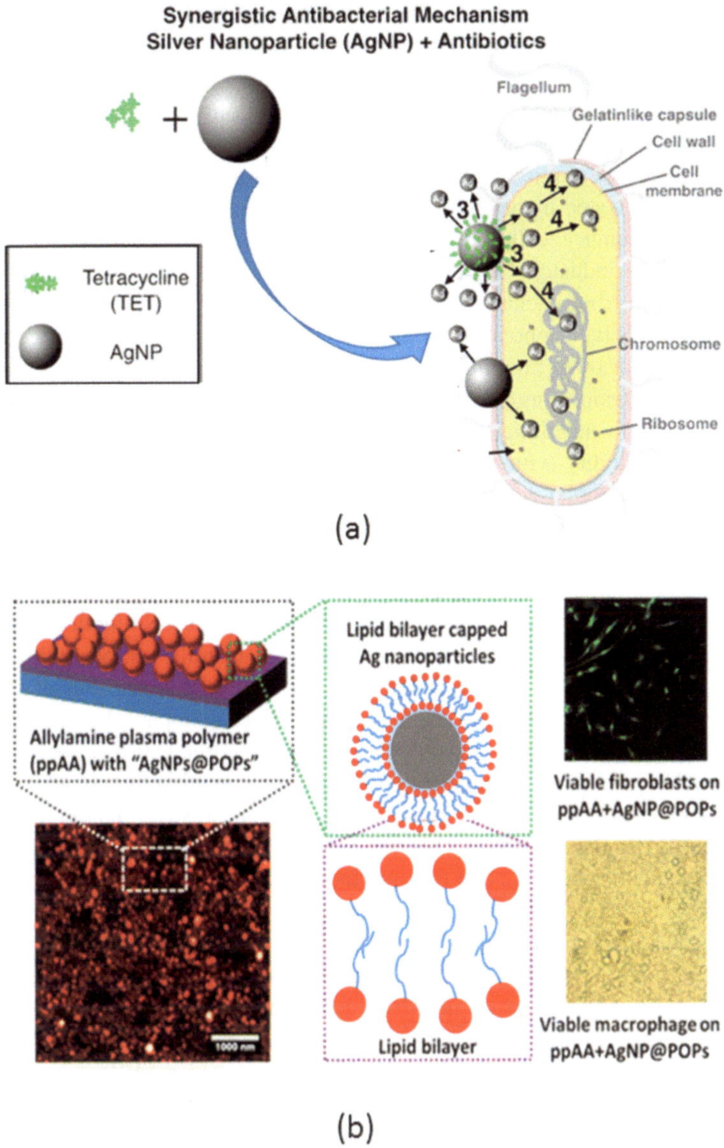

Fig. 1.4 (**a**) Synergistic effect of silver nanoparticles complexed with antibiotics on bacteria. (**b**) Anti-biofilm activity of silver nanoparticles encapsulated with a lipid 1-palmitoyl-2-oleoyl-sn-glycero-3-phospho-L-serine (POP) and coated with plasma polymer allylamine (ppAA). Adapted with permission from reference Deng et al. (2016) and Taheri et al. (2015). Copyright (2016 and 2015) American Chemical Society (ACS)

1.7 Silver Nanoparticles as a Probe for Detecting Contaminants/Adulterants in Food

Using silver for food and beverage storage applications is old as any civilization. Several ancient societies stored water and wine in vessels made of silver. Silver's broad spectrum of antimicrobial property and low cost rendered its extensive use as a disinfectant of water in developing countries (Gangadharan et al. 2010). Silver nanoparticles being excellent antimicrobial agent are also being widely used in food packaging industry. The technology involves use of silver nanoparticles as ingredient in making packaging material that renders food preservation for longer time duration. Silver nanoparticles are also being incorporated in refrigerators to avoid bacterial or other microbial growth (Chaloupka et al. 2010). One of the major advantages of using inorganic nanoparticles is that they can be incorporated with polymers to form functional antimicrobial material. AgNPs based nanocomposites offer greater stability with slower rate of silver ion release that is important for sustained antimicrobial activity. Silver nanoparticles are not only limited to food packaging and preservation, they are also being used as probes for detection of food contamination. Several methods of detecting melamine have been published; however, only few are able to detect it below 50 nM level that too by applying cumbersome procedure and using sophisticated instruments. Sulfanilic acid capped silver nanoparticles were used to detect melamine in milk samples with limit of detection (LOD) of 10.6 nM which indicates its high sensitivity (Song et al. 2015). Apart from spherical nanoparticles, anisotropic silver nanoparticles are also being used as a potent candidate for detecting food contamination. Due to anisotropy, these nanoparticles possess numerous hotspots generating maximum signals with minimum background signals. The recent development of silver nanorods array as substrates for Raman scattering to detect toxins, food borne pathogens, pesticides, and adulterants has been a boon towards food safety management. Silver nanorods synthesized using oblique angle deposition technique determines the nanorod length, density, and tilting angle which determines the surface-enhanced Raman scattering (SERS) performance (Das et al. 2019) as shown in Fig. 1.5. The spectral data obtained for the silver nanorods substrates containing pathogenic bacteria are studied for identification of the strain causing food poisoning/contamination. In a recent study, silver nanoparticles based filter paper was employed as SERS substrate which was able to bring down the detection limit for rhodamine 6G and rhodamine B to picomolar and nanomolar concentrations. To validate its utility in practical field, the substrate was subjected to detect illicit dyes used on vegetables and contaminants in pond, rain, and tap water with high reproducibility (Pan et al. 2015).

1.8 Application of Platinum Nanoparticles in Healthcare

Platinum nanoparticles are not as much popular candidates for therapeutics when compared to gold and silver nanoparticles or the derivate of its bulk metal cisplatin. Cisplatin is readily and extensively used during chemotherapy with several other

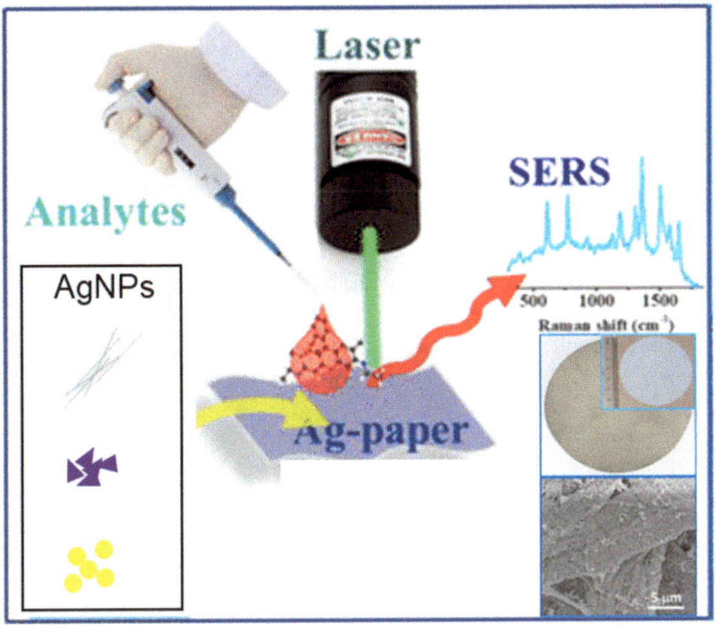

Fig. 1.5 Silver nanoparticles based cellulose paper for detection of illicit dyes used on vegetables using SERS technique. Adapted with permission from reference Das et al. (2019). Copyright (2019) American Chemical Society (ACS)

drugs administered in cycles. However, platinum nanoparticles show high catalytic activity and are being widely used in synthesizing multifunctional nanoparticles for healthcare applications (Pedone et al. 2017; Zhao et al. 2019). In vitro diagnostics (IVD) has been employing gold nanoparticles for decades because of their superior physico-chemical properties, better colloidal stability, and colorimetric results which are visible through naked eyes. However, the issue regarding sensitivity often obscures the precision of results. Therefore, researchers have circumvented the limitation by engineering gold nanoparticles possessing dual functionality by coating them with ultrafine platinum layer (i.e. Au@PtNPs). The Au@PtNPs retain the properties of both the elements in nano-regime. The dual functionalities, plasmonics (from Au) and catalysis (from Pt), offer two distinct alternatives for detection: one due to color produced by plasmonic (less sensitive) and second is the high sensitive color due to catalysis of chromogenic substrates (highly sensitive), resulting in "on demand" tunable detection performance (Gao et al. 2017). Multifunctional nanoparticles are considered to be superior nanomaterials enabling therapy and diagnosis simultaneously for biomedical applications. Platinum, cobalt, and gold were used to synthesize multifunctional magnetic nanoparticles of superior quality which exhibits high magnetic resonance and plasmonic property (Katagiri et al. 2019). The gold core of the Co-Pt@AuNPs retained its plasmonic property along with the ability to generate local heat during photothermal therapy in vivo. In a

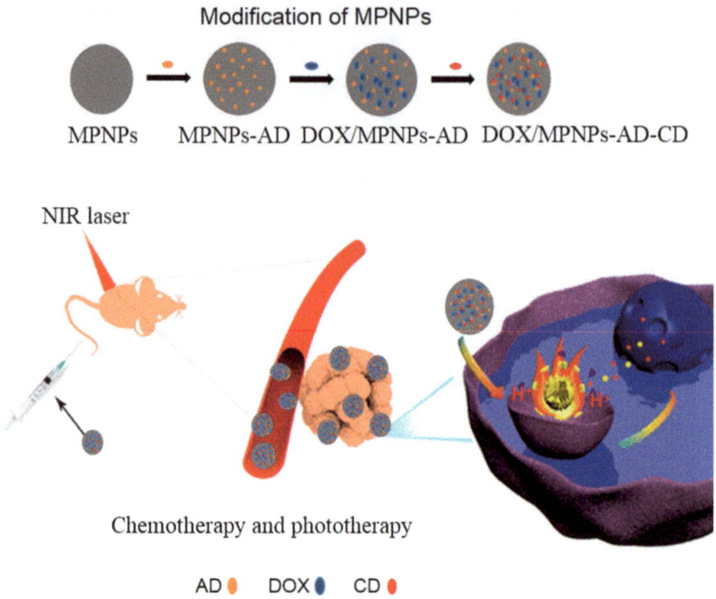

Fig. 1.6 Doxorubicin (DOX) and β-cyclodextrin (β-CD)-capped mesoporous platinum nanoparticles (MPNPs) (DOX/MPNPs-AD-CD) for combinatorial phototherapy and chemotherapy. Adapted with permission from reference Zhao et al. (2019). Copyright (2019) American Chemical Society (ACS)

recent work it was reported that the mesoporous platinum nanoparticles were used as a drug cargo for combined chemo-photothermal treatment of cancer (Zhao et al. 2019) (Fig. 1.6).

1.9 Platinum Nanoparticles as a Probe for Detecting Contaminants/Adulterants in Food

Platinum nanoparticles are one of the noble metal nanoparticles known for their high catalytic activity. Likewise its counterpart noble metals, platinum nanoparticles exhibit plasmon bands in the UV-region. The development of biosensors for detection of food contaminants and adulterants also employed platinum nanoparticles as one of the component. A sensitive optical biosensor for the detection of *Salmonella typhimurium* was based on gold@platinum nanoparticles (Au@PtNPs) aided with a passive 3D micromixer for rapid detection (Zheng et al. 2019) as shown in Fig. 1.7. The function of the micromixer is to separate 99% of the targeted bacteria magnetically within 10 min from the sample. The detection limit of the optical biosensor was 17 CFU/mL. The biosensor could also be modified by using different antibodies conjugated to Au@PtNPs for the specific epitope in bacteria. Platinum nanoparticles are not only limited to detection of microbial contamination but also being used in intelligent packaging of food items. Intelligent packaging refers to a new class of

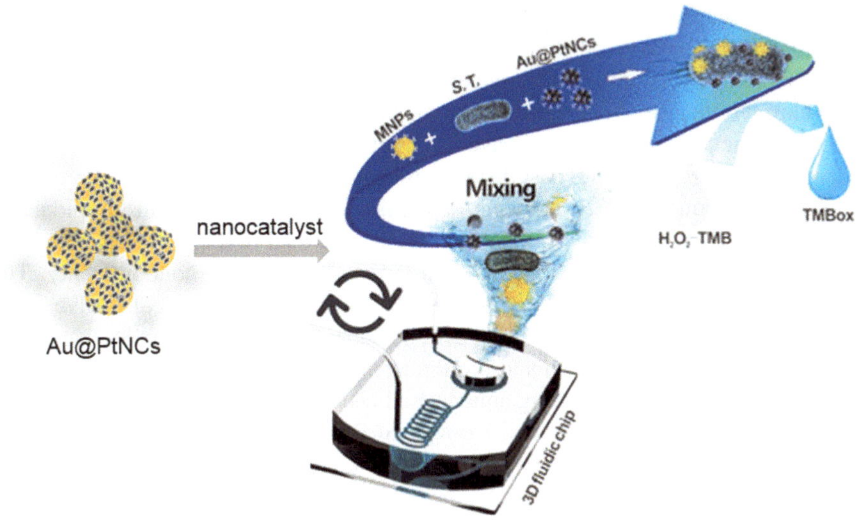

Fig. 1.7 Detection of *Salmonella typhimurium* in food by using gold–platinum as nanocatalysts in a fluidic chip. Adapted with permission from reference Zheng et al. (2019). Copyright (2019) American Chemical Society (ACS)

package that allows the condition of food contents to be monitored from production line to consumers. The quality of packaged food is also determined by pH as it is one of the indirect measures of food spoilage. The volatile compounds such as ammonia, trimethylamine, and dimethylamine are the byproducts of bacterial contamination which subsequently alter the pH (towards basic) of packaged foods. Thioglycolic acid (TGA) stabilized platinum nanoparticles were synthesized and the absorption spectra were studied under different pH. The change in spectra indicated the pH change of the food items, hence helps in monitoring the food quality. Although very few works have been published regarding application of platinum nanoparticles for food safety management but with such high catalytic activity, it will come up with pronounced wide range of applications.

1.10 Conclusion

The unique optical and chemical property of engineered plasmonic metal nanoparticles makes them the ideal candidates for enabling them as molecular nanoprobes in healthcare and food safety management application. The excellent optical enhancement and ligand binding capability due to surface chemistry of metal nanoparticle surfaces provide robust sensing and detection capability. The applications related to health care and food safety management are increasing tremendously in global market. This requires precise detection methods for either qualitative or quantitative analysis. Isotropic and anisotropic metal nanoparticles (gold/silver/platinum) can be tailored for specific end-user applications by changing their size, shape, and surface chemistry. Any subtle changes in the size and morphology will change the SPR of metal nanoparticles that may have impact on cellular uptake, biodistribution, and clearance in vivo. In this chapter, we have focused on the impact of the metal nanoparticles on food safety and health care. Anisotropic and isotropic metal nanoparticles behave differently and specifically which are dependent on the size, shape, and surface chemistry. The initial section of the chapter described various synthetic strategies/approaches of synthesizing anisotropic metal nanoparticles. Wet chemical approach is flexible and widely used method for synthesizing anisotropic nanoparticles in terms of yield and monodispersity. The chapter also emphasized on various healthcare applications of gold, silver, and platinum nanoparticles. Gold and silver nanoparticles of anisotropic shapes are used as diagnostic probes and anti-biofilm agents, respectively. Silver nanoparticles being known for its outstanding antibacterial activity are being used extensively in food industry for packaging and storage. SERs-based detection probes for food contaminants are being developed recently using gold and silver nanoparticles. Gold and silver nanoparticles are the most widely used nanoprobes as compared to platinum nanoparticles in food and biological applications. However, the antioxidant and catalytic property of platinum nanoparticles enables them as potential candidate for sensing and drug delivery. This chapter provided an overview of the current state-of-the-art techniques based on metal nanoprobes and their use as sensing probes for health care and food safety management. The chapter emphasizes on the need for such plasmonic metal nanoprobes that could benefit human health and may translate laboratory research to end-consumer applications.

AgNPs synthesis using organisms	Stabilizing/capping agents	Anti-inflammatory agents produced	References
Terminalia sp.	Polysaccharides, polyphenolic, protein, and flavonoids	Reactive oxygen species (ROS)	de Araujo et al. (2019)
Fruit extract from *Garcinia Mangostana*	Secondary metabolites (flavonoids, tannins, anthraquinones, saponins, ascorbic acid, carbohydrates, and phenols)	Cytokines	Shanmuganathan et al. (2019)
Leaf extract from *Pteristripartita*	Phenolics, terpenoids, flavonoids, proteins, tannins, and glycosides	Serotonin, histamine, and prostaglandins	Baskaran et al. (2016)
Leucas aspera	Alkaloids, flavonoids, terpenoids,, phenol,, phytosterol, tannins, carbohydrates, saponins, aliphatic amines, alkyl halides	Indomethacin	Suganya et al. (2014)

Acknowledgement We are thankful to the esteemed editors for the invitation to contribute a book chapter in the book titled: "Bio-nano Interface: Applications in Food, Healthcare and Sustainability." We acknowledge CSIR-Indian Institute of Toxicology Research (CSIR-IITR),

Lucknow, as a platform for overwhelming support. The book chapter has been approved by the publication committee of the CSIR-IITR and the manuscript number for communication is 3652. Nabojit Das is thankful to CSIR for providing the fellowship.

References

Abbasi E, Milani M, Fekri Aval S, Kouhi M, Akbarzadeh A, Tayefi Nasrabadi H, Nikasa P, Joo SW, Hanifehpour Y, Nejati-Koshki K, Samiei M (2016) Silver nanoparticles: synthesis methods, bio-applications and properties. Crit Rev Microbiol 42(2):173–180

Albanese A, Tang PS, Chan WC (2012) The effect of nanoparticle size, shape, and surface chemistry on biological systems. Annu Rev Biomed Eng 14:1–16

Ali MA, Eldin TAS, Moghazy GE, Tork IM, Omara II (2014) Detection of E. coli O157: H7 in feed samples using gold nanoparticles sensor. Int J Curr Microbiol App Sci 3(6):697–708

Baskaran X, Vigila AVG, Parimelazhagan T, Muralidhara-Rao D, Zhang S (2016) Biosynthesis, characterization, and evaluation of bioactivities of leaf extract-mediated biocompatible silver nanoparticles from an early tracheophyte, Pteris tripartita sw. Int J Nanomed 11:5789

Burrows ND, Harvey S, Idesis FA, Murphy CJ (2017) Understanding the seed-mediated growth of gold nanorods through a fractional factorial design of experiments. Langmuir 33(8):1891–1907

Chaloupka K, Malam Y, Seifalian AM (2010) Nanosilver as a new generation of nanoproduct in biomedical applications. Trends Biotechnol 28(11):580–588

Chen L, Huang B, Qiu X, Wang X, Luque R, Li Y (2016) Seed-mediated growth of MOF-encapsulated Pd@ag core–shell nanoparticles: toward advanced room temperature nanocatalysts. Chem Sci 7(1):228–233

Chi H, Liu B, Guan G, Zhang Z, Han MY (2010) A simple, reliable and sensitive colorimetric visualization of melamine in milk by unmodified gold nanoparticles. Analyst 135(5):1070–1075

Das D, Senapati S, Nanda KK (2019) "Rinse, repeat": an efficient and reusable SERS and catalytic platform fabricated by controlled deposition of silver nanoparticles on cellulose paper. ACS Sustain Chem Eng 7(16):14089–14101

Dasgupta S, Auth T, Gompper G (2014) Shape and orientation matter for the cellular uptake of nonspherical particles. Nano Lett 14(2):687–693

de Araujo AR, Ramos-Jesus J, de Oliveira TM, de Carvalho AMA, Nunes PHM, Daboit TC, Carvalho AP, Barroso MF, de Almeida MP, Plácido A, Rodrigues A (2019) Identification of Eschweilenol C in derivative of Terminalia fagifolia Mart. and green synthesis of bioactive and biocompatible silver nanoparticles. Ind Crop Prod 137:52–65

Delman J, Yang M (2012) A value chain gone awry: implications of the "tainted Milk scandal" in 2008 for political and social Organization in Rural China. In: Organizing rural China–rural China organizing, 205-221. Rowman & Littlefield, Lanham, MD

Deng H, McShan D, Zhang Y, Sinha SS, Arslan Z, Ray PC, Yu H (2016) Mechanistic study of the synergistic antibacterial activity of combined silver nanoparticles and common antibiotics. Environ Sci Technol 50(16):8840–8848

Duncan TV (2011) Applications of nanotechnology in food packaging and food safety: barrier materials, antimicrobials and sensors. J Colloid Interface Sci 363(1):1–24

Fong KE, Yung LYL (2013) Localized surface plasmon resonance: a unique property of plasmonic nanoparticles for nucleic acid detection. Nanoscale 5(24):12043–12071

Gangadharan D, Harshvardan K, Gnanasekar G, Dixit D, Popat KM, Anand PS (2010) Polymeric microspheres containing silver nanoparticles as a bactericidal agent for water disinfection. Water Res 44(18):5481–5487

Gao Z, Ye H, Tang D, Tao J, Habibi S, Minerick A, Tang D, Xia X (2017) Platinum-decorated gold nanoparticles with dual functionalities for ultrasensitive colorimetric in vitro diagnostics. Nano Lett 17(9):5572–5579

Herrera GM, Padilla AC, Hernandez-Rivera SP (2013) Surface enhanced Raman scattering (SERS) studies of gold and silver nanoparticles prepared by laser ablation. Nano 3(1):158–172

Huang X, El-Sayed MA (2011) Plasmonic photo-thermal therapy (PPTT). Alexandria J Med 47 (1):1–9

Kalyane D, Raval N, Maheshwari R, Tambe V, Kalia K, Tekade RK (2019) Employment of enhanced permeability and retention effect (EPR): nanoparticle-based precision tools for targeting of therapeutic and diagnostic agent in cancer. Mater Sci Eng C 98:1252–1276

Kang J, Zhang Y, Li X, Miao L, Wu A (2016) A rapid colorimetric sensor of clenbuterol based on cysteamine-modified gold nanoparticles. ACS Appl Mater Interfaces 8(1):1–5

Katagiri M, Cuya Huaman JL, Matsumoto T, Suzuki K, Miyamura H, Balachandran J (2019) Magneto-Plasmonic co@ Pt@ au nanocrystals for biosensing and therapeutics. ACS Appl Nano Mater 3(1):418–427

Kim M, Lee JH, Nam JM (2019a) Plasmonic Photothermal nanoparticles for biomedical applications. Adv Sci 6(17):1900471

Kim S, Seo J, Park HH, Kim N, Oh JW, Nam JM (2019b) Plasmonic nanoparticle-interfaced lipid bilayer membranes. Acc Chem Res 52(10):2793–2805

Ko J, Lee C, Choo J (2015) Highly sensitive SERS-based immunoassay of aflatoxin B1 using silica-encapsulated hollow gold nanoparticles. J Hazard Mater 285:11–17

Kumar A, Das N, Satija NK, Mandrah K, Roy SK, Rayavarapu RG (2020) A novel approach towards synthesis and characterization of non-cytotoxic gold nanoparticles using taurine as capping agent. Nano 10(1):45

Kumar A, Kim A, Nam JM (2016) Plasmonically engineered nanoprobes for biomedical applications. J Am Chem Soc 138(44):14509–14525

Landers TF, Cohen B, Wittum TE, Larson EL (2012) A review of antibiotic use in food animals: perspective, policy, and potential. Public Health Rep 127(1):4–22

Li WR, Xie XB, Shi QS, Zeng HY, You-Sheng OY, Chen YB (2010) Antibacterial activity and mechanism of silver nanoparticles on Escherichia coli. Appl Microbiol Biotechnol 85 (4):1115–1122

Lin J, Huang Z, Wu H, Zhou W, Jin P, Wei P, Zhang Y, Zheng F, Zhang J, Xu J, Hu Y (2014) Inhibition of autophagy enhances the anticancer activity of silver nanoparticles. Autophagy 10 (11):2006–2020

Lohse SE, Burrows ND, Scarabelli L, Liz-Marzán LM, Murphy CJ (2014) Anisotropic noble metal nanocrystal growth: the role of halides. Chem Mater 26(1):34–43

Manshian BB, Himmelreich U, Soenen SJ (2017) Standard cellular testing conditions generate an exaggerated nanoparticle cytotoxicity profile. Chem Res Toxicol 30(2):595–603

Martinez-Gutierrez F, Boegli L, Agostinho A, Sánchez EM, Bach H, Ruiz F, James G (2013) Anti-biofilm activity of silver nanoparticles against different microorganisms. Biofouling 29 (6):651–660

Mieszawska AJ, Mulder WJ, Fayad ZA, Cormode DP (2013) Multifunctional gold nanoparticles for diagnosis and therapy of disease. Mol Pharm 10(3):831–847

Moorthi PV, Balasubramanian C, Mohan S (2015) An improved insecticidal activity of silver nanoparticle synthesized by using Sargassum muticum. Appl Biochem Biotechnol 175 (1):135–140

Mubarak Ali D, Thajuddin N, Jeganathan K, Gunasekaran M (2011) Plant extract mediated synthesis of silver and gold nanoparticles and its antibacterial activity against clinically isolated pathogens. Colloids Surf B Biointerfaces 85(2):360–365

Mustafa DE, Yang T, Xuan Z, Chen S, Tu H, Zhang A (2010) Surface plasmon coupling effect of gold nanoparticles with different shape and size on conventional surface plasmon resonance signal. Plasmonics 5(3):221–231

Nel AE, Mädler L, Velegol D, Xia T, Hoek EMV, Somasundaran P, Klaessig F, Castranova V, Thompson M (2009) Understanding biophysicochemical interactions at the nano–bio interface. Nat Mater 8(7):543–557

Pan Y, Guo X, Zhu J, Wang X, Zhang H, Kang Y, Wu T, Du Y (2015) A new SERS substrate based on silver nanoparticle functionalized polymethacrylate monoliths in a capillary, and it application to the trace determination of pesticides. Microchim Acta 182(9–10):1775–1782

Paramasivam G, Kayambu N, Rabel AM, Sundramoorthy AK, Sundaramurthy A (2017) Anisotropic noble metal nanoparticles: synthesis, surface functionalization and applications in biosensing, bioimaging, drug delivery and theranostics. Acta Biomater 49:45–65
Pedone D, Moglianetti M, De Luca E, Bardi G, Pompa PP (2017) Platinum nanoparticles in nanobiomedicine. Chem Soc Rev 46(16):4951–4975
Puschner B, Reimschuessel R (2011) Toxicosis caused by melamine and cyanuric acid in dogs and cats: uncovering the mystery and subsequent global implications. Clin Lab Med 31(1):181–199
Rajakumar G, Rahuman AA (2011) Larvicidal activity of synthesized silver nanoparticles using Eclipta prostrata leaf extract against filariasis and malaria vectors. Acta Trop 118(3):196–203
Reguera J, Langer J, de Aberasturi DJ, Liz-Marzán LM (2017) Anisotropic metal nanoparticles for surface enhanced Raman scattering. Chem Soc Rev 46(13):3866–3885
Schleh C, Semmler-Behnke M, Lipka J, Wenk A, Hirn S, Schäffler M, Schmid G, Simon U, Kreyling WG (2012) Size and surface charge of gold nanoparticles determine absorption across intestinal barriers and accumulation in secondary target organs after oral administration. Nanotoxicology 6(1):36–46
Shah M, Badwaik VD, Dakshinamurthy R (2014) Biological applications of gold nanoparticles. J Nanosci Nanotechnol 14(1):344–362
Shahwany A, Tawfeeq HK, Hamed SE (2016) Antibacterial and anti-biofilm activity of three phenolic plant extracts and silver nanoparticles on Staphylococcus aureus and Klebsiella pneumoniae. J Biomed Biotechnol 4:12–18
Shanmuganathan R, Karuppusamy I, Saravanan M, Muthukumar H, Ponnuchamy K, Ramkumar VS, Pugazhendhi A (2019) Synthesis of silver nanoparticles and their biomedical applications-a comprehensive review. Curr Pharm Des 25(24):2650–2660
Song J, Wu F, Wan Y, Ma L (2015) Colorimetric detection of melamine in pretreated milk using silver nanoparticles functionalized with sulfanilic acid. Food Control 50:356–361
Stoehr LC, Gonzalez E, Stampfl A, Casals E, Duschl A, Puntes V, Oostingh GJ (2011) Shape matters: effects of silver nanospheres and wires on human alveolar epithelial cells. Part Fibre Toxicol 8(1):36
Suganya G, Karthi S, Shivakumar MS (2014) Larvicidal potential of silver nanoparticles synthesized from Leucas aspera leaf extracts against dengue vector Aedes aegypti. Parasitol Res 113(3):875–880
Taheri S, Cavallaro A, Christo SN, Majewski P, Barton M, Hayball JD, Vasilev K (2015) Antibacterial plasma polymer films conjugated with phospholipid encapsulated silver nanoparticles. ACS Biomater Sci Eng 1(12):1278–1286
Wang Y, He J, Liu C, Chong WH, Chen H (2015) Thermodynamics versus kinetics in nanosynthesis. Angew Chem Int Ed 54(7):2022–2051
Xie X, Liao J, Shao X, Li Q, Lin Y (2017) The effect of shape on cellular uptake of gold nanoparticles in the forms of stars, rods, and triangles. Sci Rep 7(1):1–9
Yang L, Zhou Z, Song J, Chen X (2019) Anisotropic nanomaterials for shape-dependent physicochemical and biomedical applications. Chem Soc Rev 48(19):5140–5176
Yang X, Yang M, Pang B, Vara M, Xia Y (2015) Gold nanomaterials at work in biomedicine. Chem Rev 115(19):10410–10488
You SM, Luo K, Jung JY, Jeong KB, Lee ES, Oh MH, Kim YR (2020) Gold nanoparticle-coated starch magnetic beads for the separation, concentration and SERS-based detection of E. coli O157: H7. ACS Appl Mater Interfaces 12(16):18292–18300
Zagorovsky K, Chan WC (2013) A plasmonic DNAzyme strategy for point-of-care genetic detection of infectious pathogens. Angew Chem Int Ed 52(11):3168–3171
Zhao W, Li Z, Yang H, Ren C, Lv F, Gao S, Ma H, Jin Y, Ge K, Liu D, Zhang J (2019) Mesoporous platinum nanotherapeutics for combined chemo-photothermal cancer treatment. ACS Appl Bio Mater 2(8):3269–3278
Zhao Y, Wang Y, Ran F, Cui Y, Liu C, Zhao Q, Gao Y, Wang D, Wang S (2017) A comparison between sphere and rod nanoparticles regarding their in vivo biological behavior and pharmacokinetics. Sci Rep 7(1):1–11

Zheng L, Cai G, Qi W, Wang S, Wang M, Lin J (2019) Optical biosensor for rapid detection of Salmonella typhimurium based on porous gold@platinum Nanocatalysts and a 3D fluidic Chip. ACS Sensors 5(1):65–72

An Introduction to Different Methods of Nanoparticles Synthesis

2

Rohit Pritam Das and Arun Kumar Pradhan

Abstract

On-going research in nanotechnology has confirmed a diversity of methods to synthesize nanoparticles (NPs) from a various range of materials. Nanotechnology is considered as a growing sector which brings together a multidisciplinary fields including biology, chemistry, physics, medicine, etc. with certain benefits. It includes ceramics, metals, semiconductors, polymers, metal oxides, etc. NPs have distinctive structural physicochemical and morphological characteristics based on their origin and synthesis methods. These are important in a wide variety of applications concomitant to electronic, optoelectronic, optical, electrochemical, environment. and biomedical fields. With the help of nanotechnology the major beneficial effects have been seen in agriculture, industrial sector and much more. This current review comes up with key applications of nanoparticles and extensive overview on various physical, chemical and bio-assisted methods. With all its emerging aspects nanoparticle has shown a huge upcoming opportunity in near future.

Keywords

Nanoparticles · Ball milling · Sputtering · Laser ablation · Sol-gel

R. P. Das · A. K. Pradhan (✉)
Centre for Biotechnology, Siksha 'O' Anusandhan (Deemed to be University), Bhubaneswar, Odisha, India
e-mail: arunpradhan@soa.ac.in

M. Arakha et al. (eds.), *Bio-Nano Interface*,
https://doi.org/10.1007/978-981-16-2516-9_2

2.1 Introduction

Nanotechnology is the combination of two words that is 'nano' and 'technology'. This deals with very small particles of less than 100 nm size. NPs with a wide range of utility in various emerging fields such as medicine, energy, environment, textile, etc. play an important role in this modern era. It is clear from the recent studies that in nano-level various substances behave significantly different. Various nanotechniques helped researchers to overcome the limitation of traditional methods. For example, quantum mechanics, a branch of physics which deals with the study of subatomic particles such as neutron, proton and electron in a nanoscale level. In the context of chemical properties, a high surface area exposure makes NPs as a suitable catalyst for the chemical reaction. A free movement of the atoms within or around the other atoms creates a chemical zone where chemical properties of the atoms can be altered. NPs with many potential health benefits such as disease treatment such as cancer, better imaging and diagnostic equipment draw the attention of medical researchers. Scientists are trying to develop methods through which NPs are delivered directly to specific cells, which seems to be beneficial for the cancer treatment where radio therapy as well as chemotherapy can also cause damage to healthy cells. Properties like physical, chemical, biological, electrical and optical helped to improve the technology. NP shows diverse applications only due to its surface properties and small dimension. The process of synthesis nanoparticles has been based upon 2 approaches that are 'bottom-up' and 'top-down'. Bottom-up approach is basically used for the preparation of nanoparticles (Fig. 2.1). In bottom-up certain physical as well as chemical effort has been used to mobilize the basic units of nanoparticles in to complex structure. An implementation of this bottom-up approach is carried out from biological surrounding where nature utilized chemical force to build up all essential structure needed by life. A wide variety of bottom-up approaches such as condensation of atomic vapours as well as coalescence of liquid atom has been developed for producing nanoparticles. The basic goal of this

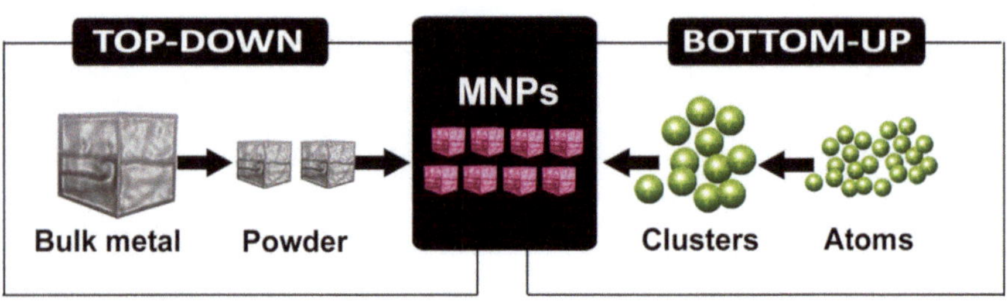

Fig. 2.1 Schematic representation of top-down and bottom-up procedure (https://nanotechnologyjd.weebly.com/manufacturing-processed.html)

approach is to synthesize nanoparticle in specific nanosizes for all particles by maintaining the uniqueness in functionality.

The diverse applications of NPs include:

- Drug discovery (Coto-García et al. 2011).
- Biosensor (Salata 2004).
- Bioimaging (Das et al. 2009).
- Molecular tagging (Becheri et al. 2008).
- Food technology (Taheri et al. 2014).
- Textile manufactureing (Gadomskii and Kharitonov 2004).
- Quantum computers (Nagavarma et al. 2012).
- Quantum laser (Abhilash 2010).

Now the second approach as per the name suggests top-down, i.e. from larger to smaller. Either chemical or fabrication technique is used here. Chemical procedure has been implemented by etching the metal by acids to obtain the required particles. The synthesis procedure is based on three types that are

(a) Physical method for synthesis of nanoparticles.
(b) Chemical method for synthesis of nanoparticles.
(c) Biological method for synthesis of nanoparticles.

2.2 Physical Method for Synthesis of Nanoparticle

In this section we shall discuss upon some physical methods which deal with high energy radiation, condensation, as well as thermal energy for the synthesis of nanoparticles. Some of the highly considerable methodologies for the physical synthesis of nanoparticles are:

high energy ball milling, laser ablation, electrospraying, inert gas condensation, physical vapour deposition, laser pyrolysis, flash spray pyrolysis and melt mixing.

2.2.1 High Energy Ball Milling

In high energy ball milling (HEBM) process the moving balls are used to break the chemical bonds by utilizing their kinetic energy on the milling substances which leads to structural changes and chemical reaction at normal room temperature (Chen et al. 1999 and Gleize et al. 1994). Tan Xing et al. have reported that ball milling technology is utilized as a green synthesis method to synthesize nitrogen doped carbon particle which may lead as an electrochemical catalysis (Xing et al. 2013). HEBM is a mechano-chemical synthesis process as it utilizes local temperature more than 1000 times as well as pressure (Dhand et al. 2015). With hydrophobic as well as hydrophilic properties biosurfactants play a critical role in nanoparticle synthesis. Recently surfactants mediated ball milling considered as efficient methods for NPs synthesis. The electrostatics forces generated by surfactants prohibit the particles from being agglomeration which dramatically enhances the particle size. Various types of ball milling methods such as vibrator mills, planetary ball mills, attritor mills, tumbler ball mills were adapted for a high throughput synthesis (He et al. 1996). Out of which tumbler ball milling methods are more economical than HEBM method (Koch 1998). Vibrator and planetary mills have known for their ability to produce small amount of NPs less than 100 g within a minimum time span (Ullah et al. 2014).

2.2.2 Inert Gas Condensation

Inert gas condensation method deals with the inert gases such as (helium or argon) and also with liquid nitrogen acting as a substrate holder for preparation of NPs. By utilizing this protocol Ward et al. *used to synthesize Mn NPs* (Chen et al. 2007). *For silver as well as platinum NPs synthesis the inert gas condensation is considered to be highly efficient method* (Ward et al. 2006). *With the help of combination of plasma*-gas-condensation-type cluster deposition apparatus, inert gas condensation (IGC) is acheived (Maicu et al. 2014). Benelmekki et al. have mentioned in his report that the metallic dielectric multicore-shell synthesis of NPs was carried with the help of IGC method (Gracia-Pinilla et al. 2010). The NPs were deposited in Si shell, which plays an important role in preventing the oxidation of nanoparticles and maintaining their aggregation.

2.2.3 Physical Vapour Deposition (PVD)

Physical vapour deposition is a collective synthesis process of NPs which deals with the vacuum deposition technique having three basic fundamental steps. First one is vaporization of metal particle from solid. Then in second step transportation of the vaporized particle. Final step belongs to nucleation and formation of thin layer films of NPs (Benelmekki et al. 2015).

2.2.3.1 Sputtering

Sputtering is working on the vacuum-based technology which deals with momentum transfer principle, which usually used to deposit films and NPs. The basic mechanism of this technique relies upon several processes. With the help of collision methods the inert gas plasma is produced within the electrodes. Then the ions present in the plasma pushed up towards the target with the help of energy present between electrodes. This energy strikes the target that leads to ejection of material from the target followed by the transportation and deposition of the NPs to the substrate. Another upgraded version of this method is magnetron sputtering which overcomes the former one by archiving the two advantages, i.e. higher deposition rate and prevention of target overheating and damage. Higher deposition rate is just because of the magnetic field generated by magnetron sputtering. The path of electron became curved due to which more ionization of gases occurs and causing more ions heat to the target, thus enhancing the ion deposition.

Veith et al. (2007) were able to generate Au NPs (less than 3 nm) deposition with basic support materials such as WO_3 and carbon surface. Bouchat et al. reported that synthesis of both metallic and non-metallic NPs such as titanium oxide (TiO2), silver (Ag), gold (Au), yttrium (Y), carbon (C), cobalt, (Co) and iron (Fe) has been carried out with the help of magnetron sputtering (Bouchat et al. 2013).

2.2.3.2 Electron Beam Evaporation (EBE)

This vacuum-based PVD process is usually considered for the production of thin layers as well as NPs. EBE system is composed of electron filament which required continuous current supply for keep it as working condition resulting generation of electron beams. These electron beams then used to target materials through magnets. EBE method has been implemented as development of 2D and 3D metal patterning (Rumi et al. 2008). This methods were also implemented for synthesis of Au NPs and platinum (Pt) NPs on multiwall carbon nanotubes to produce composite electrodes for sensors and energy storage applications (Hsieh et al. 2010). By the help of facile EBE process, fabrication of antibacterial Ag NPs was done on diameter-controlled TiO2 nanotubes (Uhm et al. 2013). It is reported that fabricated Ag NPs decorated three dimensional graphene (GR) scaffolds for electrochemical application (Bello et al. 2014).

2.2.3.3 Laser Ablation (LA) and Pulse Laser Deposition (PLD)

Through this laser ablation technique particles are evaporated from a solid source by utilizing high power laser beam (Chen and Yeh 2002). On the basis of utilization LA is divided into two categories, i.e. continuous laser or pulsed laser. LA produces workable approach to help in production of polymeric materials. In his report Kris et al. have reported a micro lens assembly that utilizes LA technique with the help of scanning the polymer surface to achieve a lens shape with the optimal focal distance and diameter (Naessens et al. 2001). In the context of NPs production, Lee et al. reported the synthesis of FeCo NPs in an inert gas atmosphere by PLD (Happy et al. 2004). Also Ong et al. (2008) have emphasized on the properties of FeCo NPs such as pulses number, temperature and gas pressure which are based on morphology and size of NPs synthesized by PLD. Again it is seen that in the presence of higher argon gas pressure, the average particle size has increased. At the same time some morphological changes have been observed with the increasing number of pulses. The interconnected chains were converted to fibrous. Similarly, in context to pressure, a significant changes has also been reported. In low pressure, NPs are shown as a floccules-like Nano-networks while in high pressure, as chain-like network. Andrea et al. also mentioned about the deposition of nanostructured silver thin films which consist of arrays of nanoparticles with the help of PLD in an argon atmosphere (Andrea et al. 2009). The size and morphology of Ag NPs are controlled by optimizing the Ar pressure and number of pulses.

2.2.3.4 Vacuum Arc (VA)

Vacuum arc is a vacuum-based PVD process in which the arc is used to vaporize the material for the synthesis of metallic, ceramic and composite NPs and films. With the help of plasma arc process, Mg-Al alloy was synthesized (Akbari et al. 2015). In another work, Lei et al. (2007) have reported his work for the synthesis of intermetallic Fe-Sn NPs done with the help of using vacuum arc. Amaratunga et al. (1996) have examined some excellent properties of thin film carbon NPs with the help of high pressure arc discharge process.

2.2.4 Laser Pyrolysis

The CO_2 laser pyrolysis technique is based on vapour phage synthesis (Amato et al. 2013). A large category of oxide NPs like TiO_2, SiO_2, Al_2 and O_3 are synthesized by this process. Further using laser pyrolysis technique some non-oxide compounds like Si, $S_{i3}N_3$, MoS_2 and ternary composites like Si, C, N are also synthesized.

2.2.5 Flame Spray Pyrolysis (FSP)

Among all the flame aerosol technologies flame spray pyrolysis is the latest one (Teoh et al. 2010). It is a high combustion method where the precursor was found in liquid form. Usually in an organic solvent, high combustion enthalpy, i.e. higher than 50% of total energy of combustion also observed. The liquid precursor can follow the droplet-to-particle route or gas-to-particle route; however, latter results in more homogenous morphologies and size. Sokolowski et al. first utilized this strategy to synthesize Al_2O_3 NPs by combusting an ultrasonically dispersed spray of aluminium acetylacetonate in benzene–ethanol solvent mixture (Sokolowski et al. 1977).

2.2.6 Electrospraying Technique

Electrospraying technique is quite similar with the electrospinning, only differentiate with the product produced by it. An electrochemical device that mix a solution containing selected polymer and the solvent is taken up by the syringe. Then a high voltage is applied to the capillary tip that results in the production of charged droplets. The solvent is evaporated on its way to the counter electrode and particles are then collected as the end product. This approach was well known for providing flexibility and control on surface parameters (Sridhar et al. 2015). Deppert et al. have studied the deposition of distinct NPs and the formation of gold NPs with the help of electrospraying technique (Böttger et al. 2007). A lipid based delivery system was reported by Sridhar et al. in his comprehensive review over electrospraying technique (Sridhar and Ramakrishna 2013).

2.2.7 Melt Mixing Technique

This technique utilizes mechanical mixing of polymers with modified version of nanofilters by extrusion or kneading and less commonly by injection moulding technique (Karak 2009). This method is an oldest technique to archive the desired materials. This is the most commonly used mechanical process because it is environment friendly and it is also well suited for current industrial practices (Krause et al. 2009). Melt mixing techniques are utilized to prepare nanocomposite NPs of conducting polymer polypyrrole with polypropylene (Sevil and Zuhal 2010).

2.3 Chemical Method for Synthesis of Nanoparticle

As per physical methods NPs are also synthesized by chemical methodology. Some of the commonly used chemical methods are Sol–gel method, micro-emulsion technique, hydrothermal synthesis, polyol synthesis, chemical vapour synthesis and plasma enhanced chemical vapour deposition.

2.3.1 Sol-Gel Methods

Sol-gel method is the combination of two components that are 'sol' a colloidal suspension of solid particle with liquid and 'gel' liquid containing polymers. Thus, this process includes the creation of 'sols' in the liquid that lead to the formation of a network of discrete particles or network polymers by the connection of sol particles (Jianing et al. 1990). The first step in formation of gel phage is hydrolysis and condensations. In hydrolysis water is used to disintegrate the bonds of precursor followed by condensations that lead to formation of nanoparticles. ZnO NPs were synthesized by this method using zinc acetate dehydrate and triethanolamine (TEA) as the precursors. Goncalves et al. demonstrated the one pot synthesis of CdSe NPs using sol–gel method and realized quantum size effect in fabricated NPs, which is important for developing advanced optoelectronic devices (Goncalves et al. 2014).

2.3.2 Micro-emulsion Technique

This technique is made up of three components, i.e. water body called as polar phage, second non-polar phase that deals with the hydrocarbon liquid or oil and finally surfactant. An interfacial layer is created by surfactants which separate aqueous and organic phage. This layer acts as a steric barrier that prevent the coalescence of the droplets (Solanki and Murthy 2011). Micro-emulsion technique was not progressed much in terms of designing organic NPs due to phase separation constraint and thus only few reports are available (Malik et al. 2012). Mostly through micro-emulsion technique synthesis of inorganic nanomaterials such as metal NPs (Au, Pt, Pd), semiconducting metal sulphite NPs (CdS, PbS, CuS, Cu2S and CdSe), metal salt NPs ($BaCO_3$, $CaCO_3$ and $SrCO_3$), metal oxide NPs (ZrO_2, TiO_2, SiO_2, GeO_2 and Fe_2O_3), magnetic NPs ((Mn, Zn)Fe_2O_4, (Ni, Zn)Fe_2O_4, $ZnFe_2O_4$ and $BaFe_{12}O_{19}$) and composite NPs (CdS–TiO_2, CdS–ZnS, CdS–SnO_2) is carried out. Reduction strategy was used for the synthesis of metal nanomaterials. Water-to-surfactant molar ratio (u), type of continuous phase, metal ion concentration, type and concentration of the reducing agent, structure and amount of the surfactant used are some of the key parameters that control the metal NPs synthesis effectively.

A reverse micelle micro-emulsion strategy was reported by Boutonnet et al. (1982) The synthesis of Pt., Pd, Rh and Ir NPs employing H_2PtCl_6, Pd $(NH_2)_4Cl_2$, $RhCl_2$ and $IrCl_3$, respectively, as the precursors with hydrazine (N_2H_4) and active hydrogen as the reducing agent.

2.3.3 Hydrothermal Synthesis

This technique is well known for the fabrication of NPs of metal oxide, iron oxide and lithium iron. This technique is based on two types of working processes such as batch hydrothermal and continuous hydrothermal process. In continuous hydrothermal process high rate of reaction was conducted within a very short time period. This is better than batch hydrothermal process (Hayashi and Hakuta 2010). This technique provides advantage on synthesis of NPs by optimizing size, morphology, composition and surface chemistry (Abedini et al. 2013).

2.3.4 Polyol Synthesis

In this technique, metal containing compounds are used as reaction medium which plays as a reducing agent and complex agent as well as dissolved stabilizing agents (Rahman and Green 2009). A wide range of metal based NPs such as Ag, Pt, Pd, Pt, Cu, ZnO is synthesized by this technique. Xia et al. reported the synthesis of Pt NPs using polyol process at varying molar ratios of $NaNO_3$ and H_2PtC_{16} (Herricks et al. 2004).

2.3.5 Chemical Vapour Deposition (CPD)

This process is carried out for the deposition of solid films from vapour phase under very high temperature condition. The films produced by this procedure consist of some ultrafine particle. This technique is also termed as chemical vapour reaction (CVR) also termed as chemical vapour condensation (CVC) and chemical vapour preparation (CVP). The precursor required for this synthesis is present in three different conditions (solid phase, liquid phase and gas phase) are converted to vapour form in the reactor under nucleation process. Jin et al. reported the synthesis of Cr doped ZnO (CZO) NPs using CVS process (Jin et al. 2007). Suffner et al. demonstrated the synthesis of fluorine doped SnO2 (FTO) NPs using CVS method

(Suffner et al. 2010). Further Lee et al. (2009) reported the fabrication of fullerene-likeWS2 NPs using CVS process.

2.3.6 Plasma Enhanced Chemical Vapour Deposition (PECVD)

This technique is also addressed as plasma assisted chemical vapour deposition (PACVD) and is generally considered for the deposition of thin films. A variety of NPS are also been synthesized with the help of this technique. This mechanism also includes vacuum processing system, gaseous precursor, power supply and heater. As mentioned before in CVD process the synthesis of thin film requires a high temperature but in PECVD process the requirement of temperature is comparatively less. Carbon NPs and hydrogenated SiNPs synthesis were also reported by MPECVED (Vach et al. 2006).

2.4 Biological Method for Synthesis of Nanoparticle

Biological synthesis of nanoparticle is also termed as green synthesis due to its environment friendly, cost effective and low toxic effect in the process of fabrication and synthesis of NPs. In these methods biological active agents such as bacteria, fungi, viruses, yeast are taking part in the synthesis process of metal oxide NPs. Biological assisted method is broadly divided into three categories such as (1) biogenic synthesis using microorganisms, (2) biogenic synthesis using biomolecules as the templates, (3) biogenic synthesis using plant extracts.

2.4.1 Biogenic Synthesis Using Microorganisms

Bio-reactors are primarily used for the synthesis of NPs. Prokaryotic agents such as bacteria, actinomycetes, fungi, algae and yeast are extensively used for this bio-reactor. Microorganisms grab target ions from their environment and then turn the metal ions into the element metal through enzymes generated by cellular activities. On the basis of location of NPs this process also can be classified into two categories as intracellular and extracellular. In the intracellular process the metal ions are transported into microbial cells to form NPs with the help of enzymes. In extracellular process, the synthesis of NPs involved in trapping of metal ions on the surface of the cells and reducing ions in the presence of enzyme (Zhang et al. 2011). Bacteria use proteins, enzymes, reducing agents to reduce metal ions. Bacterium *S. aureus* was used for the extracellular synthesis of bioactive Ag NPs with antimicrobial activities (Nanda and Saravanan 2009). Babu and Gunasekaran (2009) have reported about a new hyper metal resistance *Bacillus cereus* that form intracellular cereus crystalline Ag NPs. This system is also reported for its ability to synthesize semiconducting CdS nanocrystals (Nanda and Saravanan 2009). Sharma et al. explained the hidden potential of marine bacteria for the synthesis of AuNPs

(Sharma et al. 2012). In the context of eukaryotic microorganism Kowshik et al. reported the extracellular production of silver NPs with the help of silver tolerant yeast strain (MKY3) (Kowshik et al. 2003).

2.4.2 Biomolecules as Templates to Design Nanoparticles

Various biomolecules such as nucleic acid, membranes, viruses were used as templates for the NPs synthesis. DNA has well known for its huge attraction with transition metal ions. The process involves a reduction Au leading with the formation of Au atoms and the metal cluster develops into Au NPs (Zinchenko et al. 2014). Kundu et al. reported the synthesis of wide variety of NPs and nanoparticles assemblies using DNA as the templates or organic scaffolds (Kundu 2013; Majumdar et al. 2013; Kundu and Nithiyanantham 2014). Attempts were also made to design organosols of Os NPs and b-MnO2 NPs using DNA as scaffolds via homogenous reduction route with their promising applications in catalytic hydrogenation and oxidative polymerization of pyrrole, respectively (Anantharaj et al. 2014; Ede et al. 2015a). DNA has also been used as scaffolds to grow self-assembled NiWO4, ZnWO4 and MnWO4 nanoparticles with different morphologies (Nithiyanantham et al. 2014; Ede et al. 2015b; Nithiyanantham et al. 2015).

2.4.3 Biogenic Synthesis Using Plant Extracts

Nanoparticle synthesis in context to plant product such as plant extract, plant biomass is a very effective as well as eco-friendly process. These methods were frequently used to synthesize NPs of metal, bimetallic alloys and metal oxide (Iravani 2011). Akhtar et al. (2013) have adequately demarcated various plant biometabolites that could help in the preparation of NPs based on their valuable role as reducing agents and capping agents. Qu et al. (2011) synthesized wurtzite ZnO NPs from Zn-hyperaccumulator plant Sedum alfredii with mean size of 53.7 nm.

2.5 Conclusion

Scientists are continuously trying to investigate the novel and advanced approach for synthesis of nanoparticles. The optimal size as well as morphology of NPs play important roles on selection of NPs for various applications. To develop an ideal quality of NPs certain parameters such as ideal size and morphology are maintained. In this review we demonstrate distinct overview regarding different strategies for the synthesis of NPs. This review article may able to provide comprehensive information about different synthesis process of NPs, which could be beneficial for the researchers. Being a leading nanomaterial in targeted drug delivery and nanomedicines, precise size with specified surface characteristics are the key desirables during nanoparticle synthesis. Recently nanotechnology has become a key concept for its enormous technological ability, low toxic effect, cost effective characteristics. In this review electrospraying technique is introduced as a dry technique to design a polymeric nanoparticles except using a toxic chemical. It provides prospects to develop nanoformulations incorporating drugs/growth factors/biomolecules with versatile structural designs like core–shell NPs, hybrid NPs, composite NPs, etc., that will be useful in drug delivery, tissue engineering, biosensors.

References

Abedini A, Daud AR, Hamid MA, Othman NK, Saion E (2013) A review on radiation-induced nucleation and growth of colloidal metallic nanoparticles. Nanoscale Res Lett 8(1):474

Abhilash M (2010) In silico analysis of cranberry proanthocyanidins epicatechin (4beta-8, 2beta-0-7) as an inhibitor for modelled afimbrial adhesin virulence protein of uropathogenic Escherichia coli. Int J Pharm Bio Sci 1(1):1–7

Akbari MK, Derakhshan R, Mirzaee O (2015) A case study in vapor phase synthesis of mg–Al alloy nanoparticles by plasma arc evaporation technique. Chem Eng J 259:918–926

Akhtar MS, Panwar J, Yun YS (2013) Biogenic synthesis of metallic nanoparticles by plant extracts. ACS Sustain Chem Eng 1(6):591–602

Amaratunga GA, Chhowalla M, Kiely CJ, Alexandrou I, Aharonov R, Devenish RM (1996) Hard elastic carbon thin films from linking of carbon nanoparticles. Nature 383(6598):321–323

Amato R, Falconieri M, Gagliardi S, Popovici E, Serra E, Terranova G, Borsella E (2013) Synthesis of ceramic nanoparticles by laser pyrolysis: from research to applications. J Anal Appl Pyrolysis 104:461–469

Anantharaj S, Nithiyanantham U, Ede SR, Kundu S (2014) Osmium organosol on DNA: application in catalytic hydrogenation reaction and in SERS studies. Ind Eng Chem Res 53 (49):19228–19238

Andrea C, Neri F, Ossi PM, Santo N, Trusso S (2009) The controlled pulsed laser deposition of Ag nanoparticle arrays for surface enhanced Raman scattering. Nanotechnology 20(24):245606

Babu MG, Gunasekaran P (2009) Production and structural characterization of crystalline silver nanoparticles from Bacillus cereus isolate. Colloids Surf B: Biointerfaces 74(1):191–195

Becheri A, Dürr M, Lo Nostro P et al (2008) Synthesis and characterization of zinc oxide nanoparticles: application to textiles as UV-absorbers. J Nanopart Res 10:679–689. https://doi.org/10.1007/s11051-007-9318-3

Bello A, Fabiane M, Dodoo-Arhin D, Ozoemena KI, Manyala N (2014) Silver nanoparticles decorated on a three-dimensional graphene scaffold for electrochemical applications. J Phys Chem Solids 75(1):109–114

Benelmekki M, Vernieres J, Kim JH, Diaz RE, Grammatikopoulos P, Sowwan M (2015) On the formation of ternary metallic-dielectric multicore-shell nanoparticles by inert-gas condensation method. Mater Chem Phys 151:275–281

Böttger PM, Bi Z, Adolph D, Dick KA, Karlsson LS, Karlsson MN, Wacaser BA, Deppert K (2007) Electrospraying of colloidal nanoparticles for seeding of nanostructure growth. Nanotechnology 18(10):105304

Bouchat V, Moreau N, Colomer JF, Lucas S (2013) On some applications of nanoparticles synthesized in the gas phase by magnetron discharges. J Surf Eng Mater Adv Technol 3:184–189. https://doi.org/10.4236/jsemat.2013.33025

Boutonnet M, Kizling J, Stenius P, Maire G (1982) The preparation of monodisperse colloidal metal particles from microemulsions. Colloids Surf 5(3):209–225

Chen CN, Chen YL, Tseng WJ (2007) Surfactant-assisted de-agglomeration of graphite nanoparticles by wet ball mixing. J Mater Process Technol 190(1–3):61–64
Chen Y, Fitz Gerald JD, Williams JS, Bulcock S (1999) Synthesis of boron nitride nanotubes at low temperatures using reactive ball milling. Chem Phys Lett 299:260–264
Chen YH, Yeh CS (2002) Laser ablation method: use of surfactants to form the dispersed ag nanoparticles. Colloids Surf A Physicochem Eng Asp 197(1–3):133–139
Coto-García AM, Sotelo-González E, Fernández-Argüelles MT, Pereiro R, Costa-Fernández JM, Sanz-Medel A (2011) Nanoparticles as fluorescent labels for optical imaging and sensing in genomics and proteomics. Anal Bioanal Chem 399(1):29–42
Das M, Saxena N, Dwivedi PD (2009) Emerging trends of nanoparticles application in food technology: safety paradigms. Nanotoxicology 3(1):10–18
Dhand C, Dwivedi N, Loh XJ, Ying AN, Verma NK, Beuerman RW, Lakshminarayanan R, Ramakrishna S (2015) Methods and strategies for the synthesis of diverse nanoparticles and their applications: a comprehensive overview. RSC Adv 5(127):105003–105037
Ede SR, Anantharaj S, Nithiyanantham U, Kundu S (2015b) DNA-encapsulated chain and wire-like β-MnO 2 organosol for oxidative polymerization of pyrrole to polypyrrole. Phys Chem Chem Phys 17(7):5474–5484
Ede SR, Ramadoss A, Nithiyanantham U, Anantharaj S, Kundu S (2015a) Bio-molecule assisted aggregation of ZnWO4 nanoparticles (NPs) into chain-like assemblies: material for high performance supercapacitor and as catalyst for benzyl alcohol oxidation. Inorg Chem 54 (8):3851–3863
Gadomskii ON, Kharitonov YY (2004) Quantum computer based on activated dielectric nanoparticles selectively interacting with short optical pulses. Quantum Electron 34(3):249
Gleize P, Schouler MC, Gadelle P, Caillet M (1994) Growth of tubular boron nitride filaments. J Mater Sci 29(6):1575–1580
Goncalves LF, Kanodarwala FK, Stride JA, Silva CJ, Pereira MR, Gomes MJ (2014) One-pot synthesis of CdSe nanoparticles exhibiting quantum size effect within a sol–gel derived ureasilicate matrix. J Photochem Photobiol A Chem 285:21–29
Gracia-Pinilla M, Martínez E, Vidaurri GS, Pérez-Tijerina E (2010) Deposition of size-selected cu nanoparticles by inert gas condensation. Nanoscale Res Lett 5(1):180–188
Happy RS, Rawat RV, Ramanujan PL, Patran A, Tan TL, Springham SV, Lee S (2004) 31st EPS conference on plasma Phys. London, 28G, 1–4
Hayashi H, Hakuta Y (2010) Hydrothermal synthesis of metal oxide nanoparticles in supercritical water. Materials 3(7):3794–3817
He Y, Schwarz RB, Archuleta JI (1996) Bulk glass formation in the Pd–Ni–P system. Appl Phys Lett 69(13):1861–1863
Herricks T, Chen J, Xia Y (2004) Polyol synthesis of platinum nanoparticles: control of morphology with sodium nitrate. Nano Lett 4(12):2367–2371
Hsieh TF, Chuang CC, Chou YC, Shu CM (2010) Fabrication of nanoparticles on vertically aligned multi-wall carbon nanotubes by e-beam evaporation. Mater Des 31(4):1684–1687
Iravani S (2011) Green synthesis of metal nanoparticles using plants. Green Chem 13 (10):2638–2650
Jianing Z, Hans R. Sol-gel science: the physics and chemistry of sol-gel processing. C. J. Brinker, G. W. Scherer, Academic Press, Cambridge, MA, 1990
Jin W, Lee IK, Kompch A, Dörfler U, Winterer M (2007) Chemical vapor synthesis and characterization of chromium doped zinc oxide nanoparticles. J Eur Ceram Soc 27(13–15):4333–4337
Karak N (2009) Fundamentals of polymers: raw materials to finish products. PHI Learning, New Delhi, India
Koch CC (1998) Intermetallic matrix composites prepared by mechanical alloying – a review. Mater Sci Eng A Struct Mater 244:39–48
Kowshik M, Ashtaputre S, Kharrazi S, Vogel W, Urban J, Kulkarn SK, Paknikar KM (2003) Extracellular synthesis of silver nanoparticles by a silver-tolerant yeast strain MKY3. Nanotechnology 14:95–100

Krause B, Pötschke P, Häußler L (2009) Influence of small scale melt mixing conditions on electrical resistivity of carbon nanotube-polyamide composites. Compos Sci Technol 69 (10):1505–1515
Kundu S (2013) Formation of self-assembled ag nanoparticles on DNA chains with enhanced catalytic activity. Phys Chem Chem Phys 15(33):14107–14119
Kundu S, Nithiyanantham U (2014) DNA-mediated fast synthesis of shape-selective ZnO nanostructures and their potential applications in catalysis and dye-sensitized solar cells. Ind Eng Chem Res 53(35):13667–13679
Lee DW, Tolochko OV, Turaev FR, Kim D, Kim BK (2009) Synthesis and characterization of WS2 nanoparticles by chemical vapor condensation. Nanosci Nanotechnol 9:1–6
Lei JP, Dong XL, Zhu XG, Lei MK, Huang H, Zhang XF, Lu B, Park WJ, Chung HS (2007) Formation and characterization of intermetallic Fe–Sn nanoparticles synthesized by an arc discharge method. Intermetallics 15(12):1589–1594
Maicu M, Schmittgens R, Hecker D, Glob D, Frach P, Gerlach G (2014) Synthesis and deposition of metal nanoparticles by gas phase condensation (GPC) process. J Vac Sci Technol A 32:02B113
Majumdar D, Singha A, Mondal PK, Kundu S (2013) DNA-mediated wirelike clusters of silver nanoparticles: an ultrasensitive SERS substrate. ACS Appl Mater Interfaces 5(16):7798–7807
Malik MA, Wani MY, Hashim MA (2012) Microemulsion method: a novel route to synthesize organic and inorganic nanomaterials: 1st Nano update. Arab J Chem 5(4):397–417
Naessens K, Van Daele P, Baets R (2001) Direct writing of microlenses in polycarbonate with excimer laser ablation. SPIE Int Soc Opt Eng 4426:124–127
Nagavarma BV, Yadav HK, Ayaz AV, Vasudha LS, Shivakumar HG (2012) Different techniques for preparation of polymeric nanoparticles-a review. Asian J Pharm Clin Res 5(3):16–23
Nanda A, Saravanan M (2009) Biosynthesis of silver nanoparticles from Staphylococcus aureus and its antimicrobial activity against MRSA and MRSE. Nanomedicine 5(4):452–456
Nithiyanantham U, Ede SR, Anantharaj S, Kundu S (2015) Self-assembled NiWO4 nanoparticles into chain-like aggregates on DNA scaffold with pronounced catalytic and supercapacitor activities. Cryst Growth Des 15(2):673–686
Nithiyanantham U, Ede SR, Kesavan T, Ragupathy P, Mukadam MD, Yusuf SM, Kundu S (2014) Shape-selective formation of MnWO 4 nanomaterials on a DNA scaffold: magnetic, catalytic and supercapacitor studies. RSC Adv 4(72):38169–38181
Ong PL, Mahmood S, Zhang T, Lin JJ, Ramanujan RV, Lee P, Rawat RS (2008) Synthesis of FeCo nanoparticles by pulsed laser deposition in a diffusion cloud chamber. Appl Surf Sci 254 (7):1909–1914
Qu J, Luo C, Hou J (2011) Synthesis of ZnO nanoparticles from Zn-hyperaccumulator (Sedum alfredii Hance) plants. Micro Nano Lett 6(3):174–176
Rahman P, Green M (2009) The synthesis of rare earth fluoride based nanoparticles. Nanoscale 1 (2):214–224
Rumi M, Barlow S, Wang J, Perry JW, Marder SR (2008) Two-photon absorbing materials and two-photon-induced chemistry. In: Photoresponsive polymers I. Springer, Berlin, pp 1–95
Salata OV (2004) Applications of nanoparticles in biology and medicine. J Nanobiotechnol 2(1):3
Sevil B, Zuhal K (2010) Synthesis and characterization of polypyrrole nanoparticles and their nanocomposites with poly (propylene). In: Macromolecular symposia 2010 Sep, vol 295. Wiley, Weinheim, pp 59–64
Sharma N, Pinnaka AK, Raje M, Ashish FN, Bhattacharyya MS, Choudhury AR (2012) Exploitation of marine bacteria for production of gold nanoparticles. Microb Cell Factories 11(1):86
Sokolowski MA, Sokolowska A, Michalski A, Gokieli B (1977) The "in-flame-reaction" method for Al2O3 aerosol formation. J Aerosol Sci 8(4):219–230
Solanki JN, Murthy ZV (2011) Controlled size silver nanoparticles synthesis with water-in-oil microemulsion method: a topical review. Ind Eng Chem Res 50(22):12311–12323
Sridhar R, Lakshminarayanan R, Madhaiyan K, Barathi VA, Lim KH, Ramakrishna S (2015) Electrosprayed nanoparticles and electrospun nanofibers based on natural materials:

applications in tissue regeneration, drug delivery and pharmaceuticals. Chem Soc Rev 44 (3):790–814
Sridhar R, Ramakrishna S (2013) Electrosprayed nanoparticles for drug delivery and pharmaceutical applications. Biomatter 3(3):e24281
Suffner J, Ágoston P, Kling J, Hahn H (2010) Chemical vapor synthesis of fluorine-doped SnO 2 (FTO) nanoparticles. J Nanopart Res 12(7):2579–2588
Taheri S, Cavallaro A, Christo SN, Smith LE, Majewski P, Barton M, Hayball JD, Vasilev K (2014) Substrate independent silver nanoparticle based antibacterial coatings. Biomaterials 35 (16):4601–4609
Teoh WY, Amal R, Mädler L (2010) Flame spray pyrolysis: an enabling technology for nanoparticles design and fabrication. Nanoscale 2(8):1324–1347
Uhm SH, Song DH, Kwon JS, Lee SB, Han JG, Kim KM, Kim KN (2013) E-beam fabrication of antibacterial silver nanoparticles on diameter-controlled TiO2 nanotubes for bio-implants. Surf Coat Technol 228:S360–S366
Ullah M, Ali M, Hamid SB (2014) Surfactant-assisted ball milling: a novel route to novel materials with controlled nanostructure-a review. Rev Adv Mater Sci 1:37
Vach H, Brulin Q, Chaâbane N, Novikova T, i Cabarrocas PR, Kalache B, Hassouni K, Botti S, Reining L. (2006) Growth dynamics of hydrogenated silicon nanoparticles under realistic conditions of a plasma reactor. Comput Mater Sci 35(3):216–222
Veith GM, Lupini AR, Pennycook SJ, Villa A, Prati L, Dudney NJ (2007) Magnetron sputtering of gold nanoparticles onto WO3 and activated carbon. Catal Today 122(3–4):248–253
Ward MB, Brydson R, Cochrane RF (2006) Mn nanoparticles produced by inert gas condensation. J Phys Conf Ser 26(1):296
Xing T, Sunarso J, Yang W, Yin Y, Glushenkov AM, Li LH, Howlett PC, Chen Y (2013) Ball milling: a green mechanochemical approach for synthesis of nitrogen doped carbon nanoparticles. Nanoscale 5(17):7970–7976
Zhang X, Yan S, Tyagi RD, Surampalli RY (2011) Synthesis of nanoparticles by microorganisms and their application in enhancing microbiological reaction rates. Chemosphere 82(4):489–494
Zinchenko A, Miwa Y, Lopatina LI, Sergeyev VG, Murata S (2014) DNA hydrogel as a template for synthesis of ultrasmall gold nanoparticles for catalytic applications. ACS Appl Mater Interfaces 6(5):3226–3232

Classification, Synthesis and Application of Nanoparticles Against Infectious Diseases

3

Bhabani Shankar Das, Ankita Das, Abhisek Mishra, and Manoranjan Arakha

Abstract

Nowadays, resistant bacterial strains have become major threats to medical science, causing various infectious disease conditions against different traditional antibiotics. Medical world is also looking for potential antimicrobial agents against different resistant bacterial strains to replace conventional antibiotics. Due to advance physico-chemical properties, nanoparticles have pulled tremendous recognition from different groups of research for their advance uses in various streams of medical science. Unfortunately, currently engineered nanoparticles became a major challenge towards researcher for using nanoparticles against drug-resistant bacterial species. Although, there are several physical and chemical synthesis approaches to engineer nanomaterials, nevertheless most of them are not eco-friendly. Hence, different research groups have adopted green synthesis approaches for synthesis of nanoparticles. In this context, we highlights about different synthesis ways such as physical as well as chemical synthesis approaches to engineered nanoparticles along with green synthesis method, an eco-friendly approach in comparison to other conventional methods. Nanoantibiotics also deliver a new way of escaping present antimicrobial discovery model and also holds potential to become a new bioweapon toward

B. S. Das · A. Mishra
Institute of Technical Education and Research, Siksha 'O' Anusandhan (Deemed to be University), Bhubaneswar, Odisha, India

A. Das
Symbiosis Institute of Health Sciences (SIHS), Symbiosis International (Deemed University), Pune, Maharashtra, India

M. Arakha (✉)
Centre for Biotechnology, Siksha 'O' Anusandhan (Deemed to be University), Bhubaneswar, Odisha, India
e-mail: marakha@soa.ac.in

M. Arakha et al. (eds.), *Bio-Nano Interface*,
https://doi.org/10.1007/978-981-16-2516-9_3

bacterial resistant strains. Therefore, in this chapter we have discussed application of nanoparticle as a novel antibiotics with their antibacterial mechanism, could be potential solutions against different infectious diseases caused by various microorganisms.

Keywords

Nanoparticles · Green synthesis · Bacteria · Antibacterial activity · Infectious diseases · Nanoantibiotics

3.1 Introduction

Human vision and thoughts often give rise to newer scientific and technological tools. From such visions, nanotechnology came in twenty-first century, which created a latest milestone in the history of science and technology. Nanotechnology signifies a new revolutionary pathway for creation of new technological tools by using materials at nanometer scale (10^{-9} m), which is one billion times smaller than a meter. However, the term "nano" comes from a Greek word "nano," signifies "very small in size," which has been widely used for manipulation of materials at atomic scale (Rai et al. 2009). One of the great Nobel physicists Richard P Feynman introduced the concept of nanotechnology in 1959 by delivering a speech on "There's plenty of room at the bottom" on December, 1959 at the conference of (Feynman 1960) American Physical Society. (Feynman 1960). Nanotechnology has numerous advantages in various fields of science and technology, which includes the chemical, physical and also biological systems at nanoscales varying from single molecules or atoms to submicron in dimensions (Prince et al. 2016; Arakha et al. 2017; Arakha et al. 2016). Additionally, it has also the capability to change our views and perceptions and to provide us with the ability to solve global problems using different nanomaterials. Nowadays, there are numerous nanotechnological products available in the market and also research work is going on formulating novel nanoparticles, which could be used for human welfare. Recently, nanotechnology has become a multidisciplinary research area gaining fundamental understanding of the optical, magnetic and mechanical effects of nanostructures for wide range of applications (Nasrollahzadeh et al. 2019). Nowadays, researchers are trying to engineer nanomaterials with advanced physico-chemical properties for different applications in the field of pharmaceutical and biological sciences (Arakha et al. 2015a; Arakha et al. 2015b; Tiwari et al. 2018; Yadav et al. 2018).

Nanoparticles, being the building blocks of nanotechnology, can be found in nature as two types: natural nanoparticles and fabricated (man-made) nanoparticles. Natural nanoparticles are mainly found in dust, soil, fine sand, viruses, and volcanic ash, whereas fabricated (man-made) nanoparticles basically found in cosmetics items, sports items, plastics, textiles, gold and silver metals (Madeła et al. 2016). Further fabricated nanoparticles are divided into two sub-categories: that is incidental nanoparticles as well as engineering nanoparticles. Incidental nanoparticles arise from gasoline, fumes, smoke, fire and diesel engines, etc., whereas engineering nanoparticles contain different types of nanotubes, nanowires, quantum dots, etc. (Rai et al. 2009; Sau and Rogach 2010).

3.2 Classification of Nanoparticles

3.2.1 Dimensionality

Based on their dimensions, the nanomaterials are categorized as zero, one, two, and three dimensions. Such dimensionality represents the sum of degree of freedom in momentum of materials (Madeła et al. 2016).

Zero-dimensional nanomaterials are fixed in length, height and breadth at a particular point such as composite, particles array and core shell nanoparticles. Some useful applications of these types of nanomaterials are semiconductor, quantum dots, supercapacitor used in daily life. However, one-dimensional nanomaterials possess only single parameter. Different forms of nanomaterials are nano tubes, nano rods, nano wires, etc. These nanomaterials are used in various fields such as computer chips, eyeglass antireflection and rough coating, etc. Additionally, two-dimensional nanomaterials possess only length and breadth. For example, carbon coated nanobelts, fiber-layer films, and carbon coated nanoplates, which are widely used in solar photovoltaic, mobile phone, electronic structures, and semiconductor membranes, etc. Three-dimensional nanomaterials possess all the specifications such as length, height, and breadth. For example, fullerenes, colloidal particles, heterolayers, which are used in photonic crystals and pharmaceuticals.

3.2.2 Morphology

Nanoparticles are based upon different shape, structure and size. According to the morphology, nanoparticles are of various shape, structure, size and categorized having high and the low aspect ratio. High-aspect-ratio nanoparticles involve nanotubes and nanowires of different shapes, such as helices, zigzags, chains. Low aspect ratio includes circular, cubic, oval, helical, prism, or pole (Buzea et al. 2007).

3.2.3 Composition

A single substance or a mixture of many materials can be used for nanoparticles. Natural nanoparticles are mostly clusters of materials with various compositions that can easily be synthesized by different methods by single composite materials (Buzea et al. 2007).

3.2.4 Agglomeration and Uniformity

Nanomaterials can exist, depending on their chemistry and electromagnetic characteristics, like dispersed aerosols, colloids, suspensions, or agglomerates. Nanoparticles can act as larger particles in an agglomerate state, based on the size of agglomerate (Buzea et al. 2007). Therefore, all the shape, size, dimension, surface reactivity, and agglomeration of nanoparticles must be taken into considered for formulating new nanoparticles (Buzea et al. 2007).

3.3 Classification Based on Different Types of Nanomaterials

3.3.1 Inorganic Nanoparticles

Inorganic nanoparticles have raised the interest of many groups conducting research because of their advanced physico-chemical properties, small size and the surface plasmon resonance property. The size of inorganic nanoparticles mainly depends on concentration of metal salts, temperature, and rate of chemical reactions in medium.

3.3.2 Metal Oxide and Metal Nanoparticles

Metal oxide and metal nanoparticles are usually categorized under inorganic nanoparticles. Metal nanoparticles are the nanomaterials which consist of only one element. Such nanoparticles are synthesized by destructive or constructive methods from the metals to the nanometric sizes of metal nanoparticles (Salavati-Niasari et al. 2008). These nanoparticles are important for bio-labeling applications because of their excellent luminescent properties, producing luminescent patterns and fluorescent resonances (Salavati-Niasari et al. 2008). Metal nanoparticles are developed by simple techniques including bio-assisted synthesis, hydrothermal synthesis, and a microwave-assisted synthesis in the form of colloidal solutions. The characteristics of these nanoparticles include size from 1 to 100 nm, high surface area, high pore size, surface charge density and surface charge, crystallinity, etc. (Salavati-Niasari et al. 2008). Aluminum (Al), cadmium (Cd), cobalt (Co), copper (Cu), gold (Au), iron (Fe), lead (Pb), silver (Ag), and zinc (Zn) are widely used for metal nanoparticles synthesis. The bimetallic nanoparticle groups (e.g., Ni-Cu, Pd-Pt) mostly used as core-shells as well as in alloy structures, which are also used in metal nanoparticles (Zaleska-Medynska et al. 2016). Additionally, because of advanced mechanical, catalytic, antimicrobial, optical, anti-cancer and anti-viral properties, metal nanoparticles are extremely useful for various biomedical applications.

One of the most stable natural compounds is metal oxides. They are made up of by combining electrode charges of metal cation and oxide anioin with in a metal reaction processes (Yin and Tang 2016). The polar surfaces have anionic oxygen and are insoluble in most organic solvents due to tight bonds between metal and oxygen.

(Yin and Tang 2016). Metal oxide nanoparticles have attracted the attention of scientific groups for various approaches because of their advanced physico-chemical properties like supercapacitors, fuel cells, sensors, catalysis, flame retardants, good band gap/tunable band gap, chemical stability, outstanding thermal properties, etc. (Ray and Kuruma 2020). Metal oxide nanoparticles have been synthesized to enhance their properties, such as iron nanoparticles (Fe) that immediately oxidize iron oxide (Fe_2O_3) at room temperature when oxygen is present which increases their reactivity with regard to Fe nanoparticles (Tai et al. 2007). Some important metal oxide nanoparticles are Cerium oxide (CeO_2), Iron oxide (Fe_2O_3), Aluminum oxide (Al_2O_3), Magnetite (Fe_3O_4), Silicon dioxide (SiO_2), Titanium oxide (TiO_2), Zinc oxide (ZnO) (Tai et al. 2007). Metal oxide nanoparticles are widely used in different areas like in water purification, nutrition, pharmaceutical, electricity and environmental remediation, etc. (Kumar and Ray 2018).

3.3.3 Organic Nanoparticles

Organic nanoparticles are usually biodegradable and non-toxic in nature. Organic nanoparticles are usually known for dendrimers, micelles, liposomes, synthetic molecules, lightweight polymers and ferritin, etc. (Ealias and Saravanakumar 2017). Particles like micelles and liposomes have a hollow core, which are often called nanocapsules. They are also prone to thermal as well as electromagnets like heat and light (Tiwari et al. 2008). Liposomes are vesicles of phospholipids and are made up of only lipid compounds. Most liposomes are between 50 and 100 nm in size, while other liposomes are between 100 and 800 nm. Generally, liposomes are made up of amphiphilic compounds and they are spherical in shape. Additionally, liposomes are biocompatible, biodegradable, flexible, non-toxic, non-immunogenic, and effective at trapping (Akbarzadeh et al. 2013). Micelles are made up of amphiphilic molecules (polymers or lipids) and are between 10 and 100 nm in size. Depending on the environment, they only provide hydrophilic and hydrophobic surfaces. For example, in aqueous mediums, only hydrophilic groups are exposed as well as hide their hydrophobic parts inside the structure. Additionally, micelles have high drug strength, long circulation, and high biological stability (Orive et al. 2009). Dendrimers are highly branched and one or more nuclei have been developed. The scale of such nanoparticles can be regulated by the number of generations developed in the cores and its size is about less than 10 nm. Dendrimers are one-disperse polymer system synthesized with three structural parts such as surface, core, and branch which are regulated by polymerization (Adair et al. 2010). Specific methodologies were used to made polymeric nanoparticles such as rapid expansion into the supercritical solution (RESS), rapid expansion into a solvent (RESOLV) (Rapid Expansion of a Supercritical solution into a liquid solvent), and polymerization of emulsion free surfactants (Rao and Geckeler 2011).

3.3.4 Carbon Nanoparticles

Carbon has a unique property that forms covalent bonds with rest carbon atoms in various hybridization states such as SP, SP^2, and SP^3 to form different structures of long chains and small molecules. Carbon is a flexible material with different chemical, optical, mechanical, and electrical properties that can be modified by the structure and its surface chemistry (Georgakilas et al. 2015). Nanoparticles are made of entirely carbon materials which are called as carbon nanoparticles. It may be categorized into fullerenes, graphenes, carbon nanotubes, carbon nanofibers, black carbon as well as activated carbon in nano size form.

3.4 Synthesis of Nanoparticles

Numerous methods for synthesis of nanostructures have been used over the last two decades. They are designed to manufacture nanostructures for particular applications. Various techniques such as top-down and bottom-up methods are available for synthesizing nanoparticles.

3.4.1 Top-Down Method

Top-down method involves breaking down of larger molecules into single atomic form according to the need. This approach is suitable for the creation of structures at nanometer scale. This technique is mostly suitable for the synthesis of organized and integrated nano structures, most of which are used in electronic circuit (Mohan et al. 2018). The benefit of this approach is the ability to integrate the appropriate entity or a function at an exact location as the process is regulated externally.

3.4.2 Bottom-Up Method

In the bottom-up method, single molecules and atoms are integrated to produce nanostructures. During this process, nano structures are synthesized by connecting atoms together that form new substrates. They are further stacked on each other, and formed new nano structures. This technique involves creating buildings of molecular structure using atom by atom method (Lee and Jun 2019). These are best suited for assembly and establishment of short-range structures at the nanometer (nm) scale. Bottom-up approach synthesis is cost effective in comparison to top-down approach. This is indeed a very powerful way to create identical atomic structures, though the man-made materials that have been created so far are far simpler than the complex structure (Lee and Jun 2019).

3.5 Physical Methods for Synthesis of Nanoparticles

3.5.1 Mechanical Milling/Ball Milling of Nanoparticles

Nanomaterials are formed in a mechanical device, generically referred as ball milling. High energy ball milling (HEBM) is the most popular method used for nanomaterials synthesis (El-Eskandarany 2013). In this process the mechanical forces involved in high energy ball milling helps in breaking the bulk materials into nano form. For HEBM, the samples of the starting material are prone to extreme deformation, cold work hardening, and eventual fragmentation (Mukasyan et al. 2015). It is the way nanocrystals from metal oxides like cerium (CeO2) and zinc oxide (ZnO) are prepared. It can also be used to manufacture carbon and boron nitride nanotubes. In the ball milling process, other nanoparticles such as SiO_2, Al_2O_3, TiO_2, and nanosilica are also used (Cao 2007). The purpose of using this technique is to bring the particle size into nanoform. Mechanical methods have been used to manufacture amorphous and nanocrystalline materials as well as metal and non-metal nanocomposite materials by milling process. It is a more economical method for large-scale production of nanosized particles.

3.5.2 Laser Ablation

The process of laser ablation depends on the physical characteristics of the metals and the medium of the atmosphere. It works by synthesizing nanoparticles by the use of laser ablating techniques in a solid target particularly in a gaseous as well as the liquid environment which was further collected in nanopowder or a colloidal solution as nanoparticles. The principle of the laser ablation process is to remove surface atoms by the evaporation process. It is a rapid and easy method for the synthesis of nanoparticles. A wide variety of nanoparticles such as semiconductor quantum dots, metal and metal oxides, carbon nanoparticles, nanowires as well as core–shell nanoparticles are synthesized using this process (Kim et al. 2017). This method uses laser technology as a primary source to ablaze solid materials. At the time of synthesis, enormous energy is used to evaporate the light-absorbing material at a certain point on a solid surface. Laser ablation can generate high purity nanoparticles as the purity of the particles depend upon the purity of targeted as well as non-contaminated gas and liquid media from the reactor. The production of high purity nanoparticles in quantum size is achieved by rapid vapor technique (Kim et al. 2017). Characteristics of nanoparticles specifically size, shape, composition and structure, the size distribution of each target material depend on the laser parameters used for ablation such as wavelength, pulse replicate rate, pulse width, pulse power, and affluence along with on the environment in which the generation occurs like in vacuum, gas, or liquid. In this process, there is no need for long reactions, high temperatures, or synthetic multi-step chemicals. Hence, it is an eco-friendly and laboratory safe method for nanomaterial synthesis because it does not use any hazardous and highly flammable chemicals (Semaltianos 2010).

3.5.3 Sputtering

Sputtering is generally a Physical Vapor Deposition (PVD) vacuum process which is mostly used for depositing films and nanoparticle formation from materials. Sputtering is typically performed under high vacuum pressure and dry conditions. The basic principle behind sputtering is bombardment of energetic ions (usually argon ions) from the plasma of a gaseous discharge on a target (cathode of the discharge) which creates vapor. Radiofrequency (RF) controller, DC, pulsed DC are used to make these materials using the sputtering method (Lugscheider et al. 1998). Sputtering is a suitable method to make nanoparticles with materials having high melting points. There are three types of sputtering, i.e. RF sputtering, magnetron sputtering and ion beam sputtering. Recently, two new methods are used for the deposition of thin films which are used in the superconducting and magnetic film. The advantages of the sputtering method are controlled film thickness, high purity, low temperature, high speed and desirable adhesion, easy to operate, and environmentally friendly. The main benefit of the method is that highly pure and even thin films are created and are designed to sputter both metal (DC sputtering) and semiconductor (RF sputtering). In the last few years, nanostructures and nanoparticles were also used in this technique (Siegel et al. 2011). Using this process, nanoparticles like Au, Ag, and Cu as well as metal nanoparticles such as Palladium (Pd), Platinum (Pt), Rhodium (Rh), Iridium (Ir), Ruthenium (Ru), Tungsten (W), Molybdenum (Mo), Niobium (Nb), Titanium (Ti), Indium (In), Tin (Sn), as well as Zirconium (Zr) and oxide nanoparticles like titanium oxide (TiO_2), copper oxide (CuO), and tantalum oxide (Ta_2O_5) are synthesized. The sputtering method is a highly efficient method for the synthesis of high purity and high-performance materials. It also provides new research opportunities, especially through the synthesis of liquid colloidal nanoparticles and nanoclusters (Ishida et al. 2017).

3.6 Chemical Methods for Synthesis of Nanoparticles

3.6.1 Sol–Gel Technique

Sol–gel technique is one of the well developed synthetic technique which is used to prepare new nanoparticles like metal and metal oxide. In this technique, sol is considered as colloidal solution of solids suspended in a liquid phase, whereas gel is considered as solid macromolecule submerged in a solvent (Gonçalves and Marques 2019). The initial stage of this method is the starting material into a sol, which is a colloidal solution for the production of a gel (Fig. 3.1). This gel consists of individual particles and commonly used precursors in this sol–gel process are metal oxides and chlorides (Ramesh 2013). Such precursors are hydrolyzed and condensed to colloid formation by shaking, stirring and sonication methods. This method is often favored because of its economic viability and low temperature method that gives us control over the nanoparticles product composition. By the sol–gel method

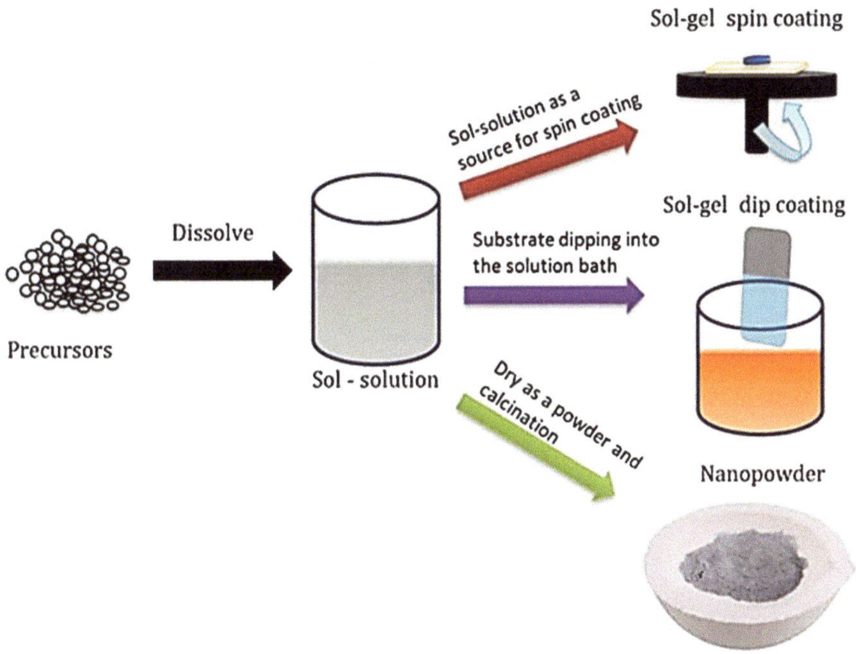

Fig. 3.1 Nanoparticles formed by sol–gel process. Adapted with permission from Ref (Thiagarajan et al. 2017)

silver, titanium oxide, gold, zinc and silica nanoparticles are synthesized like SiO2, TiO2, and ZrO2 (Epifani et al. 2000)

3.6.2 Micro-Emulsion Technique

Micro-emulsion is one of the most recent and ideal techniques for synthesis of inorganic nanoparticles. For the nanoparticles synthesis, this technique is used along with three different unmixable fluids such as oil in water, water in oil, and water in supercritical CO2 (Fig. 3.2) (Simonazzi et al. 2018). This micro-emulsion has three components in a single phase such as water, oil, and surfactant. Generally, water and oil are usually immiscible but they are miscible if the surfactant is used and also capable to overcome the interfacial tension between those fluids. Micro-emulsion is made up of surfactant aggregates in the ranges 1 Nm to 100 Nm. It constitutes mainly in three phases such as water, oil, and surfactant that influences the aggregate geometry.

When water is the bulk fluid, the micro-emulsion is considered to be oil along with water (O/W), and oil contains less in quantity, with minimal quantities of surfactant. Similarly, if oil is the bulk fluid, it is considered to be water along with oil (W/O), and water contains less quantity. From this combination in this process micelles are formed which are made up of combination of oil in water and surfactant

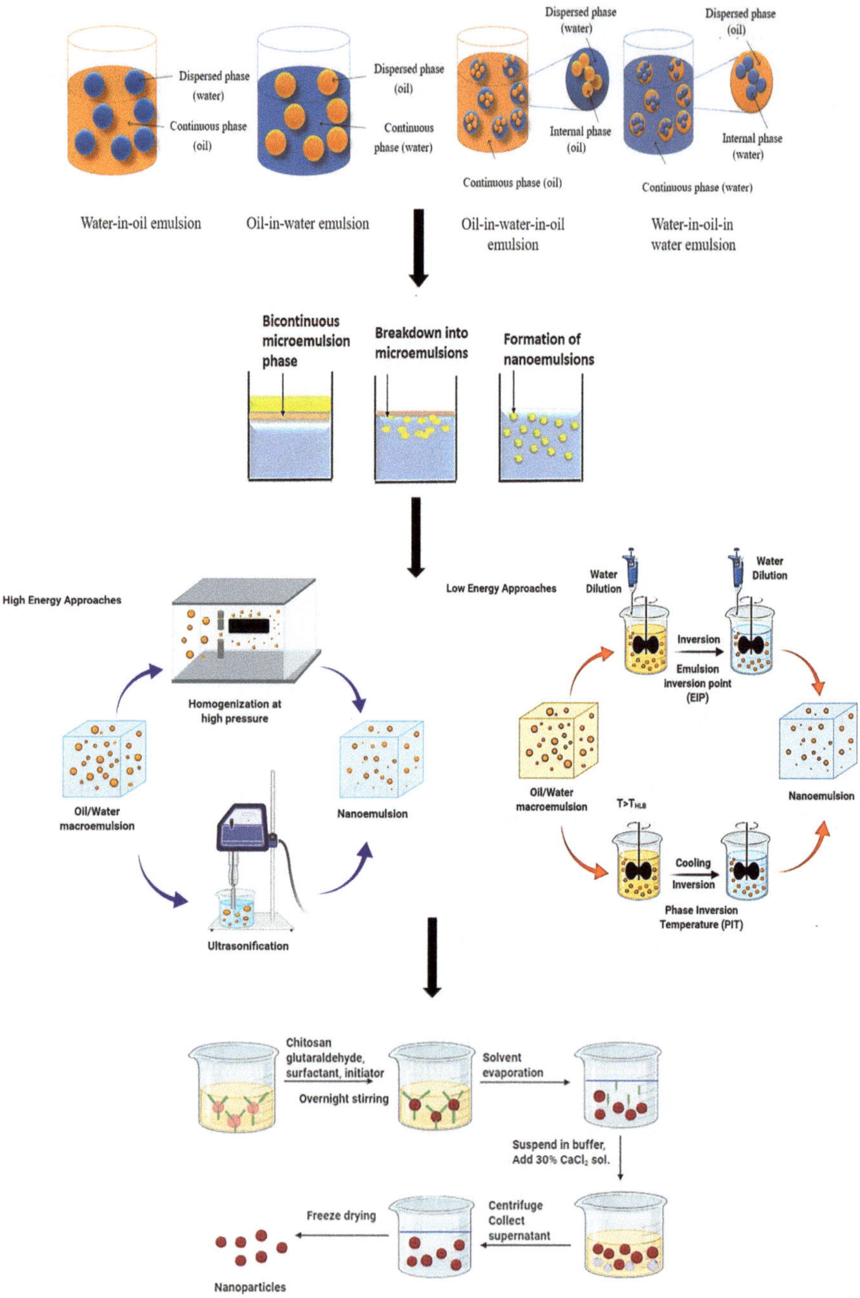

Fig. 3.2 Nanoparticles formed by micro-emulsion process. Adapted with permission from Ref (Yahaya Khan et al. 2014; Aswathanarayan and Vittal 2019; Gupta et al. 2016; Chandra Hembram et al. 2016)

which gives an aggregate shape to minimize free energy (Ghorbani 2014). For the synthesis of metal, metal oxide, magnetic and composite nanoparticles, this technique is applied.

3.6.3 Electrochemical Technique

Here, electricity is used as controlling power in this electrochemical synthesis process for the creation of nanoparticles. In this technique, nanoparticles are synthesized when electric current is passed between electrodes and electrolyte interface. Advantages of the electrochemical techniques include avoiding vacuum systems which are used in physical techniques, reasonable cost, quick operation, high durability, easy availability of equipment and instruments, less pollution (pure product,) and environmentally safe process (environment-friendly). Scientists have been carried out research work on this electrochemical technique to advance the basic understanding and industrial applications, but many aspects of this technique are still under review (Murty et al. 2013).

3.7 Green Synthesis Approaches for Synthesis of Nanoparticles

Some physical and chemical approaches have been well established for synthesis of different nanomaterials. These approaches have limitations with the use of toxic chemical agents, hazardous material processing, costly chemicals, and high energy use. Various chemical methods have been established for nanoparticle synthesis; however, these processes have some limitations like product stability, aggregation of particles, and control of crystal growth for better exposure for long term. Additionally, physical processes require the use of expensive machines, high temperature and pressure, large space required for system setup (Chandra Hebram et al. 2016). Nanoparticles applications are limited, because of the use of toxic chemicals and organic solvents during the process of synthesis. Compared with traditional approaches, environmental friendly techniques for nanoparticle synthesis must be needed. Therefore, nowadays the green synthesis methods are considered as alternate source for the nanoparticle synthesis (Musarrat et al. 2011).

The green synthesis approaches are non-toxic, environmentally friendly and limited the use of reducing and stabilizing agents, hence they have drawn the attention of scientist (Jegadeeswaran et al. 2012; Nayak et al. 2016; Panda et al. 2016; Sharma et al. 2018). In addition, the biological methods are used for nanoparticles synthesis using biological agents such as plants, fungi, algae, actinomycetes, yeast, bacteria, and viruses (Arakha and Jha 2018; Gaidhani et al. 2013; Mohanpuria et al. 2008; Mukherjee et al. 2001; Nagajyothi and Lee 2011; Nayak et al. 2018; Thakkar et al. 2010). It is reported that metal ions reduction rate by different biological agents such as plant extracts, bacteria, fungi is rapid at normal pressure and ambient temperature in comparison to other conventional methods

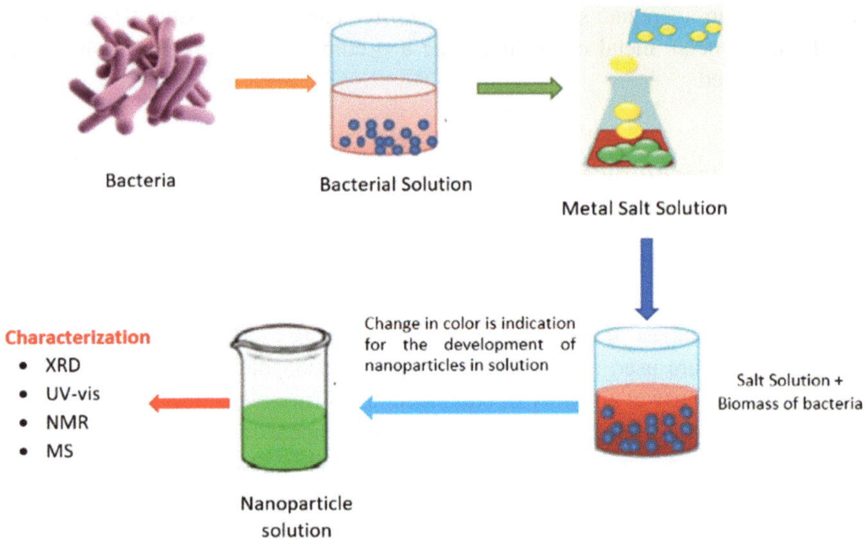

Fig. 3.3 Bacterial synthesis of nanoparticle. Adapted with permission from Ref (Gour and Jain 2019)

(Singh et al. 2018). However, among different microorganisms, bacteria have attracted various researchers for synthesis of different nanoparticles since they are resistant to different concentrations of heavy metals (Panda et al. 2016). Also, bacteria possess higher catalytic activity, larger surface area, easy to control and also help to improve the enzymatic and metal salt reaction for nanoparticles synthesis (Seo et al. 2006). Some bacterial species also developed specific defense mechanism to decrease toxicity of heavy metals ions (Durán et al. 2011). Figure 3.3 demonstrates the synthesis of nanoparticles using bacteria adopting green synthesis methods.

3.8 Nanoparticles Synthesis Using Bacteria

Bacteria have significant ability to resist metal ions, as well as adapt to utmost environmental conditions, hence considered to be potential candidate for nanoparticle synthesis. Hence, bacteria are considered as biofactory for the various nanoparticles synthesis (Choi et al. 2018). Bacteria are also used as modern realistic agents for green synthesis techniques and nanoparticles are synthesized using the microbial enzymes, vitamins, polysaccharides, proteins present in the bacteria (Iravani 2014). It is reported that bacteria can synthesize nanoparticles using both intracellular and extracellular mechanism (Das et al. 2017; Nayak et al. 2016). But mechanism behind intracellular and extracellular production of nanoparticles varies from microbes to microbes. Since the cell wall of bacteria is negatively charged so it attracts the positive metal ions and hence forms an electrostatic interaction. This prompts the enzymes of bacterial cell wall to reduce the metal ions into nanoparticles. This process is known as intracellular mechanism. The so formed nanoparticles can be diffused across the cell membrane to the solution. However, in the extracellular process, nitrate reductase enzyme helps to reduce the metal ions into nanoparticles (Das et al. 2017) (Fig. 3.4).

From extensive study, Nayak et al. also have proposed an intracellular synthesis of silver nanoparticles utilizing bacteria *Bacillus thuringiensis*. They have reported that the whole synthesis process has two important events such as the Ag (I) reduction into elemental silver Ag(0) as well as the formation of the nanocrystals by capping the silver elements by proteins available in bacterial cells. However, in the process of exocytosis/cellular efflux system, the synthesized nanoparticles will be secreted out of the bacterial cell membrane (Nayak et al. 2016). Hence, different scientific groups have synthesized different nanoparticles like gold, silver, sulfide, cadmium, copper, etc. using different bacteria (Table 3.1).

3.9 Application of Nanoparticles

3.9.1 Nanoparticles as Novel Antibiotics

Nowadays, treatment of bacterial infection using different conventional antibiotics is becoming a major challenge due to expression of resistant factors of bacteria. Bacterial genes are expressed based on environmental factors which lead to utilization of various biochemical pathways to escape the lethal action of antibiotics. Some bacteria like vancomycin-resistant *Enterococcus*—VRE, carbapenem-resistant *Enterobacteriaceae*—CRE, methicillin-resistant *Staphylococcus aureus*—MRSA, and multi-drug resistant *Mycobacterium tuberculosis*—MDR-TB are becoming resistant to conventional antibiotics and hence becoming a serious threat to human mankind throughout the globe. Bacteria exhibit resistance to many antibiotics by different mechanisms such as modifying the cell components and certain enzymes which degrade or modify the antibiotics, such as aminoglycosides and β-lactamases, drug influx through porin channel becomes poor, genetic alteration, etc. The majority of bacteria living in biofilm form also relate to specific species that help to interact with each other and the environment. Biofilms are microbial aggregates, primarily based on a solid surface and on extracellular material, i.e. extracellular polymeric substances (EPS) (Romero et al. 2010). But the expression of EPSs makes the connection irreversible that bacteria travel reversibly on the surface. The synthesis of the bacterial flagellum is blocked when the bacteria are settled down, and then bacteria reproduce rapidly, thereby leading to a biofilm. In the above process, bacteria are trapped and form a barrier that can resist antibiotics and lead to chronic systemic infections.

The rampant use of antibiotics has brought numerous health hazards, like superbugs which do not react with medicinal products as well as epidemics with no defense mechanism for killing bacterial infections. The inefficient conventional approach of drug delivery often contributes to weak therapeutic indexes and low bioavailability of drugs. Keeping all of things, antimicrobial nanoparticles as well as drug carriers of

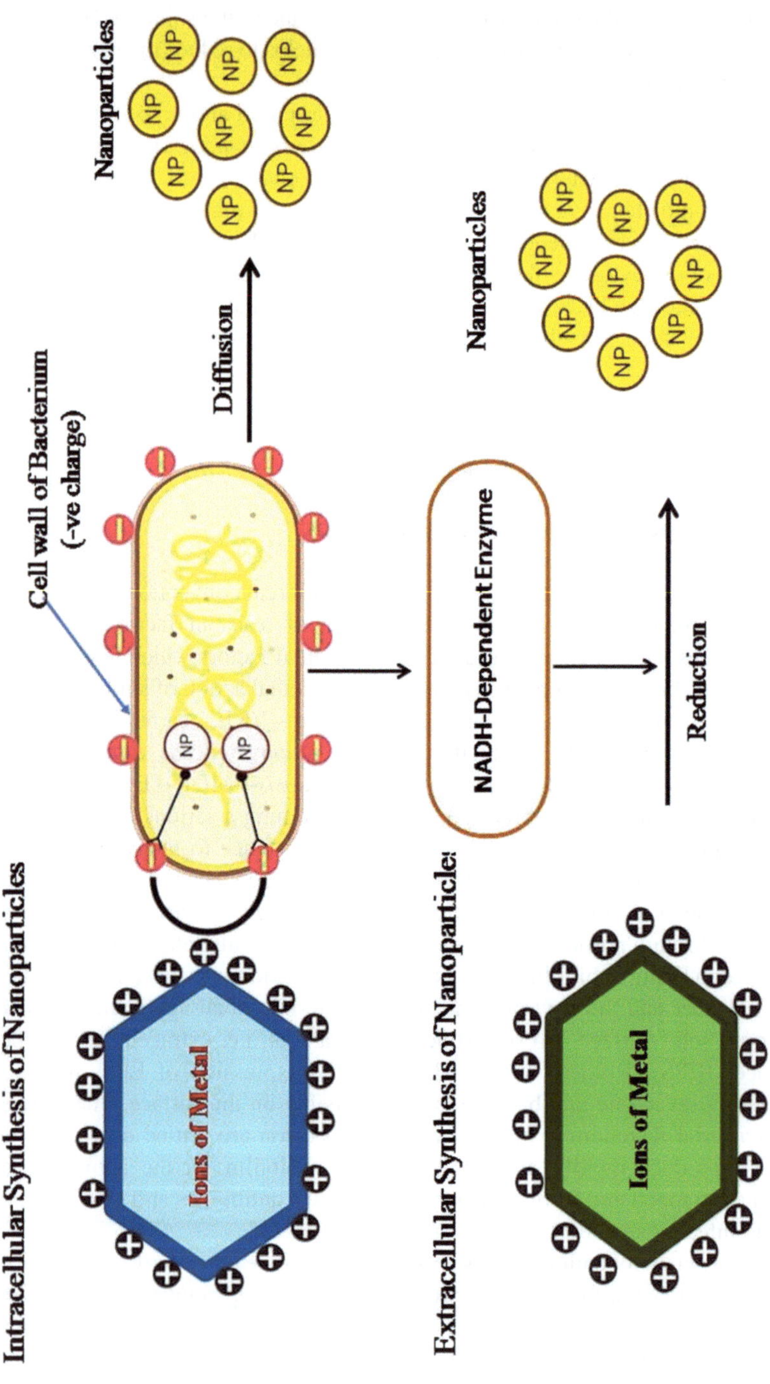

Fig. 3.4 The extracellular and intracellular mechanism for nanoparticles

Table 3.1 Bacteria synthesizing different types of nanoparticles

Bacteria	Nanoparticles	Size (nm)	References
Escherichia coli VM1	Silver	10–15	Maharani et al. (2016)
Weissella oryzae	Silver	10–30	Singh et al. (2016)
Bacillus methylotrophicus	Silver	10–30	Wang et al. (2016)
Bacillus cereus NCIM 2458	Copper	26–27	Tiwari et al. (2016)
Deinococcus radiodurans	Gold	43	Li et al. (2016a)
Bacillus cereus	Gold	20–50	Pourali et al. (2017)
Kinneretia THG-SQI4	Silver	15–25	Singh et al. (2017)
Pseudomonas putida	Silver	6–10	Gopinath et al. (2017)
Nostoc sp. HKAR-2	Silver	51–100	Sonker et al. (2017)
Shewanella oneidensis	*Cadmium sulfide*	15	Wang et al. (2018)
Streptomyces xinghaiensis	*Silver*	64	Wypij et al. (2018)
Botryococcus braunii	Silver	40–90	Arya et al. (2019)

nanosized have developed as potential effective agents those fight against the infectious diseases. Due to the ultra-smaller form and advanced physico-chemical properties, nanoparticles have the potential to kill the bacteria (Niemeyer 2001). Metal and metal oxide nanoparticles act as antimicrobial agents. These nanoparticles work on different antimicrobial mechanisms against different pathogens.

The increased utilization of nanomaterials especially in biomedical science increased the demand to investigate the antibacterial mechanisms of nanoparticles (Bayda et al. 2018). Different forms of nanoparticles reveal various mechanisms to overcome resistance against bacteria. The metabolic behavior of the bacteria also changes with nanoparticles by electrostatic interaction, van der Waals interaction, hydrophobic interactions, and receptor–ligand interaction (Choi et al. 2017). Such nanoparticles can also cross the barrier of bacterial membrane and assemble with metabolic pathways that affect the shape and function of the cell membrane of nanoparticles. Eventually, these nanoparticles interact with the basic components of bacterial cells that cause oxidative stress, modifications in gene expression, permeability, protein deactivation, various alterations of proteins, and enzyme inhibition (Yang et al. 2009). The following mechanisms are widely suggested in the current research which include oxidative stress, non-oxidative processes, and release of metal ions which is shown in Fig. 3.5.

Nanoparticles are considered to be the best possible therapeutic drug carriers by passive targeting or active targeting for effective therapy against drug resistant bacteria (Zaidi et al. 2017). Alternatively, nanoparticles are also recognized to cause genotoxicity, cytotoxicity, immunotoxicity based on the synthetic technique, size dependent, nanoparticles concentration, and mechanism of action against bacteria (Bertrand and Leroux 2012; Schrand et al. 2010). Therefore, as antibiotics carriers, nanoparticles can be good choice for delivery of antibiotic to clear toxicity issue. Although, the applications of nanoparticles as antibiotics are not explored completely, however, there is limited study on the biosynthesis of combinations of nanoparticles, which is important in the resistance era to accept toxic free antimicrobials (Smekalova et al. 2016; Hwang et al. 2012). The following table (Table 3.2) shows the

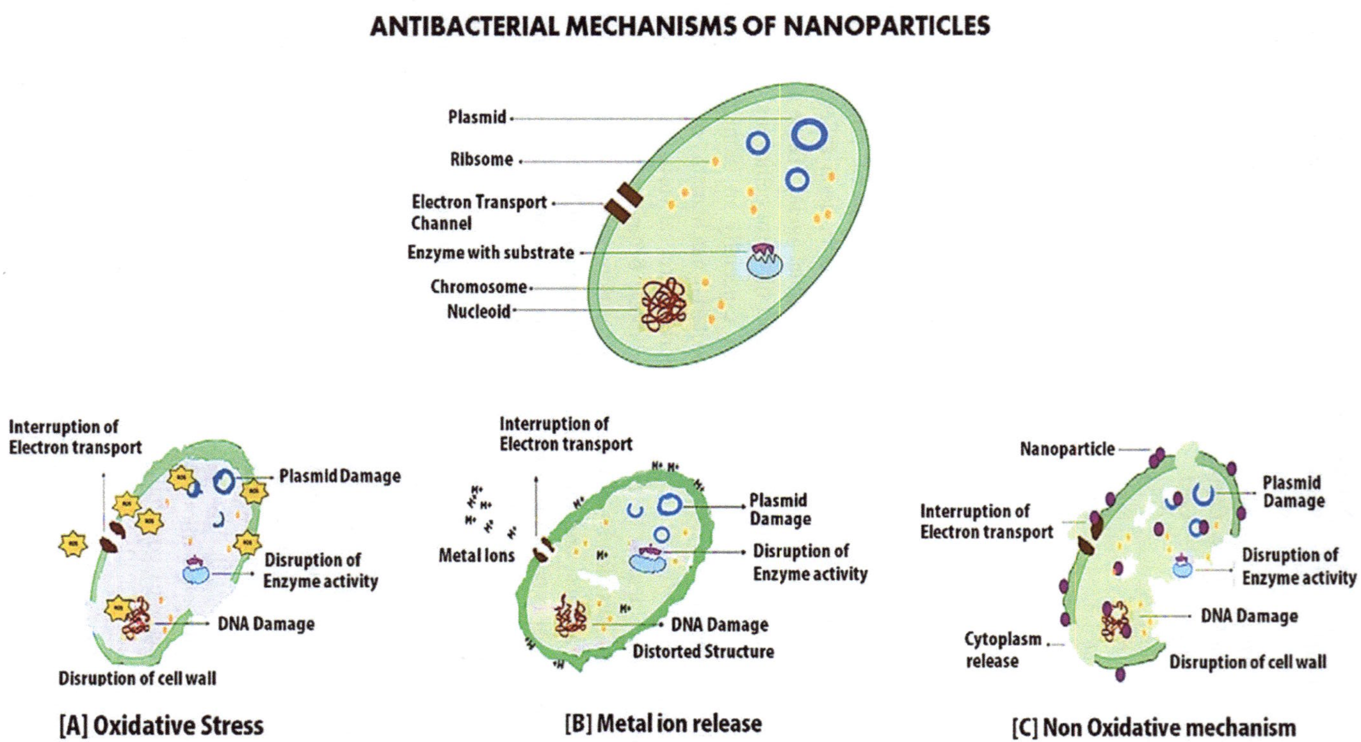

Fig. 3.5 Diagrammatic representation of antibacterial mechanism of nanoparticles. Adapted with permission from Ref (Zaidi et al. 2017)

Table 3.2 Enhanced antimicrobial activity of different nanoparticles upon conjugation with different antibiotics against different target microorganisms

Nanoparticle	Targeted microorganism	Conjugation/coating	Mode of study	References
Gold nanoparticle	*Pseudomonas aeruginosa*	Antimicrobial peptide esculentin – 1a (1–21) NH2	In vitro	Casciaro et al. (2017)
Gold nanoparticle	*Klebsiella pneumonia*	Chlorhexidine (au CHX)	In vitro	Ahmed et al. (2016)
Gold nanoparticle	Methicillin-resistant *Staphylococcus aureus*	Amoxicillin conjugation	In vitro and in vivo	Kalita et al. (2016)
Gold nanoparticle	*Pseudomonas aeruginosa, Salmonella typhimurium, Listeria monocytogenes, Streptococcus aureus, Escherichia coli*	Chitosan-streptomycin conjugation	In vitro	Mu et al. (2016)
Gold nanoparticle	Methicillin-resistant *Staphylococcus aureus*	IgG conjugation	In vitro	Panáček et al. (2016)
Silver nanoparticles	*Acinetobacter baumannii*	Citrate capping	In vitro and in vivo	Wan et al. (2016)
Silica-gentamycin (SG) nanoparticles	*Staphylococcus aureus*	Incorporation within a gelatin matrix and cross-linking on microarc oxidized titanium	In vitro	Wang et al. (2017)
Mesoporous silica nanoparticle (MSN)	*Francisella tularensis*	Functionalized with disulfide snap-tops.	In vivo	Lee et al. (2016)
Nanoporous silica nanoparticles (NPSNPs)	*Streptococcus mutans* and *Staphylococcus aureus*	Modified with poly (4-vinylpyridine)	In vitro	Sambhy et al. (2008)
Porous silicon nanoparticles (pSiNPs)	*Pseudomonas aeruginosa and Staphylococcus aureus*	Metallic silver deposition, TAT peptide conjugation	In vitro and in vivo	Kim et al. (2016)
Quantum dots and magnetic beads	*Brucella spp.*	Polyclonal antibody (IgG and IgY) conjugation	In vitro	Song et al. (2017)
Quantum dots	*Escherichia coli, Streptococcus aureus, Bacillus*	Spermidine-capping	In vitro and in vivo	Li et al. (2016b)

(continued)

Table 3.2 (continued)

Nanoparticle	Targeted microorganism	Conjugation/coating	Mode of study	References
	subtilis, *Pseudomonas aeruginosa*, and methicillin-resistant *Streptococcus aureus (MRSA)*			
Silver quantum dots	*Mycobacterium smegmatis* and *Mycobacterium bovis*	Transferrin conjugation	In vitro and ex vivo	Mahajan et al. (2012)
Multi-walled carbon nanotubes	Methicillin-resistant *Staphylococcus aureus*	IgG conjugation	In vivo	Mocan et al. (2016)
Single walled carbon nanotubes with antimicrobial peptide	*Staphylococcus aureus*, *Streptococcus pyogenes*, *Salmonella enterica*, and *Escherichia coli*	Silver coating	In vitro	Chaudhari et al. (2016)
Core–shell magnetic nanoparticle (MNPs)	Methicillin-resistant *Staphylococcus aureus* (MRSA) and multi-drug resistant *Pseudomonas aeruginosa* (MDR)	Conjugated antibacterial peptides cathelicidin LL-37, synthetic ceragenins CSA-13 and CSA-131, vancomycin and colistin	In vitro	Niemirowicz et al. (2016)

antimicrobial activity of different nanoparticles upon conjugation with different antibiotics against different target microorganism.

3.9.2 Nanoparticles as Therapeutic Agents Against Infectious Diseases

Metals and metal oxides nanoparticles could be used as effective antimicrobials primarily because of their small size and advanced chemical as well as the physical properties such as high reactivity and improved surface to volume ratio (Weir et al. 2008). Different researchers have investigated and demonstrated their ability as nanoantibiotics to control infectious diseases. Nanoantibiotics are nanomaterials having capacity to boost the effectiveness of existing antibiotics and improve their antimicrobial activity against resistant bacteria. These nanoantibiotics also have an effective cargo transport system, which can directly target drug-resistant bacteria, for the transportation of cargo into bacteria, with gene editing of the cargo causing RNA (CRISPRCas9 RNA Guide) or gene silencing (SiRNA).

A chitosan generally found in the shellfish outer skeleton provides a wide variety of antimicrobial mechanisms which is used as a delivery vehicle. Gene manipulation and RNA (RNAi) interference enable bacteria, through silencing their resistance genes, to be tuned to antimicrobial substances. The combination of gene editing or RNAi with nanoantibiotics reveals a novel approach by Professor Kwon. Using this synergistic approach, he is able to destroy multitasking bacteria. The benefit of these nanoantibiotics is that natural microbes have no resistance to them so far. They do not affect human cells. Nanoantibiotics provides a new approach circumventing the existing antimicrobial discovery paradigm and a new mechanism for outbreaks in microbes which have not yet been detected in microbes, which could result in a longer term solution for the development of drug resistance formation.

3.10 Conclusion

Nanotechnology is a multidisciplinary field, where nanoparticles are considered as building blocks. Here we reviewed different mechanisms for nanoparticles synthesis. Although there are enormous physical and chemical synthesis approaches for nanoparticles, most of the methods are not eco-friendly. Hence, the later part of the chapter discussed about an environmental friendly method for synthesis of nanoparticles, i.e. green synthesis method. In this context, the chapter discussed about green synthesis of nanoparticles by bacteria as well as possible mechanism behind it. Additionally, the paper highlights about potential of some bacteria for nanoparticles synthesis. Antibacterial activity of nanoparticles is also discussed briefly with possible mechanisms, which could be used as potential therapeutic agents against many infectious diseases.

References

Adair JH, Parette MP, Altınoglu EI, Kester M (2010) Nanoparticulate alternatives for drug delivery. ACS Nano 4(9):4967–4970

Ahmed A, Khan AK, Anwar A, Ali SA, Shah MR (2016) Biofilm inhibitory effect of chlorhexidine conjugated gold nanoparticles against Klebsiella pneumoniae. Microb Pathog 98:50–56

Akbarzadeh A, Rezaei-Sadabady R, Davaran S, Joo SW, Zarghami N, Hanifehpour Y, Samiei M, Kouhi M, Nejati-Koshki K (2013) Liposome: classification, preparation, and applications. Nanoscale Res Lett 8(1):102

Arakha M, Jha S (2018) Interfacial phenomena on biological membranes. Springer

Arakha M, Borah SM, Saleem M, Jha AN, Jha S (2016) Interfacial assembly at silver nanoparticle enhances the antibacterial efficacy of nisin. Free Radic Biol Med 101:434–445

Arakha M, Pal S, Samantarrai D, Panigrahi TK, Mallick BC, Pramanik K, Mallick B, Jha S (2015a) Antimicrobial activity of iron oxide nanoparticle upon modulation of nanoparticle-bacteria interface. Sci Rep 5:14813

Arakha M, Roy J, Nayak PS, Mallick B, Jha S (2017) Zinc oxide nanoparticle energy band gap reduction triggers the oxidative stress resulting into autophagy-mediated apoptotic cell death. Free Radic Biol Med 110:42–53

Arakha M, Saleem M, Mallick BC, Jha S (2015b) The effects of interfacial potential on antimicrobial propensity of ZnO nanoparticle. Sci Rep 5:9578

Arya A, Mishra V, Chundawat TS (2019) Green synthesis of silver nanoparticles from green algae (Botryococcus braunii) and its catalytic behavior for the synthesis of benzimidazoles. Chem Data Collect 20:100190

Aswathanarayan JB, Vittal RR (2019) Nanoemulsions and their potential applications in food industry. Front Sustain Food Syst 3(95):2571–2581

Bayda S, Hadla M, Palazzolo S, Riello P, Corona G, Toffoli G, Rizzolio F (2018) Inorganic nanoparticles for cancer therapy: a transition from lab to clinic. Curr Med Chem 25 (34):4269–4303

Bertrand N, Leroux J-C (2012) The journey of a drug-carrier in the body: an anatomo-physiological perspective. J Control Release 161(2):152–163

Buzea C, Pacheco II, Robbie K (2007) Nanomaterials and nanoparticles: sources and toxicity. Biointerphases 2(4):MR17–MR71

Cao W (2007) Synthesis of nanomaterials by high energy ball milling. Skyspring Nanomaterials, Houston, TX

Casciaro B, Moros M, Rivera-Fernández S, Bellelli A, Jesús M, Mangoni ML (2017) Gold-nanoparticles coated with the antimicrobial peptide esculentin-1a (1-21) NH2 as a reliable strategy for antipseudomonal drugs. Acta Biomater 47:170–181

Chandra Hembram K, Prabha S, Chandra R, Ahmed B, Nimesh S (2016) Advances in preparation and characterization of chitosan nanoparticles for therapeutics. Artif Cells Nanomed Biotechnol 44(1):305–314

Chaudhari AA, Deb Nath S, Kate K, Dennis V, Singh SR, Owen DR, Palazzo C, Arnold RD, Miller ME, Pillai SR (2016) A novel covalent approach to bio-conjugate silver coated single walled carbon nanotubes with antimicrobial peptide. J Nanobiotechnol 14(1):58

Choi H-J, Pammi S, Park B-J, Eom J-H, An H, Kim HY, Kim M, Seol D, Kim Y, Yoon S-GJ (2017) Resistance against water and acid water (pH= 4.0) via Al-doped ZnO thin films for environmentally friendly glass panels. J Alloys Comp 719:271–280

Choi Y, Park TJ, Lee DC, Lee SY (2018) Recombinant *Escherichia coli* as a biofactory for various single-and multi-element nanomaterials. Proc Natl Acad Sci 115(23):5944–5949

Das RK, Pachapur VL, Lonappan L, Naghdi M, Pulicharla R, Maiti S, Cledon M, Dalila LMA, Sarma SJ, Brar SK (2017) Biological synthesis of metallic nanoparticles: plants, animals and microbial aspects. Nanotechnol Environ Eng 2(1):18

Durán N, Marcato PD, Durán M, Yadav A, Gade A, Rai M (2011) Mechanistic aspects in the biogenic synthesis of extracellular metal nanoparticles by peptides, bacteria, fungi, and plants. Appl Microbiol Biotechnol 90(5):1609–1624

Ealias AM, Saravanakumar M (2017) A review on the classification, characterisation, synthesis of nanoparticles and their application. IOP Conf Ser Mater Sci Eng 2017:032019

El-Eskandarany MS (2013) Mechanical alloying: for fabrication of advanced engineering materials. University Press of Mississippi, Jackson, MS

Epifani M, Giannini C, Tapfer L, Vasanelli L (2000) Sol–gel synthesis and characterization of Ag and Au nanoparticles in SiO2, TiO2, and ZrO2 thin films. J Am Ceram Soc 83(10):2385–2393

Feynman RP (1960) There's plenty of room at the bottom. California Institute of Technology, Engineering and Science magazine, Pasadena, CA

Gaidhani S, Singh R, Singh D, Patel U, Shevade K, Yeshvekar R, Chopade BA (2013) Biofilm disruption activity of silver nanoparticles synthesized by Acinetobacter calcoaceticus PUCM 1005. Mater Lett 108:324–327

Georgakilas V, Perman JA, Tucek J, Zboril R (2015) Broad family of carbon nanoallotropes: classification, chemistry, and applications of fullerenes, carbon dots, nanotubes, graphene, nanodiamonds, and combined superstructures. Chem Rev 115(11):4744–4822

Ghorbani HR (2014) A review of methods for synthesis of Al nanoparticles. Orient J Chem 30 (4):1941–1949

Gonçalves GAB, Marques P (2019) Nanostructured materials for treating aquatic pollution. Springer, Cham

Gopinath V, Priyadarshini S, Loke MF, Arunkumar J, Marsili E, MubarakAli D, Velusamy P, Vadivelu J (2017) Biogenic synthesis, characterization of antibacterial silver nanoparticles and its cell cytotoxicity. Arab J Chem 10(8):1107–1117

Gour A, Jain NK (2019) Advances in green synthesis of nanoparticles. Artif Cells Nanomed Biotechnol 47(1):844–851

Gupta A, Eral HB, Hatton TA, Doyle PS (2016) Nanoemulsions: formation, properties and applications. Soft Matter 12(11):2826–2841

Hwang IS, Hwang JH, Choi H, Kim K-J, Lee DG (2012) Synergistic effects between silver nanoparticles and antibiotics and the mechanisms involved. J Med Microbiol 61(12):1719–1726

Iravani S (2014) Bacteria in nanoparticle synthesis: current status and future prospects. Int Sch Res Notices 2014:359316

Ishida Y, Corpuz RD, Yonezawa T (2017) Matrix sputtering method: a novel physical approach for photoluminescent noble metal nanoclusters. Acc Chem Res 50(12):2986–2995

Jegadeeswaran P, Shivaraj R, Venckatesh R (2012) Green synthesis of silver nanoparticles from extract of Padina tetrastromatica leaf. Dig J Nanomater Biostruct 7(3):991–998

Kalita S, Kandimalla R, Sharma KK, Kataki AC, Deka M, Kotoky J (2016) Amoxicillin functionalized gold nanoparticles reverts MRSA resistance. Mater Sci Eng C 61:720–727

Kim M, Osone S, Kim T, Higashi H, Seto TJ (2017) Synthesis of nanoparticles by laser ablation: a review. KONA Powder Part J 2017:9

Kim T, Braun GB, Z-g S, Hussain S, Ruoslahti E, Sailor M (2016) Composite porous silicon–silver nanoparticles as theranostic antibacterial agents. ACS Appl Mater Interfaces 8 (44):30449–30457

Kumar N, Ray SS (2018) Synthesis and functionalization of nanomaterials. In: Processing of polymer-based nanocomposites. Springer, Cham, pp 15–55

Lee BY, Li Z, Clemens DL, Dillon BJ, Hwang AA, Zink JI, Horwitz MA (2016) Redox-triggered release of moxifloxacin from mesoporous silica nanoparticles functionalized with disulfide snap-tops enhances efficacy against pneumonic tularemia in mice. Small 12(27):3690–3702

Lee SH, Jun B-H (2019) Silver nanoparticles: synthesis and application for nanomedicine. Int J Mol Sci 20(4):865

Li J, Li Q, Ma X, Tian B, Li T, Yu J, Dai S, Weng Y, Hua Y (2016a) Biosynthesis of gold nanoparticles by the extreme bacterium Deinococcus radiodurans and an evaluation of their antibacterial properties. Int J Nanomed 11:5931

Li YJ, Harroun SG, Su YC, Huang CF, Unnikrishnan B, Lin HJ, Lin CH, Huang CC (2016b) Synthesis of self-assembled spermidine-carbon quantum dots effective against multidrug-resistant bacteria. Adv Healthc Mater 5(19):2545–2554

Lugscheider E, Bärwulf S, Barimani C, Riester M, Hilgers HJ (1998) Magnetron-sputtered hard material coatings on thermoplastic polymers for clean room applications. Surf Coat Technol 108:398–402

Madeła M, Neczaj E, Grosser A (2016) Fate of engineered nanoparticles in wastewater treatment plant. Inżynieria i Ochrona Środowiska 19(4):577–587

Mahajan SD, Aalinkeel R, Law W-C, Reynolds JL, Nair BB, Sykes DE, Yong K-T, Roy I, Prasad PN, Schwartz SA (2012) Anti-HIV-1 nanotherapeutics: promises and challenges for the future. Int J Nanomed 7:5301

Maharani V, Sundaramanickam A, Balasubramanian T (2016) In vitro anticancer activity of silver nanoparticle synthesized by Escherichia coli VM1 isolated from marine sediments of Ennore southeast coast of India. Enzyme Microb Technol 95:146–154

Mocan L, Ilie I, Tabaran FA, Iancu C, Mosteanu O, Pop T, Zdrehus C, Bartos D, Mocan T, Matea C (2016) Selective laser ablation of methicillin-resistant staphylococcus aureus with IgG functionalized multi-walled carbon nanotubes. J Biomed Nanotechnol 12(4):781–788

Mohan C, Carvajal-Millan E, Ravishankar C (2018) Research methodology in food sciences: integrated theory and practice. CRC Press, Boca Raton, FL
Mohanpuria P, Rana NK, Yadav SK (2008) Biosynthesis of nanoparticles: technological concepts and future applications. J Nanopart Res 10(3):507–517
Mu H, Liu Q, Niu H, Sun Y, Duan J (2016) Gold nanoparticles make chitosan–streptomycin conjugates effective towards gram-negative bacterial biofilm. RSC Adv 6(11):8714–8721
Mukasyan AS, Rogachev AS, Aruna ST (2015) Combustion synthesis in nanostructured reactive systems. Adv Powder Technol 26(3):954–976
Mukherjee P, Ahmad A, Mandal D, Senapati S, Sainkar SR, Khan MI, Ramani R, Parischa R, Ajayakumar P, Alam M (2001) Bioreduction of AuCl4− ions by the fungus, Verticillium sp. and surface trapping of the gold nanoparticles formed, Angewandte Chemie international edition, vol 40(19), pp 3585–3588
Murty B, Shankar P, Raj B, Rath B, Murday J (2013) Textbook of nanoscience and nanotechnology. Springer, Cham
Musarrat J, Dwivedi S, Singh BR, Saquib Q, Al-Khedhairy AA (2011) Microbially synthesized nanoparticles: scope and applications. In: Microbes and microbial technology. Springer, pp 101–126
Nagajyothi PC, Lee KD (2011) Synthesis of plant-mediated silver nanoparticles using dioscorea batatas rhizome extract and evaluation of their antimicrobial activities. J Nanomater 2011:573429
Nasrollahzadeh M, Sajadi MS, Atarod M, Sajjadi M, Isaabadi Z (2019) An introduction to green nanotechnology. Academic Press, Cambridge, MA
Nayak PS, Arakha M, Kumar A, Asthana S, Mallick BC, Jha S (2016) An approach towards continuous production of silver nanoparticles using Bacillus thuringiensis. RSC Adv 6 (10):8232–8242
Nayak PS, Pradhan S, Arakha M, Kumar D, Saleem M, Mallick B, Jha S (2018) Silver nanoparticles fabricated using medicinal plant extracts show enhanced antimicrobial and selective cytotoxic propensities. IET Nanobiotechnol 13(2):193–201
Niemeyer CM (2001) Nanoparticles, proteins, and nucleic acids: biotechnology meets materials science. Angew Chem Int Ed 40(22):4128–4158
Niemirowicz K, Piktel E, Wilczewska AZ, Markiewicz KH, Durnaś B, Wątek M, Puszkarz I, Wróblewska M, Niklińska W, Savage PB (2016) Core–shell magnetic nanoparticles display synergistic antibacterial effects against Pseudomonas aeruginosa and Staphylococcus aureus when combined with cathelicidin LL-37 or selected ceragenins. Int J Nanomed 11:5443
Orive G, Anitua E, Pedraz JL, Emerich DF (2009) Biomaterials for promoting brain protection, repair and regeneration. Nat Rev Neurosci 10(9):682–692
Panáček A, Smékalová M, Večeřová R, Bogdanová K, Röderová M, Kolář M, Kilianová M, Hradilová Š, Froning JP, Havrdová M (2016) Silver nanoparticles strongly enhance and restore bactericidal activity of inactive antibiotics against multiresistant Enterobacteriaceae. Colloids Surf B Biointerfaces 142:392–399
Panda S, Yadav KK, Nayak PS, Arakha M, Jha S (2016) Screening of metal-resistant coal mine bacteria for biofabrication of elemental silver nanoparticle. Bull Mater Sci 39(2):397–404
Pourali P, Badiee SH, Manafi S, Noorani T, Rezaei A, Yahyaei B (2017) Biosynthesis of gold nanoparticles by two bacterial and fungal strains, Bacillus cereus and Fusarium oxysporum, and assessment and comparison of their nanotoxicity in vitro by direct and indirect assays. Electron J Biotechnol 29:86–93
Prince A, Sandhu P, Ror P, Dash E, Sharma S, Arakha M, Jha S, Akhter Y, Saleem M (2016) Lipid-II independent antimicrobial mechanism of nisin depends on its crowding and degree of oligomerization. Sci Rep 6(1):1–15
Rai M, Yadav A, Gade A (2009) Current trends in phytosynthesis of metal nanoparticles. Crit Rev Biotechnol 29(1):78–78
Ramesh S (2013) Sol-gel synthesis and characterization of nanoparticles. J Nanosci 2013:929321

Rao JP, Geckeler KE (2011) Polymer nanoparticles: preparation techniques and size-control parameters. Prog Polym Sci 36(7):887–913

Ray SS, Kuruma M (2020) Halogen-free flame-retardant polymers: next-generation fillers for polymer nanocomposite applications, vol 294. Springer, Berlin

Romero D, Aguilar C, Losick R, Kolter R (2010) Amyloid fibers provide structural integrity to Bacillus subtilis biofilms. Proc Natl Acad Sci 107(5):2230–2234

Salavati-Niasari M, Davar F, Mir N (2008) Synthesis and characterization of metallic copper nanoparticles via thermal decomposition. Polyhedron 27(17):3514–3518

Sambhy V, Peterson BR, Sen A (2008) Antibacterial and hemolytic activities of pyridinium polymers as a function of the spatial relationship between the positive charge and the pendant alkyl tail. Angew Chem Int Ed 47(7):1250–1254

Sau TK, Rogach AL (2010) Nonspherical noble metal nanoparticles: colloid-chemical synthesis and morphology control. Adv Mater 22(16):1781–1804

Schrand AM, Rahman MF, Hussain SM, Schlager JJ, Smith DA, Syed AF (2010) Metal-based nanoparticles and their toxicity assessment. Wiley Interdiscip Rev Nanomed Nanobiotechnol 2 (5):544–568

Semaltianos N (2010) Nanoparticles by laser ablation. Crit Rev Solid State Mater Sci 35 (2):105–124

Sharma M, Nayak PS, Asthana S, Mahapatra D, Arakha M, Jha S (2018) Biofabrication of silver nanoparticles using bacteria from mangrove swamp. IET Nanobiotechnol 12(5):626–632

Seo WS, Lee JH, Sun X, Suzuki Y, Mann D, Liu Z, Terashima M, Yang PC, McConnell MV, Nishimura DG (2006) FeCo/graphitic-shell nanocrystals as advanced magnetic-resonanceimaging and near-infrared agents. Nat Mater 5(12):971–976

Siegel J, Lyutakov O, Rybka V, Kolská Z, Švorčík V (2011) Properties of gold nanostructures sputtered on glass. Nanoscale Res Lett 6(1):1–9

Simonazzi A, Cid AG, Villegas M, Romero AI, Palma SD, Bermúdez JM (2018) Nanotechnology applications in drug controlled release. In: Drug targeting and stimuli sensitive drug delivery systems. Elsevier, Amsterdam, pp 81–116

Singh H, Du J, Yi T-H (2017) Kinneretia THG-SQI4 mediated biosynthesis of silver nanoparticles and its antimicrobial efficacy. Artif Cells Nanomed Biotechnol 45(3):602–608

Singh P, Kim YJ, Wang C, Mathiyalagan R, Yang DC (2016) Weissella oryzae DC6-facilitated green synthesis of silver nanoparticles and their antimicrobial potential. Artif Cells Nanomed Biotechnol 44(6):1569–1575

Singh J, Dutta T, Kim K-H, Rawat M, Samddar P, Kumar P (2018) 'Green'synthesis of metals and their oxide nanoparticles: applications for environmental remediation. J Nanobiotechnol 16 (1):1–24

Smekalova M, Aragon V, Panacek A, Prucek R, Zboril R, Kvitek L (2016) Enhanced antibacterial effect of antibiotics in combination with silver nanoparticles against animal pathogens. Vet J 209:174–179

Song D, Qu X, Liu Y, Li L, Yin D, Li J, Xu K, Xie R, Zhai Y, Zhang H (2017) A rapid detection method of Brucella with quantum dots and magnetic beads conjugated with different polyclonal antibodies. Nanoscale Res Lett 12(1):179

Sonker AS, Pathak J, Kannaujiya VK, Sinha RP (2017) Characterization and in vitro antitumor, antibacterial and antifungal activities of green synthesized silver nanoparticles using cell extract of Nostoc sp. strain HKAR-2. Canadian J Biotechnol 1(1):26

Tai CY, Tai C-T, Chang M-H, Liu H-SJ (2007) Synthesis of magnesium hydroxide and oxide nanoparticles using a spinning disk reactor. Ind Eng Chem Res 46(17):5536–5541

Thakkar KN, Mhatre SS, Parikh RY (2010) Biological synthesis of metallic nanoparticles. Nanomed Nanotechnol Biol Med 6(2):257–262

Thiagarajan S, Sanmugam A, Vikraman D (2017) Facile methodology of sol-gel synthesis for metal oxide nanostructures. In: Recent applications in sol-gel synthesis. Intechopen, pp 1–17

Tiwari A, Prince A, Arakha M, Jha S, Saleem M (2018) Passive membrane penetration by ZnO nanoparticles is driven by the interplay of electrostatic and phase boundary conditions. Nanoscale 10(7):3369–3384

Tiwari DK, Behari J, Sen P (2008) Application of nanoparticles in waste water treatment 1. World Appl Sci J 3(3):417–433

Tiwari M, Jain P, Hariharapura RC, Narayanan K, Bhat U, Udupa N, Rao JV (2016) Biosynthesis of copper nanoparticles using copper-resistant Bacillus cereus, a soil isolate. Process Biochem 51 (10):1348–1356

Wan G, Ruan L, Yin Y, Yang T, Ge M, Cheng X (2016) Effects of silver nanoparticles in combination with antibiotics on the resistant bacteria Acinetobacter baumannii. Int J Nanomed 11:3789

Wang C, Kim YJ, Singh P, Mathiyalagan R, Jin Y, Yang DCJ (2016) Green synthesis of silver nanoparticles by Bacillus methylotrophicus, and their antimicrobial activity. Artif Cells Nanomed Biotechnol 44(4):1127–1132

Wang J, Wu G, Liu X, Sun G, Li D, Wei H (2017) A decomposable silica-based antibacterial coating for percutaneous titanium implant. Int J Nanomed 12:371

Wang L, Chen S, Ding Y, Zhu Q, Zhang N, Yu SJ (2018) Biofabrication of morphology improved cadmium sulfide nanoparticles using Shewanella oneidensis bacterial cells and ionic liquid: for toxicity against brain cancer cell lines. J Photochem Photobiol B Biol 178:424–427

Weir E, Lawlor A, Whelan A, Regan F (2008) The use of nanoparticles in anti-microbial materials and their characterization. Analyst 133(7):835–845

Wypij M, Czarnecka J, Świecimska M, Dahm H, Rai M, Golinska P (2018) Synthesis, characterization and evaluation of antimicrobial and cytotoxic activities of biogenic silver nanoparticles synthesized from Streptomyces xinghaiensis OF1 strain. World J Microbiol Biotechnol 34(2):23

Yadav KK, Arakha M, Das B, Mallick B, Jha S (2018) Preferential binding to zinc oxide nanoparticle interface inhibits lysozyme fibrillation and cytotoxicity. Int J Biol Macromol 116:955–965

Yahaya Khan M, Abdul Karim ZA, Hagos FY, Aziz ARA, Tan IM (2014) Current trends in water-in-diesel emulsion as a fuel. Sci World J 2014:2356–6140

Yang W, Shen C, Ji Q, An H, Wang J, Liu Q, Zhang Z (2009) Food storage material silver nanoparticles interfere with DNA replication fidelity and bind with DNA. Nanotechnol Environ Eng 20(8):085102

Yin H, Tang Z (2016) Ultrathin two-dimensional layered metal hydroxides: an emerging platform for advanced catalysis, energy conversion and storage. Chem Soc Rev 45(18):4873–4891

Zaidi S, Misba L, Khan AU (2017) Nano-therapeutics: a revolution in infection control in post antibiotic era. Nanomed Nanotechnol Biol Med 13(7):2281–2301

Zaleska-Medynska A, Marchelek M, Diak M, Grabowska E (2016) Noble metal-based bimetallic nanoparticles: the effect of the structure on the optical, catalytic and photocatalytic properties. Adv Colloid Interface Sci 229:80–107

Nanotechnology in Food Science

4

Rina Ningthoujam, Barsarani Jena, Sabita Pattanayak,
Santwona Dash, Manasa Kumar Panda, Rajendra Kumar Behera,
Nabin Kumar Dhal, and Yengkhom Disco Singh

Abstract

The approach of nanotechnology has made rapid changes in different fields of science. In food science, this advanced technology improves the system of processing, packaging, storage, transport and further operation of investment and market value. Nanotechnology in food science results in an advantage over conventional and other methods of food processing. The nano based particles and materials increase mechanical strength, barrier properties, help in the detection of pathogens in food and alert the status of food. Its contribution to the food industry is extending globally. Its main role in the food industry also includes the extension of shelf life; reduce deterioration, maintaining quality, and improvement in food value addition. With this novel technology, we expect to supply and fulfil the food requirements of hunger mouths occurring due to the increase in population. It will reduce the wastage of post-harvest loss of agriculture and

R. Ningthoujam
Department of Vegetable Science, College of Horticulture and Forestry, Central Agricultural University, Pasighat, Arunachal Pradesh, India

B. Jena · S. Pattanayak · S. Dash
Department of Botany, School of Applied Sciences, Centurion University of Technology and Management, Bhubaneswar, Odisha, India

M. K. Panda · N. K. Dhal
Environment and Sustainability Department, CSIR-Institute of Minerals and Materials Technology, Bhubaneswar, Odisha, India

R. K. Behera
School of Life Sciences, Sambalpur University, Burla, Sambalpur, Odisha, India

Y. D. Singh (✉)
Biomaterials and Bioprocessing Research Laboratory, Department of Post Harvest Technology, College of Horticulture and Forestry, Central Agricultural University, Pasighat, Arunachal Pradesh, India

M. Arakha et al. (eds.), *Bio-Nano Interface*,
https://doi.org/10.1007/978-981-16-2516-9_4

horticulture produces. Soon in the future, we believe that the rapid development of this technology will handle the safeguarding of food and plays a pivotal role in the development of food science and its related fields.

Keywords

Nanotechnology · Food industry · Food processing · Food safety

4.1 Introduction

Nanotechnology is a technology of science which deals with nanoparticles like atoms, molecules and macromolecules within the size of approximately 1–100 nm to manipulate the system and generate objects of new attributes. Since these nanoparticles are very small, they have unique properties due to high surface to volume ratio and novel physiochemical properties like colour, solubility, strength, magnetic and thermodynamics. They have more tendencies to react differently in environment when compared to larger particles. Their colour change because of different sizes of particle reflects and absorbs light differently as they have different optical properties. Nanotechnology can revolutionize the industrial sector both in developing and developed countries to accept the new challenges for development of future material science and its allied branches. The potential impacts of nanotechnology can be applied in different fields such as manufacturing industries, pharmaceuticals, transportation, electronics, environment, agriculture, etc.

In sector of agricultural sciences and related industries, this new modern technology has the ability to change and increase the production and quality of food and fulfil the growing world food needs. Its application in agriculture and food industry was first directed by the United States Department of Agriculture (USDA) roadmap published in 2003. This novel technology plays a very important role as a new tool in cellular and molecular biology, and it can also be used as a new material to identify different plant pathogens. Other applications of nanotechnology in agriculture and food industry include detection of disease, targeted treatment, improve absorption of nutrients in plant, fight diseases and withstand environmental stress. Use of this technology in other sector of agriculture like food science brings opportunity to overcome the problems of post-harvest products. It has much application in processing, storing, packaging and transport of agricultural products. It provides new approaches for the selection of raw materials and enhances the quality of processed products. In food industry, the demand of nanoparticles is increasing as it is reported that many of them contain essential elements and found to be non-toxic (Roselli et al. 2003). It is also observed that in high temperature and pressure they remain stable (Sawai 2003). The use of nanoparticles in food science not only significantly changes the quality of food but also brings safety and benefits to our health. With this new modern technology, many researchers, organizations and industries are trying to develop novel methods which can be directly applied in food science (Dasgupta et al. 2015).

Before the emergence of nanotechnology in food science, we depend on the processing practices on traditional methods and a modified form of conventional methods. The spoliation in food is due to the invasion of microorganisms. Different conventional methods are used for the detection and control of bacteria and fungus. Some of the methods are culture-dependent microbiological methods, culture-independent detection techniques, nucleic acid sequence-based detection techniques, detection of pathogens by biochemical characterization, immunological detection techniques, application of biosensors and instrumental techniques for detection of bacteria, spectroscopic techniques, and terahertz radiation methods. In olden times to control the microbial growth and preservation in food, we used traditional methods such as freezing, chilling, nutrient restriction, water activity reduction, acidification, pasteurization, and fermentation. Some other modern methods were developed due to the change in civilization. They are food processing techniques by microwave and radio frequency, direct steam injection, ohmic heating, irradiation, etc. The utilization of conventional techniques sometimes adversely affected the organoleptic properties of food and decreases the acceptability by a consumer. Some methods of conventional are time-consuming, difficult in culturing; required more labour, low sensitivity and sometimes they give negative results. To cope with all these limitations, we incorporate nanotechnology in food science and show some positive results. Application of nanotechnology in food science can be of mainly two methods that encompass the nanostructured ingredients and nano sensing. The application of nanoparticles in food processing can be achieved in form of encapsulation, emulsions, biopolymer matrices, simple solutions, and in association with colloidal substances to make efficient delivery system. The nanostructured food ingredients are usually applied for food processing and food packaging. In food processing they are being used to improve the taste, texture, flavour, food additives, anti-caking agents, antimicrobial agents, and as fillers for improving mechanical strength. In food packaging they are used to modify the packaging material to enhance its durability and prevent degradation or increase the shelf life of food. Nanosensors in food are often related to its quality control and safety often used in (Ezhilarasi et al. 2013). The nanosensors are being used to check the contamination, mycotoxins and microorganisms in food. It is also noticed that nanobiosensors have been used for detection of pathogen in processing plants which makes attentive to consumers, procedures and distributors on safety status of food (Baeumner 2004; Cheng et al. 2006). In processing of food, the nanocarriers act as delivery system and carry the food additives in food products. The nanocarriers are nanomaterials which are being used as a transport module for other substances and can be classified as organic based, inorganic based or combination of both. The organic nanocarriers include polymeric nanoparticles, lipid-based nanoparticles, dendrimers and carbon-based nanoparticles, whereas inorganic nanocarriers are usually derived from metallic compounds such as quantum dots (Salehi et al. 2020). The size of nanocarriers directly affects the delivery of bioactive compounds to different sites in body, as it was observed that only submicron nanoparticles get absorbed efficiently compare to the larger size micro-particles (Ezhilarasi et al. 2013). The delivery compounds should have properties to assist the active compound exactly at targeted location,

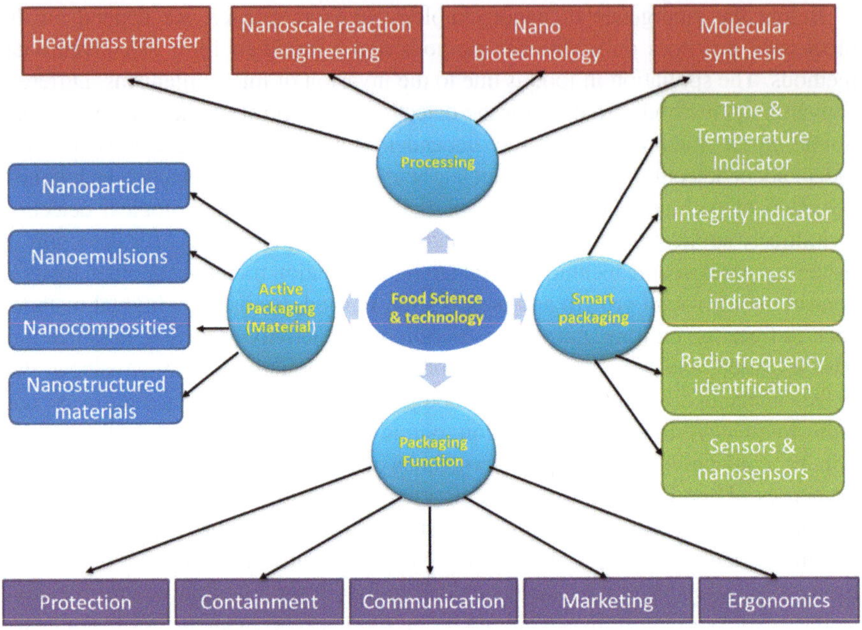

Fig. 4.1 Nanotechnology application in food science and technology

time and specific rate. The active compound should ensure to remain and maintain efficiently at appropriate levels for longer period of time. In food packaging use of nano-based particles has shown several advantages than the followed conventional methods. This novel technology improved physical performance of food, nanoparticles act as antimicrobial agent and use of inorganic nanoparticles in low concentration acts as anti-bacterial agent which are more stable in extreme conditions. Packaging of food using biosensor technology screens pathogens as the sensors are highly sensitive and they can also be used as indicator to check the changes in environmental conditions such as humidity and temperature in storage rooms. The biosensors have possible application in various nanostructures like thin films, nanorods, nanoparticles and nanofibres (Jianrong et al. 2004). Thin nano sensor chips immobilized with specific antibodies, antigens or protein molecules give signals on detecting target molecules (Subramanian 2006). Nowadays, many scientists, researchers and authorities are working to standardize this advance technology, as the importance and popularity in food science and food industry are rising up over the past few decades (reference). In near future, nanotechnology will become one of the promising powerful tools to revolutionize conventional food science and its industry (Fig. 4.1).

4.2 Nanotechnology

According to scientific survey, nanoscience and nanotechnology are rising materials having extensive activities. American physicist Richard. P. Feynman presented his historical talk on "There's Plenty of Room at the Bottom". The thought of the talk encouraged others to bring new exposure in nanotechnology. The method which enables the utilization of each single atoms and molecules through suitable tools and techniques has been explained by Faynman (Anjum et al. 2016; Feynman 1995). Until 1980s the term nanotechnology was unknown, when Eric Drexler published his first paper on the term nanotechnology (Eric Drexler 1986). The word "Nano" is derived from the Greek word "Nanos" which means dwarf or extremely small and the term "Nanotechnology" was first described by the late Norio Taniguchi in 1974 (Guhan Nath et al. 2014).

4.2.1 Types of Nanotechnology

Nanotechnology accurately means any technology on a nanoscale that has been used in the real life. Nanotechnology is defined as the production and application of physical, chemical and also biological systems at scales ranging from each single atoms or molecules to submicron dimensions, in addition to the assimilation of the resulting nanostructures into larger systems (Singh et al. 2021). In this era of nanotechnology, they are categorized as wet, dry and computational approaches.

4.2.1.1 Wet Nanotechnology

Wet technology is related with component of living organisms such as enzymes, membranes, tissues and other cellular compounds (Chung et al. 2016). Wet technology is most often known as "bottom-up" method of synthesis in which nanoparticles are accumulated atom-by-atom through a procedure of nucleation (Lateef et al. 2019). For example, nanoliposomes are used in encapsulation and release of food materials. They improve the bioavailability, stability and shelf life of ingredients. The use of liposomes and nanoliposomes in food science also includes delivery of flavours, nutrients and prevention from microbes against contamination (Reza et al. 2008).

4.2.1.2 Dry Nanotechnology

The dry technology included physical chemistry approach on the production of inorganic products like carbon, silicon, etc. (Chung et al. 2016). At the same time dry process of synthesis of nanoparticles is also frequently termed as "top-down" technology approach, where nanoscale materials are produced by breaking down bulk components into nanoparticles (Rai et al. 2016). The inorganic nanoparticles are usually used for packaging, food additives and food analysis. The metal-based nanoparticles have shown a positive impact when applied in food science. Ag, Cu, ZnO, CuO and TiO_2 prevent from growth of bacteria on packaging, storage and transportation of food (Le Guével 2017).

4.2.1.3 Computational Nanotechnology

And the last one which is known as computational technology deals with replications of nanometer-sized structures (Sinha et al. 2009). It is concerned with the development and use of computer for analysing, predicting and properties of system relevant to nanomaterials and nanotechnology. Abundant process for the synthesis of nanoparticles such as sono-chemical, laser irradiation, laser ablation, sputter deposition, solvothermal and also biological techniques is developed. But yet biological process of synthesis is most useful and broadly acceptable technology (Ojo et al. 2016). So, the utilization of green engineering, green chemistry and sustainability codes to eliminate or decrease the purpose and production of lethal substances in the area of nanotechnology are termed as "green nanotechnology" (Elegbede and Lateef 2019). The nanosensors are devices developed by integrating various nanostructured materials and biological receptors. They can detect quickly, have high sensitivity and specificity (Gálvez et al. 2007). They monitored adulterants, toxins, food borne pathogens, chemicals and pesticides (Rai et al. 2012).

The benefits of nanotechnology in food industry have been reported with better characteristics and new functionalities (Panda et al. 2020b). Different results of nanotechnology like nanoencapsulation, nanosensors, nanotube, nanofibre, etc. are applied in various dimensions of food industry. Nanoencapsulation helps in steady release of fertilizers, pesticides and herbicides more efficiently (Panda et al. 2020a). It releases slowly at the targeted area at precise dose. This method makes less wastage of agrochemical products which leads to decrease the pollution of our environment. Nanotechnology can be applied from the production phase of a crop. With help of nanotechnology in agriculture, great advantages in growth and production, detection of pest and disease and allelopathy have observed. The nanosensor reduces the damage in crops by their rapid and early detection of plant diseases. The use of this novel technology is also a very great demand in techniques of post-harvest management and processing. Since the crops of agriculture and horticulture are perishable in nature proper management after harvesting is ideal. The implementation of nanotechnology in agriculture and horticulture crops during processing and preservation helps in protection by making a barrier against food borne microorganisms. It is revealed that aroma of food can be maintained and haves the properties to use as food additives (McClements et al. 2016). Various post-harvest management operations of crops like processing, packaging, storage, transportation, etc. have been revolutionized after the incorporation of nanotechnology. Its functionality and development have discussed more in details in the following write ups.

4.3 Nanotechnology in Food Packaging

Packaging is an operational process, followed after the crops are harvested. It is practiced for the purpose of transporting the food to distant places, easy handling, maintaining quality of food, extending shelf life, protection from external injuries and preventing contamination by microorganisms. In conventional method of packaging, materials like corrugated boxes, boxboard or paperboard cartons, paper bags and sacks are commonly used. Another method of packaging is modified atmosphere packaging (MAP). In this method, fresh produces are stored, packed with plastic films which restrict the transmission of respiratory gases. This results in accumulation of carbon dioxide and reduction of oxygen around the produce to increase their storage or shelf life. As new technology is approaching in recent years, application of nanotechnology in food packaging is also a method which is being followed rapidly. It is a method like modified atmosphere packaging but here, nanoparticles are being used to develop new plastic or foil or modified packaging materials for food packaging. They act as a barrier to prevent the penetration of oxygen into the food. Nanomaterials in packaging and coating of food products can enhance the quality produce (Pinto et al. 2013). Use of nanoparticles as coating material for fruits has shown to prevent the loss of weight and shrinkage issues (Predicala 2009). Nanotechnology in packing material increases the resistance properties like mechanical, thermal, chemical and microbial by affecting its permeability to gas or water vapour. Contamination in food can be detected by developing active antimicrobial and anti-fungal sensing and signalling nanosensors for microbiological and biochemical changes (Joseph and Morrison 2006). In active anti-microbe packaging, it is in contact with the food product or headspace inside to retard the growth of microbes on surface of food (Soares et al. 2009). According to Bradley et al. (2011) and Tan et al. (2013) silver, copper, chitosan and metal oxide like titanium oxide or zinc oxide are some nanoparticles which possess anti-bacterial properties. Other forms of nanomaterials like nanocomposites and nanolaminates for food packaging have also been used to provide resistance against heat and mechanical shock. Use of these materials also helps in preserving quality and extension of shelf life of food. Nanocomposites are materials that incorporate polymers with inorganic solids at nanometric scale and are being used for packaging and coating or as ingredients. Polymers like low density polyethylene, gelatine, isotactic polypropylene and polylactic acid are generally used as nanocomposites. Additional to this, chitosan, polystyrene and polyvinylprolidone are also recorded as nanocomposite films in combination with copper or zinc oxide to deactivate food pathogen (Jin et al. 2009; Li et al. 2009; Oehlke et al. 2014). Metal-based nanocomposite like silver is utilized widely as an antimicrobial agent in the food industry and its application is also suggested to be safe in food packaging (Addo et al. 2015). Nanocomposite was first produced by incorporating organic substances with inorganic substances to bring improvement in mechanical strength, toughness, electrical and thermal conductivity. Nanocomposites are also used for coating plastic films like polyethylene terephthalate (PET) which has shown better performance than others in terms of transparency and drawn more attention towards silica and alumina coated food packaging films (Moore 1999). In horticultural food industry and value-added food items such as chocolates, bakery products and candies, nanolaminates are utilized in form of edible coatings and foaming. Nanolaminates are thin films of two or more layers made up of ultrafine-grained solids that exhibit a high concentration of interface defects; proteins, polysaccharides and lipids are used for preparing the coating films. The coating layers provide a barrier between food particle and external factors like moisture, environment, light, heat, etc. to retain the nutritional value of the food

(Cha and Chinnan 2004; Ponce et al. 2008; Rhim 2004). Nanolaminate as coating material in food packaging is one of the novel technologies of nanoscience which is widely used because of their very fine fragile nature. To extend or prolong the shelf life of food, use of nanoparticles in ethylene absorbent is also being employed. Increase in amount of ethylene gives an undesirable effect in some horticultural crops. So, use of this absorbent aid in absorption of ethylene gas and other volatile compounds produces by fruits or vegetables while transporting or storing. It significantly decreases the spoilage of economic crops. Thereby, it reduces loss in supplying chain and in return gives economic benefits to suppliers and retailers.

4.4 Nanotechnology Against Food Deterioration

Food deterioration is the spoilage of food where it becomes unfit for market and human consumption. It results in change of texture, flavour, decrease in nutrients and quality. It is caused by different factors such as growth of microorganisms in the food, denaturation of enzymes, damage in food cause by insects and rodents, due to environmental factors (temperature, humidity, heat and cold), and time. To control this problem, we need to follow some mandatory methods like inactivation of enzymes, removal of microorganisms and strictly regulating the methods of pre- and post-harvest management. Traditional or home methods to avoid the food from spoilage have also been practice from olden times. Some of the old methods that are widely followed to preserve food are drying, curing, fermenting, pickling, use of dry salt or sugar and smoking. Advance methods like irradiation, canning, pasteurization or sterilization, freezing and preservation by chemicals are also some useful techniques to stop degradation of food. Recently, many researchers are working on modern technology to fight against deterioration of food and it includes application of nano based technology. Use of nanosensors helps to assist the changes occurring in the food which may leads to spoilage. They respond to the change in gaseous composition like hydrogen, hydrogen sulphide, nitrogen oxide, sulphur oxide and ammonia. They also have the ability to detect surrounding environmental factors of the food. In agriculture and food industry, nanosensors help in detecting many undesirable factors. The nanosensors are installed in the packaging material and screen out the microorganisms that try to infest the food. It also helps in pointing out the changes in quality of food, indicating it by different colours. There are several types of nano based sensors. They are array biosensors, nanoparticle-based sensors, nanoparticles in solution, electronic noses, nano test strips and nanocantilevers (Tang et al. 2009). Array biosensors are used for detection of high or low molecular weight toxins (Frances et al. 2003). It can be successfully applied to detect content of toxin, bacteria and virus in food products. Electronic noses act similarly like human nose. Several chemical sensors are used in electronic sensors and are connected with processing data system (Sozer and Kokini 2009). Electronic tongues which have similar functions to electronic sensors have been also reported by Yuan et al. (2008). This electronic tongue sensor gives signal by changing the colour when it is in contact with any spoiled food and makes us aware that food is not suitable for consumption. Nano test strips are strips that help in determining contamination of food and it is easy and quick method for commercialization. Electrochemical biosensors can be also used as tool to monitor undesirable substances and pathogens in food. This analytical tool is also used for checking adulterants in food and beverages (Garcia et al. 2010). It is reported from University of Manitoba in Winnipeg that incorporation of nanotechnology with microelectrons can generate a tiny sensor which help to monitor the early grain spoilage. Nanobarcodes in food product are also used to screen the quality (Sonkaria et al. 2012). Nanobarcodes are of nano size barcode encoded with different element and it includes different sizes, shapes and colour. Nanobarcodes can be used like conventional barcode which helps in monitoring and controlling of food quality of agriculture produce (Prasanna 2007). Another nano based technology called nanoencapsulation also slows down deterioration of food. It is the enclosing of food products with nanoparticles to prevent degradation until it reaches the targeted site. After it is capsulated in food materials, it acts as a barrier for exchange of moisture and gas which in return helps in extension of shelf life of food. It also maintains the colour, flavour, texture, enzymes and responds as anti-browning agents. The encapsulation can be done by covering with thin films, adding an outer layer and by micro dispersion. The coating of food with nanoparticles prevents it from entry of microorganism and microbial spoilage. The utilization of this advance technology helps in protection of bio-active compounds like vitamins, minerals carbohydrates, proteins and anti-oxidants (Sekhon 2010). The production technique of nanocapsulation includes emulsification, solvent extraction, high pressure, homogenization, coacervation, nanoprecipitation, inclusion complexation and supercritical anti-solvent precipitation (Reis et al. 2006; Solans et al. 2005; Zuidam and Shimoni 2010). Coating of fresh foods with mixture of gelatin and nano crystals of cellulose, chitosan, nano silica or incorporating together has been found significantly effective in preventing deterioration of fresh foods (Medeiros et al. 2014). During manufacturing and storing, spoilage of food is generally observed. It is believed that utilization of nanocochleates can control the degradation of micronutrients and anti-oxidants in food. Nanocochleates respond as delivery system and have ability to trap the desire molecule. In advent of nanoscience, it has shown many advantages over conventional methods to control food deterioration and brings assurance about food safety.

4.5 Nanotechnology for Food Storage

Food is essential for the growth and to prolong life; with providing nutrients like carbohydrate, fats, proteins, vitamins and minerals (Pal 2017). In this phase of time nanotechnology is applied for the storage, packaging, processing, transportation, and for some other securities; also, for food industry—nanostructured materials, nanoorganic materials and metal oxide with bioactive elements are used (Pradhan et al.

2015). Contamination of food through chemical, physical and biological means put direct impact on the food chain (Pal 2017). So, it becomes very important to protect food from any type of contamination. Nanotechnology provides effective packaging, advanced packaging and creative packaging methods. In food industries, food packaging involves primal utilization of nanotechnology which aids in food safety and maximizes food shortage. In case of distribution system and encapsulation, nanotechnology is extensively used to provide facilities like edible coating and high barrier packaging as a result of which food can be stored for longer period of time against moisture, gas and lipid (Kumar et al. 2016). Different types of nanotechniques like nano-smart dust, metal-based nanosensors, nanobarcodes, nanobiosensors, abuse indicators, full-temperature history indicator, reflective-interferometry, array biosensors, nano test strips, carbon black and polyaniline, single walled carbon nanotubes, and DNA are applied for food packaging (Pradhan et al. 2015). Nanosensors like electronic noses, nanocantilevers, array biosensors, nanoparticle in solution, nanoparticles-based sensors and nano test strips are used in food packaging as it has a good role in finding out internal and external condition of food products.

4.6 Nanotechnology in Food Pathogen Detection

Food borne infections are caused by germs like various bacteria, parasites, viruses that contaminate the food products. Chemicals and harmful toxins are also responsible for food borne diseases. So, to control and avoid the food borne diseases excellent detection method is required as a result of which food will remain safe and occurrence of food borne diseases can be reduced (Law et al. 2015). Nanotechnology is widely used in food microbiology, as nanosensors for microbial detection, antimicrobial activity of nanoparticles and nanomaterials used for food packaging (Nasr 2015). Micro fluidic assays, biosensors, fluorescent nanoparticles, magnetic nanoparticles, nanosensors, silica nanoparticles, nanoparticle-established assays, nano diagnostic approaches are some of the important pathogen detection techniques (Bulbul et al. 2015). This technology has better reasons from conventional methods in terms of cost efficiency. The low cost nanosensors help in screening of different pathogens present in food during packaging operation (Ranjan et al. 2014). The alternative form of nanobiosensor known as nano bioluminescent spray can detect the strains of pathogen in food products by producing a visual glow. The visual glow in spray is due to the presence of different magnetic nanoparticles. For example, iron oxide is used for isolation of DNA for milk pathogen like *Listeria monocytogenes*. Gold nanoparticles are used for detection of *E. coli, Staphylococcus aureus, Vibrio parahaemolyticus, Salmonella enteric and Salmonella typhi* in foods. The magnetic nanoparticles are also used for detection of toxic substances found in food. Gold, iron oxide and super paramagnetic nanoparticles are used for detection of aflatoxins B1 and aflatoxins M1. Zinc oxide nanoparticles are used in analyte detection of Ochratoxin-A. The incorporation of silicon nanosensors and proteins also has the property to detect pathogen in liquid products by vibrating at different frequencies

(Wang et al. 2011). Additionally, the success research report of array based immunosorbent assay when coupled with liposomal nanovesicles on identification of *E. coli, L. monocytogenes and Salmonella spp.* was revealed (Chen et al. 2011).

4.6.1 Gold NPs

For microbial detection and identification, gold NPs have significant role in organic or aqueous solvent. Due to the small size of NPs (20–60 nm diameter), NPs are used in drug delivery, DNA detection, environmental detection, biomarkers and as chemical sensor (Primec 2016).

4.6.2 Magnetic NPs

Magnetic nanoparticles are used for immunomagnetic separation and to control sensing platforms (Hayat et al. 2013). Due to the large surface area, it controls and improves the immobilization. For food pathogen detection, magnetic microsphere immunoassay is first proposed by Kim et al. (Kim et al. 2010).

4.6.3 Biosensors

To study and detect biological organisms and events, biosensor is used (Syed et al. 2014). This device can isolate tissues, enzymes, immunosystems, organelles or whole cell. Biosensors are mainly divided into two types: direct detection sensors and indirect detection sensors. Amperometric immunosensor, potentiometric, piezoelectric, colorimetric immunosensor, electrochemical immunosensor, surface plasmon resonance are some biosensors used for pathogen detection (Senturk et al. 2018).

4.7 Implication and Perspective

According to the study of researchers and scholars it is revealed that application of nanotechnology in food science is successful in many ways. As an emerging technology it is valued in food industry and related field has improved. Food nanotechnology is being applied directly or indirectly in packaging, processing, storage, food security, against deterioration, and detection of pathogen in food. In agriculture sector its potentiality is utilized in development of precision farming. Nanotechnology shows various benefits in development of functional foods, nutraceuticals and other food products. Through this new technology smart and active packaging of food, tracking devices and targeted delivery of essential compounds in food have successfully employed. Their role in food industry proved great potential over conventional techniques. It protects the sensitive food ingredients by applying nanoencapsulation techniques and improves the stability of bioactive compounds. This new tool can increase the production of agriculture and horticulture crops, have ability to check fertility status of soil, and protect plant from insects, pests and diseases. Monitoring of undesirable bacteria like *Escherichia coli* and *Staphylococcus aureus* in foods is done successfully by this novel technology. According to the observation of many scientists, nano-based particles of metal and metal oxide are considered as safe materials in food industry. Therefore, it is used as food preservatives. However, some negative impacts and limitations of nanoscience in food industry have also been reported. It is believed that the incorporation of nanoparticles in food can sometime cause health threat and prove toxic to the environment. When nanomaterial present in food additive comes in direct contact with human organs, it brings health hazards and nanoemulsions in high amount cause biological changes in our body (Cushen et al. 2012). Nanoparticles sometime show effects in the cellular components of the biological system. It is also revealed that they get attached in the cells of immune system (Jordan et al. 2005). Nanocarriers containing high amount of organic solvents and emulsifiers can have toxic effects on human body (McClements and Rao 2011). According to World Health Organization (WHO), Food and Drug Administration (FDA) and European Food Safety Authority (EFSA) solvents and emulsifiers are toxic but using it below a certain dose is considered to be safe. Therefore, several researches and regulatory observations to protect from adverse effect of nanoscience and to reduce the risk in human health and environment are required. Moreover, to exploit it in food science a scientific and practicable protocol is necessary.

4.8 Conclusion

In food science, nanotechnology plays a great role right from pre-harvest of crops to post-harvest operational methods. Nanotechnology is a revolutionary science with providing scope in different fields like in bio-engineering, nano-fabrics, optical engineering, nano-devices, nanobiotechnology, defence, medicine and drug. Nanotechnology helps in food packaging, processing and preservation directly or indirectly in many ways. It protects our foodstuffs from moisture, gases and lipids which results in extension of shelf life. It helps in a proper way to get good result food material and reached in safe and hygienic condition by delivering the bioactive compounds. We can conclude that the advance method of this technology in food science has great potential to change the system of food industry in positive effects. After observing its wide scope in food science and other related fields, it is very necessary for more exploration and scientific research.

Acknowledgments The authors are grateful to respective institutions for support. Yengkhom Disco Singh is thankful to the honourable Vice Chancellor of Central Agricultural University, Imphal, Manipur, India for providing the facilities.

References

Addo NS, Thomas TA, Begley TH, Noonan GO et al (2015) Characterisation and potential migration of silver nanoparticles from commercially available polymeric food contact materials. Food Addit Contam Part A 32(6):1003–1011

Anjum S, Abbasi BH, Shinwari ZK et al (2016) Plant-mediated green synthesis of silver nanoparticles for biomedical applications: challenges and opportunities. Pak J Bot 48 (4):1731–1760

Baeumner A (2004) Nanosensors identify pathogens in food. Food Technol 58(8):52–56

Bradley EL, Castle L, Chaudhry Q et al (2011) Applications of nanomaterials in food packaging with a consideration of opportunities for developing countries. Trends Food Sci Technol 22:603–610

Cha D, Chinnan M (2004) Biopolymer based antimicrobial packaging. Crit Rev Food Sci Nutr 44 (4):223–237

Chen S, Ma L, Yuan R, Chai Y, Xiang Y, Wang C et al (2011) Electrochemical sensor based on Prussian blue nanorods and gold nanochains for the determination of H_2O_2. Eur Food Res Technol 232:87–95

Cheng Q, Li C, Pavlinek V, Saha P, Wang H et al (2006) Surface- modified antibacterial TiO_2/ag + nanoparticles: preparation and properties. Appl Surf Sci 252:4154–4160

Chung IM, Park I, Seung-Hyun K, Thiruvengadam M, Rajakumar G et al (2016) Plant-mediated synthesis of silver nanoparticles: their characteristic properties and therapeutic applications. Nanoscale Res Lett 11(1):40

Cushen N, Kerry J, Morris M, Cruz-Romero M, Cummins E et al (2012) Nanotechnologies in the food industry- recent developments, risks and regulation. Trends Food Sci Technol 24:30–46

Dasgupta N, Ranjan S, Mundekkad D, Ramalingam C, Shanker R, Kumar A et al (2015) Nanotechnology in agrofood: from field to plate. Food Res Int 69:381–400

Elegbede JA, Lateef A (2019) Green nanotechnology in Nigeria: the research landscape, challenges and prospects. Ann Sci Technol 4(2):6–38

Eric Drexler K (1986) Engines of creation: the coming era of nanotechnology. Anchor Book, New York, pp 1–10

Ezhilarasi PN, Karthik P, Chhanwal N, Anandharamakrishnan C et al (2013) Nanoencapsulation techniques for bioactive components. Food Bioprocess Technol 6:628–647

Feynman RP (1995) No ordinary genius: the illustrated Richard Feynman. WW Norton & Company, New York

Frances SL, Chris RT, Lisa CS, Kim ES, Yura S, Joel PG et al (2003) Array biosensor for detection of toxins. Anal Bioanal Chem 377(3):469–477

Gálvez A, Abriouel H, López RL, Omar NB et al (2007) Bacteriocin-based strategies for food biopreservation. Int J Food Microbiol 120(1–2):51–70

Garcia M, Forbe T, Gonzalez E et al (2010) Potential applications of nanotechnology in the agro food sector. Ciȩncia e Tecnologia de Alimentos 30(3):573–581

Guhan Nath S, Aaron SI, Raj AAS, Ranganathan TV et al (2014) Recent innovations in nanotechnology in food processing and its various applications–a review. Int J Pharm Sci Rev Res 29 (2):116–124

Jianrong C, Yuqing M, Nongyue H, Xiaohua W, Sijiao L et al (2004) Nanotechnology and biosensors. Biotechnol Adv 22:505–518

Jin H, Xia F, Jiang C, Zhao Y, Lin HE et al (2009) Nanoencapsulation of lutein with hydroxypropylmethyl cellulose phthalate by supercritical anti solvent. Chinese J Chem Eng 17(4):672–677

Jordan J, Jacob KI, Tannenbaum R, Sharaf MA, Jasiuk I et al (2005) Experimental trends in polymer nanocomposites. Mater Sci Eng A 393(1–2):1–11

Joseph T, Morrison M (2006) Nanotechnology in agriculture and food: A nano forum report, Institute of nanotechnology. http://www.nanoforum.org

Lateef A, Elegbede JA, Akinola PO, Ajayi VA et al (2019) Biomedical applications of green synthesized-metallic nanoparticles: a review. Pan Afr J Life Sci 3:157–182

Law JWF, Ab Mutalib NS, Chan KG, Lee LH et al (2015) Rapid methods for the detection of food borne bacterial pathogens: principles, applications, advantages and limitations. Front Microbiol 5:770

Le Guével X (2017) Overview of inorganic nanoparticles for food science applications. In: Nanotechnology in agriculture and food science. Wiley, Hoboken, NJ, pp 197–208

Li H, Li F, Wang L, Sheng J, Xin Z, Zhao L, Xiao H, Zheng Y, Hu Q et al (2009) Effect of nano packing on preservation quality of Chinese jujube. Food Chem 114(2):547–552

McClements DJ, DeLoid G, Pyrgiotakis G, Shatkin JA, Xiao H, Demokritou P et al (2016) The role of the food matrix and gastrointestinal tract in the assessment of biological properties of ingested engineered nanomaterials (iENMs): state of the science and knowledge gaps. NanoImpact 3:47–57

McClements DJ, Rao J (2011) Food grade nanoemulsions: formulation, fabrication, properties, performance, biological fate and potential toxicity. Crit Rev Food Sci Nutr 51(4):285–330

Medeiros BGDS, Souza MP, Pinheiro AC, Bourbon AI, Cerqueira MA, Vicente AA et al (2014) Physical characterisation of an alginate/lysozyme nano laminate coating and its evaluation on 'Coalho' cheese shelf life. Food Biopro Technology 7:1088–1098

Moore S (1999) Nanocomposite achieves exceptional barrier in films. Mod Plast 76(2):31–32

Oehlke K, Adamiuk M, Behsnilian D, Graf V, Mayer ME, Walz E, Greiner R et al (2014) Potential bioavailability enhancement of bioactive compounds using food grade engineered nanomaterials. Food Funct 5:1341–1359

Ojo SA, Lateef A, Azeez MA, Oladejo SM, Akinwale AS, Asafa TB, Yekeen TA, Akinboro A, Oladipo IC, Gueguim-Kana EB, Beukes LS et al (2016) Biomedical and catalytic applications of gold and silver-gold alloy nanoparticles biosynthesized using cell-free extract of Bacillus Safensis LAU 13: antifungal, dye degradation, anti-coagulant and thrombolytic activities. IEEE Trans Nanobiosci 15(5):433–442

Panda MK, Panda SK, Singh YD, Jit BP, Behara RK, Dhal NK (2020a) Role of nanoparticles and nanomaterials in drug delivery: an overview. In: Advances in pharmaceutical biotechnology. Springer, Singapore, pp 247–265

Panda MK, Singh YD, Behera RK, Dhal NK (2020b) Biosynthesis of nanoparticles and their potential application in food and agricultural sector. In: Green nanoparticles. Springer, Cham, pp 213–225

Pinto RJB, Daina S, Sadocco P, Neto CP, Trindade T et al (2013) Antibacterial activity of nanocomposites of copper and cellulose. Biomed Res Int 6:280512. https://doi.org/10.1155/2013/280512

Ponce AG, Roura SI, Del Valle CE, Moreira MR et al (2008) Antimicrobial and antioxidant activities of edible coatings enriched with natural plant extracts: in vitro and in vivo studies. Post-Harvest Biol Tec 49:249–300

Pradhan N, Singh S, Ojha N, Shrivastava S, Barla A, Rai V, Bose S et al (2015) Facets of nanotechnology as seen in food processing, packaging and preservation industry. Biomed Res Int 2015:365672

Prasanna BM (2007) Nanotechnology in agriculture. ICAR National Fellow Division of genetics IARI, New Delhi

Predicala B (2009) Nanotechnology: potential for agriculture. Prairie swine Centre INC University of Saskatchewan Saskatoon SK 123-13

Rai M, Deshmukh SD, Ingle AP, Gupta IR, Galdiero M, Galdiero S et al (2016) Metal nanoparticles: the protective nanoshield against virus infection. Crit Rev Microbiol 42(1):46–56

Rai M, Gade A, Gaikwad S, Marcato PD, Durán N et al (2012) Biomedical applications of nanobiosensors: the state-of-the-art. J Braz Chem Soc 23(1):14–24

Ranjan S, Dasgupta N, Chakraborty AR, Samuel SM, Ramalingam C, Shanker R, Kumar A et al (2014) Nanoscience and nanotechnologies in food industries: opportunities and research trends. J Nanopart Res 16:2464

Reis CP, Neufeld RJ, Ribeiro AJ, Veiga F et al (2006) Nanoencapsulation I: methods for preparation of drug loaded polymeric nanoparticles. Nanomedicine 2(1):8–21

Reza MM, Johnson C, Hatziantoniou S, Demetzos C et al (2008) Nanoliposomes and their applications in food nanotechnology. J Liposome Res 18(4):309–327

Rhim JW (2004) Increase in water vapour barrier property of biopolymer based edible films and coatings by compositing with lipid materials. J Food Sci Biotechnol 13(4):528–535

Roselli M, Finamore A, Garaguso I, Britti MS, Mengheri E et al (2003) Zinc oxide protects cultured enterocytes from the damage induced by *Escherichia coli*. J Nutr 133:4077–4082

Salehi B, Prado-Audelo D, María L, Cortés H, Leyva-Gómez G, Stojanović-Radić Z et al (2020) Therapeutic applications of curcumin nanomedicine formulations in cardiovascular diseases. J Clin Med 9(3):746

Sawai J (2003) Quantitative evaluation of antibacterial activities of metallic oxide powders (ZnO, MgO and CaO) by conductimetric assay. J Microb Method 54:117–182

Sekhon BS (2010) Food nanotechnology- an overview. Nanotechnol Sci Appl 3:1–15

Singh YD, Ningthoujam R, Panda MK, Jena B, Babu PJ, Mishra AK (2021) Insight from Nanomaterials and Nanotechnology towards covid-19. Sens Int 100099

Sinha S, Pan I, Chanda P, Sen SK et al (2009) Nanoparticles fabrication using ambient biological resources. J Appl Biosci 19:1113–1130

Soares NFF, Silva CAS, Santiago-silva P, Espitia PJP, Goncalves MPJC, Lopez MJG et al (2009) Active and intelligent packaging for milk and milk products. In: Engineering aspects of milk and dairy products. Routledge, London, pp 155–174

Solans C, Izquierdo PJ, Nolla J, Azemar N, Garcia-Celma MJ et al (2005) Nano-emulsions. Curr Opin Colloid Interface Sci 10(3–4):102–110

Sonkaria S, Hoon SA, Khare V et al (2012) Nanotechnology and its impact on food and nutrition. Recent Pat Food Nutr Agric 4(1):8–18

Sozer N, Kokini JL (2009) Nanotechnology and its applications in the food sector. Trends Biotechnol 27(2):82–89

Subramanian A (2006) A mixed self assembled monolayer based surface Plasmon immunosensor for detection of *E. coli 0157H7*. Biosens Bioelectron 7:998–1006

Tan H, Ma R, Lin C, Liu Z, Tang T et al (2013) Quaternized chitosan as an antimicrobial agent: antimicrobial activity, mechanism of action and biomedical applications in orthopaedics. Int J Mol Sci 14:1854–1869

Tang D, Sauceda JC, Lin Z, Ott S, Basova E, Goryacheva I, Biselli S, Lin J, Niessner R, Knopp D et al (2009) Magnetic nano gold microspheres based lateral form immunodipstick for rapid detection of alphatoxin B_2 in the food. Biosens Bioelectron 25:514–518

Wang L, Luo J, Shan S, Crew E, Yin J, Zhong C-J, Wallek B, Wong SS et al (2011) Bacterial inactivation using silver-coated magnetic nanoparticles as functional antimicrobial agents. Anal Chem 83:8688–8695

Yuan Y, Gao Y, Zhao J, Mao L et al (2008) Characterisation and stability evaluation of β carotene nano emulsions prepared by high pressure homogenization under various emulsifying conditions. Food Res Int 41(1):61–68

Zuidam NJ, Shimoni E (2010) Overview of microencapsulates for use in food products or processes and methods to make them. In: Encapsulation technologies for active food ingredients and food processing. Springer, Cham, pp 3–29

Reis CP, Neufeld RJ, Ribeiro AJ, Veiga F et al (2006) Nanoencapsulation I. Methods for preparation of drug-loaded polymeric nanoparticles. Nanomedicine 2(1):8–21

Reza, MM, Johnson C, Hatziantoniou S, Demetzos C et al (2008) Nanoliposomes and their applications in food nanotechnology. J Liposome Res 18(4):309–327

Facets of Nanotechnology in Food Processing, Packaging and Safety: An Emerald Insight

5

Sushrirekha Das, Smita Hasini Panda, Mahidhar Bolem, Nityasundar Pal, Biswa Ranjan Samantaray, and Hrudayanath Thatoi

Abstract

In recent times, accelerated advancement of nanotechnology plays a great impact on food science and food industry with booming investment and market share. This review is an attempt to present and analyze the applications of nanotechnology towards current and emerge use in food industries as food formulations, processing, and storage. In order to produce real, affordable and delicious fresh food items, the food industry appeals for new technologies that are essential for maintaining market leadership. Nanotechnology is a technology that has the hidden potential to revolutionize the food industry. The uses of nanoparticles are not only in the field of food science but also in diagnosis of diseases and as drug and gene therapy, tumor control, detection of pathogens and proteins, tissues and cell engineering, DNA probing and pest control in agricultural fields. The technologies also introduced novel integration technologies, micro-and nanoencapsulation and enzyme encapsulation. Regulatory structures that are able to handle any risks associated with nanofoods and the use of nanotechnologies in the food industry are urgently needed.

S. Das · M. Bolem · N. Pal
Marsco Aqua Clinics Pvt. Ltd., (AOC), Balasore, India

S. H. Panda
Department of Zoology, North Orissa University, Baripada, India

B. R. Samantaray
Krishi Vigyan Kendra, Mayurbhanj, Shamakhunta, Odisha, India

H. Thatoi (✉)
Department of Biotechnology, North Orissa University, Baripada, India

M. Arakha et al. (eds.), *Bio-Nano Interface*,
https://doi.org/10.1007/978-981-16-2516-9_5

Keywords

Food technology · Encapsulation · Nanotechnology · Packaging · Food safety

5.1 Introduction

The Greek word "Nano" means extremely small and the word was coined by Prof. Norio Taniguchi of Tokyo Science University. At present nanotechnology is focused on the production, manipulation and use of materials at the nanometre scale for advanced biotechnology (Vadlapudi and Bullaya 2014). In current scenario, it is needed to develop reliable, nontoxic, clean and eco-friendly experimental protocols for the synthesis of nanoparticles of controlled size and shape. Further it is necessary to elaborate the technology in a consolidated way with an approach that provides the current trends of research on the biosynthesis of nanoparticles. Biosynthesis involves using eco-friendly microorganisms, such as actinomycetes, fungus, plants, viruses and yeast (Shah et al. 2011) (Fig. 5.1).

In food nanotechnology, the first step of revolution in food processing and improvement in quality has emerged from Pasteurization process which was first introduced by Pasteur to kill the spoilage bacteria (1000 nanometers) of foods (Chellarama et al. 2014). Recent times, there is a great demand of functional foods because growing realization links between pivotal diet and human health (Alexandra & Alexandru 2017). This has led to the development of new category of foods, the so-called functional foods which have bioactive compounds without affecting the sensory perception of the consumer thereby improving the uptake of certain components (Harjinder 2016). Nowadays, consumers are demanding functional

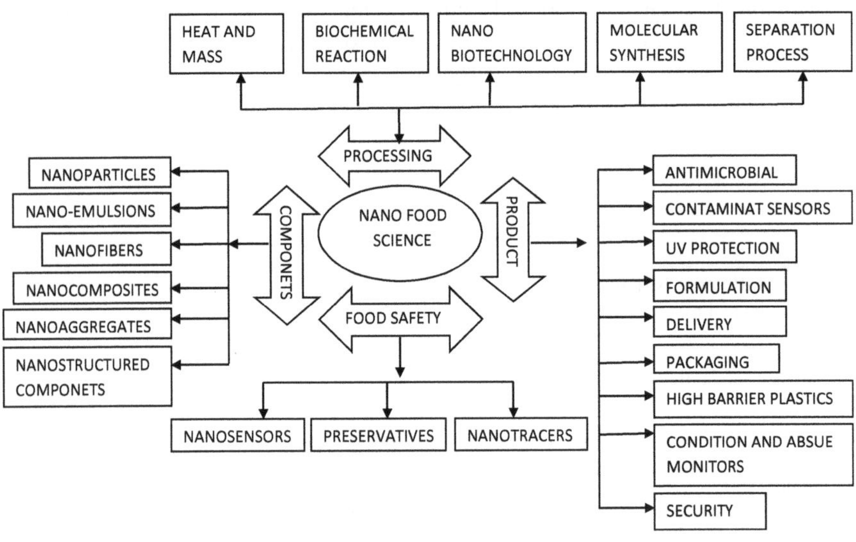

Fig. 5.1 Potential application of nanotechnology in food industry

foods which contain pro- and prebiotics but the territory of availability is limited to certain food products, mostly fermented milk products, because of the actual technological barriers concerning the processing of those bioactive substances (Kang Qixin 2016). For example, in the case of the yoghurt preparations, there are large fluctuations and poor viability of probiotic bacteria and especially of bifidobacteria during product storage, or in their passage through the digestive tract after consumption (He et al. 2015). Nanotechnology plays an important role in food technology, demand in food industry is the latest properties of nanomaterials. Thus, industries are demanding new techniques such as micro- and nanoencapsulation that would provide promising prospects for improved active research area in bioactive substances (Helmut 2004). Different types of functional nanostructures are used as building blocks to create novel structures. These include: nanoliposomes, nanoemulsions, nanoparticles and nanofibers. In the past two decades food nanotechnology has been rapidly growing because nanoscale structures have shown unique functionalities that improve sensorial, physical, chemical, biological, antimicrobial, nutritional and healthfulness properties of food products (Pan and Zhong 2016). The use of nanotechnology in food sectors includes nanosensors, tracking devices, targeted delivery of required components, food safety, new product development, precision processing, smart packaging, etc (Mahendra 2017). Nanotechnology has the potential to combat food security, disease treatment methods, new tools for molecular and cellular biology, new material for pathogen detection and protection of the environment (Dasgupta et al. 2015). Nanostructuring also adds value to traditional materials by enhancing their mechanical strength, superconductivity and ability to incorporate and efficiently deliver active substances into biological systems, at low costs and with limited environmental impact (Alam et al. 2016) The current work's exposure is to draw on modern techniques for innovative practical packaging that give answer to current barrier in the various fields of food science and food microbiology. It is confront by Lagaron et al. that the development of these ingenious concepts can be lugged by (1) integration of micro- and nanoencapsulation and controlled release of bioactive components or nanocomponents from biodegradable and/or sustainable packaging systems of these active substances either in the packaging and/or within foods and (2) packaging provided with enzymatic activity exerting a health-promoting benefit through transformation of specific food-borne components (Ishrat et al. 2016) (Tables 5.1 and 5.2).

The current review focuses on the development of such novel functional hybrid food/packaging systems including prebiotics, probiotics, phytochemicals, marine oils, lactose free foods, encapsulated vitamins, bioavailable flavonoids, etc. This will provide alternative, more efficient and in some cases, unique industrial means to provide foods with improved impact on human health upon consumption in the field of food microbiology (Bellah et al. 2012). Nanomaterials are customarily classified as various types such as (1) Nanoparticles (NPs), (2) Nanoclays (NCs) and (3) Nanoemulsions (NEs).

Table 5.1 List of microorganisms synthesizing metal nanoparticle

Microorganism	Nanoparticle	Size in (nm)	References
Pseudomonas stutzeri AG259	Silver	200	Venkataraman et al. (2011)
Lactobacillus strains	Silver	500	
Bacillus megaterium	Silver	46.9	
Klebsiella pneumoniae	Silver	50	
Bacillus licheniformis	Silver	50	
Corynebacterium sp.	Silver	10–15	
Bacillus subtilis	Silver	5–60	
Geobacter sulfurreducens	Silver	200	
Morganella sp.	Silver	20 ± 5	
Escherichia coli	Silver	1–100	
Proteus mirabilis	Silver	10–20	
Bacillus sp.	Silver	5–15	
Bacillus cereus	Silver	4–5	
Staphylococcus aureus	Silver	1–100	
Lactic acid bacteria	Silver	11.2	
Brevibacterium casei	Silver	50	
Enterobacter cloacae	Silver	50–100	Sukumaran and Eldho (2012)
Proteus mirabilis	Silver	10–20	Sukumaran and Eldho (2012)
Pseudomonas aeruginosa	Copper	50–150	Ratnika et al. (2012)
Bacillus subtilis	Gold	5–25	Abhilash et al. (2011)

5.2 Nanoparticles

Nanoparticles are very small in size, i.e. below 100 nm in at least one dimension. The term nanoparticles was coined by Walstra in the year 2003. In the past two decades food nanotechnology has been rapidly growing because structures of the nanoscale showed unique features which improve the nutritional and health properties of foods (Pan and Zhong 2016). Nanoparticles can be grouped into different types based on their ability to carry different reactions with different ingredients and environmental conditions. NP can be used as a UV protection barrier in food packaging. Depending on the chemical characteristics, Nanoparticles can be divided into two broad categories: (a) organic and (b) inorganic.

5.2.1 Organic Nanoparticles

Organic nanoparticles (ONPs) are produced from organic compounds that have potential for use in foods to provide novel functionalities (Duncan 2011; Yada et al. 2014). ONPs also known as nanocapsules used as vehicles for delivery of essential nutrients or pharmaceuticals. There are six classical methods for the preparation of nanocapsules: nanoprecipitation, emulsion–diffusion, double

Table 5.2 Different nano techniques and nano systems used in food industries

Delivery system	Characteristics	Applications
Liposomes	Biocompatible and biodegradable; suitable for DNA and RNA delivery into mammalian cells	Medical and pharmaceutical applications; easy to synthesize and easy to storage
Carbon nanotubes	Gene and drug delivery; nontoxic in mammalian cells and tissues	Gene delivery for disease therapy
Gold, silver magnetic	Biodiagnostics, imaging, gene and drug delivery for disease	Gold nanoparticles are nontoxic; non-opsogenic; magnetic hyperthermia therapy used for treatment of cancer cells
Fullerenes	Carbon nanoclusters: used in different ways such as making it water soluble	Protective barrier bound to DNA expanding chances of incorporation into chromosomes
Adenovirus	Gene therapy, low pathogenicity for human	Induces strong immune response
Adeno associated virus	Non-pathogenic and toxicity; long-term transgene expression	No specific integration
Lentivirus	Carry large gene inserts, infect non-dividing cells	Biosafety problems
Oncoviruses	Specific gene therapy; infect only dividing cells	Possible recombination with endogenous human retroviruses
Dendrimers	Encapsulation of genes, highly payload delivery	Act as multifunctional delivery system
Nanoemulsions	Stable dispersion, 100 nm size, uses in different lipids and emulsifiers	Delivery and stabilization of lipophilic compounds
Microemulsion	Mixture of water, oil and surfactants, size ranges 5–100 nm	Solubilization and delivery of hydrophobic and hydrophilic compounds

emulsification, emulsion–coacervation, polymer coating and layer-by-layer (Mora-Huertas et al. 2010). Generally, organic nanoparticles can be fabricated through two main methods, i.e., "top-down" and ''bottom-up" approaches (Forough and Farhadi 2010). In the traditional solid-state processing of the products, top-down routes are used. The first phase is based on the bluck content and making it smaller, splitting large particles with the use of physical processes such as crushing, grinding and grinding. In bottom-up methods, physical, chemical, and biological principles are used to build ONPs from particles or molecules. Both these methods differ in the magnitude of mechanical energy used to produce ONPs; top-down and bottom-up approaches are also called high-energy and low-energy methods, respectively (Abeyasinghe et al. 2016). These two methods can be used in combination by forming structures using a low-energy method and then reducing size using a high-energy method. A wide range of materials are available in NOs, for example, food additives (benzoic acid, citric acid, ascorbic acid) and supplements (vitamins A and E, isoflavones, beta-carotene, lutein, omega-3 fatty acids and coenzyme-Q).

They are used to enhance the nutrient value of food systems through improvement or alteration of food functionality (Arruebo et al. 2009).

5.2.2 Inorganic Nanoparticles (INP)

INP are also undergo several methods for production, such as gas phase INP synthesis method and liquid phase INP synthesis method which are further classified into different methods. Gas phase INP synthesis methods have mainly three types for INP synthesis which are named as flamed spray synthesis (Stark and Pratsinis 2002), laser-induced gas evaporation synthesis (Ullmannet al. 2002) and plasma-based synthesis (Kinemuchi et al. 2002). Liquid phase INP synthesis methods may further categorized into co-precipitation method (Cushing et al. 2004) and sol–gel approach (Hench and West 1990). Inorganic ingredients used in food, e.g. titanium dioxide, a food colourant. For long time preservation of prepared foods, storage containers/ utensils (food contact materials) will be embedded on INP. Food packaging and storage include engineered nanomaterials (ENMs) of transition metals, such as silver and iron; alkaline earth metals, such as selenium and silicates.

5.3 Nanoclays (NCS)

These particles are occurring naturally by means of aluminium silicate, composed of fine-grained minerals and having sheet-like geometry (Joseph and Morrison 2006). The sheet-structured silicates are referred as phyllosilicates (Aftab et al. 2016). They are low cost and eco-friendly materials. There are several NCs products in the market, e.g. Imperm, Aegis, Durethan, etc. Imperm (from Nanocor Inc.) is used in multi-layer polyethylene bottles and sheets for food and beverage packaging to minimize the loss of CO_2 from the drink and the ingress of O_2 into the bottle, thus keeping beverages fresher and extending shelf life. They provide an attractive alternative (Jayanta and Kwang-Hyun, 2017).

5.4 Nanoemulsions (NES)

The lipid phase consists of two phases, i.e. an aqueous phase and lower phase. In an aqueous phase, oil droplet being surrounded by a thin interfacial layer consisting of emulsifier molecules, called nanoemulsion (NE) (Tadros et al. 2004). High-energy or low-energy methods are used for production of NEs. NEs are highly stable to gravitational separation because the small particle size effects dominate gravitational forces. Encapsulation and delivery of functional compounds is one of the major fields of nanotechnology applied to food industry. Center for Biological Nanotechnology, 2001 reported antimicrobial NEs can be used in for decontamination of food equipment, packaging or food.

5.5 Preparation and Factor Affecting Biosynthesis of Nanoparticles

Generally, nanomaterials can be fabricated through two main methods, i.e., "top-down" and "bottom-up" approaches (Forough and Farhadi 2010). Chemical reduction method is the most common synthetic pathway for metal nanoparticles synthesis (Pal et al. 2007). In the case of nanoparticles (NPs) synthesis, bottom-up (or self-assembly) procedures involve a homogeneous system wherein catalysts (e.g. reducing agents, enzymes) are producing nanostructures affected by catalyst properties, reaction media and conditions (e.g. solvents, stabilizers, temperature) (Deepali et al. 2015). The physicochemical properties, surface and morphological characteristics of nanoparticles will influence their fate activity, transport and toxicity (Contescu and Putyera 2009). Synthesis of nanoparticles by high-energy physical processes for researchers to scan for biologically arbitrating methods using toxic chemical compounds (Gopinath and Velusamy 2013). Biologically synthesized nanoparticles are cost-effective, easily sealed up and eco-friendly and finally it is economically demanded (Mahasneh 2013). There are some methods used in nanoparticle preparation which are governed with good advantages and highly positive scale up such as (1) Nanoprecipitation of polymers, (2) Polymerization of alkylcyanoacrylates, (3) Interfacial polycondensations, (4) Formation of polyelectrolyte complexes, (5) Nano from neutral nanogels, (6) Ionic gelation, (7) Gelation of emulsion droplets, (8) Emulsification reverse salting-out, (9) Colloidal mill, (10) Natural organisms (Fungi) and (11) Natural organisms (Bacteria) (Nagavarma et al. 2015).

The biosynthesis of nanoparticles will be affected by some factors such as temperature, reaction time, reactant concentration, scalability, pH, etc. Size, shape, stability and physiological properties also affect in the biosynthesis, some nanoparticles have also been found to be toxic due to their factors such as composition, size, shape and surface chemistry (Kulkarni and Muddapur 2014). Presence of these toxic formation agents on the synthesized nanoparticles has prevented their clinical and biomedical application (Shah et al. 2011).

5.6 Characterization of Nanoparticles

In chemical and biological synthesis of nanoparticles, the aqueous state of metal ion precursors from metal salts will reduce and as a result a colour change will occur, this identification of colour change is the first qualitative detection of nanoparticles (Poinern et al. 2013). After the reaction, nanoparticles can be separated from the colloidal solution by centrifugation and then examined using advanced characterization techniques. The analytical approaches have been dived into three groups; separation techniques, imaging techniques and characterization techniques. These techniques include UV-visible spectroscopy, dynamic light scattering (DLS), atomic force microscopy (AFM), transmission electron microscopy (TEM), scanning electron microscopy (SEM), energy dispersive spectroscopy (EDS), powder X-ray diffraction (XRD), Fourier transform infrared spectroscopy (FT-IR) and Raman spectroscopy. AFM, SEM and TEM are considered as the direct methods for obtaining data from images of the nanoparticles. UV-Vis, DLS, XRD, EDS, FT-IR and Raman spectroscopies are considered for determining the composition, structure, crystal phase and properties of nanoparticles (Feldheim and Foss 2002; Jayanta and Kwang 2014). Characterization of nano delivery system is an essential part of understanding the benefits, as well as detection of toxicity in food.

5.7 Nanotechnology in Food Microbiology

Nanotechnology has prospective revolution in food industry. It has been used in innovative development of biosensors for detection of pathogens and chemical contaminants. This new technology also raises a serious concern about toxicological aspects of nanoparticles in food, with emphasis on the risk assessment and safety issues. Current research demand in food industries is encapsulation of probiotics and prebiotics to improve shelf life of the product. Probiotics are comprised of living microorganisms, when consumed in adequate amounts confer a health benefit on the host (FAO/WHO 2002). The live probiotics of different species subsume in foods in the form of fermentated milk, cheese, puddings fortified juices and curd (Nurit et al. 2015). Main aim of the nanoencapsulation is to develop probiotic bacterial preparations that could be delivered to certain parts of gastro intestinal tract where the probiotic bacteria interact with specific receptors and also to maintain good population (Jun et al. 2016). After encapsulation of probiotic bacteria into the products, the encapsulated forms of ingredients maintain longer shelf life of the product. The preparatory designer probiotics may act as de novo vaccines, which have the potentiality to modulate immune response (Timothy et al. 2016). Different varieties of probiotic nano foods are found in market, as capsules, liquids and powders. Dr. Kim's probiotic nanofood and calcium combination reduces the risk of osteoporosis (Kim et al. 2007). Fermented beverages like kefir, tempeh and kimchi have good populations of probiotics that can be used in diet.

5.8 Nanoencapsulation and Microencapsulation

Nanoencapsulation is walled material with Nano sizes. Nanoencapsulation provides several benefits such as protection against oxidation, pH triggered controlled release, change in flavour character, ease of handling and bioavailability and efficacy. Nanoencapsulation is a technology to pack substances in miniature making use of techniques such as nanocomposite, nano emulsification and nano structuration (Barrangou et al. 2011). These are providing the final product of functionality that includes controlled release of the core. By using this technique the protection of bioactive compounds, such as vitamins, antioxidants, proteins, and lipids as well as carbohydrates may be achieved for the production of functional foods with enhanced functionality and stability. Nanoencapsulation can make significant saving for formulations and can reduce the amount for active ingredients needed (Augustin and Hemar 2009).

Nano capsules are involved in the delivery of the desired components and entrapment of the odour and unwanted components in the food and thereby resulting in the preservation of the food. Nano capsules carry the food supplements via the gastrointestinal tract and which leads to increased bioavailability of the substance. There are six basic ways for the preparation of nano capsules. These are namely as nanoprecipitation, emulsions diffusion, double emulsification, emulsion-concentration, polymer coating and layer-by-layer. Nano-coating used as carrier functional ingredient during nanoencapsulation process and the nanoparticle is growing interest for the use of encapsulated nanomaterials in agriculture for production of nano formulated agrochemicals (Dingman 2008). In animal feeds, nanocapsules can be used to incorporate nano additives, antimicrobials and detoxifying compounds. Now nanocapsules and nanoemulsions have been used in production of nanopesticide. In the food industry, the nanocapsules techniques have increased because the materials can be protected from mixture, heat or other extreme conditions thus enhancing their stability. Nano encapsulation technologies have the potential to meet food industry challenges concerning the effective delivery of health functional ingredients and controlled release of flavour compounds. Lipid based nanoencapsulation system can enhance the performance of antioxidants by improving their solubility, bioavailability and preventing their unwanted interactions with other food components. The main lipid based nanoencapsulation systems that can be used for the protection and delivery of foods. Nanoliposome technology presents exciting opportunities for the area food technologists such as encapsulation and controlled release of food materials. The delivery vehicle for lipid soluble ingredients, protection from degradation during processing, controlled site specifics release, compatible with other food constituents, greater residence time and greater absorption nanomaterials are the several advantages for nanoencapsulation ingredients materials. Nanoencapsulation incorporated in fermented milk, yoghurts, cheese, puddings and fruit based drinks used to develop designer probiotics bacterial preparation. These probiotics bacterial preparation may act as vaccines which are responses immune modulating.

Microencapsulation is defined as a technology for packaging liquids, solids or gaseous substance in miniature. The sealed capsules that can release their contents at controlled rates under specific conditions. Encapsulation involves the incorporation of food ingredients and other materials in small capsules (Etheridge et al. 2013). For this technique, the applications increased in the food industry because the encapsulation materials can be protected from moisture, heat and other conditions to enhancing their stability and maintaining viability. Protein, fats, starches, alginates and lipid materials can be employed as encapsulating materials.

5.9 Nanoemulsions and Microemulsions

In pharmaceuticals, cosmetic and as models for biological membranes areas, the microemulsions have numerous applications. However in food, the microemulsions are limited by the type of surfactants that are used to facilitate their formation (Campos et al. 2011). Microemulsions are thermodynamically stable because of the mixture of water, oil and amphiphilic molecules which assemble spontaneously into nanometre-scale droplets. Microemulsion systems cannot be diluted, so the microemulsion system breaks down at increased water concentration. This restricts the application of some microemulsions in food systems (Monaliben et al. 2015). The microemulsions system potentially is used for the encapsulation of oil-soluble bioactive in food system. Nanoemulsions are only metastable, i.e. non-equilibrium systems with a tendency to separate into the constituent phases over a period of time. Nanoemulsions are optically transparent or translucent and are thermodynamically unstable systems (Chuesiang et al. 2018). In compared with microemulsions, nanoemulsions may possess a higher kinetic stability against creaming, flocculation and coalescence due to mainly the characteristic narrow droplet size distribution, low viscosity and Brownian motion between the droplets dominating over their low gravitational separation force. The main limitations for developing food nanoemulsions are the high cost of production and finding suitable food grade surfactants (Monaliben et al. 2015).

5.10 Nanofood Market

The worldwide sales of nanotechnology products in the food and beverage packaging sector increased from US$ 150 million in 2002 to US$ 860 million in 2004, reach to US$ 20.4 billion by 2010 (Helmut 2004). The consulting firm Cientifica has estimated the food applications of nanotechnologies around $410 million, food processing US$100 million, food ingredients US$100 million.

A comprehensive analysis of the worldwide state of investigational and approved nanomedicine products as of January 2012 (Volkmar et al. 2014) has identified 67 commercialized nanodevices and 33 marketed nanotherapeutics. A total of 25 devices and 122 therapeutics currently in development accounted for 789 ongoing clinical trials. Global market for nanocomposite, nanoparticles, and nano clays and nanotubes is about $1.6 billion in 2016 and it will reach $5.3 billion by 2021. The global nanotechnology market should reach $90.5 billion by 2021 from $39.2 billion in 2016 at a compound annual growth rate (CAGR) of 18.2%, from 2016 to 2021. The global market for nanofiber product reached $203.2 million and $276.8 million in 2013 and 2014, respectively. This market is projected to grow from $383.7 million in 2015 to nearly $2.0 billion in 2020, representing a compound annual growth rate (CAGR) of 38.6% between 2015 and 2020.

5.11 Food Processing Using Nanotechnology

The food processing system includes control of chemical contaminants, microbiological hazards and pathogens in order to promote food safety. For improving food safety, scientists are continuing their research process of investigating the use of nanoparticles as antibacterial agents. Recently, bulk amounts of SiO_2 and TiO_2 oxides have been permitted as food additives (marketed as E551 and E171, respectively). *Salmonella* spp. is recognized as one of the main causes of foodborne illnesses in humans worldwide. Silver nanoparticles (AgNPs) are known to be highly toxic to Gram-negative and Gram-positive microorganisms, including multidrug resistant bacteria. Silver has properties to kill pathogens biofilm in mammalian cells and tissues which is low toxic (Beyth et al. 2015). Therefore, silver nanoparticles are being considered as an important means of overcoming the growing problem of antibacterial resistance. Nutraceuticals value of nanoparticles containing nanosized ingredients and additives such as vitamins, antimicrobials, antioxidants and preservatives are currently available for enhanced taste, absorption and bioavailability (Momin et al. 2013). Some nutraceuticals incorporated in the carriers include lycopene, beta-carotenes and phytosterols are used in healthy foods to prevent the accumulation of cholesterol. The US based Oilfresh Corporation has introduced a nanofrying technique where a nanoceramic product has been marketed which reduces oil use in restaurants and fast food shops (Momin et al. 2015).

5.12 Packaging Techniques Using Nanotechnology

The innovative packaging solution is the reduction of spoilage. Packaging is the technology of producing, processing and shipment of food products for distribution, storage, sale and use. The food products are more secure through the use of nano sensors for pathogen and contaminant detection. The antimicrobial agents are being infused in to storage containers to retard bacterial growth and follow for longer storage of foods. Silver is a good and well known antimicrobial agent for food packaging (Carmen 2014). The 24-h growth of bacteria was reduced by over 98 percent because of the silver nanomaterials was developed with the help of mechanical and thermal properties (Woodrow Wilson International Centre for Sholars 2006). Nanocomposites are bonded nanoparticles in polymers to enhance properties like lightweight, better recyclability, spoilage and flavour issues (Pehanich 2006). Nowadays the nanocomposite materials are widely used in beer bottles (A to Z of Nanotechnology 2006).

In this era, the nanotechnology techniques are rapidly used in the food industry. Food packaging with nanotechnology techniques is the earliest commercial application in food sector. In commercial, about 400–500 nano-packaging products are estimated to be used and also the nano-packaging manufacture of 25% of all food packaging is predicted for next decade (Reynolds 2007). The release of antimicrobials, antioxidants, enzymes, nutraceuticals and flavours to extend shelf life can also be designed with the help of nano-packaging (Cha and Chinnan, 2004).

The pipeline and some antimicrobial films of food packaging are the new exciting nanotechnology products which are now in market to improve the shelf life of food and dairy products (El Amin 2007). Nanotechnology has been considered to have high potential for food packaging applications very early on. The companies are producing nanotechnology packaging materials for improving food safety and extend the life of food and dairy products. The important parts of food industry are food packaging and monitoring which are related with development and nanotechnology research (Broody 2003). Nanocomposites improved the food packaging by maintaing the mechanical strength, heat resistance, reduce weight and improve barrier against carbon dioxide, oxygen, ultraviolet radiation, moisture of food packaging (Gomez et al. 2014). Various food packaging techniques using nanoparticles have been discussed below.

5.12.1 Nano-Coatings

The nano-coatings are very efficient for keeping out oxygen and retaining carbon dioxide in packaging technology such as oxygen scavengers. A nano-coating is an aqueous based barrier coating. The use of aqueous based barrier coating food packaging provides an oxygen barrier and gas barrier. The widely used of waxy coating is now available in market for some foods such as apples and cheeses. Edible coating and films are developed with the help of Nano technology application, which is used on wide variety of foods and that edible coatings serve as moisture, lipid and gas barriers (Dammak et al. 2017). These edible coatings are invisible for the human eye. An edible antibacterial Nano-coating can be applied directly in bakery goods, which is developed in early 2007 by the U.S company Sono-Tec Corporation (El Amin 2007). In food packaging, the nano-coating techniques are improved day by day.

5.12.2 Nanolaminates

Nanolaminate films are suitable for use in the food industry. Nanolaminate exhibits a high concentration of interface defects and also its properties depend on their thickness and composition. For the preparation of edible coatings and films over conventional technology, the Nanolaminates have a number of important applications within the food industry. Nanolaminates also have some good advantage for the preparations. Nanolaminate coating could be created from proteins, polysaccharides, lipids by using the simple processing operation like as washing and dipping. A nanolaminate consists of two or more layer and bonded to each other with chemically and physically. For the creation of different layers, variety of different adsorbing substances could be used. The different layers are including natural polyelectrolytes, charged lipids and colloidal particles. These layers are possible to incorporate active functional agents in to the field. These functional agents would increase the shelf life and quality of coated foods. The nanolaminates are used as coatings which are attached to food surface because of thin nature (Kotov 2003).

5.12.3 Nano Crystal

The dairy and food packaging materials are incorporating as well as embedding with nano clays and nanocrystals. The advantages of nanocrystal and clay nanocomposite in the packaging material improved light in weight, shutter proof, heat resistance and shelf life (Ravichandran 2010). The nanocrystal is composed of atoms either a single or poly crystalline arrangement and having at least one dimension smaller than 100 nm.

5.12.4 Nanomaterials

Carbon nanotubes are ideal carriers for delivery of biomolecules by means of their low toxicity and easy penetration to cell membranes in which induced cell death by increasing membrane permeability (Khodakovskaya 2009). The polymer nonclay nanocomposite technology has been proven to improve the mechanical, barrier and thermal properties of polymers for packaging application. The nanomaterial will be analysed by the latest innovation in food packaging. The development of the new polymer nanomaterials is completely monitoring of safety and quality based food packaging in the food packaging industry.

5.12.5 Biobased Packaging

Biodegradable plastics are polymeric materials. It is at least one step in the degradation process. The biodegradation leads to fragmentation or disintegration of the plastics with no toxic or environmentally harmful residue in a perfect condition of moisture, temperature and oxygen availability (Chandra and Rustagi 1998). Biodegradable polymers are specific type of polymer which breaks down after its intended purpose to result in natural byproducts such as gases (CO_2, N_2), water, biomass, and inorganic salts. Biodegradable polymers have a long history and many are natural products. Biodegradable polymers are classified into natural polymers, synthetic polymers and modified natural polymers. The composition, structure, properties and applicability are the main factors that are described for the biodegradable polymers. The industries are facing environmental problems in the present era, and the polymer does not degrade under standard environmental conditions. So polymers play a vital role in biological environment degradation. Using polymers can solve the environmental problems. Recently, the preparation and characterization of various kinds of biodegradable polymers, i.e., nanocomposites are suitable for a wide range of applications (SinhaRay and Bousmina 2005). Starch and derivatives, polylactic acid (PLA), poly (butylene succinate) (PBS), polyhydroxybutyrate (PHB) and aliphatic polyester as PCl are the important ingredients for packaging.

5.12.6 Smart Packaging

In this smart packaging system, a technique known as bio-metrix has been developed, which evolves a multi detective test, i.e., food expert ID. This test was therefore met with various food scares for nanosurveillance (Mahmoud 2015). A nanotech company known as pSi Nutria develops various nanobased tracking technologies, which involve detection of pathogens and monitoring foods through ingestible Biosilicon. Nowadays smart packaging are used in Nestel, British air ways, Monoprix supermarket, to detect colour change in various foods, where food has been contaminated by food-borne pathogens. Here, packaging includes electronic noses and tongues, engineered nanosensors. The above nanosensors are developed by Kraft along with Rutgers University (U.S.) (Momin et al. 2013).

5.13 Role of Nanosensor in Food Safety

A nanosensor is a device consisting of an electronic data processing part and a sensing layer or part, which can translate a signal such as light, or the presence of an organic substance or gas into an electronic signal structured at the nanometre scale (Arfat et al. 2016). Nanosensors are small, portable, rapid response and processing. Nanosensors provide increased security in food industries in manufacturing, processing and shipping of food products. A number of biosensors, sensors based on nanoparticles, nanocantilevers, nano-test strips, solution nanoparticles, electronic noses, etc., have been used as nanosensors for the detection of pathogens, pollutants, chemicals and food analysis (Valdes et al. 2009). EU-funded Biofinger project has developed portable biosensor Bio-Nano and Micro Electro Mechanical systems (BioMEMS) using nanocantilevers for detection of biological entities, chemicals and toxins (Joseph and Morrison 2006). The European Union regulations for food safety have recommended that, for the introduction of new nanotechnology, specific safety standards and testing procedures are required. In India food safety regulations are introduced but not adequate to monitoring safety of nanoparticles. Biotechnology has designed sensors which give increased sensitivity to environmental changes. Nanosensors are also used by the store keepers to detect expiry dates of food items (Kim et al. 2007). Carbon nano tubes used as nanosensor device are small enough to trap and measure individual proteins or small molecules. Nanosensors work by initiating enzymatic reactions or by using nano-engineered branching molecules called dendrimers as probes to bind to target chemicals and proteins. Pathogen and contaminate detection is possible with increased sensitivity and decreased response time due to nanosensors. The nanotechnology has developed a test for detection of *S. aureus* based on the use of magnetic beads, which can detect most of *S. aureus* strains at a very low viable count. This will result in improving food safety. Similar results were also obtained for *Listeria monocytogenes*. Aflatoxins are carcinogenic, mutagenic, teratogenic and immunosuppressive substances, which are produced as secondary metabolites by the fungi *Aspergillus flavus* and A. Parasiticus growing on a variety of food products.

5.14 Future Trends and Perspectives of Nanotechnology

Nanotechnology has been reported to have a wide range of applications in the field of food industry from food analysis, packaging to storage. Technology helps in identifying bacterial population with possible carcinogenicity and other harmful activities. Research and development in response to consumer preferences gave rise to active, intelligent and bio active food packaging techniques that are purely innovative. These innovative packaging technologies contributed toward the enhancement of food quality, safety, feasibility and bioactivity of functional components. Nano-biosensors play an important role, in detection of food contamination levels and safety increasing its shelf life. Food products and ingredients like milk proteins, cereals, breads and polysaccharide have developed maillard conjugates with very good emulsifying properties are been used in nanoencapsulating heat sensitive bioactives including probiotics (Kolsoom et al. 2014). Beverages, meat, sausages, cheese and other foods based on enzymatic hydrolysis in milk globular protein, α-lactoalbumin development of carbon nanotubes showing antimicrobial properties are being used in food packaging. Hydrophobic healthy foods in the form of nanocomplexes incorporated into soft drinks, ice creams and chocolates. Nanotechnology develops new technique, i.e. nanocharcoal absorbant, which decolourizes the food products. Besides this, nanotechnology also uses nanomaterials found in nanofood products and includes various metal oxides and metals like silver and iron complexes as antimicrobial and antioxidant supplements, calcium and magnesium salt complexes used in health supplements, selenium used as an additive in green tea products, etc. Cola-tasting nano milk and fat reduced nano mayonnaise are the two examples of nanofood products developed. Nano-packaging materials are bound with the polymer matrix to increases its mechanical strength and makes it impermeable to gases, volatile components such as flavours or moisture. Organic nanomaterial has been developed, is used in food/feed industry to increase its consumption and improve bioavailability of various minerals, vitamins and antioxidants of our body. For example, a naturally occurring nanomaterial is lycopene a type of carotenoid in tomato. Nanotechnology developed nano bio-fertilizers and other new agricultural techniques for good quality of crops. Nanolaminate films used as edible coating over a variety of foods like vegetables, fruits, meat, chocolates, candies, etc. Biopolymeric nanoparticles like polylactic acid (PLA) helps in encapsulation. Use of nano composites materials for coating plastic, a future alternative to silica and alumina coating as it is biodegradable and helps to reduce packaging wastes. Nanotechnology helps in reducing agro wastes and converts it into a high quality bio-fuels.

There is a lot of research going on in recent times to produce nanocapsules containing vitamins that will be discharged when nano sensors detect a lack of vitamin in your body. After this research it could bring about a super vitamin storage framework in human body system that conveys the supplements require by the body, when needed. Nanocapsule contains flavour or colour enhancers which when inserted in the food. The technique has not been distributed, so it will be interesting to perceive how this specific trick is accomplished.

References

Abeyasinghe N, Kumar S, Sun K et al (2016) Enhanced emission from single isolated gold quantum dots investigated using two-photon-excited fluorescence near-field scanning optical microscopy. J Am Chem Soc 138(50):16299–16307

Abhilash RK et al (2011) Microbial synthesis of iron-based nanomaterials—a review. Bull Mater Sci 34(2):191–198

Aftab A, Ismat FR, Usman S et al (2016) Application of nanotechnology in agriculture and food science. World J Pharm Sci 4(7):45–54

Alexandra EO, Alexandru MG (2017) Nanotechnology applications in food, flavour, stability, nutrition and safety. www. elesvieor.com

Arfat YA, Benjakul S, Prodpran T et al (2016) Physico-mechanical characterization and antimicrobial properties of fish protein isolate/fish skin gelatin-zinc oxide (ZnO) nanocomposite films. Food Bioprocess Technol 9:101–112

Arruebo M, Valladares M, González-Fernández Á (2009) Antibody-conjugated nanoparticles for biomedical applications. J Nanomater 2009:37

Augustin MA, Hemar Y (2009) Nano- and micro-structured assemblies for encapsulation of food ingredients. Chem Soc Rev 38:902–912

Barrangou R, Lahtinen SJ, Ibrahim F, Ouwehand AC (2011) Genus *Lactobacillus*. In: Lactic acid Bacteria: microbiological and functional aspects. CRC Press, London, pp 77–91

Bellah MM, Christensen SM, Iqbal SM (2012) Nanostructures for medical diagnostics. J Nanomater 2012:2

Beyth N, Houri-Haddad Y, Domb A et al (2015) Alternative antimicrobial approach: nano-antimicrobial materials. https://doi.org/10.1155/2015/246012

Brody AL, Bugusu B, Han JH et al (2008) Innovative food packaging solutions. J Food Sci 73:107–116

Campos CA, Gerschenson LN, Flores SK (2011) Development of edible films and coatings with antimicrobial activity. Food Bioprocess Technol 4:849–875

Carmen L, Simone B, Veronica C et al (2014) Antibacterial activity of silver nanoparticles: sensitivity of different *Salmonella* serovars. Front Microbiol 5:227

Cha D, Chinnan M (2004) Biopolymer-based antimicrobial packaging: a review. Crit Rev Food Sci Nutr 44:223–237

Chandra R, Rustgi R (1998) Biodegradable polymers. Prog Polym Sci 23:1273–1335

Chellarama C, Murugaboopathib G, Johna AA et al (2014) Significance of nanotechnology in food industry. APCBEE Procedia 8:109–113

Chuesiang P, Siripatrawan U, Sanguandeekul R et al (2018) Optimization of cinnamon oil nanoemulsions using phase inversion temperature method: impact of oil phase composition and surfactant concentration. J Colloid Interface Sci 514:208–216

Contescu CI, Putyera K (2009) Dekker encyclopedia of nanoscience and nanotechnology I-VI. ISBN 13 978-0-8493-9639-7

Cushing BL, Kolesnichenko VL, O'Connor CJ (2004) Recent advances in the liquid-phase syntheses of inorganic nanoparticles. Chem Rev 104:3893–3946

Dammak I, de Carvalho RA, Trindade CSF et al (2017) Properties of active gelatin films incorporated with rutin-loaded nanoemulsions. Int J Biol Macromol 98:39–49

Dasgupta N, Ranjan S, Mundekkad D et al (2015) Nanotechnology in agrofood: from field to plate. Food Res Int 69:381–400

Deepali S, Suvardhan K, Krishna B (2015) Biogenic synthesis of nanoparticles: a review. Arab J Chem 12(8):3576–3600

Dingman J (2008) Nanotechnology: its impact on food safety. J Environ Health 70:47–50

Duncan TV (2011) Applications of nanotechnology in food packaging and food safety: barrier materials, antimicrobials and sensors. J Colloid Interface Sci 363:1–24

El Amin A (2007) Nanoscale particles designed to block UV light. Food Production Daily com Europe. 18 October 2007

Etheridge ML, Campbell SA, Erdman AG (2013) The big picture on nanomedicine: the state of investigational and approved nanomedicine products. Nanomedicine 9:1–14

FAO / WHO Expert Consultation (2002) Evaluation of health and nutritional properties of powder milk and live lactic acid bacteria. Córdoba, Argentina: Food and Agriculture Organization of the United Nations and World Health Organization 2002

Feldheim DL, Foss CA (2002) Metal nanoparticles: synthesis, characterization and applications. Marcel Dekker Inc, New York

Forough M, Farhadi K (2010) Biological and green synthesis of silver nanoparticles. Turkish J Eng Env Sci 34:281–287

Gomez EJ, Lopezde DC, Hernandez-M P et al (2014) Advances in antioxidant active food packaging. Trends Food Sci Technol 35:42–51

Gopinath V, Mubarak Ali D, Priyadarshini S et al (2012) Biosynthesis of silver nanoparticles from Tribulus terrestris and its antimicrobial activity: a novel biological approach. Colloids Surf B: Biointerfaces 96:69–74

Harjinder S (2016) Nanotechnology applications in functional foods; opportunities and challenges. Prev Nutr Food Sci 21(1):1–8

He X, Aker WG, Fu PP et al (2015) Toxicity of engineered metal oxide nanomaterials mediated by nano-bio-ecointeractions: a review and perspective. Environ Sci Nano 2:564

Hench LL, West JK (1990) The sol–gel process. Chem Rev 90:33–72

Helmut K (2004) Nanotechnology in food and food processing industry worldwide 2008–2010-2015 report. Springer, Zurich, Switzerland

Ishrat M, Gulzar AN, Shuaib MD et al (2016) Novel food packaging technologies: innovations and future prospective. J Saudi Society Agri Sci 11:1658–1177

Jayanta KP, Kwang HB (2014) Green Nanobiotechnology: factors affecting synthesis and characterization techniques. J Nanomater 2014-10(3):507–517

Jayanta KP, Kwang HB (2017) Antibacterial activity and synergistic antibacterial potential of biosynthesized silver nanoparticles against foodborne pathogenic bacteria along with its anticandidal and antioxidant effects. Front Microbiol. https://doi.org/10.3389/fmicb.2017.00167

Joseph T, Morrison M (2006) Nanotechnology in agriculture and food: a nanoforum report. Institute of Nanotechnology. http://www.nanoforum.org

Jun L, Cecilia Ying JS, Nikki L et al (2016) Probiotics modulated gut microbiota suppresses hepatocellular carcinoma growth in mice. Proc Natl Acad Sci U S A 113(9):1306–1315

Kang P, Qixin Z (2016) Organic nanoparticles in foods: fabrication, characterization, and utilization. Annu Rev Food Sci Technol 7:111–111

Khodakovskaya M, Enkeleda D, Meena M et al (2009) Carbon nanotubes are able to penetrate plant seed coat and dramatically affect seed germination and plant growth. ACS Nano 3(10):3221–3227

Kim JS, Kuk E, Yu KN et al (2007) Antimicrobial effects of silver nanoparticles. Nanomed Nanotechnol Biol Med 3:95–101

Kinemuchi Y, Ikeuchi T, Suzuki T et al (2002) Synthesis of nanosize Pzt powders by pulsed wire discharge. IEEE Trans Plasma Sci 30:1858–1862

Kolsoom P, Rosita J, Karimi G et al (2014) Effect of probiotics supplementation on bone mineral content and bone mass density. Sci World J 6:12. https://doi.org/10.1155/2014/595962

Kotov NA (2003) Layer-by-layer assembly of nanoparticles and nanocolloids: intermolecular interactions, structure and materials perspective. In: Decher G, Schlenoff JB (eds) Multilayer

thin films: sequential assembly of nanocomposite materials. Wiley-VCH, Weinheim, pp 207–243
Kulkarni N, Muddapur U (2014) Biosynthesis of metal nanoparticles: a review. J Nanotech:510246
Mahasneh AM (2013) Bionanotechnology: the novel nanoparticles based approach for disease therapy. Jordan J Biol Sci 6:246–251
Mahendra P (2017) Nanotechnology: a new approach in food packaging. J Food Microbiol Saf Hyg 2:2
Mahmoud MB (2015) Nanotechnology in food industry; advances in food processing, packaging and food safety. Int J Curr Microbiol App Sci 4(5):345–357
Momin JK, Joshi BH (2015) Nanotechnology in foods. In: Rai M, Ribeiro C, Mattoso L, Duran N (eds) Nanotechnologies in food and agriculture. Springer, Switzerland, pp 3–24
Momin JK, Jayakumar C, Prajapati JB (2013) Potential of nanotechnology in functional foods. Emirates J Food Agricul 25(1):1019
Monaliben S, Derek F, Shashi S, Suraj KT, Gérrard EJP (2015) Green synthesis of metallic nanoparticles via biological. Materials 8:7278–7308
Mora-Huertas CE, Fessi H, Elaissari A (2010) Polymer-basednanocapsules for drug delivery. Int J Pharm 385:113–142
Nagavarma BVN, Hemant KSY, Ayaz A et al (2015) Different techniques for preparation of polymeric nanoparticles-a review. In: Alarcon EI et al (eds) Silver nanoparticle applications, engineering materials. Springer, Switzerland. https://doi.org/10.1007/978-3-319-11262-6_2
Nurit B, Yael HH, Avi D et al (2015) Antibacterial activity of dental composites containing quaternary ammonium polyethylenimine nanoparticles against *Streptococcus mutans*. Biomaterials 27(21):3995–4002
Pan K, Zhong Q (2016) Organic nanoparticles in foods: fabrication, characterization, and utilization. Annu Rev Food Sci Technol 7:245–266
Pehanich M (2006) Small gains in processing and packaging. Food Process 11:46–48
Poinern GEJ, Le X, Chapman P et al (2013) Green biosynthesis of gold nanometre scale plate using the leaf extracts from an indigenous Australian plant Eucalyptus macrocarpa. Gold Bull 46:165–173
Ratnika V, Seema B et al (2012) A review: biological synthesis of silver and copper nanoparticles. Nano Biomed Eng 4(2):99–106
Ravichandran R (2010) Nanoparticles in drug delivery: potential green nano biomedicine applications. Int J Nanotechnol Biomed 1:108–130
Shah M, Fawcett D, Sharma S et al (2011) A novel green one-step synthesis of silver nanoparticles using chitosan: catalytic activity and antimicrobial studies. Appl Nanosci 4:113–119
Sinha Ray S, Bousmina M (2005) Biodegradable polymers and their layered silicate nanocomposites: in greening the 21st century materials world. Prog Mater Sci 50(8):962–1079
Stark WJ, Pratsinis SE (2002) Aerosol flame reactors for manufacture of nanoparticles. Powder Technol 126:103–108
Sukumaran P, Eldho KP (2012) Silver nanoparticles: mechanism of antimicrobial action, synthesis, medical applications, and toxicity effects. Int Nano Lett 2:32
Tadros T, Izquierdo P et al (2004) Formation and stability of nano-emulsions. Adv Colloid Interf Sci 108–109:303–318
Timothy WY, Elif F, Kendra AT et al (2016) Microencapsulation in alginate and chitosan microgels to enhance viability of *Bifidobacterium longum* for Oral delivery. Front Microbiol 7:494
Vadlapudi V, Bullaya D (2014) Review: green synthesis of silver and gold nanoparticles. Middle-East J Sci Res 19(6):834–842
Valdes MG, Gonzalez ACV, Calzon JA et al (2009) Analytical nanotechnology for food analysis. Microchim Acta 166:1–19
Volkmar W, Tracy KP, Nicole M (2014) Nanopharmaceuticals (part 1): products on the market. Int J Nanomed 9:4357–4373
Yada RY, Buck N, Canady R et al (2014) Engineered nanoscale food ingredients: evaluation of current knowledge on material characteristics relevant to uptake from the gastrointestinal tract. Compr Rev Food Sci Food Saf 13:730–744

Nanotechnology and Its Potential Application in Postharvest Technology

6

Barsarani Jena, Rina Ningthoujam, Sabita Pattanayak, Santwona Dash, Manasa Kumar Panda, Bimal Prasad Jit, Mohinikanti Das, and Yengkhom Disco Singh

Abstract

Due to ever-increasing human population, there is a huge need of enough supply of food in the near future to overcome food crisis. Now, only increasing sufficient amount of food is not enough to stand alone to resolve the problems, otherwise the postharvest loss which accounts up to 33% of total agricultural produce will be a hampered in the process. Hence, many combinatorial efforts using novel scientific technologies such as biotechnology, nano biotechnology, nanoparticles and other relevant technologies are being fostered to tackle the problems. Nanotechnology as its unique properties can be explored to minimize the postharvest losses. Nanoparticles such as gold, silver, zinc and other metallic nanoparticles

B. Jena · S. Pattanayak · S. Dash
Department of Botany, School of Applied Sciences, Centurion University of Technology and Management, Bhubaneswar, Odisha, India

R. Ningthoujam
Department of Vegetable Science, College of Horticulture and Forestry, Central Agricultural University, Pasighat, Arunachal Pradesh, India

M. K. Panda
Environment and Sustainability Department, CSIR-Institute of Minerals and Materials Technology, Bhubaneswar, Odisha, India

B. P. Jit
School of Life Sciences, Sambalpur University, Jyoti Vihar, Burla, Sambalpur, Odisha, India

M. Das
Department of Botany, College of Basic Science and Humanities, Odisha University of Agriculture and Technology, Bhubaneswar, Odisha, India

Y. D. Singh (✉)
Biomaterials and Bioprocessing Research Laboratory, Department of Post Harvest Technology, College of Horticulture and Forestry, Central Agricultural University, Pasighat, Arunachal Pradesh, India

M. Arakha et al. (eds.), *Bio-Nano Interface*,
https://doi.org/10.1007/978-981-16-2516-9_6

are often showing potential results in controlling the micro-organism deterioration and increasing the shelf life period of highly perishable crops. Nanotechnology is a useful technology that can be applied to all branches of the science and technology. This chapter focuses on packaging of crops using nanotechnology to increase its shelf life. It also discussed about different nanomaterials playing together in the food to safeguard from micro-organism and other pathogens especially while storage. It is concluded that there is need to explore on using of nanoparticles in postharvest management of crops to enhance their shelf life period. It also concludes that there is need of substantial research on novel technologies in the food and agricultural sectors.

Keywords

Nanomaterials · Postharvest loss · Shelf life · Food · Perishable crops

6.1 Introduction

In current days, nanotechnology application in different sectors is progressively improving by mitigating side effects. It plays major role in fundamental science, material science, engineering and agricultural science. To solve recent problems facing in agriculture, nanotechnology can be used which occupies the confluence of biotechnology, bioengineering and nano biology (Yadollahi et al. 2009). The prospective of this technology is well achieved by many industries in preparing and developing commercial products which are widely available in the market. Besides its multifaceted applications, some of the basic fields where it can play major role include electronics, photonics, pharmaceuticals and cosmetics, food and textiles (Abbas et al. 2009). It is being seen that nanotechnology is vastly used in our day to day life and this will bring change in the society. Firstly, in 2003, the USA, Department of Agriculture had utilized the nanoparticles in agriculture and food industry as their roadmap agenda (He et al. 2018; US DOA 2003). Then, thereafter, gradually the momentum of nanoparticle application has increased. Analysis on the topic has taken off over the last decade. It mainly disguises the all aspect of food and agriculture industry, irrigation/water filtration, food processing and packaging, animal feed and aquaculture (Dasgupta and Ranjan 2018; Peter et al. 2016; Finglas et al. 2014; Bryksa and Yadav 2012; Sozer and Kokini 2009). In the midst of high demands of food and its products and by looking to the ever-increasing human population, horticulture science which mainly focuses on art and cultivation of and handling of fruits, vegetables, flowers, orchids, ornamental plants will provide in supplying sufficient and healthy food in the near future. However, a problem of postharvest lost in horticultural and agricultural produce dissociates the food supply and quality food materials reached to the customer. This can be minimized by applying the biotechnology and nanotechnology in postharvest lost. Nanotechnology can play major role in extension of shelf life of horticultural crops, control from growth of micro-organism, food packaging and quality management, nanochips in labelling of the packaged food products and so on (Yadollahi et al. 2009). This chapter emphasizes on some current outcomes on the utilization of nanotechnology for horticultural crops postharvest shelf life improvement.

6.2 Nanomaterials

The word nano comes from a Greek work meaning dwarf which signifies the one billionth of a metre, i.e. 1 nm = 10^{-9} m (Taniguchi 1974). Nanotechnology is a branch of science which deals with the nanoparticles of range between 1 nm and 100 nm (Salehi et al. 2020). In twenty-first century, nanomaterials have gained significant momentum worldwide due to their peculiar properties and patterns in contrast to conventional products (Camargo and Smith 2009; Panigrahi 2013; Panda et al. 2020a). The nanomaterials show variation in their physical, chemical and biological properties from atoms, molecules and bulk materials within the dimensions of 100 nm (Panigrahi 2013). Nanomaterials act like a small crossover in the middle of atomic structures and bulk materials. These bulk commodities possess persistent physical properties in behalf of atomic structures with random atoms that are situated in the space one by one and tries to associate with each other across the atomic boundaries but the nanomaterials are aligned in a crystalline lattice that are in conjunction with a single atom (Sharma et al. 2009). Nanomaterials are having some exceptional characteristics that differentiate them from bulk materials including larger aspect of energy, larger fragment of surface atoms, dimensional captivity and lower deformity (Cao and Wang 2004).

Nanomaterials are generally classified into two groups corresponding to organic nanomaterial and inorganic nanomaterial. The organic nanomaterials encompass carbon nanoparticle while inorganic nanomaterials comprised of magnetic, noble (gold, silver) and semiconductor nanomaterials, namely zinc oxide, titanium oxide (Pradhan 2013). In comparison to organic nanomaterials, inorganic nanomaterials are observed to be more useful for its better quality, products and multipurpose function (Panigrahi 2013). Nanomaterials can exist with other materials of various chemical natures serving as metal oxides, polymers, silicates, non-oxide ceramics, organic carbon and biomolecules with dissimilar designs like spherical, oval, cubical, helical, prism, cylindrical and tubes (Akinsiku 2017). Due to exclusive properties, structure, shapes and sizes of nanomaterials, it is now widely been used in many scientific fields including physics, chemistry, biology, biotechnology, microbiology, etc. (Lin and Xing 2007).

6.3 Properties of Nanomaterial

Nanoparticles are wide class of material that include particular size of one dimension less than 100 nm at least (Singh et al. 2021). Nanoparticles can exhibit different physical and chemical properties based on the type of particles and source of its synthesis (Khan et al. 2017).

6.3.1 Physicochemical Properties of Nanoparticle

The physicochemical properties of nanoparticles include large surface area, mechanical strong, optically active and chemically reactive. This unique property makes nanoparticles suitable applicant for various applications in the field of agriculture, food, environment, health care and other diagnosis purposes.

6.4 Applications of Nanotechnology

Nanotechnology is used by biologist, chemist, physicists, engineers and doctors. Generally, in agricultural field nanotechnology is used to identify and eliminate pathogens and viruses (Singh et al. 2015). In agricultural field, nanotechnology is used for the development of genetically modified crops, precision farming technique and chemical pesticides. The nanotechnology has significant effect in food industry for the development of design methods, functional materials, instrumentation and product development for food safety and packaging (Hosseini and Rezaei 2011).

6.4.1 For the Control of Disease and Pest in Plants

Nanoparticles like nano alumino-silicate, nano silver, carbon nanomaterials, magnetic nanoparticles, titanium dioxide (TiO_2), magnetic nanoparticles with particular concentration used for the control of several plant diseases caused by different pathogens (Yadollahi et al. 2009).

6.4.2 For Detecting Plant Diseases

Nanotechnology is used to detect the plant disease at its early stage. This technology is rapid, accurate and can give results within few hours. For the detection of phytobacteria nanotechnology is used as a biomarker (Khan and Rizvi 2014).

6.4.3 For the Control of Plant Diseases

Crop management can be improved and modified by following new ways of nanotechnology. By following broadcasting or spraying technique, plant protecting chemicals and plant nutrients are applied to crops. Low concentration of chemical applied to the target site to avoid problems like degradation by photolysis, leaching of chemicals, microbial degradation and hydrolysis (Singh et al. 2015).

6.5 Use of Nanoparticles to Control the Plant Diseases

Different forms of nanoparticles can be used in controlling plant disease. Nanoparticles such as silver, silica, alumina silicates and carbon are often used in controlling plant diseases (Khan and Rizvi 2014).

6.5.1 Nano-Agriculture

The use of some nanoparticles in agriculture field, which have some profitable effect to crops, is known as nanoagriculture (Srilatha 2011). Nanoparticles decrease the release of chemical level and the damage to other plant tissues. Fertilizers, pesticides and nutrients are applying to plants by drenching or spraying process. Nanotechnology increases the safety, effectiveness, patient obedience by developing new products against pests (Caraglia et al. 2011). There are some nanoparticles such as silver nanoparticles, mesoporous silica nanoparticles, carbon nanotubes, nanoemulsion, nano sensors which provide effects against pests and improve crops quality.

6.5.2 Silver Nanoparticles

Root diseases can be reduced by using silver nanoparticles and improve health and growth of several plants (Srilatha 2011).

6.5.3 Nano Sensors

Nano sensors play a role as a promising tool for the food production and in the agriculture field (Srilatha 2011). For the detection of contaminants like toxins bio-hazardous substances, viruses, bacteria, etc. nano sensors are used in food systems and agriculture.

6.5.4 Mesoporous Silica Nanoparticles

Mesoporous silica nanoparticles (MSNs) have high pore volume, high specific area and physicochemical stability so it has an excellent drug delivery system (Srilatha 2011). DNA and chemicals are delivered into isolated plant cells by MSN. This facilitated to greatly studied on functional genomic of the plants. Gene expression system of the plants can be easily tapped using with MSN. They act as carrier to foreign gene delivering at site of interest.

6.5.5 Nanoemulsion

Nanoemulsion is dispersion of two immiscible liquids stabilized by surfactant. The dispersed phase comprises size range of 5–200 nm (Zhao et al. 2014). Nanoemulsion is used in the field of agriculture, biomedical, cleaning products and cosmetic. Most of the emulsions used in agriculture are based on biological sources which may be included lipid nanocarriers, plant oils, emulsifiers, biosurfactants, cosurfactants, surface nanoemulsion, encapsulated active ingredients, etc.

6.5.6 Precision Farming

Precision farming is an integral part of crop management system to improve the production of a particular crop by using advanced technology such as remote sensing (RS), GPS and geographical information system (GIS). This technology gives the data of disparities in the field such as soil, crops incidence with disease, etc. timely to act upon it for remedies. It also give the farmers to take right decision in the right place and at right time (Shibusawa 2002). This process identifies specific difference in site of fields and adjusts management. Precision farming or precision agriculture is a process that managing crop production inputs such a fertilizer, water, seed and cut down waste (Shibusawa 2002). The collection and analysis of information become easier by following precision farming (Hakkim et al. 2016). There are some tools and equipment needed for the precision farming.

6.6 Global Positioning System (GPS)

GPS is a network of satellite system helping in recording exact positional information such as longitude, latitude and elevation of the crop field (Batte 2000). It provides the exact position of field information such as weed invasion, soil type, pest occurrence, water holes and obstructions also provide field location to farmers so that input process becomes easy (Hakkim et al. 2016).

6.6.1 Sensor Technologies

Sensor technologies generally used to measure humidity, vegetation, temperature, structure, physical character, texture, humidity, vapour, air, nutrient level, etc. (Hakkim et al. 2016).

6.6.2 Geographic Information System

GIS is a computerized map, provides information on field topography, surface drainage, soil testing, soil types, irrigation, chemical application rates and crop yield (Hakkim et al. 2016).

6.6.3 Grid Soil Sampling and Variable-Rate Fertilizer (VRT)

Grid soil sampling is one of the methods for sampling the soil at equal distance across a given land. The method is now incorporated with many advanced tools to ensure accurate variable rate. This technology can provide farmers to map his farms based on the soil colour by using aerial photography methods, topography and the past experience of the farmers to manage the farms in an effective way.

6.6.4 Rate Controllers

These are the devices used to design to control the delivery rate of chemicals, fertilizers, pesticides or other liquids in the crop fields. It is a kind of automatic tools which monitor and control the speed of the tractor/sprayers flow across the field.

6.6.5 Yield Monitor

These are the monitor used in studies and analysing of yield parameters of a particular crop. It contains several components which may include sensors, data storage device, user interface (display and key pad), main computer and other necessary components. The information generated is stored in the computer and is used for site specific evaluation and monitoring by planting variable rates of seed and applying variable rates of herbicides and pesticides.

6.6.6 Nano-Biofarming

Nanoparticle can be yielded by the following particle farming process; following harvest procedure the nanoparticles can be separated mechanically from plant tissue (Misra et al. 2013a). Nano biotechnology has the ability to increase the quality and efficiency of production and food safety. Nano biotechnology provides the new ways to study DNA, molecules, nutraceutical and pharmaceutical use (Subramanian and Tarafdar 2011). Crops yield and values of nutritional capacity increase by nanotechnology. Recycling of wastes like food, cosmetics, medicines can be recycled by following nano-biofarming technique (Agrawal and Rathore 2014). Gold particles and silver particles are found out from *Cyamopsis tetragonolobus*,

Medicago sativa, Brassica juncea, Allium sativum and *Memecylon edule* extracts (Agrawal and Rathore 2014). Nanoparticles are stored depending on the nature, shapes, sizes and species of plant tissue (Agrawal and Rathore 2014).

6.7 Nano Formulation in Packing and Quality of Food

Packaging of food product in food science is a major problem in determining and developing an effective packaging material to protect from microbial contaminants. Packaging of edible food items using films coating with cinnamon or oregano oil, or nanoparticle of zinc, calcium and other materials that can kill bacteria is highly important. Nano-fibres using by green packaging method made from lobster shells or organic corn (both antimicrobial and biodegradable) is also an effort in food safety. Bayer polymers have produced a nano-composite "hybrid system" film durethan, enriched with silicate nanoparticles, which reduces the entry of oxygen and other gasses. It prevents the food from spoiling. When this plastic is processed into a thin film and wrapped over food, it gives more benefit than normal plastics. This thin film wrapping can extend shelf life of the food materials by preventing easy contaminations of the pathogens which causes food spoiled (Misra et al. 2013b).

6.7.1 Nanotechnology for Food Packaging

It is very important to prevent food from loss by developing a proper packaging method. Nanoparticles are one of the most used advanced technologies that offer to extend shelf life of packaging foods. Packaging of food materials using nanoparticles not only enhance its shelf life but also secure the quality of the food by preventing from spoilage. It can sustain the flavour of the food for longer period. Nanomaterials are encapsulated with other composite to form nano-composite and these composites provide a barrier in between the food and outer environment thereby proofing from leakages, gases, pathogen infestation and other physiological changes. Nanotechnology incorporation in food packaging is eco-friendly and can be more safe and secure than using of chemical preservers. The nanoparticles like silver, titanium oxide, zinc oxide can be utilized to prevent from pathogens infestation (Sharma and Dhanjal 2016).

6.7.2 Nanoencapsulation

Nanoencapsulation or nanoformulation is designing of a nanocomposite consisting of nanoparticles and other materials for effective and accurate delivery of the system that involves controlled release. Nanoformulation has got many advantages over other materials due to their stability, holding capacity of volatile components, controlled moisture release, bioavailability, efficacy and targeted delivery. Nanoencapsulation can be used in postharvest disease prevention by extending the shelf life. It has been tasted safe and controlled discharges to food preservation (Sharma and Dhanjal 2016).

6.8 Safety of Nano-Packaging Material

By the experiment of research, it is projected that the intellectual packaging will allow an improved monitoring of the flow in the next generation for protection of food material. By addition of suitable nanoparticle it is easy to synthesize packaging material of intense tensile strength. The nanotechnology field ability and its benefits are needed to be explored in the concern of industry and human health. Mainly two concern of importance involves issue related to food safety and quality that may affect the consumer and second issue is related to environment safety. Bio-nanotechnology will help as a significant approach to conquer present defies that are linked with active food packaging solutions (Sharma and Dhanjal 2016).

6.9 Biosynthesis of Nanomaterials

Nanomaterials synthesized by two approaches including bottom to top and top to bottom approach. In bottom to top approach, the smaller atoms aggregate together to form a bigger nucleus and later on gives rise to nanoparticles by both chemical and biological process while in top to bottom approach, a bigger mass of particle broken down into smaller fragments by physical process (Ahmed et al. 2016). In comparison of physical and chemical methods, biological method is widely accepted for the biosynthesis of nanomaterials due to its non-toxic and environment friendly nature (Ahmed et al. 2016). The physical and chemical method involves toxic chemicals, high temperature and pressure which is hazardous for environment although biological method includes microbes like bacteria, fungi and plant extracts (Rafique et al. 2017) which is non-pathogenic, worthwhile, energy saving, safe, fast, simplistic, high yielding and bio-compatible (Mohapatra et al. 2015).

Green synthesis approach of synthesis of nanoparticles is environmentally friendly, non-toxic in nature. The source of the synthesis includes living organisms like plants, fungi, yeast, actinomycetes and bacteria (Konishi et al. 2007; Panda et al. 2020). The biosynthesis of nanomaterials using bacteria was done both intra and extra cellular (Kowshik et al. 2003). When the bacteria are subjected to higher concentration of any heavy metal, the metal ions get accumulated to the surface of gram-positive and gram-negative bacteria and then the bacteria uptakes the heavy metal through their cytoplasmic membrane (Beveridge and Fyfe 1985). Various bacterial species have been recognized for the biosynthesis of nanomaterials including *Bacillus subtilis* (Beveridge and Murray 1976), *Bacillus licheniformis* (Kalimuthu et al. 2008), *Escherichia coli* (Shahverdi et al. 2007) and so on.

A flask containing nutrient broth is prepared and bacterial strain is inoculated with it followed by continuous shaking for 12 h. Centrifugation is done after incubation period and the obtained supernatant is used for the synthesis of nanomaterial which interacts with interested salts of nanomaterials. Again, after the incubation period about 120 h, the change in biomass colour indicates the synthesis of nanomaterials (Singh et al. 2014).

Similarly, nanomaterials synthesized using fungi show interesting properties including prolonged storage and improved longevity. Many fungal species have been reported for the synthesis of nanomaterials like *Aspergillus* (Bhainsa and D'souza 2006), *Penicillium* (Mukherjee et al. 2001), *Fusarium*, etc. Fungal strains are cultured in broth media on continuous shaking in rotary shaker for 120 h. After the incubation period, the culture is centrifuged and the obtained mycelia is allowed to interact with salts of nanomaterials and kept in dark for 70–75 h for the synthesis of nanomaterials (Singh et al. 2014).

Biosynthesis of nanomaterials using plant extracts has a little different mechanism. Plant extracts contain various secondary metabolites including flavonoids and phenolic compounds which have the ability to act as reducing and capping agent for the synthesis of nanomaterials. Several studies have been reported for the synthesis of nanomaterial using plant extracts like *Taraxacum officinale* (Saratale et al. 2018), *Coriandrum sativum* (Khan et al. 2018), *Tribulus terrestris* (Gopinath et al. 2012) and showing various application as antibacterial and antifungal agents (Nelson et al. 2005).

The plant parts are collected, dried and crushed into fine powder. With the help of deionized water, plant extract is prepared followed by filtration. Then the plant extract is mixed with salt solution of interested nanomaterial. After few minutes, the change in colour of the solution indicates the synthesis of nanomaterial (Singh et al. 2014).

6.10 Postharvest Food Processing

Nanomaterials synthesized by biological approaches show better stability, solubility, eco-friendly biodegradable nanomaterials and also exhibit its potential application in postharvest technology. Postharvest technology is a process for preserving and long lasting mean life of biodegradable and decomposable food products (Pedreschi et al. 2013). Postharvest is a period of crop production instantly following harvest along with cooling, cleaning, sorting and packing (Bachmann and Earles 2000). The main aim of postharvest is to keep the products moisture free, delay in spoilage, protection from physical damages, decrease in chemical changes of the products and low temperature storage also plays a critical role in maintaining food quality and turgidity during postharvest (Beaudry 1999). The fruits and vegetables feel necessity for proper postharvest processing technologies to lessen the quantitative and qualitative damages behind harvesting (Singh et al. 2014). Due to un-implementation of postharvest processing can lead to reduction in per capita availability of crops. So, this technology is highly required for maintaining global food security by enhancing the wealth of agricultural products (Singh et al. 2014).

A numerous study has been reported which explains the application of nanomaterials in postharvest physiology (Fig. 6.1). Nanomaterials are now used in

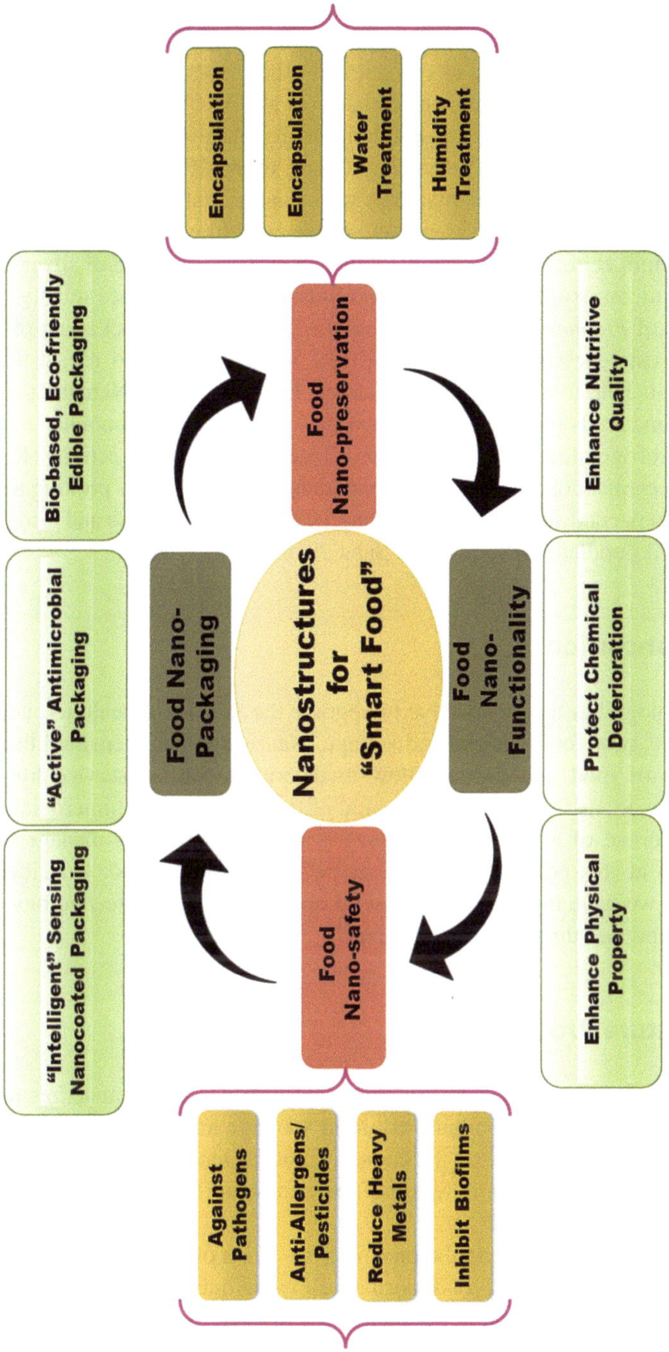

Fig. 6.1 Systematic representation of application of nanotechnology in postharvest technology

food packaging which has the capacity to withstand heat and cold. To develop the quality of food packaging, various nano tubes, nano wires and nano fibres were being implemented (Sharon et al. 2010). Durethan film made up of silicate nanoparticles blocks the entry of oxygen and other gases, thus protecting the food from spoilage by preserving moisture (Sharon et al., 2010). Cellulose nanomaterials were succeeded in increasing the storability of bananas with better moisture barricade, wettability and surface adherence (Deng et al. 2017). The combination of chitosan, silica and copper nanomaterials shows antifungal effect on *Botrytis cinerea* that causes gray mold on table grapes and hence maintains the postharvest quality of grapes (Hashim et al. 2019). A nano polymer polythene made up of silver, titanium oxide and silicon oxide nanomaterials exhibits greater appearance quality, reduced weight loss of *Flammulina velutipes*, winter mushroom after packaging of 15 days on postharvest senescence (Shi et al. 2018). Some studies also reveal that silver nanomaterials help to increase the postharvest life of cut lilies (Nemati et al. 2014). Silver nanomaterials that are used in food packaging, storage containers, chopping boards and refrigerators show tremendous effect of antimicrobial activity (Kour et al. 2014). On treatment of zinc oxide nanomaterials, the strawberries prolong their shelf life during expanded storage without changing the fruit sweetness and also decrease the microbial population (Sogvar and Saba 2016).

6.11 Conclusion

Nanotechnology has the prospective to upgrade the foods into healthier, nourishing and nutritive, new food packets and storage. Nanomaterials intensified the flavour and texture of food and do not allow to deteriorate during service life. It also preserves the goods fresh for longer period of time due to its antimicrobial activity. Effective storage of agricultural products is most important to manage the huge demand, so in this context preservation, packaging is required. Considering the matter here we have highlighted the useful application of nanotechnology in postharvest technology throughout the chapter.

6.12 Future Prospective

In future carbon nanotubes can be incorporated with packaging materials for detection of toxic proteins, microorganisms, food spoilage and to activate an intelligent packaging system. Research on the carbon nanotube nanomaterials for its potential health risk due to capability of some of the nanomaterials has the ability to cross blood–brain barrier.

Rapid increase in world population emerges a doubt of availability of food for all, so as to avoid the scarcity of food, postharvest technique should be introduced with new scientific approaches for better quality, elongation of shelf life and to diminish the food losses which can lead to rise in availability of agricultural products among peoples.

Acknowledgments YDS is very much thankful to the Vice Chancellor of Central Agricultural University, Imphal, Manipur, India for providing facilities and support.

References

Abbas KA, Saleh AM, Mohamed A et al (2009) The recent advances in the nanotechnology and its applications in food processing: a review. J Food Agric Environ 7(3–4):14–17

Agrawal S, Rathore P (2014) Nanotechnology pros and cons to agriculture: a review. Int J Curr Microbiol App Sci 3(3):43–55

Ahmed S, Ahmad M, Swami BL (2016) A review on plants extract mediated synthesis of silver nanoparticles for antimicrobial applications: a green expertise. J Adv Res 7(1):17–28

Akinsiku AA (2017) Green synthesis of capped silver nanoparticles and their hybrids: antimicrobial and optical properties (doctoral dissertation, Covenant University, Ota, Nigeria.)

Bachmann J, Earles R (2000). Postharvest handling of fruits and vegetables 1-19 ATTRA

Batte MT (2000) Factors influencing the profitability of precision farming systems. J Soil Water Conserv 55(1):12–18

Beaudry RM (1999) Effect of O_2 and CO_2 partial pressure on selected phenomena affecting fruit and vegetable quality. Postharvest Biol Technol 15(3):293–303

Beveridge TJ, Fyfe WS (1985) Metal fixation by bacterial cell walls. Can J Earth Sci 22 (12):1893–1898

Beveridge TJ, Murray RG (1976) Uptake and retention of metals by cell walls of Bacillus subtilis. J Bacteriol 127(3):1502–1518

Bhainsa KC, D'souza SF (2006) Extracellular biosynthesis of silver nanoparticles using the fungus Aspergillus fumigatus. Colloids Surf B: Biointerfaces 47(2):160–164

Bryksa BC, Yadav RY (2012) Challenges in food nanoscale science and technology. J Food Drug Anal 20:418–421

Camargo A, Smith JS (2009) Image pattern classification for the identification of disease causing agents in plants. Comput Electron Agric 66(2):121–125

Cao G, Wang Y (2004) Nanostructures and nanomaterials: synthesis, properties and applications, vol 2. World Scientific, Singapore

Caraglia M, Rosa GD, Abbruzzese A, Leonetti C (2011) Nanotechnologies: new opportunities for old drugs. The case of amino-bisphosphonates. Nanomedic Biotherapeu Dis 1:103e

Dasgupta N, Ranjan S (2018) An introduction to food grade nanoemulsions, vol 13. Springer, Singapore

Deng Z, Jung J, Simonsen J, Zhao Y (2017) Cellulose nanomaterials emulsion coatings for controlling physiological activity, modifying surface morphology, and enhancing storability of postharvest bananas (Musa acuminate). Food Chem 232:359–368

Finglas PM, Yada RY, Toldrá F (2014) Nanotechnology in foods: science behind and future perspectives. Trends Food Sci Technol 40:125–126

Gopinath V, Mubarak Ali D, Priyadarshini S, Priyadharsshini NM, Thajuddin N, Velusamy P (2012) Biosynthesis of silver nanoparticles from Tribulus terrestris and its antimicrobial activity: a novel biological approach. Colloids Surf B: Biointerfaces 96:69–74

Hakkim VA, Joseph EA, Gokul AA, Mufeedha K (2016) Precision farming: the future of Indian agriculture. J Appl Biol Biotechnol 4(06):068–072

Hashim AF, Youssef K, Abd-Elsalam KA (2019) Ecofriendly nanomaterials for controlling graymold of table grapes and maintaining postharvest quality. Eur J Plant Pathol 154 (2):377–388

He X, Deng H, Hwang HM (2018) The current application of nanotechnology in food and agriculture. J Food Drug Anal 27(1):1–21

Hosseini SM, Rezaei R (2011) Factors affecting the perceptions of Iranian agricultural researchers towards nanotechnology. Public Underst Sci 20(4):513–524

Kah M, Hofmann T (2014) Nanopesticide research: current trends and future priorities. Environ Int 63:224–235

Kalimuthu K, Babu RS, Venkataraman D, Bilal M, Gurunathan S (2008) Biosynthesis of silver nanocrystals by Bacillus licheniformis. Colloids Surf B: Biointerfaces 65(1):150–153

Khan I, Saeed K, Khan I (2017) Nanoparticles: properties, applications and toxicities. Arab J Chem 12(7):908–931

Khan MR, Rizvi TF (2014) Nanotechnology: scope and application in plant disease management. Plant Pathol J 13(3):214–231

Khan MZH, Tareq FK, Hossen MA, Roki MNAM (2018) Green synthesis and characterization of silver nanoparticles using Coriandrum sativum leaf extract. J Eng Sci Technol 13(1):158–166

Konishi Y, Ohno K, Saitoh N, Nomura T, Nagamine S, Hishida H, Takahashi Y, Uruga T (2007) Bioreductive deposition of platinum nanoparticles on the bacterium Shewanella algae. J Biotechnol 128(3):648–653

Kour H, Malik AA, Ahmad N, Wani TA, Kaul RK, Bhat A (2014) Nanotechnology–new lifeline for food industry. Critical reviews in food science and nutrition (just-accepted), 00-00

Kowshik M, Ashtaputre S, Kharraz S, Vogel W et al (2003) Nanotechnology 14:95

Lin D, Xing B (2007) Phytotoxicity of nanoparticles: inhibition of seed germination and root growth. Environ Pollut 150(2):243–250

Misra AN, Misra M, Singh R (2013a) Nanotechnology in agriculture and food industry. Int J Pure Appl Sci Technol 16(2):1

Misra R, Dhokale B, Jadhav T, Mobin SM (2013b) Donor–acceptor meso-alkynylated ferrocenyl BODIPYs: synthesis, structure, and properties. Dalton Trans 42(37):13658–13666

Mohapatra B, Kuriakose S, Mohapatra S (2015) Rapid green synthesis of silver nanoparticles and nanorods using Piper nigrum extract. J Alloys Compd 637:119–126

Mukherjee P, Ahmad A, Mandal D, Senapati S, Sainkar SR, Khan MI, Parishcha R, Ajaykumar PV, Alam M, Kumar R, Sastry M (2001) Fungus-mediated synthesis of silver nanoparticles and their immobilization in the mycelial matrix: a novel biological approach to nanoparticle synthesis. Nano Lett 1(10):515–519

Nelson D, Priscyla DM, Oswaldo LA, Gabriel IHDS, Elisa E (2005) Mechanical aspects of biosynthesis of silver nanoparticles by several Fusarium oxysporum strains. J Nanobiotechnol 3:8

Nemati SH, Esfandiyari B, Tehranifar A, Rezaei A, Ashrafi SJ (2014) Effect of nano-silver particles on postharvest life of Lilium orientaliscv.'shocking'. Int J Postharvest Technol Innovat 4 (1):46–53

Panda MK, Panda SK, Singh YD, Jit BP, Behara RK, Dhal NK (2020a) Role of nanoparticles and nanomaterials in drug delivery: an overview. In: Advances in pharmaceutical biotechnology. Springer, Singapore, pp 247–265

Panda MK, Singh YD, Behera RK, Dhal NK (2020) Biosynthesis of nanoparticles and their potential application in food and agricultural sector in green nanoparticles. Springer, Cham, pp 213–225

Panigrahi T (2013) Synthesis and characterization of silver nanoparticles using leaf extract of Azadirachta indica (doctoral dissertation)

Pedreschi R, Lurie S, Hertog M, Nicolai B, Mes J, Woltering E (2013) Post-harvest proteomics and food security. Proteomics 13(12–13):1772–1783

Peters RJ, Bouwmeester H, Gottardo S, Amenta V, Arena M, Brandhoff P, Marvin HJ, Mech A, Moniz FB, Pesudo LQ, Rauscher H (2016) Nanomaterials for products and application in agriculture, feed and food. Trends Food Sci Technol 54:155–164

Pradhan S (2013) Comparative analysis of silver nanoparticles prepared from different plant extracts (Hibiscus rosa sinensis, Moringa oleifera, Acorus calamus, Cucurbita maxima, Azadirachta indica) through green synthesis method (doctoral dissertation)

Rafique M, Sadaf I, Rafique MS, Tahir MB (2017) A review on green synthesis of silver nanoparticles and their applications. Artif Cells Nanomed Biotechnol 45:1272–1291

Salehi B, Prado-Audelo D, María L, Cortés H, Leyva-Gómez G, Stojanović-Radić Z et al (2020) Therapeutic applications of curcumin nanomedicine formulations in cardiovascular diseases. J Clin Med 9(3):746

Saratale RG, Benelli G, Kumar G, Kim DS, Saratale GD (2018) Bio-fabrication of silver nanoparticles using the leaf extract of an ancient herbal medicine, dandelion (Taraxacum officinale), evaluation of their antioxidant, anticancer potential, and antimicrobial activity against *phytopathogens*. Environ Sci Pollut Res 25(11):10392–10406

Shahverdi AR, Fakhimi A, Shahverdi HR, Minaian S (2007) Synthesis and effect of silver nanoparticles on the antibacterial activity of different antibiotics against Staphylococcus aureus and Escherichia coli. Nanomedicine 3(2):168–171

Sharma D, Dhanjal DS (2016) Bio-nanotechnology for active food packaging. J Appl Pharm Sci 6 (09):220–226

Sharma VK, Yngard RA, Lin Y (2009) Silver nanoparticles: green synthesis and their antimicrobial activities. Adv Colloid Interf Sci 145(1–2):83–96

Sharon M, Choudhary AK, Kumar R (2010) Nanotechnology in agricultural diseases and food safety. J Phytol 4:83–92

Shi C, Wu Y, Fang D, Pei F, Mariga AM, Yang W, Hu Q (2018) Effect of nanocomposite packaging on postharvest senescence of Flammulina velutipes. Food Chem 246:414–421

Shibusawa S (2002) Precision farming approaches to small-farm agriculture. Food and Fertilizer Technology Center, Taipei, Taiwan

Singh S, Singh BK, Yadav SM, Gupta AK (2015) Applications of nanotechnology in agricultural and their role in disease management. Res J Nanosci Nanotechnol 5(1):1–5

Singh V, Hedayetullah M, Zaman P, Meher J (2014) Postharvest technology of fruits and vegetables: an overview. J Postharvest Technol 2(2):124–135

Singh YD, Ningthoujam R, Panda MK, Jena B, Babu PJ, Mishra AK (2021) Insight from Nanomaterials and Nanotechnology towards covid-19. Sens Int 100099

Sogvar OB, Saba MK, Emamifar A, Hallaj R (2016) Influence of nano-ZnO on microbial growth, bioactive content and postharvest quality of strawberries during storage. Innovative Food Sci Emerg Technol 35:168–176

Sozer N, Kokini JL (2009) Nanotechnology and its applications in the food sector. Trends Biotechnol 27(2):82–89

Srilatha B (2011) Nanotechnology in agriculture. J Nanomed Nanotechnol 2(7):1000123

Subramanian KS, Tarafdar JC (2011) Prospects of nanotechnology in Indian farming. Indian J Agric Sci 81(10):887–893

Taniguchi N (1974) On the basic concept of nanotechnology. Proceeding of the ICPE

US DOA (2003) Nanoscale science and engineering for agriculture and food systems: a report submitted to cooperative state research, education and extension service. Washington, DC: Department of Agriculture; 2003. the United States Department of Agriculture: National Planning Workshop

Yadollahi A, Arzani K, Khoshghalb H (2009) The role of nanotechnology in horticultural crops postharvest management. In Southeast Asia Symposium on Quality and Safety of Fresh and Fresh-Cut Produce 875: 49–56

Zhao L, Seth A, Wibowo N, Zhao CX, Mitter N, Yu C, Middelberg AP (2014) Nanoparticle vaccines. Vaccine 32(3):327–337

Nanotechnology Mediated Detection and Control of Phytopathogens

7

Banishree Sahoo, Sandip Kumar Rath, Sailendra Kumar Mahanta, and Manoranjan Arakha

Abstract

In current environmental scenario, about 20–40% of crops are destroyed by pests and pathogens annually. Hence to control the plant pathogens, toxic pesticides are generally used which are harmful to both environment and human beings. In this context, nanotechnology provides harmless effect to pesticides as it reduces toxicity, increases solubility of less water-soluble pesticides and increases shelf-life. It also provides good impact on environment and soil. Nanoparticles are small particles which have size in between 1 and 100 nm. This chapter intends to discuss how nanoparticles can be used for the control of plant diseases either nanoparticle alone or acting as protestants or nanocarriers for insecticides, pesticides and fungicides. Nanoparticles which are synthesized by different methods can be used for agricultural applications. Nowadays although nanotechnology is progressing quickly, however its application in agricultural fields is insignificant to control pests and pathogens practically. Hence, agricultural applications can be developed by understanding the fundamental things of research and production of commercial nanoproducts to control plant disease.

B. Sahoo · M. Arakha (✉)
Center for Biotechnology, Siksha 'O' Anusandhan (Deemed to be University), Bhubaneswar, India
e-mail: marakha@soa.ac.in

S. K. Rath
University Medical Centre, George August University Gottingen, Gottingen, Germany

S. K. Mahanta
Department of Pharmacology, Nirmala College of Pharmacy, Atmakur Road, Mangalagiri, Guntur, Andhra Pradesh, India

M. Arakha et al. (eds.), *Bio-Nano Interface*,
https://doi.org/10.1007/978-981-16-2516-9_7

Keywords

Nanotechnology · Agriculture · Nanopesticides · Nanocarriers · Disease management

7.1 Introduction

Nowadays nanotechnology has become the new attraction point of research for most of the scientists. On the evening of ninth December 1959, an article entitled "There's plenty of room at the bottom" was delivered to American Physical Society by Prof. Richard Feynman (Feynman 1960). According to him, single atoms or molecules could be controlled, so that different possibilities of formulating new material could be generated, which lead the scientific community to discover the new word of "Nanotechnology". Later the term "Nanotechnology" was coined by Norio Taniguchi, Professor of Tokyo University of Science, Japan in 1974 to describe the superfine, refined particles (Bhattacharyya et al. 2009). Hence, the research or study related to the materials in the nanoscale is known as nanotechnology. The basic and fundamental unit of nanotechnology is nanoparticles. The particles whose size lies between 1 and 100 nm are called nanoparticles and mainly composed of carbon, metal, metal oxides and organic matter (Ealias and Saravanakumar 2017). Currently, nanotechnology is an exciting field for researchers as nanoparticles act as bridge between the physico-chemical gap between the atoms/molecules and bulk (macroscopic) material (Thakkar et al. 2010). This dissimilarity is due to the small size and high surface area to volume ratio of nanoparticles (Thakkar et al. 2010). So nanoparticles are used in various fields of science and technology due to these unique features. Hence researchers now have a keen interest in the synthesis of the nanoparticles using various techniques.

7.2 Synthesis of Nanoparticles

In the last decade, extensive research has resulted in enormous progress in the field of nanotechnology and also in synthesis of nanoparticles. Various methods were established to synthesize nanoparticles based on their physical, chemical, optical and mechanical properties (Cho et al. 2013). One of the most popular synthesis methods, widely known as the "top-down approach", was developed and established by Taniguchi et al. (Tarafdar and Adhikari 2015). The overall concept of this method is relatively simple and relies on the fact that most nanoparticles can be synthesized from larger molecules and later through a series of reactions can be converted into nano form (Abou El-Nour et al. 2010). Interestingly, almost 10 years after the introduction of the "top-down approach" to synthesize nanoparticles, a new concept was put forward by K. Eric Drexler termed as "bottom- up approach" (Drexled 1986). This concept envisages that nanoparticles can be synthesized from macromolecules based on the atomic and molecular composition. Most nanoparticles are synthesized by either physical and/or chemical methods. However, biological methods for synthesis of nanoparticles have gained more attention of the research community, due to its eco-friendly nature. In general eco-friendly strategy employs plant extract, bacteria, algae, fungi, etc. to synthesize different nanomaterials (Mittal et al. 2013). Furthermore, the synthesis of nanoparticles by these processes have some benefits over the chemical synthesis method, as the nanoparticles generated in such process are nontoxic by nature (Charitidis et al. 2014).

The ecological cycle is tightly regulated, with plants being placed as the primary producers to sustain the balance in the food chain. However, in the course of evolution, infections caused by plant pathogens have emerged as new players, threatening the genetic diversity and survival of plants. Plant pathogens and pests have created new issues and challenges in agriculture, resulting in decreased crop production. It is estimated that these pests and pathogens contribute to an overall 22–40% less production in crops per year globally (Worrall et al. 2018). Hence, pest management is utmost important. In layman's term pest management means use of pesticides such as insecticides, fungicides, herbicides, etc. to kill and destroy pest and increase crop yield. The use of pesticides although results in an overall increase in crop production, but comes with a price. For example, the advantages of the use of pesticides are (1) quick action for killing (2) lower in cost. On the contrary, the harmful effects of pesticides are (1) effect on the non-target organism, (2) increase in the evolution of resistant pest population, (3) adverse effect on health, (4) reduction in soil bio-diversity, (5) decrease in nitrogen fixation, etc. (Hayles et al. 2017). It is reported that 90% of pesticides are blown to the air during or after application, which causes adverse health issues (Stephenson 2003; Ghormade et al. 2011). Hence, various research groups are working to find out an alternative for the use of pesticides that could potentially be less harmful to the environment and precisely affect only the target pests in interest and enrich soil productivity.

In such a scenario, advancements in nanotechnology have emerged as a possible new strategy to overcome the traditional problems encountered in the agricultural field (Bramlett et al. 2019). Recently, nanotechnology has been implemented in plant hormone delivery, water management, seed germination, transfer of target genes and nano-sensors (Hayles et al. 2017). For the development of the new generation of pesticide, scientists are employing nanoparticles with desired size, shape and surface properties to provide better pest management (Khandelwal et al. 2016) (Fig. 7.1).

7.3 Early Detection of Phytopathogens Using Nanoparticles

Using conventional agricultural practices, it was daunting for phytopathologists to identify unknown phytopathogens causing different plant diseases. However, with the recent advancements in nanotechnology, phytopathologists can employ new strategies to detect the plant disease to increase better crop management (Rai and Ingle 2012). In this context, molecular and immunodiagnostic techniques in

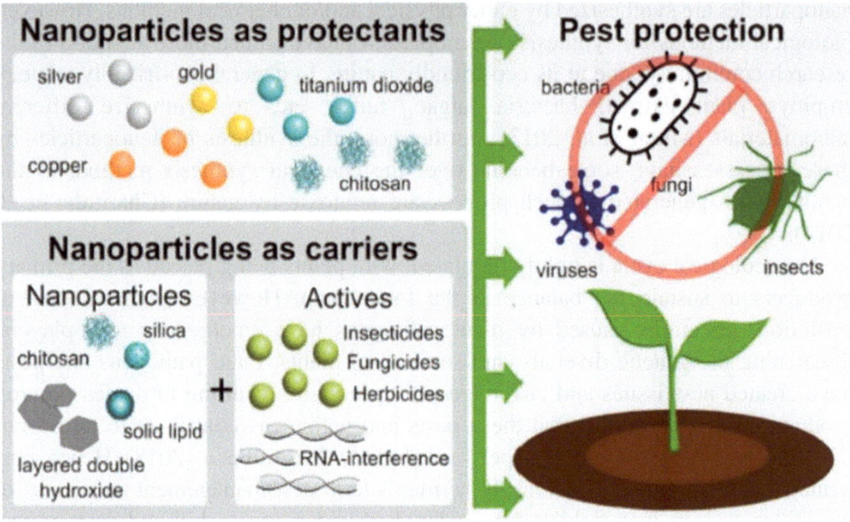

Fig. 7.1 Schematic diagram representing nanomaterial as either protectant or carrier for fungicide, insecticide or herbicide. Adapted with permission from Ref (Worrall et al. 2018)

conjunction with nanotechnology are used to characterize the pathogens for identification and detection of microorganisms. Polymerase Chain Reaction (PCR) is used for the detection of causative agent *Xanthomonas axonopodis pv. Punicae* in pomegranate (Mondal et al. 2012). ELISA test is used to detect *Xylella fastidiosa* (Sherald and Lei 1991). Advanced molecular techniques such as q-PCR is used to detect blight bacterial disease in pomegranate. These techniques have high sensitivity, reliability and specificity. Early detection of pathogens is the best method to prevent disease in plants. Phytosanitary analysis and plant quarantine etc routine surveys are performed to control the disease as these are reliable, fast and affordable process. Additionally, the lab-on-a-chip system is used for the detection of toxicity of water, nutrients in the water and control the quality of production of food (Gardeniers and Van den Berg 2004).

7.3.1 Action of Nanoparticles against Phytopathogens

7.3.1.1 Plant Disease Cycle

The process through which the development of disease occurs in plants is called the disease cycle. The events that are involved in disease development are inoculation, penetration, infection, incubation, reproduction and survival (De Wolf and Isard 2007). Pathogens are introduced into different plants by different methods. For example, some fungal pathogens release their spores into the air and these spores are spread by the air current and penetrate through the injury or wound and natural opening site of plants such as stomata and hydathodes during different environmental conditions such as moisture and temperature. When these pathogens enter into the plant tissue, they establish a parasitic relationship between pathogens and plants. Then pathogens undergo incubation or dormant period till the disease initiates (Agrahari et al. 2020). Plant pathogens reproduce sexually or asexually and survive a prolonged period till the favourable weather comes.

7.3.1.2 Host Pathogen Interaction

When the host enters into the plant tissue, immune elicitors stimulate plant defence mechanisms and hypersensitivity response occurs as reported by Stakman (Stakman 1915). Some scientists reported that host pathogen interaction is responsible for the process of apoptosis or programmed cell death in plants (Morel and Dangl 1997). Avirulence (Avr) genes are secreted by pathogens binde indirectly to the plant resistance gene (R). When both the R gene and corresponding Avr genes are present, then recognition takes place which leads to active resistance of the plant. If either Avr gene in the pathogen or R gene in the host is absent or mutated, then no recognition occurs, and the plant becomes diseased (De Wit 1995). As a result, putative reactions occur between two partners and transduction signal cascade is activated.

7.3.1.3 Generation of Reactive Oxygen Species (ROS)

Doke et al. reported that during plant–pathogen interaction, ROS is released and accumulated (Doke 1983). ROS production, otherwise known as "Oxidative burst", involves two phases. Phase I is rapid and nonspecific, while phase II is slower but yet yields a higher concentration of ROS (Wojtaszek 1997). ROS is a toxic intermediate that is generated by the reduction of molecular oxygen. Various enzymes are involved in the reaction. NADPH oxidase enzyme helps in the reduction of H_2O_2 under physiological conditions. At first, the reduction of O_2 forms superoxide anion (O^{2-}) and hydroperoxyl radical ($H_2O\bullet$), and second reduction produces hydrogen peroxide (H_2O_2), and then it reduces to form hydroxyl radical ($OH\bullet$) that is unstable, but H_2O_2 is more stable. H_2O_2 and $OH\bullet$ react with polyunsaturated lipids in the membrane and form lipid peroxide that results in the destruction of the biological membrane (Grant and Loake 2000).

7.3.1.4 Mode of Action

1. When nanoparticle is taken into the cells through the process of translocation and internalization, it helps in the degradation of intercellular ATP and DNA duplication (Lok et al. 2006).
2. Metal nanoparticles generate ROS and damage the cellular structure (Richards 1981).
3. Metal ion accumulates inside the cell and dissolves the bacterial membrane (McQuillan 2010) (Fig. 7.2).

Under stress conditions, the oxidation reactions occur in the cell, which leads to adverse effects on cell survival, signalling, death and generation of ROS (Mueller et al. 2005). Copper oxide, zinc oxide, silver nanoparticles also show antimicrobial activity against *Escherichia coli, Bacillus subtilis*, *Pseudomonas aeruginosa* and

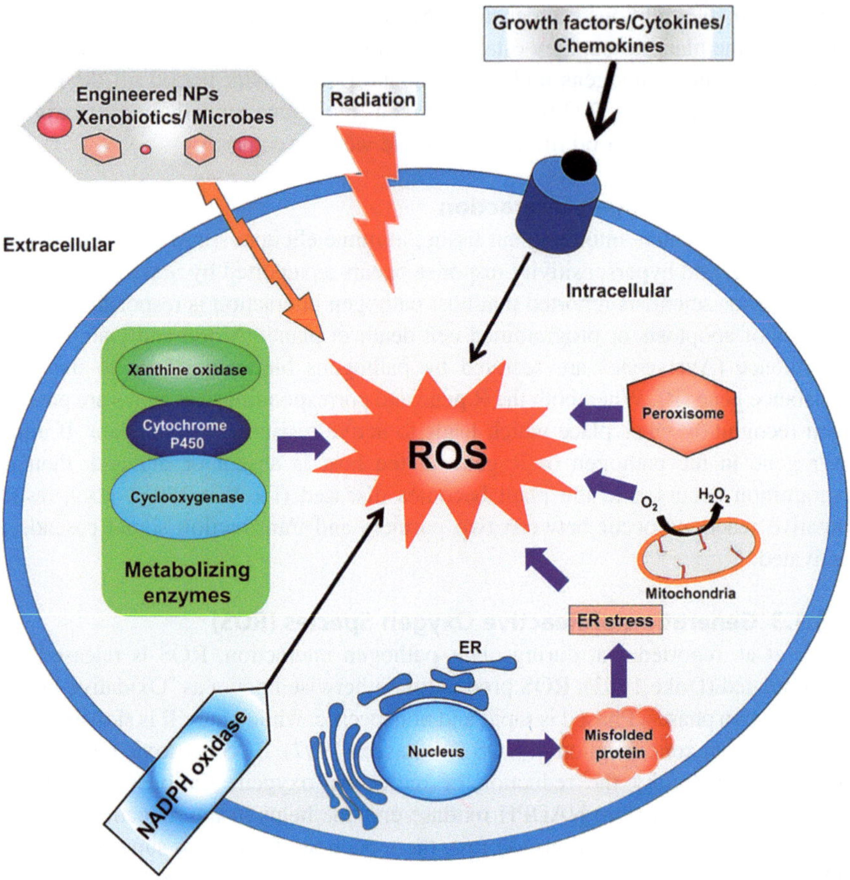

Fig. 7.2 Represents different sources of generation of ROS. Adapted with permission from Ref (Abdal Dayem et al. 2017)

Staphylococcus aureus (Viswanathan et al. 2006). Bacterial plasmolysis is a process in which subsequential degradation of components of cytoplasm and contraction of the plasma membrane take place. Upon treatment of nanoparticles with bacterial cells, nanoparticles form pores in the cell membrane, which lead to the release of intercellular glucose and trehalose into the suspension. Nanoparticles bind to the thiol groups of bacterial proteins and interfere with their activities. After binding to the cell membrane, they alter the cell electrical potential and disturbs the respiration process (Radzig et al. 2013). As a result, ROS is developed, which inhibits the respiratory enzymes (Park et al. 2009) (Fig. 7.2).

7.4 Nanoparticles in Controlling Phytopathogens

Different nanoparticles can be used to control plant pathogens. From different studies, it is reported that metalloids, metal oxides and non-metals are used as either bactericides, fungicide or nano fertilizers by suppressing foliar, stem, fruit or root rot pathogens (Kah and Hofmann 2014). To protect plants from pathogens nanoparticles can be used by two different ways: (1) Nanoparticle itself protects the production of crops (2) Nanoparticles act as a carrier for pesticides, e.g. double-stranded RNA (dsRNA) can be used by spray application, soaking into seeds, foliar tissue or roots. The advantages of nanoparticles as carrier are increased solubility of less water-soluble pesticides, decreased toxicity level, increased lifetime maintenance (Worrall et al. 2018).

Hence, this chapter focuses on nanoparticles mediated plant disease management, which can be used as protectant or carriers for insecticides, fungicides and herbicides. Although, few reports are available on prevention of plant pathogens using nanoparticles, however, the application of nanoparticles in crop pest management is not being explored well.

7.4.1 Nanoparticles Acting as Protectant

Nanoparticles are tiny materials having size in between 1 to 100 nm, and they have unique physical, chemical and biological properties in comparison to bulky material. Nanoparticles have the ability to protect the pathogen such as bacteria, virus, insect and fungus by suppressing the stem, fruit and root rot pathogen. Hence, researchers use different metals and metal oxide nanoparticles such as silver, Cu, ZnO, TiO_2, Au nanoparticles which have antifungal, antiviral and antibacterial properties (Kah and Hofmann 2014).

7.4.1.1 Ag Nanoparticle

Nowadays, the main focus is on green synthesis of Ag nanoparticles from plants, fungi, algae, etc. (Rafique et al. 2017). Initially, Ag nanoparticles were used for the control of pathogens in plants due to antimicrobial activity of Ag nanoparticles (Richards 1981). Kim et al. have investigated that double encapsulation of Ag with Ag^+ ions can be spread on the powdery mildew of roses, which can kill the fungus *Sphaerotheca pannosa* for at least 7 days (Kim et al. 2008). Lamsal et al. found that if 100 mg/mL of AgNP is applied on peppers, it suppress the Anthracnose (Lamsal et al. 2011). It was found that AgNP synthesized from the plant extract can be used for the treatment of banana at different concentrations for the post-harvest control of *Colletotrichum musae* (Jung et al. 2010). It was also found that AgNP has the antimicrobial activity against foliar fungal pathogens. When various concentrations of silver nanoparticles such as 10, 30, 50 and 100μg/mL were sprayed on the cucumber and pumpkin leaves, the suppression of powdery mildew was observed. AgNP can be used for the treatment of fungus, insects and virus (Lamsa et al. 2011). In this context, it is reported that 24μg/mL of silver nanoparticles can be used to treat total germination of *Bipolaris Sorokiniana* in greenhouse trials, which is the causative agent of spot blotch of wheat. From histochemical staining, new facts come out that due to green synthesis of nano-Ag, lignin deposits in the vascular bundles that is a unique and novel approach for disease management (Moussa et al. 2013). Soil-borne diseases which are caused by *Phytophthora parasitica, Fusarium spp.* and *Meloidogyne spp.* are suppressed by silver nanoparticle it is reported that, application of nano-silver inhibits the growth of *Sclerotium cepivorum* and *Colletotrichum Gloeasporioides* (Jung et al. 2010). Generally, nano-Ag can be used as anti-parasitic agent. Silver nanoparticle used for the inhibition of Juvelline stage of *Meloidogyne graminis* and its implementation reduces the root gall formation (Cromwell et al. 2014). Ocsoy et al. first developed a new product such as DNA-directed silver AgNP grown on graphene oxide, which has the capacity to suppress bacterial disease caused by *Xanthomonas perforans* on tomatoes. Application of 100 ppm of Ag@dsDNA@GO reduces the severity of bacterial spot disease in comparison to the conventional bactericide treatment (Ocsoy et al. 2013).

The most interesting point about AgNP is that it acts as biocontrol agent. Mallaiah et al. observed that if silver at nanoscale is combined with the biocontrol agents such as *Bacillus subtilis, Pseudomonas fluorescens, Trichoderma viride,* etc., it suppresses Fusarium wilt and increases flower yield from 5% to 12%, 14% and 15%, respectively (Mallaiah 2015). It is also helpful for the reduction of the quality of chemical and increases resistance of pathogens. From the studies, it is also reported that fluconazole has increased fungicide activity, i.e. *Alternaria alternata, Cladosporium herbarum* and *Fusarium oxysporum* (Bholay et al. 2013). But according to Gajbhiye et al., when silver nanoparticle biosynthesized from *A. Alternata* is combined with fluconazole, the antifungal activity is enhanced against plant pathogen *Phoma glomerata* (Gajbhiye et al. 2009). However, till now there is limited evidence about the defence mechanism of silver nanoparticle. Mainly silver nanoparticles provide obstacle for the production of toxicity and soil infertility.

7.4.1.2 Cu Nanoparticle

Mainly cu nanoparticles were found to have best antimicrobial activity to control plant diseases. Cu-based fungicides such as kocide 2000 35WG, kocide opti 30WG are used to treat *Phytophthora infestans* in tomato plant. Giannousi et al. monitored the symptoms of leaf lesions in every 10 days of tomato plants and found that when CuO nanoparticles at 150-340μg/ml were applied, they suppress the disease (Giannousi et al. 2013). According to Strager-Scherer et al. Cu nanoparticles, i.e. core shell cu, multivalent cu and fixed quaternary ammonium copper are used to inhibit the bacterial spot by *X. Perforans*. From the studies it was found that nano-CuO molecule can easily penetrate the bacterial membrane and leads to the busting of the cell (Strayer-Scherer et al. 2018). Copper nanoparticle acts as a potent fungicide. Some fungus such as *Fusarium solani, Fusarium oxysporum* and *Neofusicoccum sp.*, etc. invade vascular tissue of plants and block the water transport system in xylem that forms foliage wilt (Yadeta and Thomma 2013). From the studies, it was found that copper nanoparticles of different concentrations show antifungal activities against *Fusarium solani, Neofusicoccum sp.* and *Fusarium Oxysporum.* Elmer et al. (Elmer et al. 2018; Elmer and White 2016) reported that copper nanoparticles can be used as nanofertilizer. From the studies it is reported that if CuO nanoparticle is treated with tomato and eggplant, then the harvest yield increases 24% than control (Evans et al. 2007). It is reported that when 10μg/L copper nanomaterial is used in maize plant, its growth increases to 51% than the control (Adisa et al. 2019).

7.4.1.3 Zn Nanoparticle

Similar to silver and copper nanoparticle, zinc nanoparticle has antibacterial, antiviral and antifungal activity. According to Paret et al., when photocatalyst technology is combined with nanotechnology, it increases antimicrobial activity and this technique is applied for Zn nanoparticle to treat bacterial spot on leaves of rose plant which is caused by *Xanthomonas spp.* (Paret et al. 2013). It was observed that when Zn nanoparticle was sprayed on leaves of lentil plant, it helps in improving plant growth, chlorophyll, carotene content and protects the cell membrane (Siddiqui et al. 2018). Treatment of Zn nanoparticle at 1000 ppm concentration enhances seed germination, plant growth and early flowering in peanut plant compared to the 2000 ppm concentration as it shows negative and toxic effect to plant (Prasad et al. 2012). It is reported that zinc nanoparticle or Zinkicide, i.e. SG6 when sprayed on leaves of sweet orange and grape fruit, subdue cracker lesion, citrus scab (*Elsinoe fawcettii*) and melanose (*Diaporthe citri*) respectively (Graham et al. 2016). Zinc nanoparticles are also used to treat fungal pathogen (Khan et al. 2016). It is reported that ZnO nanoparticles were used to treat fungus. From the observation, it was noted that different concentrations of ZnO nanoparticle, which is synthesized from zinc acetate at 9 mM/L impede growth and morphology changes of fungus like thinning of the fibres of hyphae takes place. Antifungal activity of zinc oxide nanoparticle was noticed at 12 mM/L against *Botrytis cinerea* and *Penicillium expansum* by distorting hyphae of fungus and inhibiting Conidiophore and Conidia (He et al. 2011).

Similarly, ZnO nanoparticles can act as biocontrol agent which enhances crop health. Zn nanoparticles at 500μg/ml combined with convention fungicide tetraconazole inhibit Cercospora leaf blight of sugar beet and enhance root yield and sugar content (Dimkpa et al. 2013). However, ZnO nanoparticle upon interaction with bio-controlling agent such as *Pseudomonas chlororaphis* O6 and *Fusarium Graminearum* in vitro enhances the efficiency of bio-controlling agent (Duffy 2007). Hence it is used in plant disease management and increases the production of the crop.

7.4.2 Nanoparticles Acting as Carrier

Different nanoparticles are commonly used for drug delivery and upon conjugation with active molecules can be implemented in agriculture to treat plant pathogen.

These nanoparticles can be used as carrier in fungicide, herbicide, insecticide and RNAi inducing molecule as described below:

7.4.2.1 Chitosan Nanoparticle

Chitosan is the deacetylated form of chitin which is found in the walls of fungus and the shells of crustaceans. One of the important properties of chitosan is: it is less soluble due to its hydrophobicity in the aqueous medium; hence it is used in the drug delivery (Kashyap et al. 2015). As a result, it is mixed with organic and inorganic compounds to improve solubility (Li et al. 2011). Chitosan has functional groups such as amine and hydroxyl which are modified and interacted with chitosan molecule. It is attached to the epidermis of leave and stem for long time and facilitates the uptake of the active molecule (Malerba and Cerana 2016).

7.4.2.2 Silica Nanoparticle

Silica nanoparticle has different physical and chemical properties, synthesized using different methods (Mody et al. 2014). They have particular size, shape, crystallinity and porosity which can be used for the drug delivery purpose. Different silica NP can be produced by chemical methods such as mesoporous particle, core shell particle porous hollow silica nanoparticles (PHSN) etc. In PHSN, the pesticide is loaded into inner core to protect the active molecule and delivery to the target area. Hence silica nanoparticles are used to treat plant tolerance against biotic and abiotic stress (Barik et al. 2008).

7.4.2.3 Titanium Nanoparticle

Titanium nanoparticle is also synthesized from different methods in huge amount which can be used in various applications. TiO_2 is polymorphic in nature, is widely used to control pests, nematodes in plants.

7.5 Nanopesticides

Pesticides are used to kill pests or pathogens to protect crop and to increase the yield. It is effective to control disease in plants, but its application has adverse effect to the environment and toxicity due to bioaccumulation. Hence nanopesticides serve as a better alternate to control the pest as well as keep the environment pollution free.

7.6 Insecticides

The coating of insecticides around nanoparticles was first started in the early 2000s. Earlier studies showed that insecticides are classified into contact insecticides and systemic insecticides. Furthermore, based on different physically affected organs such as nerve, muscle, growth, respiration, midgut, it is classified into 55 chemical functional groups (Sparks and Nauen 2015). Agrochemists are now focused to increase the activity of active ingredients of insecticides by reducing the particle

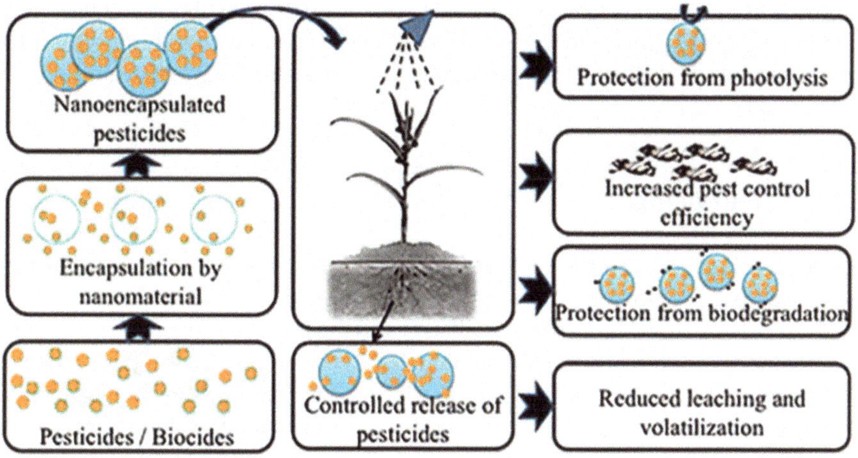

Fig. 7.3 Schematic diagram of nanoencapsulation in pesticide applications. Adapted with permission from Ref. (Nuruzzaman et al. 2016)

size or encapsulating active ingredients into the nanocapsule under normal temperature, alkaline and acidic moisture, etc. (Khan and Rizvi 2017). Hence mostly common nanoparticle carriers were chosen to target the pests or pathogens.

Generally, insecticides are less soluble in water. Hence the requirement of organic solvents is needed, but these results in increased toxicity. Nanoparticles are used to increase the solubility (Worrall et al. 2018). Besides, nanoparticles conjugated with conventional insecticides are used to inhibit the pathogens. Lu et al. reported that when the conventional thiamethoxam is loaded into dendrimer nanoparticles, it results in increased toxicity and mortality of *Hipposideros armiger* (Lu et al. 2013). Similarly, anacardic acid combined with ADH nanoparticle results in increased mortality of *S.litura* than that of anacardic acid-treated alone (Nguyen et al. 2014). Mainly nanoparticles are smaller than conventional/traditional chemicals. Hence it is important for release to the target side.

Evaporation is commonly used for the treatment of insecticides. Essential oil is used to treat but it is quickly evaporate in presence of sunlight, temperature and air (Lai et al. 2006). Hence the best method is encapsulation. If garlic essential oil is encapsulated into PEG (polyethylene glycol) and spread in harvested rice, it inhibits red flour beetles (*Tribolium castaneum*) as shown in Fig. 7.3. Imidacloprid synthesized from PEG and other aliphatic di-acids are used for encapsulation method to treat pest in different crops. Pepperman et al. reported that biodegradable microbial polymers are used to control the pests (Pepperman et al. 1991). Cameron et al. reported that there is a temperature-sensitive polymer such as intelimer, which releases pesticide according to the favourable temperature to protect the active ingredient (Cameron et al. 2018). A microcapsule is a reservoir system in which the target ingredient of pesticide is surrounded by a membrane. (Sotthivirat et al.

2007). Pest can be controlled using chitosan or alginate in this technique. SDS modified Ag/TiO_2 imidacloprid is prepared and applied on soya bean plants that are grown on soil having pH 6.2 and degraded faster within 8 days of treatment (Yan et al. 2005).

7.7 Fungicides

Fungicides are special group of pesticides that kill fungus and spores to provide improved plant protection. The word fungicide came into existence from France as early as 1938. Pierre-Marie-Alexis Millardet was the first to use fungicides to protect the vineyards from the pathogen phyllorexia. There are different types of fungus which cause disease in plants. With the advent of nanotechnology, conjugation of nanoparticles with fungicides has gained much attention. One of the recent techniques developed with the use of nanotechnology is "nano ghosts". According to Hatefalude et al. nano-sized bacterial ghost are taken from Gram-negative bacteria, attached to leave surface and improve the solubility in water of tebuconazole (Hatfaludi et al. 2004). Pyraclostrobin is another widely used fungicide that is conjugated with chitosan lactic co-polymer at different concentrations. After treatment, it inhibits *C. gossypii*, resulting in pest control within 7 days (Xu et al. 2014). Another low-soluble fungicide, Kaempferol, when loaded into chitosan or lecithin, shows 67% inhibition efficiency against *Fusarium oxysporum*. Another well-known method is encapsulation, which is mostly used as it is quick and commercial in nature. Similarly, Janatova et al. found that mesoporous silica nanoparticles (MSN) conjugated with essential oils show higher antifungal against *Aspergillus niger*, over a period of treatment of 14 days (Janatova et al. 2015).

Leaching is the most common method in which water and chemical move through the soil. It was observed that fungicide metalaxyn loaded with MNS showed an increase released rate of water. Similarly, nanoparticle encapsulated in Validamycin shows lesser efficiency than Validamycin alone (Qian et al. 2011). Kumar et.al reported that carbendazim loaded polymeric nanoparticle results in an increased rate of antifungal activity against *Fusarium oxysporum* and *Aspergillus parasiticus* than carbendazim (Kumar et al. 2017).

7.8 Herbicide

Herbicides are chemicals used to kill the herbs. Imazapic and Imazapyr are two widely used conventional herbicides (Maruyama et al. 2016). Interestingly, when these herbicides are conjugated with chitosan nanoparticle, it resulted in reduced toxicity and increased efficacy in inhibiting the *Bidens pilosa* weed. Similarly, when the conventional herbicides, i.e. Simazine and Atrazine were loaded into SLN nanoparticles it resulted in decreased toxicity and enhanced inhibition against *Raphanus raphanistrum*. (de Oliveira et al. 2015). According to Chidambaram et. al, rice husk can also be used as nanoparticle. 2,4-D is inserted into rice husk and it acts as the best herbicide agent against *Brassica species* than 2,4-D (Chidambaram

2016). Grillo et al. show that different concentrations of chitosan coated with polymeric nanoparticles help in attachment to the target plant and kill the pathogens more efficiently. These studies also show that paraquat-loaded chitosan helped to decrease toxicity against alga *Pseudokirchneriella subcapitata in* the presence of aquatic humic substances (Grillo et al. 2014).

7.9 Conclusion

Nanotechnology has emerged as a key tool in the field of agriculture and plant disease management. Conjugation of nanoparticles with pesticides or synthesis of new nanoparticles serving as biopesticides has gained much attention due to various advantages including increasing solubility of low-soluble water pesticides, precise target delivery, reduced toxicity and less pollution to the environment. The application of nanotechnology for plant disease management is yet to be explored in detail. Future works must be focussed to explore the use of nano-based pesticides to increase crop yield and productivity.

References

Abdal Dayem A, Hossain MK, Lee SB, Kim K, Saha SK, Yang G-M, Choi HY, Cho S-G (2017) The role of reactive oxygen species (ROS) in the biological activities of metallic nanoparticles. Int J Mol Sci 18(1):120

Abou El-Nour KM, Aa E, Al-Warthan A, Ammar RA (2010) Synthesis and applications of silver nanoparticles. Arab J Chem 3(3):135–140

Adisa IO, Pullagurala VLR, Peralta-Videa JR, Dimkpa CO, Elmer WH, Gardea-Torresdey JL, White JC (2019) Recent advances in nano-enabled fertilizers and pesticides: a critical review of mechanisms of action. Environ Sci Nano 6(7):2002–2030

Agrahari RK, Singh P, Koyama H, Panda SK (2020) Plant-microbe interactions for sustainable agriculture in the postgenomic era. Curr Genom 21(3):168–178

Barik T, Sahu B, Swain V (2008) Nanosilica—from medicine to pest control. Parasitol Res 103 (2):253

Bhattacharyya D, Singh S, Satnalika N, Khandelwal A, Jeon S-H (2009) Nanotechnology, big things from a tiny world: a review. Int J U-E-Serv Sci Technol 2(3):29–38

Bholay AD, Nalawade PM, Borkhataria BV (2013) Fungicidal potential of biosynthesized silver nanoparticles against phyto-pathogens and potentiation of fluconazole. World Res J Pharm Res 1(1):12–15

Bramlett M, Plaetinck G, Maienfisch PJE (2019) RNA-based biocontrols—a new paradigm in crop protection. Elsevier 6:522–527

Cameron D, Frazer E, Harvey P, Rampton M, Richardson K (2018) Researching language: issues of power and method, vol 1. Routledge, London, p 160

Charitidis CA, Georgiou P, Koklioti MA, Trompeta A-F, Markakis V (2014) Manufacturing nanomaterials: from research to industry. Manuf Rev 1:11

Chidambaram R (2016) Application of rice husk nanosorbents containing 2, 4-dichlorophenoxyacetic acid herbicide to control weeds and reduce leaching from soil. J Taiwan Inst Chem Eng 63:318–326

Cho EJ, Holback H, Liu KC, Abouelmagd SA, Park J, Yeo YJ (2013) Nanoparticle characterization: state of the art, challenges, and emerging technologies. Mol Pharm 10(6):2093–2110

Cromwell W, Yang J, Starr J, Jo Y-K (2014) Nematicidal effects of silver nanoparticles on root-knot nematode in bermudagrass. J Nematol 46(3):261

de Oliveira JL, Campos EVR, Goncalves da Silva CM, Pasquoto T, Lima R, Fraceto LF (2015) Solid lipid nanoparticles co-loaded with simazine and atrazine: preparation, characterization, and evaluation of herbicidal activity. J Agric Food Chem 63(2):422–432

De Wit PJ (1995) Fungal avirulence genes and plant resistance genes: unraveling the molecular basis of gene-for-gene interactions. In: Advances in botanical research, vol 21. Elsevier, Amsterdam, pp 147–185

De Wolf ED, Isard SA (2007) Disease cycle approach to plant disease prediction. Annu Rev Phytopathol 45:203–220

Dimkpa CO, McLean JE, Britt DW, Anderson AJ (2013) Antifungal activity of ZnO nanoparticles and their interactive effect with a biocontrol bacterium on growth antagonism of the plant pathogen Fusarium graminearum. Biometals 26(6):913–924

Doke N (1983) Involvement of superoxide anion generation in the hypersensitive response of potato tuber tissues to infection with an incompatible race of Phytophthora infestans and to the hyphal wall components. Physiol Plant Pathol 23(3):345–357

Drexled KE (1986) Engines of creation: the coming era of nanotechnology. Anchor Books, New York

Duffy B (2007) Zinc and plant disease. In: Mineral nutrition and plant disease. American Phytopathological Society, Saint Paul, MN, pp 155–175

Ealias AM, Saravanakumar M (2017) A review on the classification, characterisation, synthesis of nanoparticles and their application. In: IOP Conference Series: Materials Science and Engineering, p 032019

Elmer W, De La Torre-Roche R, Pagano L, Majumdar S, Zuverza-Mena N, Dimkpa C, Gardea-Torresdey J, White JC (2018) Effect of metalloid and metal oxide nanoparticles on Fusarium wilt of watermelon. Plant Dis 102(7):1394–1401

Elmer W, White J (2016) Nanoparticles of CuO improves growth of eggplant and tomato in disease infested soils. Environ Sci Nano 3:1072–1079

Evans I, Solberg E, Huber D (2007) Copper and plant disease. In: Mineral nutrition and plant disease. American Phytopathological Society, Saint Paul, MN, pp 177–188

Feynman RP (1960) There's plenty of room at the bottom. Eng Sci 23(5):22–36

Gajbhiye M, Kesharwani J, Ingle A, Gade A, Rai M (2009) Fungus-mediated synthesis of silver nanoparticles and their activity against pathogenic fungi in combination with fluconazole. Nanomedicine 5(4):382–386

Gardeniers J, Van den Berg A (2004) Lab-on-a-chip systems for biomedical and environmental monitoring. Anal Bioanal Chem 378(7):1700–1703

Ghormade V, Deshpande MV, Paknikar KM (2011) Perspectives for nano-biotechnology enabled protection and nutrition of plants. Biotechnol Adv 29(6):792–803

Giannousi K, Avramidis I, Dendrinou-Samara C (2013) Synthesis, characterization and evaluation of copper based nanoparticles as agrochemicals against Phytophthora infestans. RSC Adv 3 (44):21743–21752

Graham J, Johnson E, Myers M, Young M, Rajasekaran P, Das S, Santra S (2016) Potential of nano-formulated zinc oxide for control of citrus canker on grapefruit trees. Plant Dis 100 (12):2442–2447

Grant JJ, Loake GJ (2000) Role of reactive oxygen intermediates and cognate redox signaling in disease resistance. Plant Physiol 124(1):21–30

Grillo R, Rosa A, Fraceto L (2014) Poly (ε-caprolactone) nanocapsules carrying the herbicide atrazine: effect of chitosan-coating agent on physico-chemical stability and herbicide release profile. Int J Environ Sci Technol 11(6):1691–1700

Hatfaludi T, Liska M, Zellinger D, Ousman JP, Szostak M, Ambrus Á, Jalava K, Lubitz W (2004) Bacterial ghost technology for pesticide delivery. J Agric Food Chem 52(18):5627–5634

Hayles J, Johnson L, Worthley C, Losic D (2017) Nanopesticides: a review of current research and perspectives. In: New pesticides and soil sensors. Elsevier, Amsterdam, pp 193–225

He L, Liu Y, Mustapha A, Lin M (2011) Antifungal activity of zinc oxide nanoparticles against Botrytis cinerea and Penicillium expansum. Microbiol Res 166(3):207–215

Janatova A, Bernardos A, Smid J, Frankova A, Lhotka M, Kourimská L, Pulkrabek J, Kloucek P (2015) Long-term antifungal activity of volatile essential oil components released from mesoporous silica materials. Ind Crop Prod 67:216–220

Jung J-H, Kim S-W, Min J-S, Kim Y-J, Lamsal K, Kim KS, Lee YS (2010) The effect of nano-silver liquid against the white rot of the green onion caused by Sclerotium cepivorum. Mycobiology 38(1):39–45

Kah M, Hofmann T (2014) Nanopesticide research: current trends and future priorities. Environ Int 63:224–235

Kashyap PL, Xiang X, Heiden P (2015) Chitosan nanoparticle based delivery systems for sustainable agriculture. Int J Biol Macromol 77:36–51

Khan MR, Mohidin FA, Khan U, Ahamad F (2016) Native Pseudomonas spp. suppressed the root-knot nematode in in vitro and in vivo, and promoted the nodulation and grain yield in the field grown mungbean. Biol Control 101:159–168

Khan MR, Rizvi TF (2017) Application of nanofertilizer and nanopesticides for improvements in crop production and protection. In: Nanoscience and plant–soil systems. Springer, Cham, pp 405–427

Khandelwal N, Barbole RS, Banerjee SS, Chate GP, Biradar AV, Khandare JJ, Giri AP (2016) Budding trends in integrated pest management using advanced micro-and nano-materials: Challenges and perspectives. J Environ Manag 184:157–169

Kim HS, Kang HS, Chu GJ, Byun HS (2008) Antifungal effectiveness of nanosilver colloid against rose powdery mildew in greenhouses. In: Solid state phenomena. Trans Tech Publication, Zurich, pp 15–18

Kumar S, Kumar D, Dilbaghi N (2017) Preparation, characterization, and bio-efficacy evaluation of controlled release carbendazim-loaded polymeric nanoparticles. Environ Sci Pollut Res 24 (1):926–937

Lai F, Wissing SA, Müller RH, Fadda AM (2006) Artemisia arborescens L essential oil-loaded solid lipid nanoparticles for potential agricultural application: preparation and characterization. AAPS Pharm Sci Tech 7(1):E10

Lamsa K, Kim S-W, Jung JH, Kim YS, Kim KS, Lee YS (2011) Inhibition effects of silver nanoparticles against powdery mildews on cucumber and pumpkin. Mycobiology 39(1):26–32

Lamsal K, Kim SW, Jung JH, Kim YS, Kim KS, Lee YS (2011) Application of silver nanoparticles for the control of Colletotrichum species in vitro and pepper anthracnose disease in field. Mycobiology 39(3):194–199

Li M, Huang Q, Wu Y (2011) A novel chitosan-poly (lactide) copolymer and its submicron particles as imidacloprid carriers. Pest Manag Sci 67(7):831–836

Lok C-N, Ho C-M, Chen R, He Q-Y, Yu W-Y, Sun H, Tam PK-H, Chiu J-F, Che C-M (2006) Proteomic analysis of the mode of antibacterial action of silver nanoparticles. J Proteome Res 5 (4):916–924

Lu W, Lu ML, Zhang QP, Tian YQ, Zhang ZX, Xu HH (2013) Octahydrogenated retinoic acid-conjugated glycol chitosan nanoparticles as a novel carrier of azadirachtin: synthesis, characterization, and in vitro evaluation. J Polym Sci A Polym Chem 51(18):3932–3940

Malerba M, Cerana R (2016) Chitosan effects on plant systems. Int J Mol Sci 17(7):996

Mallaiah B (2015) Integrated approaches for the management of crossandra crossandra infundibuliformis l Nees wilt caused by fusarium incarnatum desm Sacc. Coimbatore

Maruyama CR, Guilger M, Pascoli M, Bileshy-José N, Abhilash P, Fraceto LF, De Lima R (2016) Nanoparticles based on chitosan as carriers for the combined herbicides imazapic and imazapyr. Sci Rep 6:19768

McQuillan J (2010) Bacterial-nanoparticle interactions. Univ Exeter 7(1):3

Mittal AK, Chisti Y, Banerjee UC (2013) Synthesis of metallic nanoparticles using plant extracts. Biotechnol Adv 31(2):346–356

Mody VV, Cox A, Shah S, Singh A, Bevins W, Parihar H (2014) Magnetic nanoparticle drug delivery systems for targeting tumor. Appl Nanosci 4(4):385–392

Mondal KK, Rajendran T, Phaneendra C, Mani C, Sharma J, Shukla R, Verma G, Kumar R, Singh D, Kumar A (2012) The reliable and rapid polymerase chain reaction (PCR) diagnosis for Xanthomonas axonopodis pv. Punicae in pomegranate. Afr J Microbiol Res 6(30):5950–5956

Morel J-B, Dangl JL (1997) The hypersensitive response and the induction of cell death in plants. Cell Death Diff 4(8):671–683

Moussa SH, Tayel AA, Alsohim AS, Abdallah RR (2013) Botryticidal activity of nanosized silver-chitosan composite and its application for the control of gray mold in strawberry. J Food Sci 78 (10):M1589–M1594

Mueller CF, Laude K, McNally JS, Harrison DG (2005) Redox mechanisms in blood vessels. Arterioscler Thromb Vasc Biol 25(2):274–278

Nguyen TNQ, Hua QC, Nguyen TT (2014) Enhancing insecticide activity of anacardic acid by intercalating it into MgAl layered double hydroxides nanoparticles. J Vietnam Environ 6 (3):208–211

Nuruzzaman M, Rahman MM, Liu Y, Naidu R (2016) Nanoencapsulation, nano-guard for pesticides: a new window for safe application. J Agric Food Chem 64(7):1447–1483

Ocsoy I, Paret ML, Ocsoy MA, Kunwar S, Chen T, You M, Tan W (2013) Nanotechnology in plant disease management: DNA-directed silver nanoparticles on graphene oxide as an antibacterial against Xanthomonas perforans. ACS Nano 7(10):8972–8980

Paret ML, Palmateer AJ, Knox GW (2013) Evaluation of a light-activated nanoparticle formulation of titanium dioxide with zinc for management of bacterial leaf spot on rosa 'Noare'. Hort Sci 48 (2):189–192

Park H-J, Kim JY, Kim J, Lee J-H, Hahn J-S, Gu MB, Yoon J (2009) Silver-ion-mediated reactive oxygen species generation affecting bactericidal activity. Water Res 43(4):1027–1032

Pepperman AB, Kuan J-CW, McCombs C (1991) Alginate controlled release formulations of metribuzin. J Control Release 17(1):105–111

Prasad T, Sudhakar P, Sreenivasulu Y, Latha P, Munaswamy V, Reddy KR, Sreeprasad T, Sajanlal P, Pradeep T (2012) Effect of nanoscale zinc oxide particles on the germination, growth and yield of peanut. J Plant Nutr 35(6):905–927

Qian K, Shi T, Tang T, Zhang S, Liu X, Cao Y (2011) Preparation and characterization of nano-sized calcium carbonate as controlled release pesticide carrier for validamycin against Rhizoctonia solani. Microchim Acta 173(1-2):51–57

Radzig M, Nadtochenko V, Koksharova O, Kiwi J, Lipasova V, Khmel I (2013) Antibacterial effects of silver nanoparticles on gram-negative bacteria: influence on the growth and biofilms formation, mechanisms of action. Colloids Surf B: Biointerfaces 102:300–306

Rafique M, Sadaf I, Rafique MS, Tahir MB (2017) A review on green synthesis of silver nanoparticles and their applications. Artif Cells Nanomed Biotechnol 45(7):1272–1291

Rai M, Ingle A (2012) Role of nanotechnology in agriculture with special reference to management of insect pests. Appl Microbiol Biotechnol 94(2):287–293

Richards R (1981) Antimicrobial action of silver nitrate. Microbios 31(124):83–91

Sherald J, Lei J (1991) Evaluation of a rapid ELISA test kit for detection of Xylella fastidiosa in landscape trees. Plant Dis 75(2):200–203

Siddiqui Z, Khan A, Khan M, Abd-Allah E (2018) Effects of zinc oxide nanoparticles (ZnO NPs) and some plant pathogens on the growth and nodulation of lentil (Lens culinaris Medik.). Acta Phytopathol Entomol Hungar 53(2):195–211

Sotthivirat S, Haslam J, Stella V (2007) Evaluation of various properties of alternative salt forms of sulfobutylether-β-cyclodextrin,(SBE) 7M-β-CD. Int J Pharm 330(1-2):73–81

Sparks TC, Nauen R (2015) IRAC: mode of action classification and insecticide resistance management. Pestic Biochem Physiol 121:122–128

Stakman E (1915) Relation between Puccinia graminis and plants highly resistant to its attack. J Agric Res 4:193–200

Stephenson GR (2003) Pesticide use and world food production: risks and benefits, vol 853. ACS Publications, Washington, DC, pp 261–270

Strayer-Scherer A, Liao Y, Young M, Ritchie L, Vallad G, Santra S, Freeman J, Clark D, Jones J, Paret M (2018) Advanced copper composites against copper-tolerant Xanthomonas perforans and tomato bacterial spot. Phytopathology 108(2):196–205

Tarafdar J, Adhikari T (2015) Nanotechnology in soil science. In: Soil science: an introduction. ICAR, New Delhi, pp 775–807

Thakkar KN, Mhatre SS, Parikh RY (2010) Biological synthesis of metallic nanoparticles. Nanomed: Nanotechnol Biol Med 6(2):257–262

Viswanathan S, Wu L-c, Huang M-R, Ho J-aA (2006) Electrochemical immunosensor for cholera toxin using liposomes and poly (3, 4-ethylenedioxythiophene)-coated carbon nanotubes. Anal Chem 78(4):1115–1121

Wojtaszek P (1997) Oxidative burst: an early plant response to pathogen infection. Biochem J 322 (3):681–692

Worrall EA, Hamid A, Mody KT, Mitter N, Pappu HR (2018) Nanotechnology for plant disease management. Agronomy 8(12):285

Xu L, Cao L-D, Li F-M, Wang X-J, Huang Q-L (2014) Utilization of chitosan-lactide copolymer nanoparticles as controlled release pesticide carrier for pyraclostrobin against Colletotrichum gossypii Southw. J Dis Sci Technol 35(4):544–550

Yadeta K, Thomma B (2013) The xylem as battleground for plant hosts and vascular wilt pathogens. Front Plant Sci 4:97

Yan J, Huang K, Wang Y, Liu S (2005) Study on anti-pollution nano-preparation of dimethomorph and its performance. Chin Sci Bull 50(2):108–112

Nanosystems for Cancer Therapy

8

Sailendra Kumar Mahanta and Manoranjan Arakha

Abstract

The chapter focuses on the various nanosystems and their applications in cancer therapy. The nanosystems comprise nanoscaled biomaterials that carry the drug molecules, proteins and genes into the tumor environment to achieve therapeutic efficacy. The current know-how of the cancer physiology and the cancer milieu facilitates designing nanosystems for cancer therapy. In terms of size, nanosystems are a thousandth part of an average human cell. Nanosystems having a size of less than 300 nm have enhanced permeability and retention in cancer cells than healthy cells.

In comparison, nanosystems less than 50 nm in size easily enter most of the healthy cells. Hence, the size of nanosystems is extremely critical for their enhanced efficacy. By exploiting their small size, nanosystems can interact with physiological molecules both outside and inside the cancer cells. Upon functionalization of nanoparticles with antibodies, drug molecules, and genes, we can have access to all the cells of the body. The system behaves like inject and forget nanodevice which can selectively target cancer cells without affecting normal cells. The nanosystems are building the roadmap for future technologies where these can be potentially used to detect cancer cells and deliver treatment at the targeted site.

S. K. Mahanta (✉)
Department of Pharmacology, Nirmala College of Pharmacy, Atmakur Road, Mangalagiri, Guntur, Andhra Pradesh, India

M. Arakha
Center for Biotechnology, Siksha 'O' Anusandhan (Deemed to be University), Bhubaneswar, India
e-mail: marakha@soa.ac.in

M. Arakha et al. (eds.), *Bio-Nano Interface*,
https://doi.org/10.1007/978-981-16-2516-9_8

Keywords

Nanosystems · Cancer · Drug delivery · Gene delivery · Functionalization · Dendrimers · Protein nanoparticles · Metal nanoparticles

8.1 Introduction

Cancer is one of the top three prevalent health issues challenging humanity. The heterogeneity of the disease and its features coupled with treatment induced drugs toxicity makes it difficult to contain and treat. The number of cancer patients is increasing rapidly, and as per the estimates projected by International Agency for Research on Cancer (IARC), we could see an increase in the number of cancer patients from 127 million new patients in 2008 to 222 million by 2030 (Bray et al. 2012). Cancer involves an abnormality in the physiological growth of cells and has the propensity to move to other parts of the body. It is a disease of heterogeneous characteristics where the cells undergo modifications in their DNA, causing aggressive growth and invasion to nearby tissues resulting in tumor formation. Current therapeutic options involve radiation therapy, immunotherapy, bone marrow transplant, cryoablation therapy along with chemotherapy. Chemotherapy can sometimes act as a double-edged sword where the Chemotherapeutic drugs can target and kill rapidly dividing healthy cells in the human body such as hair follicular cells and gastrointestinal cells resulting in loss of healthy cells in the human body.

Moreover, the tumor microenvironment is highly variable and adaptive than normal cells and prevents chemotherapeutic retention of drugs within the tumors. Due to the above reasons, gradually drug resistance creeps in the treatment process and delays recovery. The above challenges make it an uphill task for drugging the target site.

Nanosystems have emerged as vehicles for targeting tumors with a calibrated amount of therapeutic agent. Nanoparticles are defined as particles having a size of 1–100 nanometers (nm), but the broader range includes particle sizes up to 1000 nm. Particles in these ranges have improved physicochemical, electrical, surface, interaction properties, which facilitate them as nano-carriers (Albanese et al. 2012).

Nanosystems having a size of less than 300 nm can easily enter the tumor cells compared to normal cells and can cause enhanced permeation and retention in cancer cells while nanosystems less than 50 nm in size easily enter most of the cells, cross blood vessels, and can circulate throughout the body. Nanodelivery systems for therapeutic agents are being pursued worldwide due to their target specific release of the drug, improved efficacy, and minimal side effects.

8.2 Physiological Hindrances to Tumor-Specific Delivery

Intravenously administered drugs come across the first hindrance in the form of the vascular endothelial lining. Other components in the blood, such as serum and plasma proteins, protein splitting enzymes, RNA degrading enzymes, etc., also hinder the movement of the drug towards the target sites (Alonso 2004). Epithelial cells lining the gut wall act as a barrier for drugs taken orally. The drug molecules will move through the epithelial cells with the help of transcellular transport proteins. The physical and chemical characteristics of the drug molecules are detrimental in their membrane transport. Thus, we can facilitate the movement of the drug molecules across the cells lining the gut by altering the drug form, by providing a protective layer that will prevent the drug molecule from coming in contact with the gut lining or by use of molecules that would increase the porosity of the gut lining for smooth transport of drug molecules (Ganta et al. 2010).

Moreover, enzymes such as cytochrome P450 and Efflux proteins such as P glycoprotein are present in the intestinal lining, which prevent the absorption of drugs (Ganta et al. 2010; Zhang et al. 2007). In contrast to the healthy tissue, tumor microenvironment is entirely different with the presence of increased vascular networks under hypoxic and acidic conditions. The tissue matrix in tumor cells is also heterogeneous and contains inflammatory cells and fibroblasts that increase the interstitial pressure within the cells. This increase in positive pressure within the tumor environment also prevents the penetration of drug molecules deep within the tissue (Tredan et al. 2007).

Tumor-targeted delivery of chemotherapeutic drugs using nanosystems is mainly facilitated by the leaky tumor microvasculature, as shown in Fig. 8.1.

The mechanism is known as "enhanced permeability and retention" (EPR) effect (Matsumura and Maeda 1986). The pore size of leaky channels in tumor milieu varies across tumor types in the range between 300–800 nm; hence, the choice of nanosystem size is detrimental for its successful delivery (Goncalves et al. 2012). Nanosystems within a size range of 20–200 nm are known to cross the tumor milieu and remain trapped (Danhier et al. 2010; Wilhelm et al. 2016). In a rather unconventional study, Chan and Co-workers have suggested the median delivery efficiency of the nanoparticle-based systems at the tumor site is a mere 0.7% of the injected dose. Nanosystems prepared from inorganic materials have a higher chance of penetrating the tumor milieu compared to organic materials. Nanosystems having a size of less than 100 nm have higher median delivery efficiency at the tumor site compared to nanosystems of increased size. It clearly shows now that even though the pore channels in tumor vasculature are around 300–800 nm, we find nanosystems having a size less than 100 nm (0.7% median delivery) to have increased access to tumor vasculature than the nanosystems having a size greater (0.6% median deliver) than it. The charge on the nanosystem significantly affects their ability to infiltrate the tumor environment. Negatively charged nanosystems demonstrate a reduced rate of endocytosis without the clathrin-mediated endocytosis pathway. However, positively charged nanosystems internalize swiftly through the clathrin-mediated endocytosis (Harush-Frenkel et al. 2007). Moreover, nanosystems having a neutral zeta

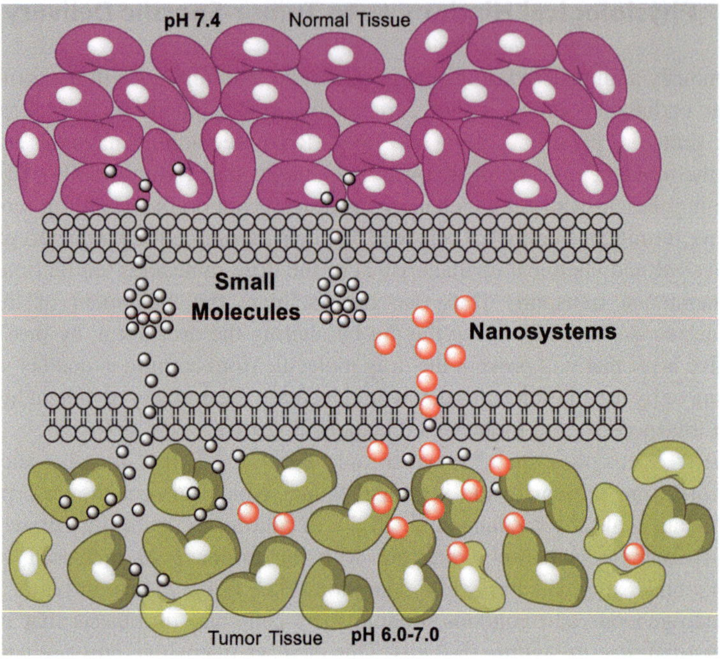

Fig. 8.1 Schematic representation of nanosystems having enhanced permeability with retention in tumor tissue

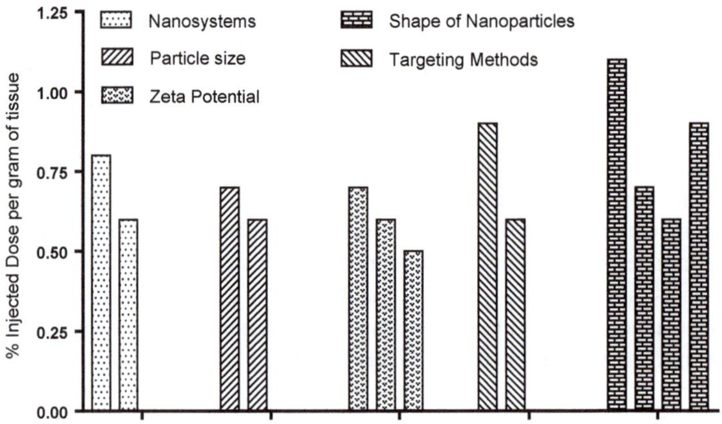

Fig. 8.2 Effect of variable parameters on % Injected Dose of nanosystems per gram of tissue

potential range (-10 mV to +10 mV) tend to have a delivery efficiency of 0.7% compared to lower (0.6%) and higher zeta potential (0.5%) over neutral zeta potential range (see Fig. 8.2). The study, however, does not take into account

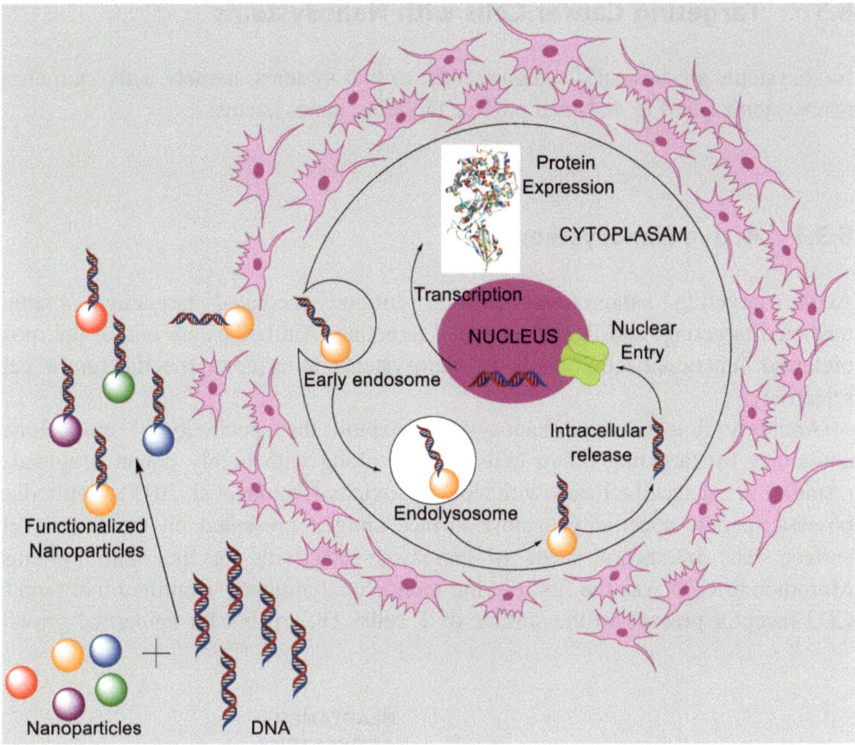

Fig. 8.3 Functionalized nanosystems for nucleic acid delivery into tumor cells

pharmacokinetics, toxicity, efficacy, and the overall impact on patient outcomes (Wilhelm et al. 2016).

However, the EPR phenomenon can be improved if the nanosystems can trick the reticulo-endothelial system (RES) and demonstrate circulation in the bloodstream for extended periods. The problem is solved by incorporating Poly(ethylene glycol) (PEG) in the nanosystem (Harush-Frenkel et al. 2007). Nanosystems without PEG functionalization can deliver drugs to organ systems where RES has a major role to play, viz. the respiratory and the hepatic system (Ayer and Klok 2017).

The specificity of the nanosystems for the tumor environment can be significantly improved by functionalization with tumor-specific ligands that bind to receptors overexpressed in tumor cells.

Functionalizing the nanosystems for active targeting to tumors, using aptamers, folic acid, nucleic acids (see Fig. 8.3), arginine-glycine-aspartic acid (RGD) tripeptides, and epidermal growth factor mAb, etc., improves target delivery.

Hence, nanosystems can score more than the other systems in using their size and physiological properties to their advantage and increase their concentration at the tumor site.

8.3 Targeting Cancer Cells with Nanosystems

Nanosystems are basically categorized in to two systems, namely active targeting nanosystems (see Fig. 8.4) and passive targeting nanosystems.

8.3.1 Active Nanosystems

Active targeting nanosystems include antibody-mediated targeting, aptamer mediated targeting, and ligand-mediated targeting. Antibodies are one of the most preferred functionalization tools in nanosystems to achieve specific tumor cell selectivity.

Antibody functionalized nanosystems exploit the specificity of monoclonal antibodies for targeting tumor cells directly along with highly potent drugloads, resulting in enhanced efficacy with reduced toxicity (Parakh et al. 2015). Antibodies possess specific specificity against surface antigen presented on the tumor cell surface. The interaction leads to increased selectivity for the nano systems. Muromonab CD3 was the first murine monoclonal antibody identified that targets CD3 receptor present on the surface of T cells. The mAb's for epidermal growth

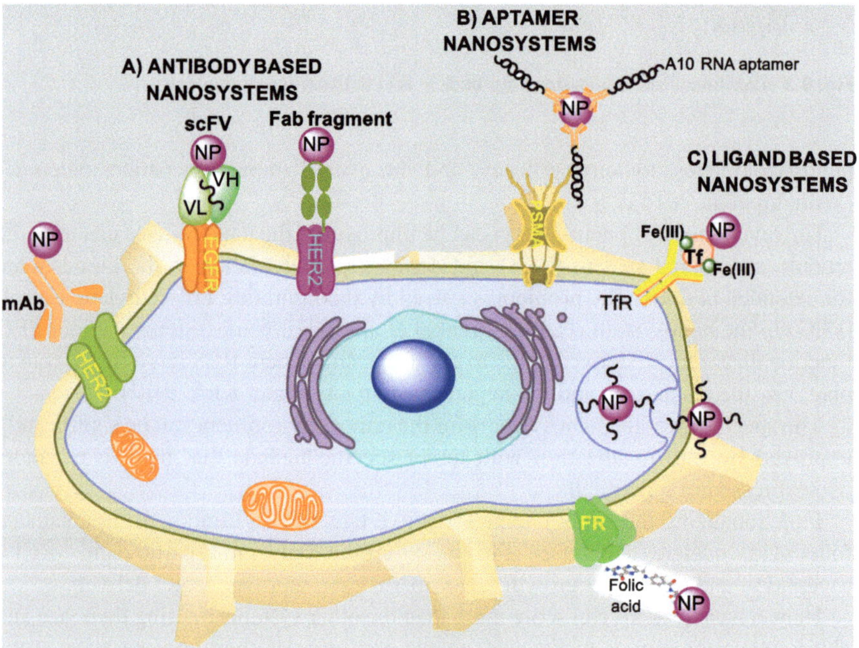

Fig. 8.4 Schematic representation of active targeting nanosystems (**a**) antibody based nanosystems (**b**) aptamer nanosystems, and (**c**) ligand based nanosystems

factor receptor (EGFR), transferrin receptor (TR), and prostate-specific membrane antigen (PSMA) are frequently used for delivery of nanosystems.

EGFR is found to be over expressed in most of the tumor tissues than normal tissues. Humanized mAb Trastuzumab which targets EGFR 2 (Her2) is one of preferred treatment options available for Her2 overexpressed breast cancer. Trastuzumab functionalization on nanosystems has been observed in a lot of experiments involving Her2 positive cancer cells (Bloise et al. 2020; Truffi et al. 2018; Di Wu et al. 2017).

Antibody fragments such as antigen-binding fragments (Fab, ~50 kDa), single-chain variable fragments (scFv ~30 kDa), and single-domain antibodies (sdAb 12–15 kDa) can be used as an alternative approach for functionalization of nanosystems. Fab and scFv are the most widely incorporated antibody fragments for functionalizing nanosystems (Byrne et al. 2008; Yu et al. 2010). Due to increased surface area of the nanosystems tagging of multiple targeting peptides on to a single nanosystem improves the loading efficiency, targeting efficiency and specificity. The Fab consists of one constant (heavy) domain and one variable (light) domain (Fig. 8.4). The scFvs are composed of a merged variable regions of constant and variable chains (Fig. 8.4, (Wang et al. 2008)). The variable section provides them with the required binding specificity (Byrne et al. 2008), whereas the absence of the fixed domains makes them non-immunogenic. Overall, in comparison with whole mAbs, antibody fragments functionalization on nanosystems reduces the challenge presented by the RES and significantly increases the residence time in the blood stream (Wang et al. 2008).

Aptamers are small single-strands of DNA or RNA molecules arranged into three dimensional (3D) structures. Their small size and low cross-sectional profile facilitates their binding to key proteins within the biological system efficiently. Aptamers are equally efficient compared to antibodies due to their high sensitivity and specificity (Ni et al. 2011). The A10 RNA aptamer binds with the extracellular binding element of PsMa and facilitates the internalization of the nanosystems as shown in Fig. 8.4 (Farokhzad et al. 2004). Cisplatin is internalized in prostate cancer cells using PsMa aptamer-tagged nanosystems (Dhar et al. 2008). Quantum dot-doxorubicin nanosystems are also functionalized with A10 RNA aptamer for tumor-specific targeted delivery (Bagalkot et al. 2007).

Transferrin (Tr) is a glycoprotein present in blood that facilitates transport of iron across the cell membrane. It binds to transferrin binding sites on the cell membrane, and causes internalization through receptor-mediated endocytosis (Fig. 8.4). The transferrin receptors on tumor cells have a 100 fold expression compared to normal cells (Danhier et al. 2010; Byrne et al. 2008). Several liposome based nanosystems using transferrin have been developed for intra-tumor delivery of doxorubicin (Li et al. 2010; Kobayashi et al. 2007; Anabousi et al. 2006). Transferrin receptor (TFr) is over expressed in tumor cells than normal cells; moreover, transferrin receptor is also overexpressed in endothelial cells of brain capillary which makes it an effective target for drugs targeting the brain. Drugs which are not able to cross the blood–brain barrier can be moved across the blood–brain barrier with the help of transferrin receptor loaded human serum albumin nanosystems (Daniels et al. 2012).

Prostate-specific membrane antigen (PSMA) is present on the outer surface of the membrane lining of prostate cancer cells and other solid tumors and is specific to tumors (Hrkach et al. 2012). PsMa-targeted dendrimer-based nanoconjugate system (see Fig. 8.4) has been used for carrying out payload delivery to tumors having PSMA on the membrane surface (Patri et al. 2004).

Folic acid is a vitamin essential for synthesizing purines and pyrimidines. Cellular uptake of folates is carried out through folate receptor (FR) (see Fig. 8.4). The FR is found to be significantly upregulated in many human tumors compared to most normal tissues (Byrne et al. 2008). FR directly transports folic acid, as well as folic acid functionalized nanosystems (Talekar et al. 2011; Hilgenbrink and Low 2005). Hence, FR is pursued as a viable target for delivering anticancer agents using nanosystems (Yu et al. 2010). Folate receptors (FR-α and β) present on the cell surface are capable enough for transferring folate into the cells. The alpha isoform is highly expressed in more than 40% of different types of human cancers, whereas FR-β is found specifically on triggered macrophages and hematopoietic cells (Xia and Low 2010). These differences in receptor distribution can be harnessed for designing folate functionalized nanosystems (Elnakat and Ratnam 2004; Pan et al. 2002).

8.3.2 Passive Targeting Systems

Passive targeting systems include dendrimers, metallic nanoparticles, metal oxide nanoparticles, magnetic nanoparticles, quantum dots, liposomes, and polymeric nanoparticles, as nanosystems for anticancer therapy.

Dendrimers are nanosized, radially symmetrical, well defined, homogeneous, monodisperse, and regularly branched macromolecules. Dendrimers usually increase in diameter linearly and develop a globular shape while generating dendrimers (Baker 2009).

They are generally synthesized from synthetic sources or amino acids, sugars, and nucleotides. Such nanosystems can be easily functionalized and conjugated to therapeutic payloads for targeted delivery. Dendrimers can carry drug molecules in their core cavities through hydrophobic contacts and various bond formations within the cavities. Dendrimer–drug conjugates are highly popular in pre-clinical development. Biocompatible dendrimers are under use for cancer treatment, as nanodelivery systems for cisplatin and doxorubicin (Szymanski et al. 2011; Kolhe et al. 2006; Gillies and Frechet 2005).

Among metal nanoparticles gold nanoparticles (GNPs) are widely preferred for drug and gene delivery as it is inert, has highly modifiable synthesis methods, increased functionalization properties, absorption coefficient is high, biocompatible, can be easily detected, and acts as a carrier for targeted delivery (Papasani et al. 2012; Shan et al. 2012; Shukla et al. 2005; Connor et al. 2005). The functionalization of GNPs with oligonucleotide results in increased uptake in cells compared to GNPs alone (Giljohann et al. 2007). GNPs are also utilized for delivery of antisense DNA into the nucleus by functionalizing antisense DNA with peptides (Liu and Franzen 2008).

PEG coating of TNF alpha functionalized GNPs enhances uptake in tumor cells and reduces the growth of tumor cells as observed in mice (Visaria et al. 2006). Thiol functionalization of PEG enhances uptake of GNPs in MC-38 colon carcinoma tumors with insignificant accumulation in healthy cells (Paciotti et al. 2004). Initiation of angiogenesis within cells is regulated by heparin-binding growth factors such as VEGF165 and bFGF; however, GNPs can bind with these growth factors, thereby preventing intracellular cytokine activity and angiogenesis (Mukherjee et al. 2005).

Magnetic nanoparticles (MNPs) in size range of 5–20 nm are prepared from magnetite or maghemite (McBain et al. 2008). Synthesis of MNPs is carried out by co-precipitation of Fe(III) and Fe(II) (Sun and Zeng 2002). Magnetic and thermal properties are unique features of MNPs apart from being biodegradable and biocompatible (El-Sherbiny et al. 2017; Laurent et al. 2011). MNPs are generally pursued as a drug or gene delivery carrier for intracellular delivery. MNPs carrying the therapeutic gene or payload of interest are injected intravenously and exposed to strong magnetic fields focussed at the site of interest in the body. Under the influence of these strong magnetic fields, the MNPs are taken up by the cells in local tissue (Dobson 2006; Wu et al. 2019). MNP-drug formulations involving doxorubicin, paclitaxel, and methotrexate are undergoing pre-clinical evaluations (Li et al. 2013; Corem-Salkmon et al. 2011; Thomsen et al. 2013). Increased targeting of tumor cells is achieved by functionalization of MNPs with folic acid since folic acid receptors are highly expressed in tumor cells (Li et al. 2013). MNPs have shown promising results in experiments involving the functionalization of MNPs with drugs such as doxorubicin and verapamil and peptides specific for tumor tissues such as cyclo (Arg-Gly-Asp-D-Phe-Lys) (Shen et al. 2013).

Zinc oxide nanoparticles (ZnONP) ZnO are considered to be a "GRAS" (generally recognized as safe) and approved by the Food and Drug Administration (FDA) with potential anticancer effect. The ZnoNP are versatile nanoparticles and can form a core, as a shell and protect drug or genes or nucleic acids within itself (Martinez-Carmona et al. 2018). ZnONP develop selective cytotoxicity towards cancer cells mediated through the development of an imbalance in zinc-dependent protein's functional activity and selective generation of reactive oxygen species in cancer cells. Zinc also plays a vital role in the maintenance of the tumor suppressor gene p53 and the caspase-6 enzyme responsible for regulating apoptosis activity (Elshama et al. 2018). Bovine alpha-lactalbumin (BLA) functionalized ZnONP have demonstrated increased toxicity against MDAMB-231 cells in a reactive oxygen species-dependent mechanism (Mahanta et al. 2017). Doxorubicin functionalized Zinc oxide quantum dots have demonstrated increased toxicity in MCF-7 cells compared to doxorubicin alone (Sharma et al. 2016). The above studies show that ZnONP can act both as a drug and gene delivery system along with its own specific toxicity against tumor cells.

Mesoporous silica nanoparticle (SiNP) is a promising carrier for gene, peptides, and drugs. It enjoys numerous advantages such as porosity for drug encapsulation, increased surface area, long-term storage, highly biocompatible (till a concentration below 2.5 nM), and low cost of scalability. SiNPs have shown promising results in drug and gene delivery in addition to their ability to silence specific protein expression (Malvindi et al. 2012). Multicomponent nanosystems comprising of PEI25k-PEG5k copolymer grafted SiNPs showed increased cellular uptake and siRNA delivery in tumor cells (Lee et al. 2011). SiNPs functionalized with PEI, and Folic acid is frequently used for targeting cancer cells (Rosenholm et al. 2009). Animal studies have confirmed that SiNPs are preferentially accumulated in tumors, and effectively deliver drugs such as camptothecin for suppressing tumor growth (Lu et al. 2010). Exploiting the high porosity properties of SiNPs, Curcumin a water-insoluble anticancer drug has been successfully delivered and found to enhance the pharmacological effect in Glioblastoma cells (Ahmadi Nasab et al. 2016).

A wide range of natural and synthetic polymers are utilized for preparing biodegradable nanoparticles such as polylactide–polyglycolide copolymers, polyacrylates, polycaprolactones, or natural polymers such as gelatin, alginate, collagen, and chitosan (Moghimi et al. 2001). These polymers in a physiological environment undergo hydrolysis to form lactic acid and glycolic acid and finally get eliminated from the body with the help of Krebs cycle. The degradation products are formed very slowly in the body without affecting the normal cellular functions of the body. PLGA nanoparticles are synthesized using the emulsion solvent evaporation technique (Jain 2000). DNA, antisense oligonucleotides, and drug molecules can be encapsulated in nanoparticles using the same technique.

PLGA nanoparticles loaded with human wild type p53 plasmid DNA demonstrated enhanced anti-proliferative activity in cancer cell lines (Prabha and Labhasetwar 2004). In a separate study, the different size of synthesized PLGA nanoparticle was administered to cancer cells, and their transfection efficiency was evaluated. The results showed that the smaller PLGA nanoparticles had a 27-fold higher transfection efficiency in COS-7 cells and a four-fold higher transfection efficiency in HEK293 in comparison to larger size of PLGA nanoparticles greater than 200 nm. The higher transfection efficiency of the small size of PLGA nanoparticles was independent of DNA functionalization. The results highlight that the smaller particle size and the uniform particle size distribution both enhance the PLGA nanoparticle-mediated gene expression (Prabha et al. 2002). PLGA nanoparticles functionalized with AS1411 (Ap), a DNA aptamer specific to nucleolin receptor (highly expressed in cancer cells and endothelial cells), were used to deliver paclitaxel in glioblastoma cell lines (Guo et al. 2011).

Chitosan is a biodegradable linear amino-polysaccharide obtained by deacetylation of chitin. It is composed mainly of 1–4 linked N-acetyl D-glucosamine and D-glucosamine subunits, (Romoren et al. 2002; Hejazi and Amiji 2003). It readily forms complex with pDNA and forms nanoparticles in the size range of 20–500 nm particles. The molecular weight of chitin and deacetylation characteristics determine the size of Chitin nanoparticles (CNP) (Illum et al. 2001). Chitosan exploits its cationic nature and forms strong electrostatic bonds with other drug molecules, DNA, and antisense oligonucleotides (Bordelon et al. 2010). Due to the above interactions CNP protect the DNA from degradation and increases its bioavailability inside the cells (Koping-Hoggard et al. 2001). In order to exploit this property of CNP, several researchers have prepared various forms of functionalized CNP such as galactosylated chitosan nanoparticles (Erbacher et al. 1998), galactosylated chitosan-graft-poly(vinylpyrrolidone) nanoparticles (PVP) (Park et al. 2004), trimethylated chitosan nanoparticles (Thanou et al. 2002), N-dodecylated chitosan nanoparticles (Li et al. 2002), deoxycholic acid altered chitosan nanoparticles (Kim et al. 2001). Doxorubicin loaded CNP have demonstrated increased toxicity against MCF-7 cells compared to Doxorubicin alone (Saeed et al. 2020). Higher transfection efficiencies were obtained with CNP functionalized with cell-penetrating peptides for the delivery of antisense oligonucleotides inside the cells (Dowaidar et al. 2018).

Protein nanosystems involving lysozyme (Mahanta et al. 2015), bovine alpha-lactalbumin (BLA) (Mahanta et al. 2017; Mahanta and Paul 2015), human serum albumin (HSA) (Li et al. 2011) are under investigation for their use as drug and gene delivery systems in cancerous cells. The albumin nanoparticle system utilizes HSA as a drug carrier for hydrophobic molecules. Albumin is physiologically known to carry hydrophobic bio-macromolecules through reversible noncovalent interactions (Hawkins et al. 2008). Albumin also binds to glycoprotein (gp60) receptor and mediates the transport of albumin-interacting molecules (Desai et al. 2006). These advantages of the albumin nanosystem facilitated the "nab" technology, a nanoparticle system comprising of albumin and Paclitaxel. The first nanosystem based product was the development of 130-nm nab-paclitaxel (Abraxane®), approved by the FDA for the treatment of breast cancer.

8.4 Future Directions

The unique properties of nanosystems make them well prepared for oncology applications. Although the use of nanosystems in medicine is relatively new, its translation into the development of formulation is promising. Nanosystem based chemotherapy is poised to influence the therapeutic outcomes in cancers in a more positive way compared to traditional chemotherapy. However, there are still areas for improvement, viz. limited clinical data and a less number of nano therapeutics receiving approval for clinical use. Still, increased clinical data are required to appreciate the nano system based therapy. The field of nanomedicine has caught up rapid pace, and the new and improved nanosystems are being developed for the treatment of cancer.

References

Ahmadi Nasab N, Hassani Kumleh H, Beygzadeh M, Teimourian S, Kazemzad M (2016) Delivery of curcumin by a pH-responsive chitosan mesoporous silica nanoparticles for cancer treatment. Artif Cells Nanomed Biotechnol 46(1):75–81

Albanese A, Tang PS, Chan WC (2012) The effect of nanoparticle size, shape, and surface chemistry on biological systems. Annu Rev Biomed Eng 14:1–16
Alonso MJ (2004) Nanomedicines for overcoming biological barriers. Biomed Pharmacother 58 (3):168–172
Anabousi S, Bakowsky U, Schneider M, Huwer H, Lehr C-M, Ehrhardt C (2006) In vitro assessment of transferrin-conjugated liposomes as drug delivery systems for inhalation therapy of lung cancer. Eur J Pharm Sci 29(5):367–374
Ayer M, Klok H-A (2017) Cell-mediated delivery of synthetic nano-and microparticles. J Control Release 259:92–104
Bagalkot V, Zhang L, Levy-Nissenbaum E, Jon S, Kantoff PW, Langer R et al (2007) Quantum dot-aptamer conjugates for synchronous cancer imaging, therapy, and sensing of drug delivery based on bi-fluorescence resonance energy transfer. Nano Lett 7(10):3065–3070
Baker JR Jr (2009) Dendrimer-based nanoparticles for cancer therapy. ASH Educ Program Book 2009(1):708–719
Bloise N, Massironi A, Della Pina C, Alongi J, Siciliani S, Manfredi A et al (2020) Extra-small gold nanospheres decorated with a thiol functionalized biodegradable and biocompatible linear polyamidoamine as nanovectors of anticancer molecules. Front Bioeng Biotechnol 8:132
Bordelon H, Biris AS, Sabliov CM, Todd MW (2010) Characterization of plasmid DNA location within chitosan/PLGA/pDNA nanoparticle complexes designed for gene delivery. J Nanomater 2011:952060
Bray F, Jemal A, Grey N, Ferlay J, Forman D (2012) Global cancer transitions according to the human development index (2008–2030): a population-based study. Lancet Oncol 13 (8):790–801
Byrne JD, Betancourt T, Brannon-Peppas L (2008) Active targeting schemes for nanoparticle systems in cancer therapeutics. Adv Drug Deliv Rev 60(15):1615–1626
Connor EE, Mwamuka J, Gole A, Murphy CJ, Wyatt MD (2005) Gold nanoparticles are taken up by human cells but do not cause acute cytotoxicity. Small 1(3):325–327
Corem-Salkmon E, Ram Z, Daniels D, Perlstein B, Last D, Salomon S et al (2011) Convection-enhanced delivery of methotrexate-loaded maghemite nanoparticles. Int J Nanomed 6:1595
Danhier F, Feron O, Preat V (2010) To exploit the tumor microenvironment: passive and active tumor targeting of nanocarriers for anti-cancer drug delivery. J Control Release 148(2):135–146
Daniels TR, Bernabeu E, Rodriguez JA, Patel S, Kozman M, Chiappetta DA et al (2012) The transferrin receptor and the targeted delivery of therapeutic agents against cancer. Biochim Biophys Acta 1820(3):291–317
Desai N, Trieu V, Yao Z, Louie L, Ci S, Yang A et al (2006) Increased antitumor activity, intratumor paclitaxel concentrations, and endothelial cell transport of cremophor-free, albumin-bound paclitaxel, ABI-007, compared with cremophor-based paclitaxel. Clin Cancer Res 12(4):1317–1324
Dhar S, Gu FX, Langer R, Farokhzad OC, Lippard SJ (2008) Targeted delivery of cisplatin to prostate cancer cells by aptamer functionalized Pt (IV) prodrug-PLGA PEG nanoparticles. Proc Natl Acad Sci 105(45):17356–17361
Di Wu MS, Xue H-Y, Wong H-L (2017) Nanomedicine applications in the treatment of breast cancer: current state of the art. Int J Nanomed 12:5879
Dobson J (2006) Gene therapy progress and prospects: magnetic nanoparticle-based gene delivery. Gene Ther 13(4):283–287
Dowaidar M, Nasser Abdelhamid H, Hallbrink M, Langel U, Zou X (2018) Chitosan enhances gene delivery of oligonucleotide complexes with magnetic nanoparticles-cell-penetrating peptide. J Biomater Appl 33(3):392–401
Elnakat H, Ratnam M (2004) Distribution, functionality and gene regulation of folate receptor isoforms: implications in targeted therapy. Adv Drug Deliv Rev 56(8):1067–1084
Elshama SS, Abdallah ME, Abdel-Karim RI (2018) Zinc oxide nanoparticles: therapeutic benefits and toxicological hazards. Open Nanomed J 5(1):16–22

El-Sherbiny IM, Elbaz NM, Sedki M, Elgammal A, Yacoub MH (2017) Magnetic nanoparticles-based drug and gene delivery systems for the treatment of pulmonary diseases. Nanomedicine 12(4):387–402

Erbacher P, Zou S, Bettinger T, Steffan AM, Remy JS (1998) Chitosan-based vector/DNA complexes for gene delivery: biophysical characteristics and transfection ability. Pharm Res 15(9):1332–1339

Farokhzad OC, Jon S, Khademhosseini A, Tran T-NT, LaVan DA, Langer R (2004) Nanoparticle-aptamer bioconjugates: a new approach for targeting prostate cancer cells. Cancer Res 64 (21):7668–7672

Ganta S, Devalapally H, Amiji M (2010) Curcumin enhances oral bioavailability and anti-tumor therapeutic efficacy of paclitaxel upon administration in nanoemulsion formulation. J Pharm Sci 99(11):4630–4641

Giljohann DA, Seferos DS, Patel PC, Millstone JE, Rosi NL, Mirkin CA (2007) Oligonucleotide loading determines cellular uptake of DNA-modified gold nanoparticles. Nano Lett 7 (12):3818–3821

Gillies ER, Frechet JMJ (2005) Dendrimers and dendritic polymers in drug delivery. Drug Discov Today 10(1):35–43

Goncalves AS, Macedo AS, Souto EB (2012) Therapeutic nanosystems for oncology nanomedicine. Clin Transl Oncol 14(12):883–890

Guo J, Gao X, Su L, Xia H, Gu G, Pang Z et al (2011) Aptamer-functionalized PEG-PLGA nanoparticles for enhanced anti-glioma drug delivery. Biomaterials 32(31):8010–8020

Harush-Frenkel O, Debotton N, Benita S, Altschuler Y (2007) Targeting of nanoparticles to the clathrin-mediated endocytic pathway. Biochem Biophys Res Commun 353(1):26–32

Hawkins MJ, Soon-Shiong P, Desai N (2008) Protein nanoparticles as drug carriers in clinical medicine. Adv Drug Deliv Rev 60(8):876–885

Hejazi R, Amiji M (2003) Chitosan-based gastrointestinal delivery systems. J Control Release 89 (2):151–165

Hilgenbrink AR, Low PS (2005) Folate receptor-mediated drug targeting: from therapeutics to diagnostics. J Pharm Sci 94(10):2135–2146

Hrkach J, Von Hoff D, Ali MM, Andrianova E, Auer J, Campbell T et al (2012) Preclinical development and clinical translation of a PSMA-targeted docetaxel nanoparticle with a differentiated pharmacological profile. Sci Transl Med 4(128):128ra39

Illum L, Jabbal-Gill I, Hinchcliffe M, Fisher AN, Davis SS (2001) Chitosan as a novel nasal delivery system for vaccines. Adv Drug Deliv Rev 51(1–3):81–96

Jain RA (2000) The manufacturing techniques of various drug loaded biodegradable poly(lactide-co-glycolide) (PLGA) devices. Biomaterials 21(23):2475–2490

Kim YH, Gihm SH, Park CR, Lee KY, Kim TW, Kwon IC et al (2001) Structural characteristics of size-controlled self-aggregates of deoxycholic acid-modified chitosan and their application as a DNA delivery carrier. Bioconjug Chem 12(6):932–938

Kobayashi T, Ishida T, Okada Y, Ise S, Harashima H, Kiwada H (2007) Effect of transferrin receptor-targeted liposomal doxorubicin in P-glycoprotein-mediated drug resistant tumor cells. Int J Pharm 329(1–2):94–102

Kolhe P, Khandare J, Pillai O, Kannan S, Lieh-Lai M, Kannan RM (2006) Preparation, cellular transport, and activity of polyamidoamine-based dendritic nanodevices with a high drug payload. Biomaterials 27(4):660–669

Koping-Hoggard M, Tubulekas I, Guan H, Edwards K, Nilsson M, Varum KM et al (2001) Chitosan as a nonviral gene delivery system. Structure-property relationships and characteristics compared with polyethylenimine in vitro and after lung administration in vivo. Gene Ther 8 (14):1108–1121

Laurent S, Dutz S, Hafeli UO, Mahmoudi M (2011) Magnetic fluid hyperthermia: focus on superparamagnetic iron oxide nanoparticles. Adv Colloid Interf Sci 166(1–2):8–23

Lee H, Sung D, Veerapandian M, Yun K, Seo SW (2011) PEGylated polyethyleneimine grafted silica nanoparticles: enhanced cellular uptake and efficient siRNA delivery. Anal Bioanal Chem 400(2):535–545
Li D, Zhang YT, Yu M, Guo J, Chaudhary D, Wang CC (2013) Cancer therapy and fluorescence imaging using the active release of doxorubicin from MSPs/Ni-LDH folate targeting nanoparticles. Biomaterials 34(32):7913–7922
Li F, Liu WG, Yao KD (2002) Preparation of oxidized glucose-crosslinked N-alkylated chitosan membrane and in vitro studies of pH-sensitive drug delivery behaviour. Biomaterials 23 (2):343–347
Li L, Fang CJ, Ryan JC, Niemi EC, Lebron JA, Bjorkman PJ et al (2010) Binding and uptake of H-ferritin are mediated by human transferrin receptor-1. Proc Natl Acad Sci 107(8):3505–3510
Li Q, Liu C, Zhao X, Zu Y, Wang Y, Zhang B et al (2011) Preparation, characterization and targeting of micronized 10-hydroxycamptothecin-loaded folate-conjugated human serum albumin nanoparticles to cancer cells. Int J Nanomed 6:397
Liu Y, Franzen S (2008) Factors determining the efficacy of nuclear delivery of antisense oligonucleotides by gold nanoparticles. Bioconjug Chem 19(5):1009–1016
Lu J, Liong M, Li Z, Zink JI, Tamanoi F (2010) Biocompatibility, biodistribution, and drug-delivery efficiency of mesoporous silica nanoparticles for Cancer therapy in animals. Small 6 (16):1794–1805
Mahanta S, Paul S (2015) Stable self-assembly of bovine α-lactalbumin exhibits target-specific antiproliferative activity in multiple cancer cells. ACS Appl Mater Interfaces 7 (51):28177–28187
Mahanta S, Paul S, Srivastava A, Pastor A, Kundu B, Chaudhuri TK (2015) Stable self-assembled nanostructured hen egg white lysozyme exhibits strong anti-proliferative activity against breast cancer cells. Colloids Surf B: Biointerfaces 130:237–245
Mahanta S, Prathap S, Ban DK, Paul S (2017) Protein functionalization of ZnO nanostructure exhibits selective and enhanced toxicity to breast cancer cells through oxidative stress-based cell death mechanism. J Photochem Photobiol B Biol 173:376–388
Malvindi MA, Brunetti V, Vecchio G, Galeone A, Cingolani R, Pompa PP (2012) SiO2 nanoparticles biocompatibility and their potential for gene delivery and silencing. Nanoscale 4 (2):486–495
Martinez-Carmona M, Gun'ko Y, Vallet-Regi M (2018) ZnO nanostructures for drug delivery and theranostic applications. Nanomaterials 8(4):268
Matsumura Y, Maeda H (1986) A new concept for macromolecular therapeutics in cancer chemotherapy: mechanism of tumoritropic accumulation of proteins and the antitumor agent smancs. Cancer Res 46(12 Pt 1):6387–6392
McBain SC, Yiu HH, Dobson J (2008) Magnetic nanoparticles for gene and drug delivery. Int J Nanomed 3(2):169–180
Moghimi SM, Hunter AC, Murray JC (2001) Long-circulating and target-specific nanoparticles: theory to practice. Pharmacol Rev 53(2):283–318
Mukherjee P, Bhattacharya R, Wang P, Wang L, Basu S, Nagy JA et al (2005) Antiangiogenic properties of gold nanoparticles. Clin Cancer Res 11(9):3530–3534
Ni X, Castanares M, Mukherjee A, Lupold SE (2011) Nucleic acid aptamers: clinical applications and promising new horizons. Curr Med Chem 18(27):4206–4214
Paciotti GF, Myer L, Weinreich D, Goia D, Pavel N, McLaughlin RE et al (2004) Colloidal gold: a novel nanoparticle vector for tumor directed drug delivery. Drug Deliv 11(3):169–183
Pan XQ, Zheng X, Shi G, Wang H, Ratnam M, Lee RJ (2002) Strategy for the treatment of acute myelogenous leukemia based on folate receptor beta-targeted liposomal doxorubicin combined with receptor induction using all-trans retinoic acid. Blood J Am Soc Hematol 100(2):594–602
Papasani MR, Wang G, Hill RA (2012) Gold nanoparticles: the importance of physiological principles to devise strategies for targeted drug delivery. Nanomedicine 8(6):804–814
Parakh S, Parslow AC, Gan HK, Scott AM (2015) Antibody-mediated delivery of therapeutics for cancer therapy. Expert Opin Drug Deliv 13(3):401–419

Park IK, Jiang HL, Cook SE, Cho MH, Kim SI, Jeong HJ et al (2004) Galactosylated chitosan (GC)-graft-poly(vinyl pyrrolidone) (PVP) as hepatocyte-targeting DNA carrier: in vitro transfection. Arch Pharm Res 27(12):1284–1289

Patri AK, Myc A, Beals J, Thomas TP, Bander NH, Baker JR (2004) Synthesis and in vitro testing of J591 antibody-dendrimer conjugates for targeted prostate cancer therapy. Bioconjug Chem 15(6):1174–1181

Prabha S, Labhasetwar V (2004) Nanoparticle-mediated wild-type p53 gene delivery results in sustained antiproliferative activity in breast cancer cells. Mol Pharm 1(3):211–219

Prabha S, Zhou WZ, Panyam J, Labhasetwar V (2002) Size-dependency of nanoparticle-mediated gene transfection: studies with fractionated nanoparticles. Int J Pharm 244(1–2):105–115

Romoren K, Thu BJ, Evensen O (2002) Immersion delivery of plasmid DNA. II. A study of the potentials of a chitosan based delivery system in rainbow trout (Oncorhynchus mykiss) fry. J Control Release 85(1–3):215–225

Rosenholm JM, Meinander A, Peuhu E, Niemi R, Eriksson JE, Sahlgren C et al (2009) Targeting of porous hybrid silica nanoparticles to Cancer cells. ACS Nano 3(1):197–206

Saeed RM, Dmour I, Taha MO (2020) Stable chitosan-based nanoparticles using Polyphosphoric acid or Hexametaphosphate for tandem ionotropic/covalent crosslinking and subsequent investigation as novel vehicles for drug delivery. Front Bioeng Biotechnol 8:4

Shan Y, Luo T, Peng C, Sheng R, Cao A, Cao X et al (2012) Gene delivery using dendrimer-entrapped gold nanoparticles as nonviral vectors. Biomaterials 33(10):3025–3035

Sharma H, Kumar K, Choudhary C, Mishra PK, Vaidya B (2016) Development and characterization of metal oxide nanoparticles for the delivery of anticancer drug. Artif Cells Nanomed Biotechnol 44(2):672–679

Shen JM, Gao FY, Yin T, Zhang HX, Ma M, Yang YJ et al (2013) cRGD-functionalized polymeric magnetic nanoparticles as a dual-drug delivery system for safe targeted cancer therapy. Pharmacol Res 70(1):102–115

Shukla R, Bansal V, Chaudhary M, Basu A, Bhonde RR, Sastry M (2005) Biocompatibility of gold nanoparticles and their endocytotic fate inside the cellular compartment: a microscopic overview. Langmuir 21(23):10644–10654

Sun S, Zeng H (2002) Size-controlled synthesis of magnetite nanoparticles. J Am Chem Soc 124 (28):8204–8205

Szymanski P, Markowicz M, Mikiciuk-Olasik E (2011) Nanotechnology in pharmaceutical and biomedical applications: dendrimers. Nano 6(06):509–539

Talekar M, Kendall J, Denny W, Garg S (2011) Targeting of nanoparticles in cancer: drug delivery and diagnostics. Anti-Cancer Drugs 22(10):949–962

Thanou M, Florea BI, Geldof M, Junginger HE, Borchard G (2002) Quaternized chitosan oligomers as novel gene delivery vectors in epithelial cell lines. Biomaterials 23(1):153–159

Thomsen LB, Linemann T, Pondman KM, Lichota J, Kim KS, Pieters RJ et al (2013) Uptake and transport of superparamagnetic iron oxide nanoparticles through human brain capillary endothelial cells. ACS Chem Neurosci 4(10):1352–1360

Tredan O, Galmarini CM, Patel K, Tannock IF (2007) Drug resistance and the solid tumor microenvironment. J Natl Cancer Inst 99(19):1441–1454

Truffi M, Colombo M, Sorrentino L, Pandolfi L, Mazzucchelli S, Pappalardo F et al (2018) Multivalent exposure of trastuzumab on iron oxide nanoparticles improves antitumor potential and reduces resistance in HER2-positive breast cancer cells. Sci Rep 8(1):6563

Visaria RK, Griffin RJ, Williams BW, Ebbini ES, Paciotti GF, Song CW et al (2006) Enhancement of tumor thermal therapy using gold nanoparticle-assisted tumor necrosis factor-alpha delivery. Mol Cancer Ther 5(4):1014–1020

Wang AZ, Gu F, Zhang L, Chan JM, Radovic-Moreno A, Shaikh MR et al (2008) Biofunctionalized targeted nanoparticles for therapeutic applications. Expert Opin Biol Ther 8 (8):1063–1070

Wilhelm S, Tavares AJ, Dai Q, Ohta S, Audet J, Dvorak HF et al (2016) Analysis of nanoparticle delivery to tumours. Nat Rev Mater 1(5):16014

Wu K, Su D, Liu J, Saha R, Wang JP (2019) Magnetic nanoparticles in nanomedicine: a review of recent advances. Nanotechnology 30(50):502003

Xia W, Low PS (2010) Folate-targeted therapies for cancer. J Med Chem 53(19):6811–6824

Yu BO, Tai HC, Xue W, Lee LJ, Lee RJ (2010) Receptor-targeted nanocarriers for therapeutic delivery to cancer. Mol Membr Biol 27(7):286–298

Zhang W, Tan TMC, Lim L-Y (2007) Impact of curcumin-induced changes in P-glycoprotein and CYP3A expression on the pharmacokinetics of peroral celiprolol and midazolam in rats. Drug Metab Dispos 35(1):110–115

Phytoplankton Mediated Nanoparticles for Cancer Therapy

9

Priyadarshini Padhi, Manasa Kumar Panda, Yengkhom Disco Singh, Kaushik Kumar Bharadwaj, Kausika Kumar Malik, and Binata Nayak

Abstract

Cancer is one of the most dreaded diseases which causes major stumbling in life expectancy around the globe. Although the primary treatment at early stages can reduce the mortality rate of cancer, the survival rate is still low and this causes a nuisance in the scientific research to explore new tools and technologies. Phytochemical based compounds mainly obtained from phytoplankton and their formulation with nanoparticles paves a new way of drug discovery to target the delivery at specific site. This can modulate the molecular pathways and cellular mechanisms involved in apoptosis, cell proliferation, migration, and invasion. However, the phytochemical alone could not be so positive to treat against cancer due to their low water solubility, low stability, poor absorption, and less site specificity. The development of nanotechnology and formulation with other drugs including phytoplankton mediated drugs, can maximize the

P. Padhi
Department of Biotechnology and Bioinformatics, Sambalpur University, Sambalpur, Odisha, India

M. K. Panda
Environment & Sustainability Department, CSIR- IMMT, Bhubaneswar, India

Y. D. Singh
Biomaterials and Bioprocessing Research Laboratory, Department of Post Harvest Technology, College of Horticulture and Forestry, Central Agricultural University, Pasighat, Arunachal Pradesh, India

K. K. Bharadwaj
Department of Bioengineering and Technology, Gauhati University Institute of Science and Technology (GUIST), Gauhati University, Guwahati, Assam, India

K. K. Malik
Centre for DNA Fingerprinting and Diagnostics, Hyderabad, Telangana, India

B. Nayak (✉)
School of Life Sciences, Sambalpur University, Sambalpur, Odisha, India

M. Arakha et al. (eds.), *Bio-Nano Interface*,
https://doi.org/10.1007/978-981-16-2516-9_9

potential of anticancer activities. Nanocarriers can enhance the stumbling block due to phytochemicals and other drugs to deliver at the site-specific.

Keywords

Cancer therapy · Phytoplankton · Bioactive molecules · Nanoformulation

9.1 Introduction

Phytoplankton is autotrophic, prokaryotic, or eukaryotic microscopic organisms under the plankton class. These organisms live in marine and fresh water ecosystems, capable of performing photosynthesis by using sunlight and water. Moreover, phytoplankton, also called microalgae, create organic carbon dioxide compounds which are dissolved into water during a cycle called carbon fastening that sustains the aquatic food web and allows aquatic life style. It is one of the world's most cosmopolitan, mostly undiscovered and diverse forms of life on the planet (Appeltans et al. 2012). Some common types of phytoplankton are diatoms, cyanobacteria, dinoflagellates, green algae, and coccolithophores.

Cyanobacteria, a comparatively more researched microalgae, originated approximately 3.5 billion years ago, having a population of 150 genera and nearly 2000 species with substantial diversity, causing a massive change in the environment by releasing oxygen from water to the atmosphere through a photochemical reaction. These prokaryotic autotrophs can exist as single-celled species or in colonies being primary first-level species in food chains in aquatic ecosystems. In fact, these prokaryotes play a major role in the marine nitrogen cycle and balance nitrogen (N) and carbon dioxide (CO_2) cycles in the biosphere (Awramik 1992; Blank and Sanchez-Baracaldo 2010; Zehr et al. 2001). Apart from their life-sustaining role, they are also a major concern for terrestrial ecosystem for the eutrophication of certain cyanobacterial species which release harmful toxins into the water bodies where they bloom (Gupta et al. 2003; Falconer and Humpage 2005; Vareli et al. 2009).

Acute liver damage, neurotoxicity, gastrointestinal disorders, and hepatic cancer are reported in humans due to cyanotoxins (Gaudin et al. 2008; Van Apeldoorn et al. 2007). Cyanotoxins identify some potential molecules having anticancer properties and can be used in cancer therapy. Among the molecules, cyanobacterial cyclopeptides is one of them, which possess significant pharmacophoric composition, which could lead to the development of a new class of anticancer therapy specifically targeting organic anion transporters (OATs) in metastatic chemotherapy-refractory cancers (Monks et al. 2007; Herfindal et al. 2009).

As a source of novel drug molecules, marine species have the immense capacity, so recently gained commercial interest. Microalgae are, in reality, a possible source of valuable pharmaceutical goods containing a variety of polyunsaturated fatty acids, fats, lipids, proteins, and other bioactive compounds. Also, microalgae molecules isolated from whole extracts and fractional extracts have been reported to have noteworthy biological activity including anticancer, anti-inflammatory, anti-oxidant, microbicidal, anti-leprosy, and anti-leishmanial function (Torres et al. 2014). These have been demonstrated in many published papers and several issued patents for chemicals obtained from marine algae. As per new advance studies in cancer science, cancer malignancy, and drug resistance remain the leading cause of death worldwide. According to the 2011–2015 National Cancer Institute studies (NCI 2020), roughly 172 of every 100,000 persons are killed each year. Cancer is one of the most common and dangerous diseases, mainly because of its treatment difficulties and ultimately leading to death. The principal source of this problem is that cancer is the result of uncontrolled replication of healthy human cells that are significantly disturbed.

In 2012, there were about 975,396 new cancer cases and 358,392 deaths from cancer among young adults around the world, equal to an aging uniform rate of 43.3 new cancer cases and 15.9 deaths from cancer per in 100,000 individuals annually (Fidler et al. 2017). Modalities of cancer care include emergency, radiation, and chemotherapy (Huang et al. 2017). The third medication type includes active therapeutic molecules with antitumor properties. Chemotherapy treatment mainly targets alkylating agents and antimetabolites that inhibit DNA synthesis (Huang et al. 2017). At present, a wide variety of anticancer drugs is required in the pharmaceutical industry. There is a substantial lack of compatibility in the effectiveness and availability of the ten highest-earning cancer drugs between the USA and Norway (Prasad et al. 2017).

Several microalgae-based substances have been extensively tested with medicinal and cancer therapeutic qualities (Campos et al. 2012; Farooqi et al. 2012). Natural materials derived from microalgae provide a large foundation for diagnosis, and experimental research has been carried out on algae related anticancer materials (Talero et al. 2015). While marine samples are difficult to obtain, a large number of marine samples have been approved as useful pharmaceutical health products (Pangestuti and Kim 2011). Numerous pharmaceutical and scientific collaborations have made significant efforts in recent years to extract and identify novel bioactive molecules from marine flora, in particular, microalgae (Pyne et al. 2017). Despite this initiative, the marine flora has remained undiscovered to some degree, and these present data are reviewed as baseline evidence to encourage further study in this area. Several natural compound-based cancer drugs have been found, but have low water solubility, making it difficult or even impossible to formulate them (Sun et al. 2014). For example, camptothecin is generally accepted as an effective in vitro anticancer agent, but its clinical use is limited because of its poor solubility (Venditto and Simanek 2010). To overcome these challenges, nanotechnology will provide an innovative approach for addressing the low solubility of hydrophobic natural marine products in aqueous medium. Nanotechnology is a branch of division to investigate the application of tiny particles like 1–100 nm (Panda et al. 2020a, b, c).

Nanomedicine has made colossal progress in formulating nanocarriers to provide tumor/cancer areas with pharmaceuticals and imaging agents. Nanomedicines deliver special benefits in terms of improved agent safety, biological activities inside a serum-rich environment, an extended period of circulation in the bloodstream, decreased allergic reactions, enhanced efficacy in permeability, retention and tumor targeting, improved release profiles, and possible application of therapeutic stimuli on demand among others (Sun et al. 2014; Zhang et al. 2017; Xia 2008; Farokhzad and Langer 2009). Hydrophobic nanoparticles encapsulation (<200 nm) allows the intravenous injection of natural algae-based drugs, as the drugs are fully soluble in water. The drug may be extracted from nanoparticles after administration of drug-loaded nanoparticles and may play a role as a disease inhibitor (Zhang et al. 2017). Nanoparticles may be made from specific inorganic (e.g., iron, silica, and carbon oxides) and organic (e.g., polymers and lipids) materials which are of differing origin. In addition, its sizes can be varied in conjunction with a variety of dimensions, ranging from a few nanometers to not more than 1 mm, its structures can be formed to smooth or rough, its plasticity can be adjusted to rigid or flexible and its surfaces can work with specific properties and interesting fields.

Their willingness to lead nanoformulations utilizing natural compounds in order to resolve sequential biological obstacles (e.g., Mononuclear Phagocytic System Sequestration, MDRs) which are important for restricting the efficacy of nanoparticles (Blanco et al. 2015; Xu et al. 2016). Furthermore, it has been shown that first nanoparticles become passively aggregate into tumors by increasing the effects of permeability and retention (EPR) ((Prabhakar et al. 2013; Millard et al. 2017; Marcazzan et al. 2018). The key benefit of nanoparticles with suitable anticancer drugs are (a) enhanced solubility for faster treatment, (b) enhanced half-life distribution throughout; (c) increased retention of treatment in target-based cancer tissues; (d) continuous and safe medication releases, and (e) decreased resistance of drugs to pump-mediated efflux.

9.2 Different Phytoplankton Mediated Nanoparticles

The potential of phytoplankton in nanomedicine development has driven a remarkable attraction for pharmaceutical research in recent years. This section presents an updated summary of current advances and opportunities in the green synthesis of nanomedicine for cancer therapy using phytoplankton. Marine organisms, especially phytoplankton are rich source of ecofriendly, non-toxic bioactive compounds with different medicinal values and production is cost-effective (Singh et al. 2015).

Different functional groups (hydroxyl, carboxyl, and amino) present in the phytoplankton act as a metal reducing agents as well as capping agents and play a major role in the green synthesis of metal nanoparticles in a single step.

For more research on the green approach of nanomaterial and nanomedicine production, we need to know about the availability, classification, and diversity of phytoplankton in the aquatic ecosystem. Approximately 70% of the earth's surface consists of marine ecosystems (Pomponi 1999) in which phytoplankton are considered as grassland of the ocean. Phytoplankton, found in both fresh water and marine ecosystems, mainly comprises of diatoms, dinoflagellates, cocolithoides, cryptophyceae, cyanophytes, and green algae. Diversity of these groups varies according to environmental conditions, though diatoms dominate in most of the cases. Figure 9.1 exhibits the diversity of phytoplankton in the marine ecosystem.

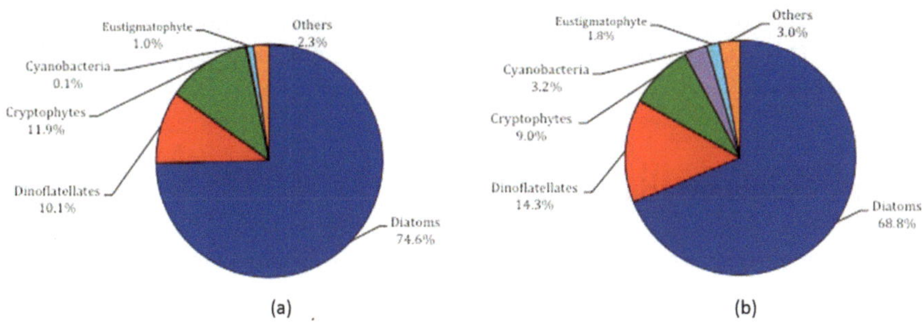

Fig. 9.1 Percentage of each type of phytoplankton in a water sample, (**a**) identified through traditional microscopy, and (**b**) analyzed with HPLC technology (Source: http://oceandatacenter.ucsc.edu/home/outreach/PhytoID_fullset.pdf)

Nature has been a relevant resource for anticancer bioactive molecules to date. According to a report by Newman and Cragg 2020, over the time frame from 01/1981 to 09/2019, about 41%, i.e., of 75 out of 185 small molecules with anticancer properties are derived directly or indirectly from naturally inspired sources. To date, many anticancer molecules have been discovered. But, due to limitations like poor water solubility and strong interaction with serum proteins leading to rapid clearance of the anticancer molecules from the target site, they are not eligible for entering clinical trials and drug formulation (Sun et al. 2014). Phytoplankton like diatoms, cyanobacteria, and coccolithophores have been used for green synthesis of nanomaterials, which are non-toxic in drug delivery than chemically synthesized nanomaterials.

9.2.1 Diatoms

It has been demonstrated that the polyethylene glycolyzed diatomite nanoparticles (DNPs) in conjugation with cell penetrating peptides (CPP) 3-aminopropyltriethoxysilane (APT), i.e., DNPs-APT-PEG-CPP have an effective cytotoxic effect on MCF-7 and MDA-MB-231 breast cancer (Terracciano et al. 2015). Recently, it has been reported that modified diatomite particles (irregular shape, approximately 350 nm in size) with a peptide-siRNA complex are efficiently uptaken by human lung epidermoid carcinoma cell line (H1355) (Schaer et al. 2007). So, this can be used as an efficient drug carrier.

9.2.2 Coccolithophores

Coccolithophores play a major role in the biomineralization of calcium carbonate and calcite nanoparticle formation. They can make complex scales of calcites ($CaCO3$) called coccoliths. The high pore density of coccoliths from *Pontosphaera japonica, Michaelsarsia elegans,* or *Calyptrolithophora papilifera* can be used for efficient drug encapsulation by suitably controlling the pore chemistry (Skeffington and Scheffel 2018). Coccoliths show high stability at very low concentration in solution. This phenomenon can be exploited for effective drug carrier (Hassenkam et al. 2011).

9.2.3 Cyanobacteria

In one report, C-Phycocyanin (C-PC), a major photosynthetic pigment of cyanobacteria, has been used to formulate nanodrug by combining C-Phycocyanin (C-PC), Carboxymethyl chitosan (CMC), and CD59-specific ligand peptide (CD59sp). This C-PC/CMC-CD59sp nanoparticle exhibits an apoptotic effect on HeLa cells (Yang et al. 2017). Organic nanomaterials like sulfated polysaccharides derived from spirulina have an anti-proliferative effect on tumor cells in vitro (Jia et al. 2008) and in vivo (Akao et al. 2009).

Another phytoplankton named *Dunaliella salina* can produce around 80 nm sized nanoparticle upon 45 min of microwave irradiation with the phytoplankton extract. This NP carries β-carotene, which is a potential anticancer agent (Lomora et al. 2019).

Though phytoplankton are rich sources of marine biomolecules, their potential in nanoformulation of anticancer drugs is largely unexplored with scarce data in our hands. It shows the need of research in-depth for more discoveries of phytoplankton mediated nanoformulation of anticancer drugs.

9.3 Strategies for Development of Phytoplankton Mediated Nanodrug Formulation for Cancer Therapy

Though phytoplankton are biologically diverse and can perform as biofactories for the production of nanoparticles and play various important roles in the formulation of nanodrugs for human life improvement, it still remains unexplored. The phytoplankton's mediated biosynthesis of nanoparticles is an emerging field for research in cancer therapeutics. The strategy for the efficient development of nanodrugs has consistently been a pillar of drug development research in cancer to improve the therapeutic performance (Bajpai et al. 2018). At present, a scanty of literature has reported the use of phytoplankton for the biosynthesis of nanoparticles in cancer therapy. Here, phytoplankton like Blue-green algae or Cyanobacteria (Cyanophyceae), Green algae (Chlorophyceae), and Diatoms, mediated different nanodrugs formulations strategies for cancer drug development are discussed. Figure 9.2 illustrates the sequential steps of phytoplanktonic nanoformulation.

9.3.1 Green Synthesis of Metallic Nanoparticles

Cyanobacteria have been harnessed for the biosynthesis of metallic nanoparticles by using metals like gold (Au), silver (Ag), gold-silver nanoalloy, and various other

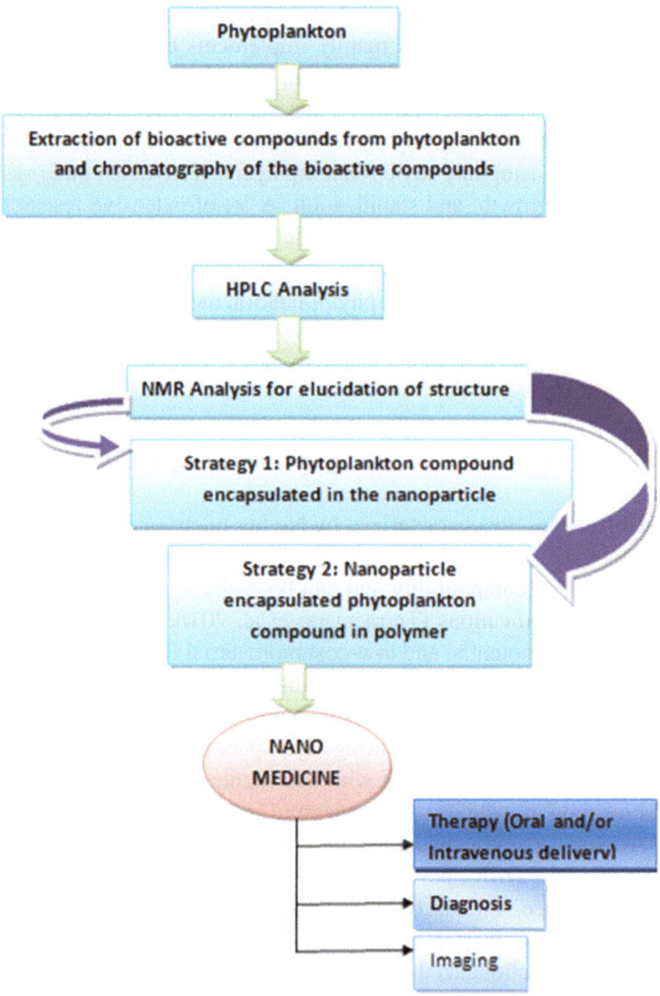

Fig. 9.2 Concept of phytoplanktonic nanoformulation: from extraction to drug delivery (Bajpai et al. 2018)

metals (Roychoudhury et al. 2016b; Sharma et al. 2016). Nanoparticles are classified into organic and inorganic. Organic nanoparticles are carbon-based, whereas metal nanoparticles of (gold (Au), silver (Ag), Copper (Cu), etc.), metal oxide nanoparticles (Zinc oxide (ZnO), titanium oxide (TiO2), SiO2, Fe2O3, etc.,), metal sulfide nanoparticles (CdS, PbS), magnetite nanoparticles, quantum dots (CdS, CdTe) are grouped as inorganic nanoparticles. In general, there are two strategies for the synthesis of metal nanoparticles, the top-down and bottom-up approaches. The top-down approach involves degrading of bulk materials into reduced nanoscale size objects, while the bottom-up approach involves assembling of atom by atom, molecule by molecule to build conformations of nanoscale. Green synthesis of metallic nanoparticles mainly implements bottom-up approaches that are ecofriendly, less toxic, cost-effective, simple, and energy-efficient (Mittal et al. 2013; Kharissova et al. 2013). The metal nanoparticles are formulated by reducing the metal-ion precursors using phytoplankton mainly in aqueous solution at room temperature without using any toxic reducing agent. It involves three steps mechanism of nucleation, growth, and stabilization. A lot of extensive research has been done in green synthesis of metallic nanoparticles, which shows antimicrobial activity, but only a few are reported showing activity against cancer cell lines. Here we have mentioned some of the reported phytoplankton mediated metallic nanoparticles against different cancer cell lines in Table 9.1. In Fig. 9.3, the simple one-step method for green synthesis of metal nanoparticles is shown.

9.3.2 Diatom Nanocarriers for Systemic Drug Delivery

Diatom silica frustules are characterized by having micro or nanoscale range porosity, high surface area, thermal stability, mechanical resistance, optical properties, non-toxicity, great biocompatibility and can be easily altered by genetic manipulation or chemical modifications (Terracciano et al. 2018; Kuppusamy et al. 2017). Thus, it seems to be a potential and low-cost biomaterial for efficient anticancer drug delivery and detection of cancer cells. The frustules of *Chaetoceros* sp. from marine diatoms with intrinsic fluorescence capacity were synthetically altered by iron oxide nanoparticles, which gives rise to a strong super-paramagnetic particle-like property. Then the surface was functionalized with Trastuzumab antibody to conjugate with HER2 receptor protein, which is overexpressed in breast cancer cells. The antibody functionalized super-paramagnetic nanoparticles of *Chaetoceros* sp. having fluorescence can separate the HER2 positive breast cancer cells (SKBR3 cell) from normal cells by fluorescence microscope (Esfandyari et al. 2020). The genetically engineered diatom *Thalassiosira pseudomonas* loading with chemotherapeutic drugs are surface-functionalized with the IgG-binding portion of protein G can selectively target and killed neuroblastoma and B-lymphoma tumor cells (Delalat et al. 2015). A novel mesoporous biodegradable silicon nanoparticle synthesized from natural diatom was found to be a potential drug carrier with a high carrying capacity of anticancer drug (doxorubicin). These nanoparticles show a sustained drug release and enhanced cytotoxicity selectively against cancer cells (Maher et al. 2016).

9.3.3 Green Carbon Nanotags for Anticancer Drug Delivery

From harmful cyanobacteria, high-quality fluorescent, photostable carbon materials called G-tags or T-tags (G-tags when conjugated with doxorubicin) are prepared. The G-tags, when conjugated with anticancer drug doxorubicin (T-tags), induce apoptosis in cancer cell lines of human hepatocellular carcinoma (HepG2) and

Table 9.1 Phytoplankton mediated green synthesis of metallic nanoparticles and their anticancer activity in various cancer cell lines

Phytoplankton name	Types	Nanoparticle type	Nanoparticle size	Morphology	Cell line	References
Trichodesmium erythraeum	Micro algae	AgNPs	26.5 nm	Crystalline, cubical	HeLa, MCF-7	Sathishkumar et al. (2019)
Lyngbya majuscula	Cyanobacteria	AgNPs	20–50 nm	Spherical	REH	Roychoudhury et al. (2016a)
Nostoc sp. Bahar M	Cyanobacteria	AgNPs	14.9 nm	Face-centered cubic crystalline	Caco-2	Hamida et al. (2020)
Desertifilum IPPAS B-1220	Cyanobacteria	AgNPs	4.5–26 nm	Spherical	MCF-7, HepG2	Hamida et al. (2020)
Nostoc linckia	Cyanobacteria	AgNPs	9.39–25.89 nm	Spherical	MCF-7	El-Naggar et al. (2017)
Oscillatoria limnetica	Cyanobacteria	AgNPs	3.30–17.97 nm	Quasi- spherical	HCT-116	Hamouda et al. (2019)
Dunaliella salina	Green microalgae	AgNPs	15.26 nm	Spherical	MCF-7	Singh et al. (2017)
Anabaena doliolum	Cyanobacteria	AgNPs	10–50 nm	Spherical	Colo205, DL	Singh et al. (2014)
Nostoc sp. HKAR-2	Cyanobacteria	AgNPs	51–100 nm	Crystalline	MCF-7	Sonker et al. (2017)

AgNPs silver nanoparticles, *MCF-7* human breast cancer, *HeLa* human cervical cancer, *REH* leukemic cell lines, *Caco-2* human colon cancer cell line, *HepG2* human liver cancer cell line, *HCT-116* human colon cancer, colo205 cell line (colon adenocarcinoma), *DL* Dalton's lymphoma

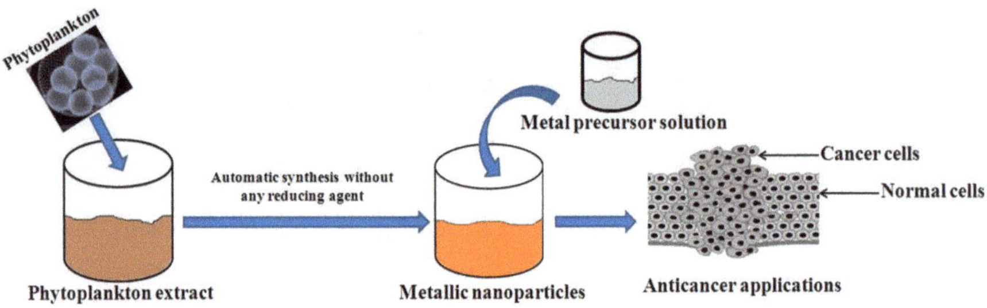

Fig. 9.3 Phytoplankton mediated single step green synthesis of metallic nanoparticles and its anticancer applications

human breast cancer (MCF-7) at a higher rate. Therefore it indicates that the loaded drug doxorubicin holds its pharmaceutical activity and releases from the T-tags after endocytic uptake in the intracellular environment. Investigation on imaging of cancer cells by G-tags is still going on (Lee et al. 2014).

9.4 Possible Future Strategies of Nanoformulation of Anticancer Drugs Isolated from Phytoplankton in Cancer Drug Development

Anticancer nanodrugs in clinical development are basically liposomes, polymeric conjugates, and polymeric nanoparticles. Liposomes having lipid bilayer membrane structure can encapsulate both hydrophilic and hydrophobic drugs. Thus it protects the encapsulating drugs during circulation in the body. Their surfaces can also be modified by functionalizing with specific ligands to target-specific cells and tissues (Torchilin 2005). Various potential anticancer compounds isolated from the cyanobacterial species are significantly active against different cancer cell lines like liver cancer (HepG-2), lung cancer (A549), colon cancer (HCT116), and breast cancer (MCF-7), however, anticancer drug of cyanobacterial origin is quite dynamic and significantly entering into clinical trials.

Anticancer drugs like Glembatumumab vedotin, Cryptophycin 1, Brentuximab vedotin 63, etc., isolated from different Cyanobacteria sp. are currently in a clinical trial or already in the market (Bajpai et al. 2018). These anticancer drugs can encapsulate into liposomes or with biodegradable polymers like polycaprolactone (PCL), polylactic acid (PLA), polylactic-co-glycolic acid (PLGA), etc., and further functionalized with specific cell surface antibody for target-specific drug delivery in specific cancer cells without damaging the healthy cells. Sulfated polysaccharides isolated from *Spirulina* showed inhibition of proliferation of human hepatoma cell line (BEL 7404) (Jia et al. 2008). Therefore, strategies can be designed to synthesize polysaccharide-based nanomaterials against various cancer cells synergist with other potential anticancer drug isolated from phytoplankton.

The nanodrug formulation strategy for cancer therapy still demands extensive research for developing an innovative approach to tackle the problems associated with the increased solubility of drug, increased time span in circulation, efficient drug delivery with enhanced drug accumulation in target cancer tissues or cells, sustained drug release, reduction in efflux pump-mediated drug resistance, alleviate toxicity to improve their clinical efficacy and optimize their therapy.

9.5 Scope of Commercialization for Nanodrug Formulation for Cancer Therapy

The scope of nanotechnology in the field of medical biology is increasing rapidly. Even though nanotechnology is 100 years old, the applications in biology are vast. Few biological applications of nanotechnology in cancer therapy include drug delivery, diagnosis, biosensor development and these are increasing day by day. Cancer is defined as the uncontrolled growth of the cell and is the leading cause of death in the world. Despite decades of knowledge shared by scientists, doctors, and pharmacists, we are still not able to eradicate cancer. This may be because of the diversity, environmental factors, and complex molecular mechanisms involved in cancer progression. Using modern technology, we have reached a stage where we can generate thousands of synthetic nano-molecules, with high purity, which can be used for cancer treatment. For practical use, these nanodrugs need proper characterization and targeting. The use of nanoformulation started in 1986. Over the past years, scientists have discovered drugs which target different pathways of the cell, like inhibitors for transcription, inhibitors of angiogenesis, and inducer of apoptosis to kill cancer cells. Due to the limitations in the standard anticancer therapy and development of multidrug resistant cancer cells, there is an urgent call to device a new way for cancer treatment. The growing number of evidences hints towards nanotechnology based treatment because of its physiochemical ability like stability, size, solubility, low toxicity, and cancer-specific internalization. The nanoformulation is based on the principle of targeting the nanomaterial to the specific organ or affected organ so that the effect of the drug will be more maximized and effective. There are varieties of synthetic and natural nanoparticle-based drugs that has been used for cancer studies. Some of the drugs that show promising results are introduced for 270 clinical trials ongoing worldwide. Although nearly 20% of the trials show clinical significance, the rest 80% trials were stopped due to lake of funding, drug sensitivity, and drug dose adjustment. However, a few numbers of nanoformulations introduce into phase-3 and 4 trial, showing promising results for cancer therapy. Recent studies are also pointing towards combinatorial therapy. Even though the results of combinatorial therapy are not so effective because of the variability of the results, but there is hope for future therapeutics. The ongoing study on nab-paclitaxel with the combination of other compounds used for metastatic breast cancer and Her2 negative breast cancer reduces metastatic malignancy. Similarly, liposomal doxorubicin is tested for ovarian and fallopian tube cancer along with carboplatin, bevacizumab, and veliparib. The detailed summary of the phase trials is summarized in the literature.

From the different studies, it is appreciable that nanoformulation has potential therapeutic importance in treating diseases. However, the importance of nanoformulation is undermined, because of the technological and biological limitations. It is palpable that nanomaterial can influence the biological material which may cause toxicity in the normal cells. Therefore, it is important to reorganize the studying protocols and an in-depth study of nanomaterials will expand its applications. In recent studies, the combination therapy for recurrent epithelial ovarian cancer treatment demonstrates that the highly active combinations are well tolerated by patients. Markedly, the use of pegylated liposomal doxorubicin along with cyclophosphamide followed by paclitaxel treatment is even safe for cardiotoxicity reported in phase-2 studies. Similarly, the treatment of paclitaxel in peritoneal malignancy does not show toxic effect with low clearance from the peritoneum. In recent years, these kind of promising results draw the attention of researchers towards the commercialization of FDA approved nanodrugs in cancer treatment. Overall, research data shows that the use of nanoformulation is not only restricted to use as a drug but also has the potential to be used for different biological developments.

9.6 Limitations of Phytoplankton Mediated Nanoparticles

The ultimate goal of green synthesis of nanoparticles is to successfully deliver the required drug to the specific target site while avoiding the accumulation of the drug at any non-specific target site with the least side effects (Terracciano et al. 2015). It is quite obvious for the immune system of the body to inactivate any nanoparticle affecting their function and elimination by macrophages, which will lead to hemolysis (Terracciano et al. 2015). Arrangement of surface cationic charges of nanoparticles also determines its accessibility to the target cell surface (Terracciano et al. 2015). Nanoparticles are also capable of forming intracellular reactive oxygen species (ROS) and apoptosis (Mu and Yan 2019). Apart from this, the required solubility for drug depends on the phase I clinical dose and the estimated human maximal tolerated dose (MTD). The human starting dose is assumed by calculating from the toxicological dose in mice or rats using a conversion formula (EORTC, N. D.D.C 1985). There are still particular challenges in enabling proper surface functionalization of the phytoplankton mediated nanoparticles to achieve the required drug loading and release within the range of specific therapeutic applications (Lomora et al. 2019). The cost-effectiveness for sustainable production of green synthesis of nanoparticles should also be investigated (Das et al. 2017).

These are the limitations in the functionality of phytoplankton-based nanomaterials, which could be addressed to successfully overcome and administrate nanomedicine.

9.7 Conclusion and Future Perspectives

Nanotechnology is showing great interest in various aspects of science including marine biology. The nanomaterials formulations with other biologically active compounds extracted from phytoplankton actually give a new line or works and seem to possess challenging therapeutic applications. Biologically active compounds from diverse group of phytoplankton such as algae and cyanobacteria have been widely considered to possess multiple functions in therapeutic applications of diseases and this field is growing industrially and biomedically. Cyanobacteria and algae are good bioresources of marine diversity with enormous biomolecules present in the form of polysaccharides, essential lipids, polyketides, vitamins, pigments, lectins, steroids, antioxidants, fibers, proteins, and halogenated compounds. These compounds are extensively used in the field of medical science, pharmaceutical science, therapeutic uses, nutraceuticals aspects, and biomedical fields. Analysis on algal metabolites has proven that this field showed significant improvement with escalating new technologies, however many challenges are need to overcome in the areas like medicine and pharmaceutical drugs when nanoformulations and commercialization are concerned. It needs to uncover the mechanism of action of phytoplankton enhanced anticancerous nanomedicine and hence a way for treating such disease.

The future perspective will be on better development of the marine drugs and commercialization using more bioactive molecules isolated from marine biodiversity encapsulated with nanoparticles to enhance targeted and site-specific delivery. Additionally, new technologies are needed to explore cancer therapy using nanoformulations and marine phytoplankton. Nano-fusion and green technologies should exploit together to work on anticancer drugs. Future works should explore on secondary metabolites of marine phytoplankton and drug discovery.

References

Akao Y, Ebihara T, Masuda H, Saeki Y, Akazawa T, Hazeki K, Hazeki O, Matsumoto M, Seya T (2009) Enhancement of antitumor natural killer cell activation by orally administered Spirulina extract in mice. Cancer Sci 100(8):1494–1501. https://doi.org/10.1111/j.1349-7006.2009.01188.x

Appeltans W, Ahyong ST, Anderson G, Angel MV, Artois T, Bailly N, Bamber R, Barber A, Bartsch I, Berta A, Błażewicz-Paszkowycz M (2012) The magnitude of global marine species diversity. Curr Biol 22(23):2189–2202. https://doi.org/10.1016/j.cub.2012.09.036

Awramik SM (1992) The oldest records of photosynthesis. Photosynth Res 33:75–89. https://doi.org/10.1007/BF00039172

Bajpai VK, Shukla S, Kang SM, Hwang SK, Song X, Huh YS, Han YK (2018) Developments of Cyanobacteria for Nano-Marine Drugs: Relevance of Nanoformulations in Cancer Therapies. Mar Drugs 16(6):179. https://doi.org/10.3390/md16060179

Blanco E, Shen H, Ferrari M (2015) Principles of nanoparticle design for overcoming biological barriers to drug delivery. Nat Biotechnol 33(9):941. https://doi.org/10.1038/nbt.3330

Blank CE, Sanchez-Baracaldo P (2010) Timing of morphological and ecological innovations in the cyanobacteria–a key to understanding the rise in atmospheric oxygen. Geobiology 8(1):1–23. https://doi.org/10.1111/j.1472-4669.2009.00220.x

Campos A, Souza CB, Lhullier C, Falkenberg M, Schenkel EP, Ribeiro-do-Valle RM, Siqueira JM (2012) Anti-tumour effects of elatol, a marine derivative compound obtained from red algae Laurencia microcladia. J Pharm Pharmacol 64(8):1146–1154. https://doi.org/10.1111/j.2042-7158.2012.01493.x

Das RK, Pachapur VL, Lonappan L, Naghdi M, Pulicharla R, Maiti S, Cledon M, Dalila LMA, Sarma SJ, Brar SK (2017) Biological synthesis of metallic nanoparticles: plants, animals and microbial aspects. Nanotechnol Environ Eng 2(1):18. https://doi.org/10.1007/s41204-017-0029-4

Delalat B, Sheppard VC, Ghaemi SR, Rao S, Prestidge CA, McPhee G, Rogers ML, Donoghue JF, Pillay V, Johns TG, Kröger N (2015) Targeted drug delivery using genetically engineered diatom biosilica. Nat Commun 6(1):1–1. https://doi.org/10.1038/ncomms9791

El-Naggar NE, Hussein MH, El-Sawah AA (2017) Bio-fabrication of silver nanoparticles by phycocyanin, characterization, in vitro anticancer activity against breast cancer cell line and in vivo cytotoxicity. Sci Rep 7(1):1–20. https://doi.org/10.1038/s41598-017-11121-3

EORTC NDDC (1985) EORTC guidelines for phase I trials with single agents in adults. Eur J Cancer Clin On 21(9):1005–1007. https://doi.org/10.1016/0277-5379(88)90349-5

Esfandyari J, Shojaedin-Givi B, Hashemzadeh H, Mozafari-Nia M, Vaezi Z, Naderi-Manesh H (2020) Capture and detection of rare cancer cells in blood by intrinsic fluorescence of a novel functionalized diatom. Photodiag Photodyn 30:101753. https://doi.org/10.1016/j.pdpdt.2020.101753

Falconer IR, Humpage AR (2005) Health risk assessment of cyanobacterial (blue-green algal) toxins in drinking water. Int J Env Res Public He 2(1):43–50. https://doi.org/10.3390/ijerph2005010043

Farokhzad OC, Langer R (2009) Impact of nanotechnology on drug delivery. ACS Nano 3(1):16–20. https://doi.org/10.1021/nn900002m

Farooqi AA, Butt G, Razzaq Z (2012) Algae extracts and methyl jasmonate anti-cancer activities in prostate cancer: choreographers of 'the dance macabre'. Cancer Cell Int 12(1):1–6. https://doi.org/10.1186/1475-2867-12-50

Fidler MM, Gupta S, Soerjomataram I, Ferlay J, Steliarova-Foucher E, Bray F (2017) Cancer incidence and mortality among young adults aged 20–39 years worldwide in 2012: a population-based study. Lancet Oncol 18(12):1579–1589. https://doi.org/10.1016/S1470-2045(17)30677-0

Gaudin J, Le Hegarat L, Nesslany F, Marzin D, Fessard V (2008) In vivo genotoxic potential of microcystin-LR: a cyanobacterial toxin, investigated both by the unscheduled DNA synthesis (UDS) and the comet assays after intravenous administration. Environ Toxicol 24(2):200–209. https://doi.org/10.1002/tox.20417

Gupta N, Pant SC, Vijayaraghavan R, Rao PL (2003) Comparative toxicity evaluation of cyanobacterial cyclic peptide toxin microcystin variants (LR, RR, YR) in mice. Toxicology 188(2–3):285–296. https://doi.org/10.1016/S0300-483X(03)00112-4

Hamida RS, Abdelmeguid NE, Ali MA, Bin-Meferij MM, Khalil MI (2020) Synthesis of silver nanoparticles using a novel cyanobacteria Desertifilum sp. extract: their antibacterial and cytotoxicity effects. Int J Nanomed 15:49

Hamouda RA, Hussein MH, Abo-elmagd RA, Bawazir SS (2019) Synthesis and biological characterization of silver nanoparticles derived from the cyanobacterium Oscillatoria limnetica. Sci Rep 9(1):1–7. https://doi.org/10.1038/s41598-019-49444-y

Hassenkam T, Johnsson A, Bechgaard K, Stipp SL (2011) Tracking single coccolith dissolution with picogram resolution and implications for CO2 sequestration and ocean acidification. Proc Natl Acad Sci 108(21):8571–8576. https://doi.org/10.1073/pnas.1009447108

Herfindal L, Kasprzykowski F, Schwede F, Lankiewicz L, Fladmark KE, Lukomska J, Wahlsten M, Sivonen K, Grzonka Z, Jastorff B, Doskeland SO (2009) Acyloxymethyl esterification of nodularin-R and microcystin-LA produces inactive protoxins that become reactivated and produce apoptosis inside intact cells. J Med Chem 52(18):5758–5762. https://doi.org/10.1021/jm900502e

Huang CY, Ju DT, Chang CF, Reddy PM, Velmurugan BK (2017) A review on the effects of current chemotherapy drugs and natural agents in treating non–small cell lung cancer. Biomedicine 7(4):23

Jia XD, Kuang XC, Pang H, Tang GF, Yu YH (2008) The effects of polysaccharide from Spirulina platensis on proliferation of BEL 7404 cells. Guangxi Med J 30:1122–1124

Kharissova OV, Dias HR, Kharisov BI, Pérez BO, Pérez VM (2013) The greener synthesis of nanoparticles. Trends Biotechnol 31(4):240–248. https://doi.org/10.1016/j.tibtech.2013.01.003

Kuppusamy P, Soundharrajan I, Srigopalram S, Yusoff MM, Maniam GP, Govindan N, Choi KC (2017) Potential pharmaceutical and biomedical applications of diatoms microalgae-an overview. Indian J Geo-Mar Sci 46(04):663–667

Lee HU, Park SY, Park ES, Son B, Lee SC, Lee JW, Lee YC, Kang KS, Kim MI, Park HG, Choi S (2014) Photoluminescent carbon nanotags from harmful cyanobacteria for drug delivery and imaging in cancer cells. Sci Rep 4:4665. https://doi.org/10.1038/srep04665

Lomora M, Shumate D, Rahman AA, Pandit A (2019) Therapeutic applications of phytoplankton, with an emphasis on diatoms and Coccolithophores. Adv Ther 2(2):1800099. https://doi.org/10.1002/adtp.201800099

Maher S, Kumeria T, Wang Y, Kaur G, Fathalla D, Fetih G, Santos A, Habib F, Evdokiou A, Losic D (2016) From the mine to cancer therapy: natural and biodegradable theranostic silicon nanocarriers from diatoms for sustained delivery of chemotherapeutics. Adv Healthc Mater 5(20):2667–2678. https://doi.org/10.1002/adhm.201600688

Marcazzan S, Varoni EM, Blanco E, Lodi G, Ferrari M (2018) Nanomedicine, an emerging therapeutic strategy for oral cancer therapy. Oral Onco 76:1–7. https://doi.org/10.1016/j.oraloncology.2017.11.014

Millard M, Yakavets I, Zorin V, Kulmukhamedova A, Marchal S, Bezdetnaya L (2017) Drug delivery to solid tumors: the predictive value of the multicellular tumor spheroid model for nanomedicine screening. Int J Nanomed 12:7993

Mittal AK, Chisti Y, Banerjee UC (2013) Synthesis of metallic nanoparticles using plant extracts. Biotechnol Adv 31(2):346–356. https://doi.org/10.1016/j.biotechadv.2013.01.003

Monks NR, Liu S, Xu Y, Yu H, Bendelow AS, Moscow JA (2007) Potent cytotoxicity of the phosphatase inhibitor microcystin LR and microcystin analogues in OATP1B1-and OATP1B3-expressing HeLa cells. Mol Cancer Ther 6(2):587–598. https://doi.org/10.1158/1535-7163.MCT-06-0500

Mu Q, Yan B (2019) Nanoparticles in Cancer therapy-novel concepts, mechanisms, and applications. Front Pharmacol 9:1552. https://doi.org/10.3389/fphar.2018.01552

National Cancer Institute (NCI) (2020) Understanding cancer. https://www.cancer.gov/aboutcancer/understanding. Accessed 18 May 2020

Newman DJ, Cragg GM (2020) Natural products as sources of new drugs over the nearly four decades from 01/1981 to 09/2019. J Nat Prod 83(3):770–803. https://doi.org/10.1021/acs.jnatprod.9b01285

Panda MK, Dhal NK, Kumar M, Mishra PM, Behera RK (2020a) Green synthesis of silver nanoparticles and its potential effect on phytopathogens. Mater Tod Proc 35:233–238

Panda MK, Panda SK, Singh YD, Jit BP, Behara RK, Dhal NK (2020b) Role of nanoparticles and nanomaterials in drug delivery: an overview. In: Advances in pharmaceutical biotechnology. Springer, Singapore, pp 247–265

Panda MK, Singh YD, Behera RK, Dhal NK (2020c) Biosynthesis of nanoparticles and their potential application in food and agricultural sector. In: Green nanoparticles. Springer, Cham, pp 213–225

Pangestuti R, Kim SK (2011) Biological activities and health benefit effects of natural pigments derived from marine algae. J Funct Food 3(4):255–266. https://doi.org/10.1016/j.jff.2011.07.001

Phtoplankton Classification, Kudela Lab, University of California, http://oceandatacenter.ucsc.edu/home/outreach/PhytoID_fullset.pdf

Pomponi SA (1999) The bioprocess-technological potential of the sea. In: Progress in industrial microbiology, vol 35. Elsevier, Amsterdam, pp 5–13. https://doi.org/10.1016/S0079-6352(99)80092-7

Prabhakar U, Maeda H, Jain RK, Sevick-Muraca EM, Zamboni W, Farokhzad OC, Barry ST, Gabizon A, Grodzinski P, Blakey DC (2013) Challenges and key considerations of the enhanced permeability and retention effect for nanomedicine drug delivery in oncology. Cancer Res 73(8):2412–2417. https://doi.org/10.1158/0008-5472.CAN-12-4561

Prasad V, De Jesús K, Mailankody S (2017) The high price of anticancer drugs: origins, implications, barriers, solutions. Nat Rev Clin Oncol 14(6):381. https://doi.org/10.1038/nrclinonc.2017.31

Pyne SK, Bhattacharjee P, Srivastav PK (2017) Microalgae (Spirulina platensis) and its bioactive molecules: review. Indian J Nutr 4(2):1–6

Roychoudhury P, Gopal PK, Paul S, Pal R (2016a) Cyanobacteria assisted biosynthesis of silver nanoparticles—a potential antileukemic agent. J Appl Phycol 28(6):3387–3394. https://doi.org/10.1007/s10811-016-0852-1

Roychoudhury P, Ghosh S, Pal R (2016b) Cyanobacteria mediated green synthesis of gold-silver nanoalloy. J Plant Biochem Biot 25(1):73–78. https://doi.org/10.1007/s13562-015-0311-0

Sathishkumar RS, Sundaramanickam A, Srinath R, Ramesh T, Saranya K, Meena M, Surya P (2019) Green synthesis of silver nanoparticles by bloom forming marine microalgae Trichodesmium erythraeum and its applications in antioxidant, drug-resistant bacteria, and cytotoxicity activity. J Saudi Chem Soc 23(8):1180–1191. https://doi.org/10.1016/j.jscs.2019.07.008

Schaer DJ, Alayash AI, Buehler PW (2007) Gating the radical hemoglobin to macrophages: the anti-inflammatory role of CD163, a scavenger receptor. Antioxid Redox Sign 9(7):991–999. https://doi.org/10.1089/ars.2007.1576

Sharma A, Sharma S, Sharma K, Chetri SP, Vashishtha A, Singh P, Kumar R, Rathi B, Agrawal V (2016) Algae as crucial organisms in advancing nanotechnology: a systematic review. J Appl Phycol 28(3):1759–1774. https://doi.org/10.1007/s10811-015-0715-1

Singh CR, Kathiresan K, Anandhan S (2015) A review on marine based nanoparticles and their potential applications. Afr J Biotechnol 14(18):1525–1532. https://doi.org/10.5897/AJB2015.14527

Singh G, Babele PK, Shahi SK, Sinha RP, Tyagi MB, Kumar A (2014) Green synthesis of silver nanoparticles using cell extracts of Anabaena doliolum and screening of its antibacterial and antitumor activity. J Microbiol Biotechnol 24(10):1354–1367. https://doi.org/10.4014/jmb.1405.05003

Singh AK, Tiwari R, Kumar V et al (2017) Photo-induced biosynthesis of silver nanoparticles from aqueous extract of Dunaliella salina and their anticancer potential. J Photochem Photobiol B Biol 166:202–211. https://doi.org/10.1016/j.jphotobiol.2016.11.020

Skeffington AW, Scheffel A (2018) Exploiting algal mineralization for nanotechnology: bringing coccoliths to the fore. Curr Opin Biotech 49:57–63. https://doi.org/10.1016/j.copbio.2017.07.013

Sonker AS, Pathak J, Kannaujiya VK, Sinha RP (2017) Characterization and in vitro antitumor, antibacterial and antifungal activities of green synthesized silver nanoparticles using cell extract of Nostoc sp. strain HKAR-2. Can J Biotechnol 1(1):26. https://doi.org/10.24870/cjb.2017-000103

Sun T, Zhang YS, Pang B, Hyun DC, Yang M, Xia Y (2014) Engineered nanoparticles for drug delivery in cancer therapy. Angew Chem Int Ed 53(46):12320–12364. https://doi.org/10.1002/anie.201403036

Talero E, García-Mauriño S, Ávila-Román J, Rodríguez-Luna A, Alcaide A, Motilva V (2015) Bioactive compounds isolated from microalgae in chronic inflammation and cancer. Mar Drugs 13(10):6152–6209. https://doi.org/10.3390/md13106152

Terracciano M, De Stefano L, Rea I (2018) Diatoms green nanotechnology for biosilica-based drug delivery systems. Pharmaceutics 10(4):242. https://doi.org/10.3390/pharmaceutics10040242

Terracciano M, Shahbazi MA, Correia A, Rea I, Lamberti A, De Stefano L, Santos HA (2015) Surface bioengineering of diatomite based nanovectors for efficient intracellular uptake and drug delivery. Nanoscale 7(47):20063–20074. https://doi.org/10.1039/C5NR05173H

Torchilin VP (2005) Recent advances with liposomes as pharmaceutical carriers. Nat Rev Drug Discov 4(2):145–160. https://doi.org/10.1038/nrd1632

Torres FA, Passalacqua TG, Velásquez AM, de Souza RA, Colepicolo P, Graminha MA (2014) New drugs with antiprotozoal activity from marine algae: a review. Rev Bras Farmacogn 24(3):265–276. https://doi.org/10.1016/j.bjp.2014.07.001

Van Apeldoorn ME, Van Egmond HP, Speijers GJ, Bakker GJ (2007) Toxins of cyanobacteria. Mol Nutr Food Res 51(1):7–60. https://doi.org/10.1002/mnfr.200600185

Vareli K, Pilidis G, Mavrogiorgou M-C, Briasoulis E, Sainis I (2009) Molecular characterization of cyanobacterial diversity and yearly fluctuations of Microcystin loads in a suburban Mediterranean Lake (Lake Pamvotis, Greece). J Environ Monit 11(8):1506–1512. https://doi.org/10.1039/B903093J

Venditto VJ, Simanek EE (2010) Cancer therapies utilizing the camptothecins: a review of the in vivo literature. Mol Pharm 7(2):307–349. https://doi.org/10.1021/mp900243b

Xia Y (2008) Nanomaterials at work in biomedical research. Nat Mater 7(10):758–760. https://doi.org/10.1038/nmat2277

Xu R, Zhang G, Mai J, Deng X, Segura-Ibarra V, Wu S, Shen J, Liu H, Hu Z, Chen L, Huang Y (2016) An injectable nanoparticle generator enhances delivery of cancer therapeutics. Nat Biotechnol 34(4):414–418. https://doi.org/10.1038/nbt.3506

Yang P, Li B, Yin QF, Wang YJ (2017) Carboxymethyl chitosan nanoparticles coupled with CD59-specific ligand peptide for targeted delivery of C-phycocyanin to HeLa cells. Tumor Biol 39(3):1010428317692267

Zehr JP, Waterbury JB, Turner PJ, Montoya JP, Omoregie E, Steward GF, Hansen A, Karl DM (2001) Unicellular cyanobacteria fix N 2 in the subtropical North Pacific Ocean. Nature 412(6847):635–638. https://doi.org/10.1038/35088063

Zhang YS, Zhang YN, Zhang W (2017) Cancer-on-a-chip systems at the frontier of nanomedicine. Drug Discov Today 22(9):1392–1399. https://doi.org/10.1016/j.drudis.2017.03.011

Terracciano M, De Stefano L, Rea I (2018) Diatoms green nanotechnology for biosilica-based drug delivery systems. Pharmaceutics 10(4):242. https://doi.org/10.3390/pharmaceutics10040242
Terracciano M, Shahbazi MA, Correia A, Rea I, Lamberti A, De Stefano L, Santos HA (2015) Surface bioengineering of diatomite based nanovectors [illegible]
[illegible]

Nanotechnology and Its Potential Implications in Ovary Cancer

10

Bimal Prasad Jit, Biswajita Padhan, and Ashok Sharma

Abstract

Prediction of an optimal diagnostic therapy for cancer patients remains a major challenge in the last few decades of research. Currently, ovary cancer has become an excruciating factor with major cause of morbidity and mortality. Owing to the complexities in the clinical presentation, origin of tumour and gene expression profiles have rendered it more difficult for accurate prognosis and diagnosis. Although last decades of research shed light on the chemotherapeutic approach, still early stage diagnosis with potential biomarker still remains question mark so far as the pathogenesis and clinical severity are concerned. Owing to the drawbacks of conventional chemotherapy, early diagnosis, discrepancy in clinical trials in context to the ovary cancer, onset of nanotechnology with novel tools can provide a potential alternative approach from biomarker detection to diagnosis in ovary cancer. In the recent past, nanotechnology has become a milestone in context to drug carriers such as polymer-drug conjugates, dendrimer, polymer micelles, carbon nanotubes, lipid/solid nanoparticles, and polymer nanoparticles have abundant benefits over conventional methods. Keeping the insight in mind the present chapter discerns potential implications of nanotechnology and its application in several dimensions of ovary cancer.

Keywords

Ovary cancer · Nanotechnology · Nanoparticles · Biomarker discovery · Receptor targeting

B. P. Jit · A. Sharma (✉)
Department of Biochemistry, All India Institute of Medical Sciences, New Delhi, India

B. Padhan
PG Department of Botany, Berhampur University, Bhanja Bihar, Odisha, India

M. Arakha et al. (eds.), *Bio-Nano Interface*,
https://doi.org/10.1007/978-981-16-2516-9_10

10.1 Introduction

Ovarian cancer considered to be the most debilitating cause of morbidity and mortality in women. The disease is ranked as seventh in worldwide and third in India as cancer causing death. The rate of incidence of the disease is highest in the populations of Central America and Northern Europe and lowest in some parts of Africa and Asia (Jemal et al. 2010). India exhibits third highest number of cases after China and USA. Due to lack of visible symptoms it remains a crucial challenge for proper diagnosis and implication of suitable drug therapy.

Ovarian cancer is observed to be the fifth most prevalent cancer among women with a lifetime risk of 1.4–1.8%. Identification of particular mutation, type of tissue affected and stage of the patients is a major challenge which determines the survival rate of the patients. Early stage detection of the disease (<III) facilitates encouraging survival rates, where as a patient diagnosed with metastatic phenotype exhibits pathophysiologic challenge and shows a five year overall survival dropping a dismal of 20% (Siegel et al. 2018). Epidemiological evidence elucidates that the overall survival rates of the ovarian cancer patients have not been improved in the last few decades, which indicates there is immense need of significant improvement in the field of ideal biomarker representing the crucial phenotype and therapeutic implications in ovarian cancer. Although there is promising approach in terms of chemotherapy for the treatment of the disease, however during the course the treatment eventually became a failure due to the onset of chemoresistance (Reid et al. 2017).

Generally there are three types of ovary cancers: Epithelial carcinoma, germ cell tumour and stromal tumour, where 90% of the ovarian cancers belong to epithelial origin, mostly outer surface layers of the ovary or fallopian tube epithelial cells. The 5-year survival rate for early stage ovarian cancer is approximately 92%; however, most patients are diagnosed with advanced stage of severity where the 5-year survival rate is only 30%. Most ovarian cancers are developed from three categories of cells: epithelial cells, sex cord-stromal cells and germ cells. Among them, epithelial ovarian cancer (EOC) accounts for 90% cases. EOCs are divided into five subtypes: (1) serous: ~50%; (2) mucinous: 5–10%; (3) endometrioid: 10–25%; (4) clear cell: 4–5% and (5) transitional cells: rare (Chen et al. 2017). Ovary cancer is the third leading cause of cancer related deaths among women in India (National Cancer Registry Report, ICMR).The phenotype is also regarded as **"silent killers"** and constitutes many sub-types, i.e. high-grade serous ovarian cancer (HGSOC), endometrioids, mucinous and clear cell tumours. HGSOC is the most prominent and major cause of morbidity and mortality which accounts for 90% of death of patients. Late-stage diagnosis and incidence of drug resistance cancer are two contributing factors which are highly crucial for the survival rate of patients with HGSOC. Variation in the incidence of ovary cancer by race and ethnicity is solely dependent on women age. It was observed that the overall survival rate is less than 30 percent (Lisio et al. 2019).

10.2 Possible Risk Factors Associated with Ovary Cancer

The incidence of ovarian cancer in India is between 0.9 and 8.4/100,000. It has been evident that the risk usually starts from age 35 and reaches a peak between the ages of 55 and 64. Although the exact cause of ovary cancer is yet to be known, several risk factors are shown to modulate the clinical severity of the disease. The potential risk factors aggravating the pathological complexity include genetic, epigenetic, reproductive history, family history, age, weight, hormonal replacement therapy (HRT), ethnicity, endometriosis, chronic inflammation, non-steroidal anti-inflammatory drugs (NSAIDS) (McLemore et al. 2009).

10.2.1 Age

Ovarian cancer is a characteristic feature of older women. Women with old age are likely to be more prone to ovary cancer than the younger. The Age Specific Incidence Rate (ASIR) for ovarian cancer shows a drift of rising disease from the age of 35 years with a highest rate of incidence between the age of 55 and 64 years (Rodriguez et al. 1995). Although onset of menarche with respect to age shows no clear association, however natural menopause found with late stage poses higher risk (Andéol et al. 1990).

10.2.2 Genetics

Inherited genetic mutations play a crucial role in ovary cancer. It has been evidence that about 20–25% of women diagnosed with ovary cancer have a hereditary tendency to develop the disease. Mutation in genes like breast cancer gene 1 (BRCA1) and breast cancer gene 2 (BRCA2) contributes 10–15% of the cases with ovary cancer (Greiser et al. 2007; Lalwani et al. 2011). About 25% of the breast cancer death in stage 1 is due to the occurrence of ovarian cancer. It was observed that the cumulative risk of ovarian cancer is 49% for the BRCA1 mutation carriers at the age of 80, whereas for BRCA2 mutation carriers it is 21%. Previously it was observed that 1–3% of the colorectal cancers are due to the occurrence of Lynch syndrome, an autosomal dominant disorder. Lynch syndrome contributes 10–15% of the total inherited ovarian cancer patients, where most of the cases are non-mucinous and present in stage I or II. It was also evident that women with Peutz-Jeghers syndrome (PJS), Nevoid basal cell carcinoma syndrome (NBCCS), Li-Fraumeni syndrome and ataxia-telangiectasia have also increased risk to develop ovary cancer (Thull and Vogel 2004).

10.2.3 Family History

Women with first degree of relatives diagnosed with ovary cancers are likely chance to develop the disease. Evidence from case control study revealed that women with a family history of ovary, breast or uterine cancer in their mother or sisters are likely chance to develop the ovary cancer (Walker et al. 2002). Reports from national cancer institute (NCI) elucidate that woman lacking a family history have a 1 in 55 chance to develop the ovary cancer, whereas the risk increases to 10 fold with known family or hereditary history (Stratton et al. 1998).

10.2.4 Ethnicity

Ethnicity does not play a significant role in modulating the onset and severity of ovary cancer. However, results from few case control study indicate that rate of epithelial ovary cancer seems to be high in white women, compared to black women (Walker et al. 2002). Women of Ashkenazi Jewish descent are characterized by BRCA1/2 mutations and thus are at a higher risk to develop ovary cancer. Recurrence of high mortality in all racial groups indicates that ethnicity does not play a crucial role in the overall survival (Liakakos 2008).

10.2.5 Reproductive History

Age of menarche and menopause, menstrual related factors, parity, pregnancy characteristics, age at child birth are some of the crucial factors affecting the occurrence and progression of the ovarian cancer. Numerous studies have put their opinion on the association of menstruation and ovulation cycle with ovary cancer (Mori et al. 1988; Kim et al. 2017). Some study shows inverse relationship between ovulation cycle and ovary cancer, whereas other strongly argues that lack of ovulation due to pregnancy or taking of oral contraceptive could pose risk to ovary cancer (Baik et al. 2007).

Menstruation at an early age (12 years) and menopause at 50 could contribute higher risk to vary cancer although fewer studies claimed no significant association. Multiple line of evidence strongly suggests that increase in the number of pregnancy is significantly associated with decrease risk of ovary cancer. However, on the contrary to the number of pregnancy, older age in pregnancy is also associated with decreased risk of the ovary cancer (Hummeida et al. 2015).

10.2.6 Gynaecological Factors

Several gynaecological factors including pelvic inflammatory disease, endometriosis, ovarian cyst and tubal ligations play an important role in the occurrence of ovary cancer. Multiple line of evidence suggests that inflammatory events in the ovary associated with repeated ovulation process release cancer cells to the surrounding tissues and contribute an increase risk in ovary cancer (Ness et al. 2000; Purdie et al. 2003; Ness et al. 2003; Lo-Ciganic et al. 2012). In addition these infectious agents like Chlamydia trachomatis also contribute to the progression of ovary cancer (Ness et al. 2003). Nulliparous women or women with hyperoestrogenism are most likely chance of development of ovary cancer. Ovarian cysts are observed to be frequent in women and it was elucidated that ovarian cyst in postmenopausal women increases the risk of progression of ovary cancer. Removal of uterus leaving the ovary or fallopian tube tied, use of oral contraceptives and breast feeding may reduce the risk of ovary cancer (Jemal et al. 2006; Sanguinete et al. 2017).

10.2.7 Hormone Replacement Therapy

Women lacking previous history of hysterectomy taking hormone therapy orally have shown to be associated with increased risk of ovary cancer. Previous study has also shown that taking oestrogen for 10 years or more may poses risk for ovary cancer; however it has no significant association with survival rate of the patients (Glud et al. 2004).

10.2.8 Lifestyle Factors

Lifestyle factors including nutrition, diet, obesity, physical activity, consumption of alcohol, caffeine, smoking play a prominent role in modulating the development of ovarian cancer. Although epidemiological evidence indicates that obesity plays a significant role in incidence of ovary cancer, however consideration of menopausal condition, hormonal stage, body mass index, waist-hip ratio is highly significant in development of the ovarian cancer. Physical activity has shown inconclusive results indicative of both positive and negative association with ovarian cancer (Momenimovahed et al. 2019).

10.3 Current Therapeutic Approach to Ovary Cancer

Despite the considerable advancement in the field medicine still ovary cancer poses a significant challenge so far as the therapeutic intervention is considered. Significant challenges have been put forwarded in the management of pathogenesis and mortality of ovary cancer. Owing to the complexities in the clinical presentation, origin of tumour and gene expression profiles have rendered it more difficult for accurate prognosis and diagnosis. Although previously some of the biomarkers used as an only clinical indicator of severity, however they are limited due to their variable degree of expression from mild to severe phenotype in ovary cancer (Jatoi et al. 2016). Ninety percent of ovarian cancers are derived from coelomic epithelium.

Germ cell tumours account for 5% of ovarian cancers and sex cord-stromal tumours approximately 7%. The common epithelial ovarian cancers (EOCs) include high-grade serous (70%), endometrioid (10%), clear cell (10%), mucinous (3%) and low-grade serous carcinomas (<5%). Difference in epidemiology, molecular profiles, clinical presentations, patterns of spread, response to chemotherapy has contributed different degree of prognosis. About 70% of patients with EOC present with advanced disease, as a result of the lack of any satisfactory screening test and specific symptoms (Chen et al. 2003).

Cytoreductive surgery encompasses the first line of treatment for ovary cancer. Although post-tumour debulking followed by intravenous cisplatin or carboplatin based chemotherapy shows good response (more than 70%), however a large fraction of the patients develops chemoresistance. Currently, doxorubicin, cisplatin, oxaliplatin, transplatin, carboplatin, paclitaxel (Taxol, TAX) have shown to be beneficial for ovary cancer. Paclitaxel has been used as a most efficient chemotherapeutic agent for relapsed ovarian cancer (Spears et al. 2019). Efforts are being made to develop new treatment options by combination of different drug approach to boost better therapeutic response. The rate of survival and clinical response towards a combination of doxorubicin, cyclophosphamide and cisplatin shows efficient response in comparison to the treatment with only cyclophosphamide or cisplatin (Chandra et al. 2019). It was also evident that patients with resistance to cisplatin show positive response when undergone combinational treatment with gemcitabine and cisplatin. Furthermore, results of randomized clinical study revealed that treatment with combination chemotherapy approach like with HEXA-CAF (hexamethyl melamine (HMMA), cyclophosphamide (CPP), methotrexate (MTX) and fluorouracil (FU) shows high survival rate in comparison to the use of single therapeutics (Corradetti et al. 2019).

Another approach elucidates that development of potential agents targeting PARP (poly ADP ribose polymerase), mTOR pathway, MAP kinase pathway has become increasingly significant in survival outcome. For example, use of olaparib, a PARP inhibitor with two or more course of platinum based chemotherapy has shown to be beneficial in high-grade serous ovary cancer and increases progression free survival. Furthermore, clinical trials with temsirolimus (mTOR inhibitor), sorafenib (kinase inhibitor) have paved the way for their implication in targeting the specific signalling pathway in ovary cancer. It was also observed that understanding the aberrant genetic mechanism and involvement of specific signalling pathway of different histological subtypes will likely to contribute the better therapeutic intervention. For example, high frequency of Ras mutation in MAP kinase signalling pathway makes them an attractive target in low-grade serous ovary cancer (Vetter and Hays 2018).

Although chemotherapy approach has shown a great promise, however association of several confounding factors like side effects, poor drug accumulation, rapid blood clearance, degradation, poor penetration capacity, drug resistance, lack of suitable ideal biomarker which could indicate the associated clinical severity of the phenotype is needed to be addressed and monitored for an efficient prognosis and diagnosis approach. Nevertheless, it is evident that the chemotherapeutic approach is limited owing to its toxicity, less efficacy and development of drug resistance.

10.4 Nanotechnology and Its Implications in Ovary Cancer

Owing to the drawbacks of conventional chemotherapy, requirement for early diagnosis, discrepancy in clinical trials in context to the ovary cancer, onset of nanotechnology with novel tools can provide a potential alternative approach from biomarker detection to diagnosis in ovary cancer. In the recent past nanotechnology has become a milestone in context to drug carriers such as polymer-drug conjugates, dendrimer, polymer micelles, carbon nanotubes, lipid/solid nanoparticles and polymer nanoparticles have abundant benefits over conventional methods (Gupta et al. 2009, 2019) (Fig. 10.1). Starting from the measurement of biomolecules in nano level by electrochemoluminiscence to biomarker detection and the cellular imaging approach has opened a new avenue in the field of disease diagnosis and preventive medicine. Currently nanoparticle platform has immensely useful in targeting several infectious, non-infectious disease including cancer (Engelberth et al. 2014).

10.4.1 Nanoformulations in Drug Delivery for Chemotherapy

Control drug release, extended circulation, less toxicity and enhanced drug protection are central to successful drug efficacy. The unique physical and chemical nature of the nano particles makes them ideal to be used in the drug delivery. Previous evidence has suggested that conjugates of nanoparticles and chemotherapeutics induce dose delivery to the tumour microenvironment (Chang et al. 2009). Previous investigation indicates doxorubicin conjugated PEGylated liposome

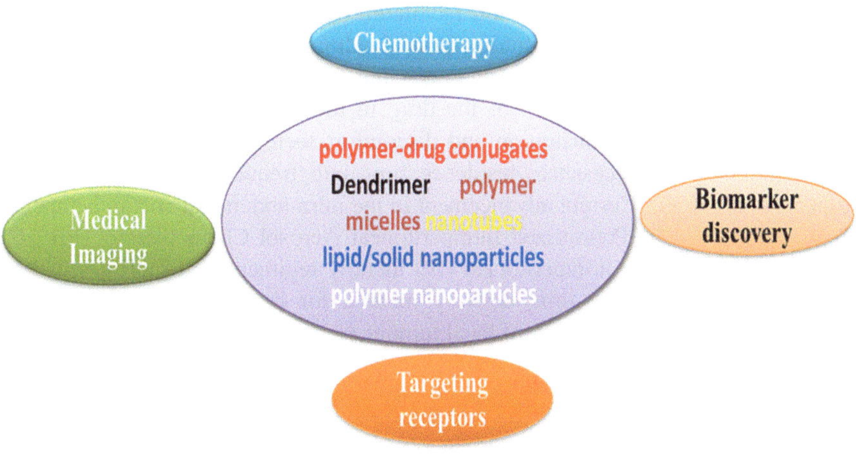

Fig. 10.1 Nanotechnology and its possible implications in ovary cancer

nanoformulations known as Doxil used in platinum resistance cancer (Gabizon et al. 1994; Ferrandina et al. 2008). Previous study in SKOV-3 cells indicates cisplatin-loaded nanoparticles having pH-sensitive poly [2-(N, N-diethylamino) ethyl methacrylate] (PDEA) dissolved at a pH < 6 characterized by rapid release of cisplatin to the cytoplasm upon integration to the acidic lysosome and thus able to overcome the chemoresistance activity (Xu et al. 2009b). Similar findings are reported in case of poly-(c-L-glutamylglutamine)–paclitaxel (PGG–PTX), nanoparticulate formulation of paclitaxel (PTX) shows better therapeutic efficacy and tumour inhibition property (Feng et al. 2004). In addition to this, paclitaxel formulations of solid lipid nano particles (SLN) have also been tested in OVCAR-3 cell lines, however the results show some cytotoxicity (Lee et al. 2012). Results of the recent study elucidated that the potential ability of PEGylated liposomal formulations of paclitaxel and cisplatin significantly reduces the growth and aggressiveness of ovary cancer (Krieger et al. 2010). Both in vivo and in vitro study revealed that the compound induces TNF mediated ERK/AKT signalling pathway and causes enhanced expression of caspase3,9, ERK and AKT thus inducing apoptosis in ovarian cancer cells (Qi et al. 2018). Nano vesicular system like lapatinib/PTX nano capsules showed an improved tumour inhibition activity in comparison to the single PTX treatment only in multidrug resistance OVCAR-3 cell lines (Vergara et al. 2012). Further advancement in nano formulation system leads to the development of three-dimensional regularly branched tree-like macromolecules called as dendrimers which have the potential anticancer activity. Previous finding revealed dendritic copolymer system with paclitaxel and cisplatin displayed significant cytotoxic activity in SKOV3 cells (Kirkpatrick et al. 2011). Furthermore, for efficient drug delivery with high local concentration in the intraperitoneal region (ip) shows significant progress due to the development of hyaluronic acid (HA)-based in-situ crosslinkable hydrogel specialized for delivery of paclitaxel (PTX) to the IP tumours (Bajaj et al. 2012). Further evidence shows that an in-situ crosslinkable hydrogel depot containing paclitaxel (PTX) nanocrystals (PNC) possess significant inhibitory effect on SKOV3 cells proliferation facilitating IP chemotherapy in ovary cancer (Sun et al. 2016). Although intraperitoneal administration of chemotherapeutic agents have shown a promising approach but the process is limited by certain drawbacks like rapid clearance of the drug in the treatment area, complications associated with i.p administration and discomfort feelings related to the catheter implantation could necessitate residence time and frequent dosing of the drug (Spears et al. 2019). Current advancement in the intra-abdominal chemotherapeutic approach, known as Pressurized Intraperitoneal Aerosol Chemotherapy (PIPAC), was introduced to the intraperitoneal (IP) therapy regimens and indicates to be a compatible and effective drug delivery method for the treatment of inoperable tumours and for the prevention of local tumour recurrence (Shariati et al. 2019).

10.4.2 Nanotechnology in Biomarker Discovery in Ovarian Cancer

It is an undoubted fact that still there is lacking of a potential specific ideal biomarker determining clinical severity of the patients in ovary caner. Past decades of research has only developed two prominent biomarker like cancer antigen-125 (CA-125) and human epididymis protein 4 (HE4) used for diagnosis of ovary cancer (Bast et al. 2009). However currently due to the advent of gold standard high-throughput technology like Systematic Evolution of Ligand by Exponential Enrichment (SELEX) or Cell-SELEX, mass spectrometry, micro array and other fields like epigenetic, cytomics provide a ray of hope and enabled the development of panel biomarker approach for early detection of ovary cancer (Yang et al. 2017).

The derisory performance of current diagnostic approaches has led to numerous investigations to identify new biomarkers and use of multiplex assays of serum protein levels (ROMA/OVA1). Recent reports indicate that in addition to CA-125, HE4 and mesothelin have some promise for use in multiplex assays to improve reliability and performance (Moore et al. 2009; Nolen and Lokshin 2013; Kimmel et al. 2012).

Recently development of screen printed gold nanoparticle electrodes for label-free impedimetric sensing and electroluminescent sensing platform with iron oxide nanoparticles labelled with a primary antibody and a dendrimer/luminol modified secondary antibody have attracted the significant attention. Both platforms have shown capable of detecting CA-125 at a very lower concentration (Ravalli et al. 2013; Li et al. 2013). Development of sandwich electrochemical aptasensor for the detection of HER2 and HER2-overexpressing breast cancer cells has been described previously. In such approach the quantitative measurement of target molecules and cells can be done using mono-antibody of HER2 (anti-HER2) and a bioconjugate of hydrazine-AuNP (Gold Nanoparticle)-Aptamer (Hyd-AuNP-Apt), by stripping voltammetry (Zhu et al. 2013). However this strategy can be implemented for detection of other biomarkers in ovary cancer. It is expected that MEMS systems, lab-on-chip, nanoparticles and different alternative sensing platforms can be further developed for screening and early detection of ovary cancer and therapeutic efficacy. Recently a novel liposome mediated nanoparticle approach was developed for blood biomarker detection in ovarian cancer patients (Hadjidemetriou et al. 2020).

10.4.3 Nanotechnology in Imaging Approach in Ovarian Cancer

The current treatment for advanced OC relies on the synergistic effect of combining surgical cytoreduction and chemotherapy. However, beside the fact that chemotherapy resistance is a major dispute in OC management, new imaging strategies are needed to target microscopic lesions and improve both cytoreductive surgery and patient outcomes (Pavlik et al. 2000).

Medical resonance imaging (MRI) is a powerful, non-invasive in vivo imaging technique with high resolution (10–100 mm) and unlimited imaging depth. The MRI signal is based on the relaxation of hydrogen nuclei (in water) after being exposed to a pulsed radiofrequency signal. Contrast in MRI can be T1 weighted (differences in the spin-lattice relaxation of tissues), or T2 weighted (difference in spin–spin relaxation). Contrast agents are administered to influence the T1or T2 relaxation parameters and enhance the imaging quality. Numerous nanoparticle constructs have been developed to provide contrast enhancement where several have been applied to cancer imaging. Iron oxide (IO) nanoparticles are frequently used for T2 imaging, whereas liposomes, micelles, dendrimers and polymeric nanoparticles incorporating paramagnetic species, such as $Gd3+$, are common for T1 imaging (Wang et al. 2011).

Pelvic transvaginal sonography (TVS) together with abdominal and pelvic transabdominal sonography is the most important procedure for the morphological evaluation of OC with the use of the Doppler and colour Doppler to study mass vascularization. Ovarian sonography can be an effective strategy to detect the changes regarding the size and adnexal architecture preceding both the development of symptoms and alterations detectable by pelvic examination (Clarke-Pearson 2009). Photoacoustic imaging (PAI), a comparatively new imaging method based on the detection of light excited ultrasound waves, may complement existing US screening techniques for improving discovery and characterization of OC. In particular, PAI might permit early detection of angiogenesis at an initial stage by allowing identification of the neovascularisation (Lao et al. 2008; Aguirre et al. 2009). Furthermore ultrasound responsive doxorubicin/curcumin co-deliver alginate nanodroplets were also evaluated for their therapeutic efficacy against multidrug resistance ovary cancer and show promising treatment efficacy (Baghbani and Moztarzadeh 2017). Evidence from clinical trials in animal model indicates dual-modality computed tomography, near-infrared fluorescence nano liposomal agent (CF800) and activable theranostic nanoplatform based on silicon naphthalocyanine polymeric NPs shows potential results in image guided surgery in ovary cancer (Zheng et al. 2015; Duffy et al. 2018). Results from the xenograft A2780 ovarian cancer model indicate potential of quantum dots (having emission about 500 nm) in targeting aptamer (MUC1) and doxorubicin for dual imaging approach (Savla et al. 2011).

10.4.4 Nanotechnology in Receptor Targeting in Ovary Cancer

Over expression of cellular proteins EGF receptor (EGFR, HER2), luteinizing hormone receptor, claudins, mucins, TAG-72, and integrins displayed by ovary cancer cells makes them ideal for drug therapeutic approach. Taking the principle of ligand–receptor interaction cancerous cells expressing any specific receptor which is absent in normal cells could be a potential therapeutic target (Liang et al. 2019). Previous reports suggest that several nano carriers with specific ligands are targeted to the tumour cells expressing the receptors. For example, gold nanoparticles, bovine serum albumin nanoparticles (BSANPs), nano gel systems are implicated in targeting the folate receptor in several ovary cancer cell lines (Zhang et al. 2004; Bhattacharya et al. 2007; Ledermann et al. 2015). Furthermore, face centred cubic (fcc)- iron-platinum (FePt) NPs have shown to preferentially bind luteinizing hormone releasing hormone receptor (LHRH) expressed by A2780 cells (Xu et al. 2009a). Assessment of Paclitaxel loaded NPs coated with anti-HER2 monoclonal antibody was evaluated in xenograft model of SKOV-3 cells (Cirstoiu-Hapca et al. 2010). Besides this OVCCAR-3 cells expressing follicle stimulating hormone receptor were targeted by poly (amidoamine) (PAMAM) dendrimers conjugated with the binding peptide domain of FSH (FSH33) (Zhang et al. 2009). Similar findings were obtained from the receptor targeting to integrin receptor, transferring receptor and other cell surface receptors indicate targeting with nanoconjugated ligands could be a novel approach for better therapeutic efficacy in ovary cancer. In addition to this markers specific to the tumour tissue could be a potential target. Previous study indicates targeting the serum marker CA-125 which is predominantly expressed by ovarian surface epithelial cells in ovary cancer is targeted with a bifunctional fusion protein construct consisting of a single-chain antibody variable domain and core streptavidin domain with biotinylated antigen and liposome shows maximum binding specificity to OVCAR-3 cells in comparison to the control (Wang et al. 2007).

10.5 Conclusion

According to of NCRP report, 2019, ovary cancer is the third most predominant form of cancer after breast and cervix cancer, among women in India. Epidemiological study revealed that by 2020 there will be around 36, 200 new ovarian cancer cases in India. Ovarian cancer is the fourth most common cancer in Indian women with an incidence of 4.9 cases per 100,000. This happens largely due to late-stage diagnosis, poor prognosis related to advanced-stage disease and high recurrence rate associated with development of chemo resistance. A suitable ideal biomarker, chemotherapeutic approach with maximum specificity and efficacy context to the different clinical representation, symptoms and degree of severity is of an immense requirement so far as the disease pathogenesis of ovary cancer is concerned. Survival statistics have not improved significantly over the last three decades and underlining the fact that improved therapeutic strategies as well as early detection require substantial improvements. Owing to the remarkable advancement in nanotechnology has paved a new avenue in context to the prognosis, diagnosis and design as well as delivery of drug candidate for better therapeutic efficacy. However more works yet to be explored in biomarker approach, non-invasive methods of prognosis, antitumour agents and early diagnosis of ovary cancer.

References

Aguirre A, Guo P, Gamelin JK, Yan S, Sanders MM, Brewer MA, Zhu Q (2009) Coregistered three-dimensional ultrasound and photoacoustic imaging system for ovarian tissue characterization. J Biomed Opt 14(5):054014

Andéol Y, Gusse M, Méchali M (1990) Characterization and expression of a Xenopus ras during oogenesis and development. Dev Biol 139(1):24–34

Baghbani F, Moztarzadeh F (2017) Bypassing multidrug resistant ovarian cancer using ultrasound responsive doxorubicin/curcumin co-deliver alginate nanodroplets. Colloids Surf B: Biointerfaces 153:132–140

Baik I, Lambe M, Liu Q, Cnattingius S, Mucci LA, Riman T, Ekbom A, Adami H-O, Hsieh C-C (2007) Gender of offspring and maternal risk of invasive epithelial ovarian cancer. Cancer Epidemiol Prev Biomark 16(11):2314–2320

Bajaj G, Kim MR, Mohammed SI, Yeo Y (2012) Hyaluronic acid-based hydrogel for regional delivery of paclitaxel to intraperitoneal tumors. J Control Release 158(3):386–392

Bast RC, Hennessy B, Mills GB (2009) The biology of ovarian cancer: new opportunities for translation. Nat Rev Cancer 9(6):415–428

Bhattacharya R, Patra CR, Earl A, Wang S, Katarya A, Lu L, Kizhakkedathu JN, Yaszemski MJ, Greipp PR, Mukhopadhyay D (2007) Attaching folic acid on gold nanoparticles using noncovalent interaction via different polyethylene glycol backbones and targeting of cancer cells. Nanomed: Nanotechnol Biol Med 3(3):224–238

Chandra A, Pius C, Nabeel M, Nair M, Vishwanatha JK, Ahmad S, Basha R (2019) Ovarian cancer: current status and strategies for improving therapeutic outcomes. Cancer Med 8(16):7018–7031

Chang B, Liu G, Xue F, Rosen DG, Xiao L, Wang X, Liu J (2009) ALDH1 expression correlates with favorable prognosis in ovarian cancers. Mod Pathol 22(6):817–823

Chen VW, Ruiz B, Killeen JL, Coté TR, Wu XC, Correa CN, Howe HL (2003) Pathology and classification of ovarian tumors. Cancer: Interdiscip Int J Am Cancer Soc 97(S10):2631–2642

Chen S, Zhang L, Yan G, Cheng S, Fathy AH, Yan N, Zhao Y (2017) Neutrophil-to-lymphocyte ratio is a potential prognostic biomarker in patients with ovarian cancer: a meta-analysis. BioMed Res Int 2017

Cirstoiu-Hapca A, Buchegger F, Lange N, Bossy L, Gurny R, Delie F (2010) Benefit of anti-HER2-coated paclitaxel-loaded immuno-nanoparticles in the treatment of disseminated ovarian cancer: therapeutic efficacy and biodistribution in mice. J Control Release 144(3):324–331

Clarke-Pearson DL (2009) Screening for ovarian cancer. N Engl J Med 361(2):170–177

Corradetti B, Pisano S, Conlan RS, Ferrari M (2019) Nanotechnology and immunotherapy in ovarian cancer: tracing new landscapes. J Pharmacol Exp Ther 370(3):636–646

Duffy MJ, Planas O, Faust A, Vogl T, Hermann S, Schäfers M, Nonell S, Strassert CA (2018) Towards optimized naphthalocyanines as sonochromes for photoacoustic imaging in vivo. Photoacoustics 9:49–61

Engelberth SA, Hempel N, Bergkvist M (2014) Development of nanoscale approaches for ovarian cancer therapeutics and diagnostics. Critical Reviews™ in Oncogenesis 19(3–4)

Feng SS, Mu L, Win KY, Huang G (2004) Nanoparticles of biodegradable polymers for clinical administration of paclitaxel. Curr Med Chem 11(4):413–424

Ferrandina G, Ludovisi M, Lorusso D, Pignata S, Breda E, Savarese A, Del Medico P, Scaltriti L, Katsaros D, Priolo D (2008) Phase III trial of gemcitabine compared with pegylated liposomal doxorubicin in progressive or recurrent ovarian cancer. J Clin Oncol 26(6):890–896

Gabizon A, Catane R, Uziely B, Kaufman B, Safra T, Cohen R, Martin F, Huang A, Barenholz Y (1994) Prolonged circulation time and enhanced accumulation in malignant exudates of doxorubicin encapsulated in polyethylene-glycol coated liposomes. Cancer Res 54(4):987–992

Glud E, Kjaer SK, Thomsen BL, Høgdall C, Christensen L, Høgdall E, Bock JE, Blaakaer J (2004) Hormone therapy and the impact of estrogen intake on the risk of ovarian cancer. Arch Intern Med 164(20):2253–2259. https://doi.org/10.1001/archinte.164.20.2253

Greiser CM, Greiser EM, Dören M (2007) Menopausal hormone therapy and risk of ovarian cancer: systematic review and meta-analysis. Hum Reprod Update 13(5):453–463

Gupta PB, Chaffer CL, Weinberg RA (2009) Cancer stem cells: mirage or reality? Nat Med 15 (9):1010–1012

Gupta PB, Pastushenko I, Skibinski A, Blanpain C, Kuperwasser C (2019) Phenotypic plasticity: driver of cancer initiation, progression, and therapy resistance. Cell Stem Cell 24(1):65–78

Hadjidemetriou M, Papafilippou L, Unwin RD, Rogan J, Clamp A, Kostarelos K (2020) Nano-scavengers for blood biomarker discovery in ovarian carcinoma. Nano Today 34:100901

Hummeida ME, Hamad K, Gadir AF, Ali AA (2015) Ovarian cancer during pregnancy: a case report and literature review. Clin Pract 5(2):727. https://doi.org/10.4081/cp.2015.727

Jatoi A, Vierkant RA, Hawthorne KM, Block MS, Ramus SJ, Larson NB, Fridley BL, Goode EL (2016) Clinical and emergent biomarkers and their relationship to the prognosis of ovarian cancer. Oncology 90(2):59–68

Jemal A, Siegel R, Ward E, Murray T, Xu J, Smigal C, Thun MJ (2006) Cancer statistics, 2006. CA: Cancer J Clin 56(2):106–130

Jemal A, Siegel R, Xu J, Ward E (2010) Cancer statistics, 2010. CA Cancer J Clin 60(5):277–300. https://doi.org/10.3322/caac.20073

Kim SJ, Rosen B, Fan I, Ivanova A, McLaughlin JR, Risch H, Narod SA, Kotsopoulos J (2017) Epidemiologic factors that predict long-term survival following a diagnosis of epithelial ovarian cancer. Br J Cancer 116(7):964–971

Kimmel DW, LeBlanc G, Meschievitz ME, Cliffel DE (2012) Electrochemical sensors and biosensors. Anal Chem 84(2):685–707

Kirkpatrick GJ, Plumb JA, Sutcliffe OB, Flint DJ, Wheate NJ (2011) Evaluation of anionic half generation 3.5–6.5 poly (amidoamine) dendrimers as delivery vehicles for the active component of the anticancer drug cisplatin. J Inorg Biochem 105(9):1115–1122

Krieger ML, Eckstein N, Schneider V, Koch M, Royer H-D, Jaehde U, Bendas G (2010) Overcoming cisplatin resistance of ovarian cancer cells by targeted liposomes in vitro. Int J Pharm 389(1–2):10–17

Lalwani N, Prasad SR, Vikram R, Shanbhogue AK, Huettner PC, Fasih N (2011) Histologic, molecular, and cytogenetic features of ovarian cancers: implications for diagnosis and treatment. Radiographics 31(3):625–646

Lao Y, Xing D, Yang S, Xiang L (2008) Noninvasive photoacoustic imaging of the developing vasculature during early tumor growth. Phys Med Biol 53(15):4203

Ledermann J, Canevari S, Thigpen T (2015) Targeting the folate receptor: diagnostic and therapeutic approaches to personalize cancer treatments. Ann Oncol 26(10):2034–2043

Lee J-Y, Kim HS, Song Y-S (2012) Genistein as a potential anticancer agent against ovarian cancer. J Tradit Complement Med 2(2):96–104

Li J, Xu Q, Fu C, Zhang Y (2013) A dramatically enhanced electrochemiluminescence assay for CA125 based on dendrimer multiply labeled luminol on Fe3O4 nanoparticles. Sens Actuators B: Chem 185:146–153

Liakakos T (2008) Early signet ring cell or diffuse-type gastric cancer: risks for endoscopic mucosal resection. Ann Surg Oncol 15(12):3620

Liang Z, Lu Z, Zhang Y, Shang D, Li R, Liu L, Zhao Z, Zhang P, Lin Q, Feng C (2019) Targeting membrane receptors of ovarian cancer cells for therapy. Curr Cancer Drug Targets 19 (6):449–467

Lisio M-A, Fu L, Goyeneche A, Gao Z-h, Telleria C (2019) High-grade serous ovarian cancer: basic sciences, clinical and therapeutic standpoints. Int J Mol Sci 20(4):952

Lo-Ciganic W-H, Zgibor JC, Bunker CH, Moysich KB, Edwards RP, Ness RB (2012) Aspirin, non-aspirin nonsteroidal anti-inflammatory drugs, or acetaminophen and risk of ovarian cancer. Epidemiology (Cambridge, Mass) 23(2):311

McLemore MR, Miaskowski C, Aouizerat BE, Chen L-m, Dodd MJ (2009) Epidemiologic and genetic factors associated with ovarian cancer. Cancer Nurs 32(4):281

Momenimovahed Z, Tiznobaik A, Taheri S, Salehiniya H (2019) Ovarian cancer in the world: epidemiology and risk factors. Int J Women's Health 11:287

Moore RG, McMeekin DS, Brown AK, DiSilvestro P, Miller MC, Allard WJ, Gajewski W, Kurman R, Bast RC Jr, Skates SJ (2009) A novel multiple marker bioassay utilizing HE4 and CA125 for the prediction of ovarian cancer in patients with a pelvic mass. Gynecol Oncol 112 (1):40–46

Mori M, Harabuchi I, Milyake H, Casagrande JT, Henderson BE, Ross RK (1988) Reproductive, genetic, and dietary risk factors for ovarian cancer. Am J Epidemiol 128(4):771–777

Ness RB, Grisso JA, Cottreau C, Klapper J, Vergona R, Wheeler JE, Morgan M, Schlesselman JJ (2000) Factors related to inflammation of the ovarian epithelium and risk of ovarian cancer. Epidemiology (Cambridge, Mass) 11(2):111–117. https://doi.org/10.1097/00001648-200003000-00006

Ness RB, Goodman MT, Shen C, Brunham RC (2003) Serologic evidence of past infection with Chlamydia trachomatis in relation to ovarian cancer. J Infect Dis 187(7):1147–1152

Nolen BM, Lokshin AE (2013) Biomarker testing for ovarian cancer: clinical utility of multiplex assays. Mol Diagn Ther 17(3):139–146

Pavlik E, DePriest P, Gallion H, Ueland F, Reedy M, Kryscio R, van Nagell J Jr (2000) Ovarian volume related to age. Gynecol Oncol 77(3):410–412

Purdie DM, Bain CJ, Siskind V, Webb PM, Green AC (2003) Ovulation and risk of epithelial ovarian cancer. Int J Cancer 104(2):228–232

Qi Z, Yin L, Xu Y, Wang F (2018) Pegylated liposomal-paclitaxel induces ovarian cancer cell apoptosis via TNF-induced ERK/AKT signaling pathway. Mol Med Rep 17(6):7497–7504

Ravalli A, Dos Santos GP, Ferroni M, Faglia G, Yamanaka H, Marrazza G (2013) New label free CA125 detection based on gold nanostructured screen-printed electrode. Sens Actuators B: Chem 179:194–200

Reid BM, Permuth JB, Sellers TA (2017) Epidemiology of ovarian cancer: a review. Cancer Biol Med 14(1):9

Rodriguez C, Calle EE, Coates RJ, Miracle-McMahill HL, Thun MJ, Heath CW Jr (1995) Estrogen replacement therapy and fatal ovarian cancer. Am J Epidemiol 141(9):828–835

Sanguinete MMM, Oliveira PHD, Martins-Filho A, Micheli DC, Tavares-Murta BM, Murta EFC, Nomelini RS (2017) Serum IL-6 and IL-8 correlate with prognostic factors in ovarian cancer. Immunol Invest 46(7):677–688

Savla R, Taratula O, Garbuzenko O, Minko T (2011) Tumor targeted quantum dot-mucin 1 aptamer-doxorubicin conjugate for imaging and treatment of cancer. J Control Release 153 (1):16–22

Shariati M, Willaert W, Ceelen W, De Smedt SC, Remaut K (2019) Aerosolization of nanotherapeutics as a newly emerging treatment regimen for peritoneal carcinomatosis. Cancers 11(7):906

Siegel RL, Miller KD, Jemal A (2018) Cancer statistics, 2018. CA: Cancer J Clin 68(1):7–30

Spears N, Lopes F, Stefansdottir A, Rossi V, De Felici M, Anderson R, Klinger F (2019) Ovarian damage from chemotherapy and current approaches to its protection. Hum Reprod Update 25 (6):673–693

Stratton JF, Pharoah P, Smith SK, Easton D, Ponder BA (1998) A systematic review and meta-analysis of family history and risk of ovarian cancer. BJOG: Int J Obstet Gynaecol 105 (5):493–499

Sun B, Taha MS, Ramsey B, Torregrosa-Allen S, Elzey BD, Yeo Y (2016) Intraperitoneal chemotherapy of ovarian cancer by hydrogel depot of paclitaxel nanocrystals. J Control Release 235:91–98

Thull DL, Vogel VG (2004) Recognition and management of hereditary breast cancer syndromes. Oncologist 9(1):13–24

Vergara D, Bellomo C, Zhang X, Vergaro V, Tinelli A, Lorusso V, Rinaldi R, Lvov YM, Leporatti S, Maffia M (2012) Lapatinib/Paclitaxel polyelectrolyte nanocapsules for overcoming multidrug resistance in ovarian cancer. Nanomed: Nanotechnol Biol Med 8(6):891–899

Vetter MH, Hays JL (2018) Use of targeted therapeutics in epithelial ovarian cancer: a review of current literature and future directions. Clin Ther 40(3):361–371

Walker GR, Schlesselman JJ, Ness RB (2002) Family history of cancer, oral contraceptive use, and ovarian cancer risk. Am J Obstet Gynecol 186(1):8–14

Wang WW-S, Das D, McQuarrie SA, Suresh MR (2007) Design of a bifunctional fusion protein for ovarian cancer drug delivery: single-chain anti-CA125 core-streptavidin fusion protein. Eur J Pharm Biopharm 65(3):398–405

Wang L, Neoh KG, Kang E-T, Shuter B (2011) Multifunctional polyglycerol-grafted Fe3O4@ SiO2 nanoparticles for targeting ovarian cancer cells. Biomaterials 32(8):2166–2173

Xu C, Yuan Z, Kohler N, Kim J, Chung MA, Sun S (2009a) FePt nanoparticles as an Fe reservoir for controlled Fe release and tumor inhibition. J Am Chem Soc 131(42):15346–15351

Xu M, Barrett SL, West-Farrell E, Kondapalli LA, Kiesewetter SE, Shea LD, Woodruff TK (2009b) In vitro grown human ovarian follicles from cancer patients support oocyte growth. Hum Reprod 24(10):2531–2540

Yang W-L, Lu Z, Bast RC Jr (2017) The role of biomarkers in the management of epithelial ovarian cancer. Expert Rev Mol Diagn 17(6):577–591

Zhang L, Hou S, Mao S, Wei D, Song X, Lu Y (2004) Uptake of folate-conjugated albumin nanoparticles to the SKOV3 cells. Int J Pharm 287(1–2):155–162

Zhang X-y, Chen J, Zheng Y-f, Gao X-l, Kang Y, J-c L, Cheng M-j, Sun H, Xu C-j (2009) Follicle-stimulating hormone peptide can facilitate paclitaxel nanoparticles to target ovarian carcinoma in vivo. Cancer Res 69(16):6506–6514

Zheng J, Muhanna N, De Souza R, Wada H, Chan H, Akens MK, Anayama T, Yasufuku K, Serra S, Irish J (2015) A multimodal nano agent for image-guided cancer surgery. Biomaterials 67:160–168

Zhu Y, Chandra P, Shim Y-B (2013) Ultrasensitive and selective electrochemical diagnosis of breast cancer based on a hydrazine–Au nanoparticle–aptamer bioconjugate. Anal Chem 85 (2):1058–1064

Wang WWS, Das D, McQuarrie SA, Suresh MR (2007) Design of a bifunctional fusion protein for ovarian cancer drug delivery: Single-chain anti-CA125 core-streptavidin fusion protein. Eur J Pharm Biopharm 65:398–405

[illegible]

Nanotechnology: An Emerging Field in Protein Aggregation and Cancer Therapeutics

11

Sonali Jena, Sonali Mohanty, Monalisha Ojha, Kumari Subham, and Suman Jha

Abstract

Nanoparticles (NPs) prominence in technological advancements in different fields of biology and medicine are for their tuneable physicochemical properties, like size and surface functionality. For the fact that NP composition can be engineered to exhibit unique functions, it has paved the path for development of novel tools and techniques with better precision and efficacy in different research fields including biomedical research. For instance, single-walled carbon nanotubes have shown an impressive potential as an efficient delivery agent for transportation of biomolecules to cells. Additionally, NPs with size-tuneable light emission property, which made it accessible to relatively non-accessible area, have been implemented to precisely deliver drug and get images of tumour sites. Thus, the primary driving forces behind the progress of nanotechnology in different fields are their surface area to volume ratio and tuneable physicochemical properties. Besides the tuneable accessible surface, NPs also has two other layers, i.e. the shell layer and the core, which have relatively less role in most of the nanotechnology-mediated biological applications. Since accessible surface area credits for most of the application, it is mostly discussed layer than other two layers. Despite knowing the physicochemical properties of NPs, understanding its behaviour to biological surfaces for its multifaceted functions becomes priority for its biological applications. Upon introduction of metallic nanoparticles into biological milieu, its surface interacts with different biological surfaces, including biomolecular surfaces of varying curvatures, to attain a stable local free energy minima state. In doing so, it forms a corona of biomolecular surfaces around it, which in turn further decides the fate of the complex like the agglomeration,

S. Jena · S. Mohanty · M. Ojha · K. Subham · S. Jha (✉)
Department of Life Science, National Institute of Technology Rourkela, Rourkela, Odisha, India
e-mail: jhas@nitrkl.ac.in

M. Arakha et al. (eds.), *Bio-Nano Interface*,
https://doi.org/10.1007/978-981-16-2516-9_11

molecular stability, core/surface dissolution rate, toxicity. Thus, the chapter evaluated different nanoparticle-based theranostic applications in protein amyloidogenesis and cancer; both are from the group of incurable human diseases.

Keywords

Nanoparticles · Protein–nanoparticle interaction · Protein aggregation · Cancer · Theragnostic

Abbreviations

APP	Amyloid precursor protein
CTC	Circulating tumour cell
CTL	Cytotoxic T lymphocytes
DMPC	Dimyristoylphosphatidylcholine
DMPG	Dimyristoylphosphatidylglycerol
DMSO	Dimethyl sulfoxide
FDA	Food and Drug Administration
hIAPP	Human islet amyloid polypeptide
MNP	Magnetic nanoparticle
MRI	Magnetic resonance imaging
NAC	Non-amyloid β-peptide component
NFT	Neurofibrillary tangles
NLS	Nuclear localization signal
NP	Nanoparticle
OI	Optical imaging
PD	Parkinson's disease
PEI	Polyethylenimine
PLGA	Poly(*n*-butyl cyanoacrylate)
QD	Quantum dot
RNAi	RNA interference
ROS	Reactive oxygen species
siRNA	Small interfering RNA
SPIONS	Supramagnetic iron oxide nanoparticle
TIIDM	Type 2 diabetes mellitus

11.1 Introduction

NPs can be categorized on the basis of their shape, size, compositions and functionalities. The metal nanoparticles, like iron, copper, gold, silver NPs, have promising applications in the field of biological science and medicine for the ease of tuning its physicochemical properties, including the surface plasmon resonance (SPR), which falls in visible region for the certain NPs (Sun and Xia 2003; Wang and Wang 2014). Metal nanoparticles, when introduced to biological milieu, interact with biological moieties with varying affinity to form a corona around nanoparticle core. The biomolecular composition of the corona remains dynamic, mostly because of varying affinity biomolecular surfaces from the local milieu for nanoparticle, as it moves through the organism system (Saptarshi et al. 2013). Thus, the interaction with corona brings changes to the physicochemical properties of nanoparticle and biomolecules' physiological function, like optical properties of gold colloid that ultimately aids in photometric-mediated applications (Raoufi et al. 2018). Similarly, fibronectin protein underwent significant conformational changes upon interaction with the gold nanoparticle (Raoufi et al. 2018). Conversely, α-synuclein interaction with ZnONP further stabilizes the protein conformation against its amyloidogenic propensity (Asthana et al. 2020). In the same line of exploring physicochemical property of NP applications, magnetic NPs, like iron oxide nanoparticle, stimulated with appropriate biomolecules such as antibodies, oligonucleotides, lectins, etc. can be adequately used to segregate biomolecules and cells of interest from complex biological systems for better diagnosis, detection and therapeutics (Bruce and Sen 2005; Lin et al. 2007a; Molday and Mackenzie 1982).

When NPs, acting as platform with multifunctional surface, recognize diversified surface of bio-macromolecules, it becomes a potential tool for regulating cellular and extracellular processes for several biological applications, like enzymatic inhibition, transcription regulation, antigen presentation, drug delivery and as sensor. Depending on the core material, the size of NPs can be adjusted from 1.5 nm to more than 10 nm which offers a suitable platform for interaction of NPs with proteins and other biomolecules (Hostetler et al. 1998). Interestingly, the particle with smaller size, i.e. higher curvature, of similar affinity to biomolecule has insignificant effect on biomolecular conformation and functions (Rabbani et al. 2014). However, surface potential difference at the interacting interface has different effect on the biomolecular dynamics and function; the higher potential difference results in stronger interaction at the interface, consequentially result in drastic conformational perturbation (Cukalevski et al. 2011). For the multifaceted consequences of NP on biomolecule surfaces, the NP–biomolecule interfacial interaction has become an interesting area of research in nanobiosciences. The details of NP–protein interaction in relation to its role in conformational changes towards amyloidogenesis have been discussed in detail, in the follow-up section of the chapter.

In addition to NP–protein interaction, multivalent surface functionality of NP has also been utilized to boost low-affinity interactions such as carbohydrate–protein interactions (Lee 1992). For instance, mannose-functionalized gold nanoparticles have been prepared and their interaction was investigated with lectin, concanavalin A (Lin et al. 2003). The bio-functionalization of NPs through non-covalent interactions offers a highly transposable NPs-mediated approaches. Coming to NP–DNA interactions, it is driven predominantly through electrostatic interactions, intercalation, groove binding and complementary single-strand DNA binding (Armitage 2005). NP interaction with DNA is strikingly analogous to protein–

DNA interactions, for example, the complementary electrostatic interaction is one such approach for the DNA assembly around histone protein, which also endorses for the high affinity NP–DNA interaction. In a study, where nano gold cluster (MMPC1) was used to recognize a 37-mer DNA duplex, cationic ligands functionalized NP surfaces offered a complementary surface for binding of the DNA backbone (Lakowicz et al. 2000; Mahtab et al. 2000; McIntosh et al. 2001). In such interaction between complementary surfaces, a particular group of nanoparticles, i.e. the NPs with photocatalytic/SPR property, has shown to produce ROS upon interaction, consequently bringing changes directly and/or indirectly in biomolecules, these changes often cause cell degeneration, as shown in Fig. 11.1 (Arakha et al. 2017).

Hence, the biomolecules interaction with different nanoparticle surfaces results in either loss of the physiological function or enhances the function/conformation of biomolecule. In earlier case, the nanoparticle surfaces need to be moderated to lower the deteriorative effect on biomolecules. Whereas the latter case offers multiple nanoparticle-mediated strategies and advantages in biomedical research like molecular and cell imaging, bioassays, biosensing, etc.

11.2 Nanoparticle-Mediated Applications in Biology and Medicine

11.2.1 Nanoparticles in Biosensor

The basic principle of biomedical diagnosis, forensic analysis and environmental monitoring is to sense biological agents, disease causing agent and toxic materials present in minute quantity in biological milieu (Otsuka et al. 2001). Since an ideal sensor typically consists of two elements, a recognition element to bind specific target and a transduction element to signal the binding event, efforts have been made to explore dual properties of photocatalytic nanoparticle, i.e. enhanced surface functionalities and photocatalytic-based cascade of signal for the binding event (De et al. 2008). For example, photocatalytic NPs such as gold NP have been adopted by Mirkin's group as bio-barcode platform for its unique surface chemistry of signal transduction amplification (Hill and Mirkin 2006; Nam et al. 2003). Nowadays with emerging sensing techniques, different NP surfaces have been fabricated to work on different sensing principles, like fluorescence sensing, colorimetric sensing and electrochemical sensing, to detect the target in real-time at a very low physiological concentration (De et al. 2008, 2009).

The Mirkin's group found that the gold nanoparticle-based bio-barcode probe can detect femtogram per millilitre of prostate-specific antigen (PSA), a common cancer biomarker that aids to indicate the existence of prostate cancer when expressed at elated levels (Lilja et al. 2008; Nam et al. 2003). Rotello's group has developed another approach using the optical properties of NP surface to detect multiple target analytes (You et al. 2007). He observed that anionic fluorescent protein that remains reversibly bound with cationic gold NP through electrostatic interactions can be

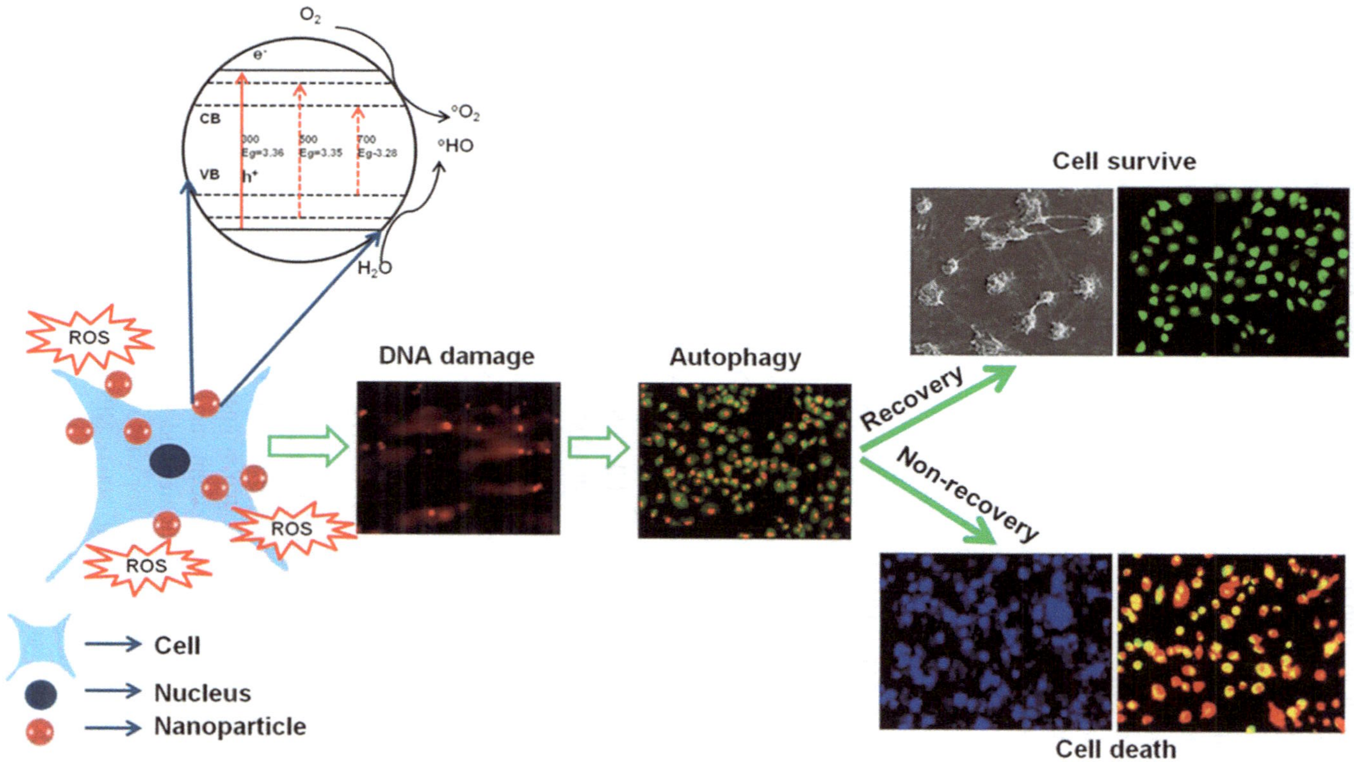

Fig. 11.1 The cytotoxic propensity of a NP depends upon its physio-chemical properties. ZnONP band gap energy is modulated by its calcination temperature which in turn regulates cell cytotoxicity. Here, the cytotoxicity of fabricated ZnONP is depicted on HT1080 cells. Narrowing the band gap of ZnONP enhances the intracellular ROS generation. This enhanced oxidative stress also causes DNA damage which is shown by comet assay and γH2AX activity assay. Some cells further try to recover the damage by undergoing $SubG_1$ phase and autophagy, which is confirmed by increased expression of LC3II and acridine orange binding assay. But the other cells that undergo non-recoverable damage, however, attain apoptotic cell death as shown by Annexin V apoptosis assay and 4,6-Diamidino-2-phenylindole dihydrochloride (DAPI) staining adopted from Arakha et al. (2017)

replaced by different anionic biomolecules (analytes) from biological milieu according to their relative affinity for the cationic nanoparticle surface, and a significant difference in fluorescence signal is observed. Hence, he coined the term of "chemical nose/tongue" for the nanoparticle-based array to detect multiple target analytes. This process is now employed for detection of specific proteins, bacteria, physicochemical differences between healthy/cancerous/metastatic human breast cells, etc. (Bajaj et al. 2009; De et al. 2009; Phillips et al. 2008). In addition to optical property of gold NP, the surface-enhanced Raman scattering (SERS) property of the particle has also been adopted for ultrasensitive detection of biomolecules such as haemoglobin, glucose, bacteria and viruses (Shafer-Peltier et al. 2003; Shanmukh et al. 2006; Xu et al. 1999; Yan et al. 2009). There has also been a rapid development to exploit the magnetic properties of certain magnetic nanoparticles (MNPs), such as SPIONs (supramagnetic iron oxide nanoparticles), for the sensors (Haun et al. 2010).

The magnetic biosensors though are one of the routinely used methods for detection of bacteria that entails the direct immunological reactions between antibodies coated magnetic NPs and surface antigens (Varshney and Li 2007; Xia et al. 2006). The Weissleder group has also applied the approach to detect *M. tuberculosis* and identified the drug-resistant strains in a period of 30 min only (Lee et al. 2009). In another study, a specific nucleic acid functionalized NP platform was used to recognize the bacteria with its 16S rRNA with a very high sensitivity, i.e. 1–2 E. coli in 10 ml of blood, and precision in determining bacterial load (Chung et al. 2013). Many other groups have also utilized functionalized MNPs to efficiently label bacteria. Metallic NPs and QDs have also been recently used for optical-based tracking of specific bacteria. Like bacteria detection, specific molecule functionalized MNPs have also been adopted to track the circulating tumour cells (CTCs) for detection and separation of the cell at early stages of cancer progress (Xu et al. 2011). The different nanoparticle-based approaches for proper detection and separation of cancer cells have been discussed in later sections of the chapter.

11.2.2 Nanoparticles in Bioimaging

In the field of molecular imaging, NPs have been widely investigated as novel labels and contrast agents. The advancement in bioimaging technology is credited to the current evolution of luminescent and magnetic NPs (Sharma et al. 2006; Tan et al. 2004), among these two luminescent nanoprobes are used for optical imaging (OI) and magnetic nanoparticles are used for magnetic resonance imaging (MRI). There also exist a two-fold NPs for concurrent imaging by OI and MRI (Kircher et al. 2003; Schellenberger et al. 2004). NPs possess many promising features for targeted imaging. Firstly, NPs due to their larger surface area are capable of delivering a large number of imaging agents at a time, leading to sensitivity improvement (Wang and Wang 2014; Welch et al. 2009). Secondly, NPs for its enhanced permeability and retention (EPR) effect can passively target tissues in vivo or increase the local concentration of contrast agents by targeted accumulation at specific sites (Wang and Wang 2014).

Mono-functionalized QDs have a wide applicability, like tracking of specific protein in cells or receptors involved in cell movement during development and metastasis (Clarke et al. 2010). Lately, QDs have also come into picture for multiplex molecular imaging of embryonic stem cells, lymph nodes, as well as tumour cells and vasculature (Ballou et al. 2007; Gao et al. 2004; Kang et al. 2009; Lin et al. 2007b). Whereas gold NPs provide a non-invasive approach for in vivo cancer imaging by means of small organic molecules as near-infrared (NIR) SERS reporters (Wang and Wang 2014). Magnetic NPs form another category of well-studied NP for molecular targeted imaging. Magnetic NP imaging systems has aided in real-time visualization of biological events, from cell migration/trafficking, enzyme activities (Bulte et al. 2001; Chien et al. 2011; Mahmoudi et al. 2013; Tung et al. 2000) and other biological interactions at molecular and cellular levels. Metal-doped ferrite NPs ($MnFe_2O_4$) have also came into picture with its advanced magnetics properties constantly being pursued for envisioning biological events (Lee et al. 2007). It can be used to make better MRI probes due to its non-toxic nature in vitro and higher magnetic susceptibility. Conjugated NPs with antibodies have shown to enhance MRI sensitivity for the detection of cancer markers (Wang and Wang 2014; Wang et al. 2015).

11.2.3 Nanoparticles in Drug Delivery

In improving some vital properties of "free" drugs, drug delivery systems (DDSs) play its part well from increasing solubility, in vivo stability, pharmacokinetics and bio-distribution to enhance the efficacy of drug (Allen and Cullis 2004). For instance, Derfus, A. M. et al. in 2007 have successfully designed multifunctional superparamagnetic NPs for far-off release of bound drugs (Derfus et al. 2007). Delivering various agents to specific organelle(s) are the present focus of many investigators. There are many molecular processes taking place in organelles that are far from the sight of understanding and demands, proper investigation has resulted in the development of potentially active delivering agents. The scope of NPs in this field is worthwhile. For proper and effective targeting, the NPs carrying oligonucleotides have to escape the endosome first. Various tools for active subcellular targeting are emerging for targeted delivery to the cytosol, nucleus, mitochondria, lysosomes and endosomes (Bareford and Swaan 2007; Lloyd 2000; Pouton et al. 2007; Vasir and Labhasetwar 2007; Yamada and Harashima 2008). Generally, there are two approaches being utilized for designing NPs for subcellular targeting: (1) Passive targeting by NPs due to its varied characteristics such as size, shape and composition (Xu et al. 2008), (2) Active targeting by functionalizing NP surfaces with the specific ligands directed on the road to the specific organelle. In one study, it was shown that the nuclear localization signal (NLS) conjugated gold NPs targets at the cancer cell nucleus to damage DNA (Kang et al. 2010). So far, for delivery to the mitochondria metal oxide or liposomal NPs have been primarily investigated. The electrostatic interactions between the NP and the organelle have also been used for the delivery to mitochondria and have been utilized for cancer research (Boddapati et al. 2008; Breunig et al. 2008). By in vitro screening of a library of NPs with varied charge and size, it has been concluded that an optimized targeted NP improves the efficiency and reduces the toxicity for cancer, Alzheimer's disease and obesity compared to non-targeted NPs or other small molecule therapeutics (Marrache and Dhar 2012).

Small interfering RNA (siRNA) is a non-coding small double stranded RNA molecule which interferes with the expression of a certain protein by RNA interference (RNAi) of respective mRNA. Hence, siRNA has been used as a biological tool to control the transcriptome of a cell that is yet to optimize for its efficient and biocompatible delivery systems. NPs, here, offer a possible solution to the obstacles in siRNA delivery. In order to transport anionic nucleic acids into cells, cationic lipid or polymeric NPs with their ability to form a condensed complex with nucleic acids have been in use (Behlke 2006). Lately, in order to deliver siRNA to silence genes of immune cells (playing pivotal roles in homeostasis and disease), NPs have been used (Peer et al. 2008). To study the cellular pathways at the single cell level in plant cells, NPs have also been used as a vehicle for delivering siRNA (Silva et al. 2010).

Hydrophobic compounds, instead of being biologically active, are not frequently utilized in biological research due to its poor solubility in aqueous medium. The widely used approaches involve the use of solvents like dimethyl sulfoxide (DMSO) or an excipient like cremophor. But their limited solubilization and frequent toxicities can complicate the biological experiments. One of the strategies to overcome the challenge of delivering these hydrophobic agents is to utilize NPs. Many NPs, especially polymeric NPs, have hydrophobic cores and are well suited for the delivery of hydrophobic agents. The use of NP delivery vehicles not only overcomes the problem of solubility of the active agent, but also protects the agent from the environment until the agent is released from NP (Wang and Wang 2014). Karve, S. et al. have shown that NP wortmannin works as an active and potent therapeutic agent in vivo in a cancer mouse model (Karve et al. 2012).

11.3 Nanoparticle–Protein Interaction and Protein Aggregation

Nanoparticles (NPs), as previously discussed in the chapter, due to their nano-size and large surface to mass ratio have relatively larger surface to interact with biomolecules in biological milieu. In a living system, the interaction of NPs with proteins serve as the basis of bio-reactivity and leads to the formation of potent nanoparticle–protein corona (NP-PC). The adsorption of protein at nano–bio interface is driven by several factors, namely NP curvature, different interactions at interface, solvation forces, etc. The formation of NP-PC is a multifarious process that depends on different factors such as characteristics of nanoparticle, amount of the interacting protein and its medium, affinity of the protein and its ability to occupy the surface of nanoparticle (Saptarshi et al. 2013). The binding of protein onto nanoparticle surface is likely to induce conformational rearrangements onto the adsorbed protein, thereby anticipated to impede them in folding pathways (Cabaleiro-Lago et al. 2010b; Fei and Perrett 2009; Kopp et al. 2017) and affect the overall bio-reactivity of nanoparticle (Saptarshi et al. 2013). Moreover, the higher curvature in nanoparticle provides additional flexibility to adsorbing protein perturbing relatively less intramolecular interactions, in comparison to the planar surfaces (Verma and Stellacci 2010). However, some proteins' multilayer adsorption and conformational rearrangement on interaction with certain nanoparticle interface lead to irreversible amyloid-like fibril formation (Colvin and Kulinowski 2007; Wagner et al. 2010; Worrall et al. 2006). These amyloid-like fibrils, directly or indirectly, are associated with degenerative diseases such as type II diabetes mellitus, Parkinson's and Alzheimer's disease (Saptarshi et al. 2013). Naturally found different polyphenols of curcumin, epigallocatechin-3-gallate and resveratrol as well as different ranges of nanoparticles (ceria, carbon nanotubes, copolymers, quantum dots, graphene sheets, iron oxide, gold, silver and polymeric NPs) have shown to moderate the fibrillation propensity of different proteins (human β2-microglobulin, α-synuclein, hIAPP, prion, amyloid-β) via conformational rearrangement upon interaction into either anti-amyloidogenic or amyloidogenic conformation (Ehrnhoefer et al. 2008; Fischer et al. 2010; Gurzov et al. 2016; Jana and Sengupta 2012; Linse et al. 2007; Mahmoudi et al. 2012; Pithadia et al. 2016). Hence, different nanoparticle–polyphenol formulations have been evaluated for more efficient anti-amyloidogenic properties, as discussed in following section.

11.3.1 Nanoparticles in Type II Diabetes Mellitus

Insulin and human islet amyloid polypeptide (amylin/hIAPP) are endocrine hormones known to regulate glucose level in blood. These two hormones are co-expressed, co-processed and co-secreted in response to same secretagogue by pancreatic β-cells. In the proteome of the pancreas, these two proteins have highest amyloidogenic propensity despite the different measures that cells have adopted to keep them in physiologically active form with the evolutionary development (Baker et al. 1988; Brange 2012; Brange et al. 1997; Michael et al. 1987). These two proteins have been predominantly observed in amyloid plaques taken from type II diabetes mellitus (TIIDM) patients. Since these amyloid plaques are known to be cytotoxic, their formation results in degradation of β-cells in pancreas and anticipated as causative agent for TIIDM occurrence in 95% of the patients (Gupta and Leahy 2014). Fibrillation of insulin is not only present in physiological condition but also can happen in the site of insulin infusions in patients, during long time storage, delivery or manufacturing of insulin. Therefore, fibrillation of insulin is one of the major concerns in the field of diabetology (Hua and Weiss 2004; Meesaragandla et al. 2020). There is no therapeutics to reverse the fibrillation of the proteins, though there are approaches to control further progress of the disease. Therefore, prevention of insulin/hIAPP fibrillation has been considered as one of the remedial perspectives for diabetes treatment (Zhang et al. 2016). The association between NPs and aggregation prone proteins has been a remarkable approach in the field of research and technology owing to its small dimension that allows easy access to cellular barriers and biological membranes (Chithrani et al. 2006; Goodman et al. 2004; Gulati et al. 2013; Verma et al. 2008). Intervention of NPs in different stages of fibrillation pathway may lead towards amyloid cytotoxicity in biological systems (Peretz et al. 2018). Some NPs such as inorganic quantum dots (Yoo et al. 2011), silver nanoparticles (AgNPs) (Anand et al. 2016), fullerenes (Kim and Lee 2003), etc. show inhibitory effect on formation of amyloid aggregates. Additionally, different NP-mediated approaches has been adopted to further enhance the anti-amyloidogenic property of phenolic compounds like resveratrol functionalized gold nanoparticles (Zhang et al. 2016). However, there are reports of different NPs, like polymeric NPs consisting of polystyrene or N-isopropylacrylamide, titanium oxide (TiO_2 NPs) (Wu et al. 2008), enhancing the fibrillation by decreasing aggregation lag time (Cabaleiro-Lago et al. 2010a). Recent studies have shown that membrane-associated fibrillation of the peptide has been restructured by almost all NPs and the lipid bilayer interaction has been strengthened (Peretz et al. 2018).

AuNPs (Gold nanoparticles) are one of the most common NPs used in the field of theranostics due to its inert, biocompatible and tunable physicochemical properties. In a study conducted by Meesaragandla et al., several biopolymers, like dextrans, chitosan, functionalized AuNPs inhibited the insulin fibrillation (Bhumkar et al. 2007; Meesaragandla et al. 2020; Wagner et al. 2010). The insulin interaction with the functionalized AuNPs restricted the conformational transition from α-helix to amyloid-like cross β-sheet structure (Meesaragandla et al. 2020). Whereas the graphene oxide nanosheet (GONP) interaction with hIAPP, stabilized the protein into coil-like conformation instead of its self-assembly into amyloid-like fibrils (Nedumpully-Govindan et al. 2016). Likewise, resveratrol (trans-3,4,5-trihydroxy-stilbene, a phenolic compound found in red wine) functionalized AuNP remodelled the membrane-associated hIAPP fibrillation, when done in presence of DMPC (dimyristoylphosphatidylcholine) and DMPG (dimyristoylphosphatidylglycerol) vesicles in vitro. It was shown that the NPs stabilized the transient helical conformation of hIAPP in presence of DMPC: DMPG vesicles, thereby inhibited the hIAPP fibrillation. Additionally, amine-functionalized polystyrene and titanium oxide nanoparticles enhanced the DMPC: DMPG vesicle-mediated hIAPP fibrillation (Peretz et al. 2018).

The bare ZnONPs (Zinc oxide nanoparticles) and AgNPs with negative surface potentials have shown to enhance the proteins fibrillation, Fig. 11.2 (Alvarez et al. 2013; Wang et al. 2017). The surface enhanced the fibrillation by inducing conformational rearrangement towards the amyloidogenic conformation upon interaction with the particle surface (Asthana et al. 2020). However, functionalizing the surface with citrate or branched polyethyleneimine significantly inhibited the protein fibrillation, largely by sequestering the protein monomers against the self-assembly into amyloid fibrils (Wang et al. 2017).

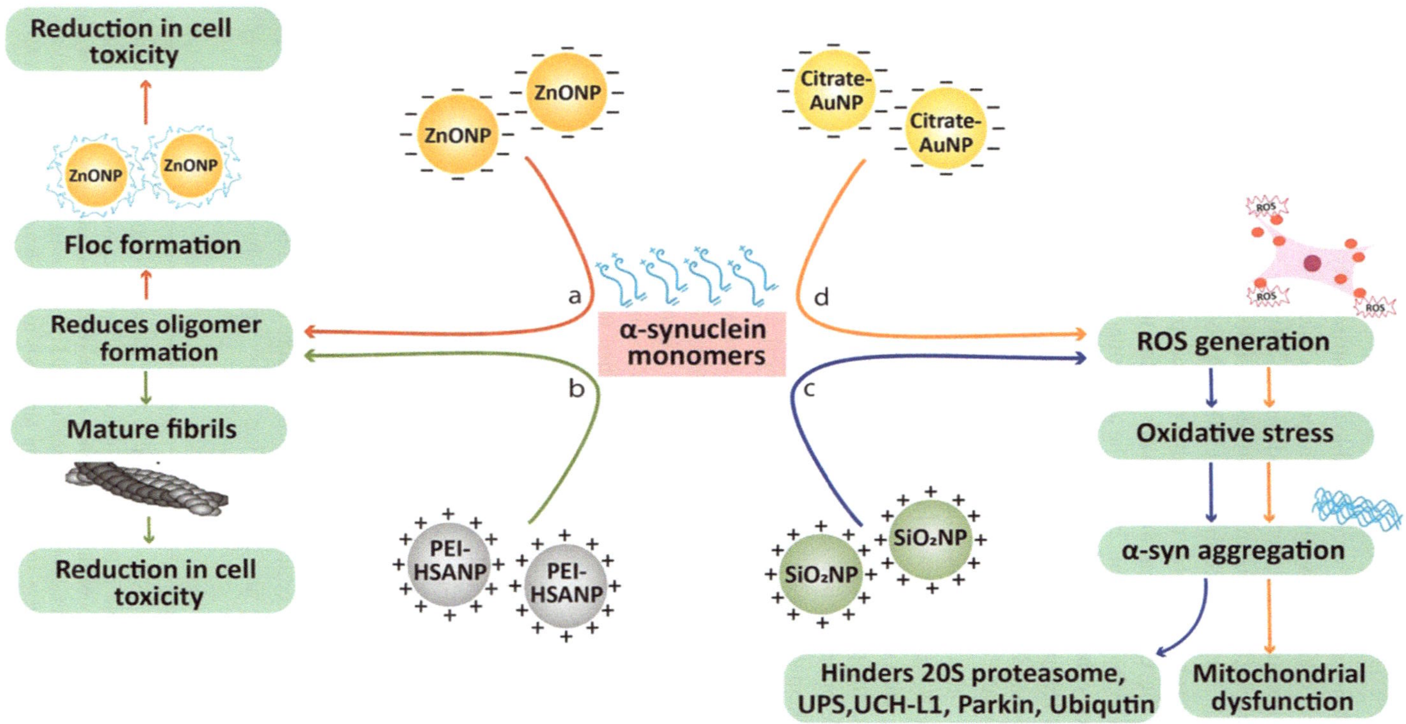

Fig. 11.2 (**a**) The interaction of negatively charged ZnONP with positively charged N-terminus of α-synuclein (α-syn) leads to formation of stable amyloidgenic conformation, i.e. flocs that are less cytotoxic in nature (Asthana et al. 2020). (**b**) The negatively charged HSA when coated with PEI renders an overall positive charge. The interaction of such NP with α-syn leads to reduction of oligomer formation and instead forms mature fibrils that are less toxic in nature and minimize cell toxicity (Mohammad-Beigi et al. 2015). (**c**) The interaction of positively charged SiO_2NP with α-syn leads to ROS generation and

11.3.2 Nanoparticles in Parkinson's Disease

Alpha-synuclein (α-syn), a small, soluble and intrinsically disordered protein extensively found in brain tissues, plays an important role in the onset and progression of Parkinson's disease (PD) (Uversky and Eliezer 2009). α-Syn consists of three main regions: an amphiphilic N-terminal part responsible for interacting with membranes (Maltsev et al. 2013; van Rooijen et al. 2009), a non-amyloid β-peptide component (NAC) region as hydrophobic region that makes up the core component of amyloid fibrils, an acidic C-terminus that remains unstructured in all forms of fibrils and monomers (van Rooijen et al. 2009). The accumulation of amyloid aggregates of the protein in lewy bodies, found in dopaminergic neurons of the midbrain, is a characteristic feature of PD (Xie and Wu 2016). Numerous NPs have been designed to transport small hydrophilic molecules across the blood–brain barrier (BBB) that offers homeostatic defence system in brain (Kanazawa et al. 2013; Patel et al. 2013). In this case, evaluation of the nanoparticle interfaces for its interaction with proteins from brain cells, i.e. α-syn, amyloid-β and tau proteins, becomes important. Several studies have shown NPs of varying sizes and structures interacting with different proteins affect the protein fibrillation propensity; however, there seems to be limited data available on interaction of α-syn with NPs and its aggregating mechanisms, as shown in schematic Fig. 11.2 (Cabaleiro-Lago et al. 2012; Huang et al. 2013; Marshall et al. 2011; Mirsadeghi et al. 2015; O'Brien et al. 2011).

Metal nanoparticles with negative surface potentials, like AuNP and ZnONP, have shown to inhibit the α-syn fibrillation propensity by either sequestering the monomeric protein (Yang et al. 2013) or by trapping it in non-amyloidogenic confirmation (Asthana et al. 2020). The predominantly positively charged N-terminus of α-syn interacts with such NP interface and intensifies the conformational entropy that kinetically confines α-syn in stable non-amyloidogenic conformation, i.e. flocs. These off-pathway aggregates significantly showed reduced cytotoxicity towards IMR32 cells (human neuroblastoma cell line) and THP-1 cells (human monocytic cell line) in vitro, which indicated the potential therapeutic approach of ZnONP in α-syn amyloidogenesis, Fig. 11.2a (Asthana et al. 2020). On the contrary, albumin-based nanoparticle with negative surface potential showed insignificant interaction with monomeric α-syn. Hence, the protein-based nanoparticle had no effect on α-syn fibrillation. Interestingly, when the surface potential was reversed by coating with positively charged polyethylenimine (PEI), it enhanced the primary nucleation and fibrillation. However, it reduced the cytotoxic nature of the resulting fibrils, Fig. 11.2b (Mohammad-Beigi et al. 2015). Thus, this protein-based non-immunogenic

Fig. 11.2 (continued) oxidative stress (Wu et al. 2011; Xie and Wu 2016). SiO2-NPs also hinder 20S proteasome activity and reduce Parkin, ubiquitin, ubiquitin carboxy-terminal hydrolase L1 (UCH-L1) protein levels in ubiquitin-proteasome system (UPS) as a result of aggregation of α-syn (Xie and Wu 2016). (**d**) Interaction of negatively charged citrate coated AuNP with α-syn leads to ROS generation and oxidative stress which in turn leads to mitochondrial dysfunction (Alvarez et al. 2013). Red: ZnONP, Green: PEI-HSANP, Blue: SiO_2NP, Orange: Citrate-AuNP

nanoparticle should also be evaluated for their use as drug carrier due to their ability to cross the blood–brain barrier, BBB (Langer et al. 2003; Mishra et al. 2006; Patel et al. 2013).

Like protein-based nanoparticle, SiO_2-NP (Silica nanoparticles) has been widely used in diagnostics, imaging and drug delivery system for many central nervous system diseases (Wang et al. 2015). Moreover, SiO_2-NP has also shown deteriorative effects on brain cells indirectly, i.e. via oxidative stress that leads to aggregation of α-syn, resulting in damage of dopaminergic neurons in the striatum, Fig. 11.2c (Wu et al. 2011; Xie and Wu 2016). Aggregation of α-syn activated the autophagy via PI3-Akt-mTOR signalling pathway in PC12 cells (dopaminergic neuron-like cell line) that can lead to neurotoxicity and development of PD. As these NPs render α-syn mediated toxicity in vitro, it develops a sense of caution over its use in biomedical field and cancer treatment in brain (Xie and Wu 2016).

11.3.3 Nanoparticles in Alzheimer's Disease

Alzheimer's disease (AD) is an age-related progressive neurodegenerative disorder that is mainly associated with aggregation of amyloid-β (Aβ) and phosphorylated tau proteins resulting in formation of amyloid plaques and intracellular neurofibrillary tangles (NFTs) (Amiri et al. 2013; Mahmoudi et al. 2013). Aβ is an amphipathic polypeptide of 36–43 amino acids and with a molecular weight of 4 kDa (Mahmoudi et al. 2013). This Aβ is generated by a cascade that consists of sequential cleavage of a precursor protein called amyloid precursor protein (APP) (Amiri et al. 2013). The Aβ self-assembly into amyloid fibril is causative agent for Alzheimer's disease. The unstructured Aβ monomer mainly consists of three regions: one highly charged stretch in the region 1–16 amino acid residues and two hydrophobic stretches in the regions of 16–20 and 30–40 residues. The hydrophobic residues are said to play a crucial part in the oligomerization, hence any perturbance in this region moderates the fibril formation (Bett et al. 2010; Liu et al. 2004). Several studies have been carried out to inhibit the aggregation by using different biological molecules such as phenols, inositols, indoles and several others (Cohen et al. 2006; Hashimoto et al. 2009; Kim and Lee 2003; Necula et al. 2007). This is where nanoparticles come into play, they can provide platform for enhanced Aβ fibrillation to decrease the cytotoxic population accumulation or/and NP surface can be moderated to influence the interactions of Aβ self-assembly in fibrils. For example, SPION with positive surface potential at lower concentrations slowed Aβ fibrillation, while at higher concentrations it enhances the fibrillation by decreasing the nucleation lag time (Cabaleiro-Lago et al. 2010b; Ikeda et al. 2006). SPION with negative surface potential or amine-functionalized tends to inhibit the fibrillation process in varying concentrations and sizes (Mahmoudi et al. 2013; Olmedo et al. 2008). Another metal oxide, TiO_2NP, surface in Wu Wei-hui et al. study also enhanced the fibrillation by decreasing the nucleation lag time. In fact, the surface enriched the pre-fibrillar protein population upon interaction with the monomeric Aβ protein (Wu et al. 2008). On contrary to NP surfaces that enhanced the fibrillation, there are also NP-based studies showing the potential inhibitory effect (Barbara et al. 2017; Cabaleiro-Lago et al. 2008; Yoo et al. 2011).

Curcuminoids, besides its anti-oxidant, anti-inflammatory properties, also have anti-amyloidogenic property. It has shown to disaggregate the Aβ fibrils and inhibit the formation of newer ones. But its solubility and bioavailability was a problem until Rouzi Barbara et al. adopted glycoprotein conjugated PLGA (poly(*n*-butyl cyanoacrylate)) nanoparticle to overcome the Curcumin bioavailability problem. Thus, this work proved the compound's potential to disintegrate Aβ fibrils and to reduce oxidative stress in the neuronal cells (Barbara et al. 2017). Another copolymers NP (N-isopropylacrylamide:N-tert-butylacrylamide, NIPAM:BAM) based approach has also been explored to inhibit the Aβ pathology. The polymeric NP inhibited the fibrillation by significantly decreasing the availability of Aβ monomer and oligomers (Cabaleiro-Lago et al. 2008). Likewise, nanogels, micelles, fullerenes and surface-modified nanogels upon interaction with Aβ peptide also stabilized the anti-amyloidogenic conformation, thereby inhibited their fibrillation and fibril-mediated toxicity (Cabaleiro-Lago et al. 2008; Ikeda et al. 2006; Kim and Lee 2003; Pai et al. 2006). N-acetyl-1-cysteine-capped quantum dots having a diameter of 3–5 nm were shown to inhibit the fibrillation of Aβ by the formation of hydrogen bonds between the NPs and Aβ protein (Xiao et al. 2010).

11.3.4 Nanoparticles in Tauopathy Disease

Tau, like α-syn protein, is an intracellular protein which plays a major role in stabilizing the microtubules in the cells (Sonawane et al. 2019; Wang and Mandelkow 2016). Tau is predominantly found in cells of CNS (Central Nervous System) and is expressed in 6 isoforms in human. It is a protein which is highly prone to phosphorylation as it consists of 85 potential threonine, serine and tyrosine phosphorylation sites. Various tauopathies are caused by aggregation of hyperphosphorylated tau proteins (Hanger et al. 2007; Tuna et al. 2009). In adult human brain, the purified tau protein shows about 10 phosphorylated residues, whereas tau isolated from tauopathic brain shows about 16 phosphorylated residues (Arai et al. 2004; Sergeant et al. 1999; Wray et al. 2008). The phosphorylation of Tau is developmentally regulated by different kinases and phosphatases. Any kind of abnormal activity in these kinases and phosphatases caused by either genetic mutations or by faulty post-translational modifications results in the hyperphosphorylation of tau, a landmark in various tauopathies including AD (Coppola et al. 2012; Gong et al. 2005; Iqbal et al. 2016). In AD, the hyperphosphorylated tau aggregates and forms paired helical filaments (PHFs) in neurons which is followed by formation of neurofibrillary tangles (Gorantla et al. 2017a, b). Tau isolated from healthy human brain is hydrophilic, unstructured and dynamic protein in solution (Mandelkow and Mandelkow 2012; Noble et al. 2013). The in vitro studies suggest that hyperphosphorylation of tau in mouse and human causes their aggregation, and dephosphorylation inhibits the aggregation and restores the normal function of tau in brain (Iqbal et al. 1998). Tau protein, although, lacks definite structure in soluble form, but attains partial β-sheet structure in its pathological state. The tau protein is composed of two main domains, namely projection and microtubule binding domain. The projection domain consists of a polyproline stretch and two inserts, while the microtubule binding domain consists of four 29 amino acid repeats. There are two hexapeptide motifs present in these repeat regions which mainly enable the transition of random coil to β-sheet, thereby causing tau protein to aggregate (von Bergen et al. 2000, 2005).

Recent studies have shown that nanoparticles have the potential to inhibit the conformational transition, thus ameliorate the tauopathy related to it. Sonwane S. K. et al. reported that metal nanoparticle coated with fungi derived protein inhibited the tauopathy to different extent by stabilizing the tau oligomers against its fibrillation. In a separate study, another metal oxide nanoparticle-mediated approach was adopted by Yoshiyama et al. to relieve the tauopathy. The group treated the tau-mutant mice with γ377-39 peptide conjugated iron oxide nanoparticle to inhibit the microglial activity; microglial activation is seen in tauopathy, and γ377-39 peptide is a known inhibitor of microglial activity. The treatment reduced the tauopathy-mediated neurodegeneration in tau-mutant mice (Adams et al. 2007; Glat et al. 2013; Luo et al. 2020; Yoshiyama et al. 2007). On the contrary, some metal nanoparticles, like Manganese oxide NPs, Nickel oxide NPs, Iron NPs, with intact surface have shown to interact with tau protein with significant or insignificant change in protein conformation. Hence, the interactions have insignificant effect on tau aggregation (Hajimohammadjafartehrani et al. 2019; Hajsalimi et al. 2018; Mehdizadeh et al. 2019).

Thus, the literature altogether, beside raising the concern over the use of unevaluated NPs for biological applications, indicated that the anti-amyloidogenic phenol/ biopolymer(s) functionalized nanoparticles has tremendous therapeutic potentials in amyloid-mediated pathologies.

11.4 Nanoparticle in Cancer

According to World Health Organization (WHO), cancer is a prime reason of death worldwide in both developing and devolved countries, approximately 9.6 million deaths in 2018, making it second most leading cause of death globally (https://www.who.int/news-room/fact-sheets/detail/cancer). About 70% deaths from cancer occur in middle- and low-income countries. Late stage presentation and failure in early stage diagnosis are the common reasons for higher death risk. Neoplasms and malignant tumours are some common name used for cancer. One of the defining characteristics of cancer is the rapid proliferation of cell that flourishes beyond their regular borderline. At later stage they can invade neighbouring parts and then transmit it to other organs, which is known as metastasis (NIH 2017; Society 2016). Lung, colorectal, pancreatic, liver and breast cancers are the common causes of cancer deaths. To turn off the occurrence of various cancers, several techniques have been demonstrated such as cancer cell screening, surgery, chemotherapy, electrosurgical excision procedure, radiation, cryotherapy and combinational therapies (Yuan et al. 2018). Traditional therapies often produce side effects and these are insufficient due to various reasons like drug resistance, chaotic and complex tumour microenvironment, bioavailability of drug, etc. In last few decades with advancement of nanotechnology, its applications for clinical analysis, imaging and therapeutics have been on fore front (Jaishree and Gupta 2012). The detection and capture of circulating tumour cells (CTCs) via NPs have drawn much attention of researchers. Additionally, nanoparticle-mediated targeted delivery of drugs in cancer therapeutics has significantly reduced the drugs dosage with more specificity, low toxicities and better bioavailability (Ferrari 2005; Nie et al. 2007; Srinivas et al. 2002; Wang et al. 2007).

11.4.1 Nanoparticles in Cancer Diagnosis

In the combat against cancer, half of the fight is won based on its early screening. Nanotechnology implements some contrast molecular agents to enable the prior and accurate diagnosis of cancerous cells as well as help in continuous monitoring of cancer patients, which has brought revolutionary changes in the field of cancer diagnosis. Although this method is not yet used clinically for early detection or diagnosis, but it is already used in various clinical practices (NIH 2017) (https://www.cancer.gov/nano/cancer-nanotechnology/detection-diagnosis). Certain nanodevices are being explored for the trap of vascular biomarkers, overexpressed proteins in malignant cells and circulating oncogenes. The biosensor designed by using nanotechnology has high sensitivity and accuracy rate. In vivo, it can detect even a single malignant cell and also has the ability of site-specific delivery of toxic drugs. For cancerous cell detection some of the nanomaterials are used, like fullerenes, quantum dots, superparamagnetic nanoparticles, dendrimers, nanodiamonds, etc. (Jaishree and Gupta 2012). These nanomaterials used specific cross-linking agents, such as particular antibodies against malignant cells for easy detection of cancer cell (Alexis et al. 2008). Individual cancer cells have some biomarkers on their surface, which provide specificity to the cancer cell. Nowadays, a novel approach for detection of biomarker quantitatively, by using lipid coated nanomaterials, i.e. quantum dots (QDs) helps/improves the histopathological diagnosis of cancer. Bruchez et al. in 1998 first used ligand conjugated CdSe QDs of semiconductor nature to image the fibroblast cells (Bruchez et al. 1998). Thereafter, it has been optimized for targeting, bioimaging, vasculature imaging and tracking of the cancerous cells (Cai and Chen 2008; Maysinger 2007; Schroeder et al. 2007; Tada et al. 2007). Like QDs, nanodiamonds have also been explored for cancer cell diagnostic for its highly biocompatible and stable photoluminescence properties (Enoki et al. 2009; Holt 2007). Cancer specific biomarker conjugated nanodiamonds show unprecedented specificity for the target cells for bioimaging. They are potential labelling and tracing agents, as they do not interrupt cell cycle and can easily conjugated with certain proteins and biomarkers. Hence, these NPs, for their optical property, can be used in bioimaging of proliferating cells in vivo, which provides important information such as tumour growth rate, angiogenesis rate, efficacy of treatment vigour of normal cells (Fu et al. 2007).

To detect and capture circulating tumour cells (CTCs) magnetic NPs have also been used, where the nanoparticle is conjugated with ligand to target and isolate the CTCs. In a preclinical study, PEGylated magnetic NPs were used for in vivo CTC detection (Galanzha et al. 2009). These magnetic nanomaterials have already been in use for magnetic resonance imaging (Wust et al. 2002). Many researchers reported that superparamagnetic nanoparticle can be associated with other nanoparticle for specific tumour targeting. This method is known as magnetic drug targeting which uses magnetic field for localization. The PEG-conjugated superparamagnetic nanoparticles have improved biocompatibility and enhanced circulation time in biological system (Harris et al. 2001). Moreover, the cancer cell-specific antibody-linked PEGylated nanoshells have also been evaluated by Hirsch et al., where the sensing is enhanced by 10 billion times and also made the approach 10,000 times more accurate than the conventional methods; for detection and targeting tumor cells. For example, the breast adenocarcinoma cells overexpressing a potent biomarker i.e., human epidermal growth factor receptor-2 are immunotargeted using nanoshells (Loo et al. 2005).

Other approaches for detection and separation of cells take into account the use of biomimetic nanotechnology. The Hong's group has developed microfluidic devices applying biomimetic approaches for the detection and separation of cells (Karnik et al. 2008).

11.4.2 Nanoparticles in Cancer Therapeutics

To turn off the occurrence of various cancers, several techniques have been demonstrated such as cancer cell screening, surgery, chemotherapy, electrosurgical excision procedure, radiation, cryotherapy and combinational therapies. Traditional therapy does not discriminate between malignant cells and healthy cells and non-selectively target malignant cells. Some factors like drug resistance, chaotic and complex tumour microenvironment, bioavailability of drugs, etc. are also responsible for the failure of traditional therapies (Jain 2001). Chemotherapy appears to be a potent therapy by using certain anticancer agents such as nitrogen mustards, Pt-based drugs and therapeutic agents like temozolomide, based upon alkylation of DNA, which have relentless side effects on cancer cells (Rasoulian et al. 2017). In radiation therapy, death of tumour cells is induced by accumulating higher amount of energy in the target site, resulting in damage of tumour cells or their vasculature (Hainfeld et al. 2006). In chemotherapy, the death of tumour cell is induced by administrating cytotoxic chemotherapeutic agents. Based upon the type of therapeutic agents used, there are different modes of action to cause cancer cell death (Crawford 2013). These two different approaches are also used combinedly for cancer treatment. Despite their outstanding clinical application, certain major drawbacks of this combined therapy are also reported (Herscher et al. 1999; Ma et al. 2003). These limitations are cytotoxic effect on normal cells, non-specific drug delivery and targeting, bioavailability of therapeutic agents, etc. Generally, the most widely used therapy which is chemotherapy showed undesired secondary responses and risks and can also able to affect bone marrow cells to reduce the immunity and increasing host susceptibility (Lebaron et al. 1988; Peppercorn et al. 2005).

To overcome these problems there is a requisite of combinational therapy, which is a combination of biocompatible nanoparticles and certain anticancer drugs. Combinational therapy is well known for its synergistic effects. In this technique, two or more therapeutic agents are used combinedly to hike the efficacy of certain chemotherapeutic agents which are used in low concentration. This technique is also trying to reduce the drug resistance by chemosensitization. Although this therapy appears to be toxic, but it can be conquered by using different therapeutic agents in lower concentration. Furthermore, this therapy may be able to check the side effects on nonmalignant cells while simultaneously having cytotoxic effects on malignant cells. This therapy is also able to diminish the possibility for drug resistance (Da Silva et al. 2016). These limitations can further be reduced through targeted delivery and increasing the local concentration of the drugs. Nanotechnology aims to mitigate the complication, coupled to diseases, at nanoscale dimensions where all biological entities coexist and cooperate. The application of nanotechnology to medicine, known as nanomedicine focuses on proposing major breakthroughs in cancer diagnosis and treatment and has the enormous potential to mitigate the problems associated with conventional therapies. Nanoparticles are capable to cross the biological barriers (bio-barriers), pile up in tumour cells and specifically target singlet cancer cells for their early detection and treatment (Wang and Thanou 2010). This interdisciplinary field combines several branches such as engineering, biology, medicine and chemistry, focusing towards major progress in early cancer detection, diagnosis and treatment.

In recent years, metal nanoparticles have been widely explored for drug delivery, targeting of diagnostic and therapeutic agents in cancer therapy (as referred in Table 11.1). Metallic NPs, being multifunctional, are supposed to have superior effects on tumour cells. The site-specific targeting feature of NPs plays a key role in cancer nanobiotechnology as they avoid hampering of normal cells and unnecessary side effects (Gindy and Prud'homme 2009). Therefore, current disease management modalities in cancer nanotechnology promise better therapy due to their tiny size and unique physicochemical properties. There are different nanomaterials, as described above for bioimaging and diagnostics, of different shapes, like nanoshells, nanotubes, nanosponges, that have been evaluated to deliver the drug to cancer site.

Nanosponges are three-dimensional mesh or scaffold like structure made up of long chain of polyester. Along with certain small molecules known as cross linkers, they acquire some further rearrangement to form spherically shaped structures with cavities. These cavities store therapeutic agents for cancer treatment and then deliver it to the target site. This microscopic sponge circulates in the neoplasm until they come across the surface to sustain delivering their cargo (Cavalli et al. 2006; Guo et al. 2008). Whereas fullerenes nanotubes, also known as carbon nanotubes (CNT), are designed by using unique atomic level arrangement of carbon in the molecular form. Ji et al. stated that a CNT device is capable of screening of single cancerous cell efficiently (Ji et al. 2010). The distinct physicochemical properties, optical properties and electrical properties of the fullerenes and some of their derivatives

Table 11.1 Metal nanoparticle-based approaches with respective mechanism that has been evaluated for the potential cancer diagnosis or therapeutics

Sl no	Metal NP	Approach	Mechanism	Outcome	References
1	Silver	Lower the number of tumour-inducing cytokines	Increases anticancer effects of therapeutic agents	Hindered the growth of fibrosarcoma in vivo	Chakraborty et al. (2016)
2	Zinc oxide	Antigen delivery to dendrimer	Augmented antigen specific CTL response	Detained growth of tumour in vivo	Cho et al. (2011)
3	Gold	Delivery of Ag or adjuvant. Photothermal therapy	Enhanced CTL responses, freed tumour antigens	Reduced and prevented tumour growth in vivo	Dreaden et al. (2011) and Almeida et al. (2011)
4	Cuprous oxide	Hamper microenvironment of tumour	Change expression profile of drosophila transcription factor	Induced systemic immunity and tumour infiltration	Kheirolomoom et al. (2015)
5	Aluminium oxide	Adjuvant	Amplifies anticancer effects of tumour cell vaccines	Reduction of tumour size, more CTL response when combinedly administrated with vaccine	Sun et al. (2010)
6	Silver (Ag)	Antigen capturing and antigen presentation to improve immune response	Delivered tumour specific protein to the antigen presenting cells	Extension of cytotoxic T cells + CD8 and enhanced both CD4+T/Treg and CD8+T/Treg ratios	Min et al. (2017)
7	Iron oxide	Shipment of protein, polarization of M1 macrophage, photothermal therapy	Enhanced proliferation of pro-inflammatory macrophages, magnetic hyperthermia	Hindered tumour growth, IONP-chaperone led to cancer cell specific CTL response	Zanganeh et al. (2016), Shevtsov et al. (2015) and Toraya-Brown et al. (2014)
8	Titanium dioxide	Stimulation of immune system by ultrasound	Enhanced pro-inflammatory cytokines and interleukins in the tumour due to ROS production	Reduced tumour growth in vivo	You et al. (2016)
9	Cobalt oxide	Antigen delivery for macrophage-based antitumour vaccine	Activation of macrophages	Enhanced Ag specific CTLs in vivo	Chattopadhyay et al. (2016)

make them popular tools in cancer diagnosis and therapy. Nowadays these are the most promising nanomaterials with the ability of both screening of tumour cells and site-specific delivery of therapeutic agents.

Liposomes are the first FDA approved nanoparticles used as a carrier to deliver the therapeutic agents. They are sized at the nanoscale dimensions and consist of a lipid bilayer enclosing the drug inside a core hydrophilic environment. In breast and ovarian cancer, liposomal doxorubicin is proved to be efficient and safe clinically (Papahadjopoulos et al. 1991; Straubinger et al. 1988). PEG coated nanoparticles show higher efficacy due to steric stabilization. This also reduces the uptake of liposome by macrophages (Papahadjopoulos et al. 1991; Woodle and Lasic 1992). Albumin taxol conjugates are the second nanoparticles to be approved by FDA. This was designed to mitigate the insolubility issues of taxol through binding of the drugs to the nanosized albumin (Damascelli et al. 2001; Desai et al. 2006; Miele et al. 2009). Currently more than 50 clinical trials of nanoparticles are ongoing for cancer. Following are the list of different metal nanoparticles which have been evaluated for its potential therapeutic approaches.

11.5 Conclusion

In conclusion, nanoparticles, like zinc oxide, iron oxide, when enter biological milieu produce reactive oxygen species (ROS) upon interaction with biomolecules leading to oxidative stress, damaging cellular organelles including lipid peroxidation and production of 4-hydroxylnonenal, malondialdehyde and other toxic aldehyde compounds (Sripetchwandee et al. 2016). These toxic elements further lead to protein damage by specific reactions and easily bypass the ubiquitin/proteasome degradation system. These misfolded and modified proteins accumulate with time and adopt a cytotoxic aggregate structure in cytoplasm, as observed in the cases of neurodegenerative disorders (Crichton et al. 2011). The level of cytotoxic-based degeneration depends on the concentration, size, shape, surface charge, functional groups and type of coating of the NP (Yarjanli et al. 2017). Another way of indirect contribution of nanoparticle in cancer and degenerative diseases is anticipated via generated ROS-mediated DNA modifications including breakdown of bases, mutations, DNA degradation, alteration in purine or pyrimidine and protein cross-linking (Birben et al. 2012). Thus, with the advancement of nanotechnology in biological sciences, it becomes pertinent to evaluate the NP surface properties and its consequence in biological milieu before its applications.

References

Adams RA et al (2007) The fibrin-derived gamma(377-395) peptide inhibits microglia activation and suppresses relapsing paralysis in central nervous system autoimmune disease. J Exp Med 204:571–582. https://doi.org/10.1084/jem.20061931

Alexis F, Rhee JW, Richie JP, Radovic-Moreno AF, Langer R, Farokhzad OC (2008) New frontiers in nanotechnology for cancer treatment. Urol Oncol-Semin Ori 26:74–85. https://doi.org/10.1016/j.urolonc.2007.03.017

Allen TM, Cullis PR (2004) Drug delivery systems: entering the mainstream. Science 303:1818–1822. https://doi.org/10.1126/science.1095833

Almeida JPM, Chen AL, Foster A, Drezek R (2011) In vivo biodistribution of nanoparticles. Nanomedicine-UK 6:815–835. https://doi.org/10.2217/Nnm.11.79

Alvarez YD, Fauerbach JA, Pellegrotti JV, Jovin TM, Jares-Erijman EA, Stefani FD (2013) Influence of gold nanoparticles on the kinetics of alpha-synuclein aggregation. Nano Lett 13:6156–6163. https://doi.org/10.1021/nl403490e

Amiri H, Saeidi K, Borhani P, Manafirad A, Ghavami M, Zerbi V (2013) Alzheimer's disease: pathophysiology and applications of magnetic nanoparticles as MRI theranostic agents. ACS Chem Neurosci 4:1417–1429. https://doi.org/10.1021/cn4001582

Anand BG, Dubey K, Shekhawat DS, Kar K (2016) Capsaicin-coated silver nanoparticles inhibit amyloid fibril formation of serum albumin. Biochemistry-US 55:3345–3348. https://doi.org/10.1021/acs.biochem.6b00418

Arai T et al (2004) Identification of amino-terminally cleaved tau fragments that distinguish progressive supranuclear palsy from corticobasal degeneration. Ann Neurol 55:72–79. https://doi.org/10.1002/ana.10793

Arakha M, Roy J, Nayak PS, Mallick B, Jha S (2017) Zinc oxide nanoparticle energy band gap reduction triggers the oxidative stress resulting into autophagy-mediated apoptotic cell death. Free Radical Bio Med 110:42–53. https://doi.org/10.1016/j.freeradbiomed.2017.05.015

Armitage BA (2005) Cyanine dye–DNA interactions: intercalation, groove binding, and aggregation. In: DNA binders and related subjects. Springer, Berlin, Heidelberg, pp 55–76

Asthana S, Bhattacharyya D, Kumari S, Nayak PS, Saleem M, Bhunia A, Jha S (2020) Interaction with zinc oxide nanoparticle kinetically traps alpha-synuclein fibrillation into off-pathway non-toxic intermediates. Int J Biol Macromol 150:68–79. https://doi.org/10.1016/j.ijbiomac.2020.01.269

Bajaj A, Miranda OR, Kim IB, Phillips RL, Jerry DJ, Bunz UHF, Rotello VM (2009) Detection and differentiation of normal, cancerous, and metastatic cells using nanoparticle-polymer sensor arrays. Proc Natl Acad Sci U S A 106:10912–10916. https://doi.org/10.1073/pnas.0900975106

Baker EN et al (1988) The structure of 2zn pig insulin crystals at 1.5-a resolution. Philos Trans R Soc B 319:369. https://doi.org/10.1098/rstb.1988.0058

Ballou B, Ernst LA, Andreko S, Harper T, Fitzpatrick JAJ, Waggoner AS, Bruchez MP (2007) Sentinel lymph node imaging using quantum dots in mouse tumor models. Bioconjugate Chem 18:389–396. https://doi.org/10.1021/bc060261j

Barbara R et al (2017) Novel Curcumin loaded nanoparticles engineered for Blood-Brain Barrier crossing and able to disrupt Abeta aggregates. Int J Pharmaceut 526:413–424. https://doi.org/10.1016/j.ijpharm.2017.05.015

Bareford LA, Swaan PW (2007) Endocytic mechanisms for targeted drug delivery. Adv Drug Deliv Rev 59:748–758. https://doi.org/10.1016/j.addr.2007.06.008

Behlke MA (2006) Progress towards in vivo use of siRNAs. Mol Ther 13:644–670. https://doi.org/10.1016/j.ymthe.2006.01.001

Bett CK, Ngunjiri JN, Serem WK, Fontenot KR, Hammer RP, McCarley RL, Garno JC (2010) Structure-activity relationships in peptide modulators of beta-amyloid protein aggregation: variation in alpha,alpha-disubstitution results in altered aggregate size and morphology. ACS Chem Neurosci 1:608–626. https://doi.org/10.1021/cn100045q

Bhumkar DR, Joshi HM, Sastry M, Pokharkar VB (2007) Chitosan reduced gold nanoparticles as novel carriers for transmucosal delivery of insulin. Pharm Res-Dordr 24:1415–1426. https://doi.org/10.1007/s11095-007-9257-9

Birben E, Sahiner UM, Sackesen C, Erzurum S, Kalayci O (2012) Oxidative stress and antioxidant defense. World Allergy Organ J 5:9–19. https://doi.org/10.1097/WOX.0b013e3182439613

Boddapati SV, D'Souza GGM, Erdogan S, Torchilin VP, Weissig V (2008) Organelle-targeted nanocarriers: specific delivery of liposomal ceramide to mitochondria enhances its cytotoxicity in vitro and in vivo. Nano Lett 8:2559–2563. https://doi.org/10.1021/nl801908y

Brange J (2012) The physico-chemical and pharmaceutical aspects of insulin and insulin preparations. Galenics of insulin. Springer-Verlag, Berlin Heidelberg. https://doi.org/10.1007/978-3-662-02526-0

Brange J, Andersen L, Laursen ED, Meyn G, Rasmussen E (1997) Toward understanding insulin fibrillation. J Pharm Sci 86:517–525. https://doi.org/10.1021/js960297s

Breunig M, Bauer S, Goefferich A (2008) Polymers and nanoparticles: intelligent tools for intracellular targeting? Eur J Pharm Biopharm 68:112–128. https://doi.org/10.1016/j.ejpb.2007.06.010

Bruce IJ, Sen T (2005) Surface modification of magnetic nanoparticles with alkoxysilanes and their application in magnetic bioseparations. Langmuir 21:7029–7035. https://doi.org/10.1021/la050553t

Bruchez M, Moronne M, Gin P, Weiss S, Alivisatos AP (1998) Semiconductor nanocrystals as fluorescent biological labels. Science 281:2013–2016. https://doi.org/10.1126/science.281.5385.2013

Bulte JWM et al (2001) Magnetodendrimers allow endosomal magnetic labeling and in vivo tracking of stem cells. Nat Biotechnol 19:1141–1147. https://doi.org/10.1038/nbt1201-1141

Cabaleiro-Lago C et al (2008) Inhibition of amyloid beta protein fibrillation by polymeric nanoparticles. J Am Chem Soc 130:15437–15443. https://doi.org/10.1021/ja8041806

Cabaleiro-Lago C, Lynch I, Dawson KA, Linse S (2010a) Inhibition of IAPP and IAPP((20-29)) fibrillation by polymeric nanoparticles. Langmuir 26:3453–3461. https://doi.org/10.1021/la902980d

Cabaleiro-Lago C, Quinlan-Pluck F, Lynch I, Dawson KA, Linse S (2010b) Dual effect of amino modified polystyrene nanoparticles on amyloid beta protein fibrillation. ACS Chem Neurosci 1:279–287. https://doi.org/10.1021/cn900027u

Cabaleiro-Lago C, Szczepankiewicz O, Linse S (2012) The effect of nanoparticles on amyloid aggregation depends on the protein stability and intrinsic aggregation rate. Langmuir 28:1852–1857. https://doi.org/10.1021/la203078w

Cai WB, Chen XY (2008) Preparation of peptide-conjugated quantum dots for tumor vasculature-targeted imaging. Nat Protoc 3:89–96. https://doi.org/10.1038/nprot.2007.478

Cavalli R, Trotta F, Tumiatti W (2006) Cyclodextrin-based nanosponges for drug delivery. J Incl Phenom Macro 56:209–213. https://doi.org/10.1007/s10847-006-9085-2

Chakraborty B, Pal R, Ali M, Singh LM, Rahman DS, Ghosh SK, Sengupta M (2016) Immunomodulatory properties of silver nanoparticles contribute to anticancer strategy for murine fibrosarcoma. Cell Mol Immunol 13:191–205. https://doi.org/10.1038/cmi.2015.05

Chattopadhyay S et al (2016) Metal based nanoparticles as cancer antigen delivery vehicles for macrophage based antitumor vaccine. Vaccine 34:957–967. https://doi.org/10.1016/j.vaccine.2015.12.053

Chien LY et al (2011) In vivo magnetic resonance imaging of cell tropsim, trafficking mechanism, and therapeutic impact of human mesenchymal stem cells in a murine glioma model. Biomaterials 32:3275–3284. https://doi.org/10.1016/j.biomaterials.2011.01.042

Chithrani BD, Ghazani AA, Chan WCW (2006) Determining the size and shape dependence of gold nanoparticle uptake into mammalian cells. Nano Lett 6:662–668. https://doi.org/10.1021/nl052396o

Cho NH et al (2011) A multifunctional core-shell nanoparticle for dendritic cell-based cancer immunotherapy. Nat Nanotechnol 6:675–682. https://doi.org/10.1038/Nnano.2011.149

Chung HJ, Castro CM, Im H, Lee H, Weissleder R (2013) A magneto-DNA nanoparticle system for rapid detection and phenotyping of bacteria. Nat Nanotechnol 8:369–375. https://doi.org/10.1038/Nnano.2013.70

Clarke S, Pinaud F, Beutel O, You CJ, Piehler J, Dahan M (2010) Covalent monofunctionalization of peptide-coated quantum dots for single-molecule assays. Nano Lett 10:2147–2154. https://doi.org/10.1021/nl100825n
Cohen T, Frydman-Marom A, Rechter M, Gazit E (2006) Inhibition of amyloid fibril formation and cytotoxicity by hydroxyindole derivatives. Biochemistry-US 45:4727–4735. https://doi.org/10.1021/bi051525c
Colvin VL, Kulinowski KM (2007) Nanoparticles as catalysts for protein fibrillation. Proc Natl Acad Sci U S A 104:8679–8680. https://doi.org/10.1073/pnas.0703194104
Coppola G et al (2012) Evidence for a role of the rare p.A152T variant in MAPT in increasing the risk for FTD-spectrum and Alzheimers diseases. Hum Mol Genet 21:3500–3512. https://doi.org/10.1093/hmg/dds161
Crawford S (2013) Is it time for a new paradigm for systemic cancer treatment? Lessons from a century of cancer chemotherapy. Front Pharmacol 4. https://doi.org/10.3389/fphar.2013.00068
Crichton RR, Dexter DT, Ward RJ (2011) Brain iron metabolism and its perturbation in neurological diseases. Monatsh Chem 142:341–355. https://doi.org/10.1007/s00706-011-0472-z
Cukalevski R, Lundqvist M, Oslakovic C, Dahlback B, Linse S, Cedervall T (2011) Structural changes in apolipoproteins bound to nanoparticles. Langmuir 27:14360–14369. https://doi.org/10.1021/la203290a
Da Silva CG, Rueda F, Lowik CW, Ossendorp F, Cruz LJ (2016) Combinatorial prospects of nano-targeted chemoimmunotherapy. Biomaterials 83:308–320. https://doi.org/10.1016/j.biomaterials.2016.01.006
Damascelli B et al (2001) Intraarterial chemotherapy with polyoxyethylated castor oil free paclitaxel, incorporated in albumin nanoparticles (ABI-007) - Phase I study of patients with squamous cell carcinoma of the head and neck and anal canal: preliminary evidence of clinical activity. Cancer 92:2592–2602. https://doi.org/10.1002/1097-0142(20011115)92:10<2592::Aid-Cncr1612>3.0.Co;2-4
De M, Ghosh PS, Rotello VM (2008) Applications of nanoparticles in biology. Adv Mater 20:4225–4241. https://doi.org/10.1002/adma.200703183
De M, Rana S, Akpinar H, Miranda OR, Arvizo RR, Bunz UHF, Rotello VM (2009) Sensing of proteins in human serum using conjugates of nanoparticles and green fluorescent protein. Nat Chem 1:461–465. https://doi.org/10.1038/Nchem.334
Derfus AM, von Maltzahn G, Harris TJ, Duza T, Vecchio KS, Ruoslahti E, Bhatia SN (2007) Remotely triggered release from magnetic nanoparticles. Adv Mater 19:3932. https://doi.org/10.1002/adma.200700091
Desai N et al (2006) Increased antitumor activity, intratumor paclitaxel concentrations, and endothelial cell transport of Cremophor-free, albumin-bound paclitaxel, ABI-007, compared with Cremophor-based paclitaxel. Clin Cancer Res 12:1317–1324. https://doi.org/10.1158/1078-0432.Ccr-05-1634
Dreaden EC, Mackey MA, Huang XH, Kang B, El-Sayed MA (2011) Beating cancer in multiple ways using nanogold. Chem Soc Rev 40:3391–3404. https://doi.org/10.1039/c0cs00180e
Ehrnhoefer DE et al (2008) EGCG redirects amyloidogenic polypeptides into unstructured, off-pathway oligomers. Nat Struct Mol Biol 15:558–566. https://doi.org/10.1038/nsmb.1437
Enoki T, Takai K, Osipov V, Baidakova M, Vul A (2009) Nanographene and nanodiamond; new members in the nanocarbon family. Chem-Asian J 4:796–804. https://doi.org/10.1002/asia.200800485
Fei L, Perrett S (2009) Effect of nanoparticles on protein folding and fibrillogenesis. Int J Mol Sci 10:646–655. https://doi.org/10.3390/ijms10020646
Ferrari M (2005) Cancer nanotechnology: opportunities and challenges. Nat Rev Cancer 5:161–171. https://doi.org/10.1038/nrc1566
Fischer M, Appelhans D, Schwarz S, Klajnert B, Bryszewska M, Voit B, Rogers M (2010) Influence of surface functionality of poly(propylene imine) dendrimers on protease resistance and propagation of the scrapie prion protein. Biomacromolecules 11:1314–1325. https://doi.org/10.1021/bm100101s

Fu CC et al (2007) Characterization and application of single fluorescent nanodiamonds as cellular biomarkers. Proc Natl Acad Sci U S A 104:727–732. https://doi.org/10.1073/pnas.0605409104

Galanzha EI, Shashkov EV, Kelly T, Kim JW, Yang LL, Zharov VP (2009) In vivo magnetic enrichment and multiplex photoacoustic detection of circulating tumour cells. Nat Nanotechnol 4:855–860. https://doi.org/10.1038/Nnano.2009.333

Gao XH, Cui YY, Levenson RM, Chung LWK, Nie SM (2004) In vivo cancer targeting and imaging with semiconductor quantum dots. Nat Biotechnol 22:969–976. https://doi.org/10.1038/nbt994

Gindy ME, Prud'homme RK (2009) Multifunctional nanoparticles for imaging, delivery and targeting in cancer therapy. Expert Opin Drug Deliv 6:865–878. https://doi.org/10.1517/17425240902932908

Glat M, Skaat H, Menkes-Caspi N, Margel S, Stern EA (2013) Age-dependent effects of microglial inhibition in vivo on Alzheimer's disease neuropathology using bioactive-conjugated iron oxide nanoparticles. J Nanobiotechnol 11. https://doi.org/10.1186/1477-3155-11-32

Gong CX, Liu F, Grundke-Iqbal I, Iqbal K (2005) Post-translational modifications of tau protein in Alzheimer's disease. J Neural Transm 112:813–838. https://doi.org/10.1007/s00702-004-0221-0

Goodman CM, McCusker CD, Yilmaz T, Rotello VM (2004) Toxicity of gold nanoparticles functionalized with cationic and anionic side chains. Bioconjugate Chem 15:897–900. https://doi.org/10.1021/bc049951i

Gorantla NV, Khandelwal P, Poddar P, Chinnathambi S (2017a) Global conformation of tau protein mapped by Raman spectroscopy. Meth Mol Biol 1523:21–31. https://doi.org/10.1007/978-1-4939-6598-4_2

Gorantla NV, Shkumatov AV, Chinnathambi S (2017b) Conformational dynamics of intracellular tau protein revealed by CD and SAXS. Meth Mol Biol 1523:3–20. https://doi.org/10.1007/978-1-4939-6598-4_1

Gulati M, Chopra DS, Singh SK, Saluja V, Pathak P, Bansal P (2013) Patents on brain permeable nanoparticles. Inida Patent

Guo LL, Gao G, Liu XL, Liu FQ (2008) Preparation and characterization of TiO2 nanosponge. Mater Chem Phys 111:322–325. https://doi.org/10.1016/j.matchemphys.2008.04.016

Gupta D, Leahy JL (2014) Islet amyloid and type 2 diabetes: overproduction or inadequate clearance and detoxification? J Clin Invest 124:3292–3294. https://doi.org/10.1172/Jci77506

Gurzov EN et al (2016) Inhibition of hIAPP amyloid aggregation and pancreatic beta-cell toxicity by OH-terminated PAMAM dendrimer. Small 12:1615–1626. https://doi.org/10.1002/smll.201502317

Hainfeld JF, Slatkin DN, Focella TM, Smilowitz HM (2006) Gold nanoparticles: a new X-ray contrast agent. Brit J Radiol 79:248–253. https://doi.org/10.1259/bjr/13169882

Hajimohammadjafartehrani M et al (2019) The effects of nickel oxide nanoparticles on tau protein and neuron-like cells: biothermodynamics and molecular studies. Int J Biol Macromol 127:330–339. https://doi.org/10.1016/j.ijbiomac.2019.01.050

Hajsalimi G, Taheri S, Shahi F, Attar F, Ahmadi H, Falahati M (2018) Interaction of iron nanoparticles with nervous system: an in vitro study. J Biomol Struct Dyn 36:928–937. https://doi.org/10.1080/07391102.2017.1302819

Hanger DP et al (2007) Novel phosphorylation sites in tau from Alzheimer brain support a role for casein kinase 1 in disease pathogenesis. J Biol Chem 282:23645–23654. https://doi.org/10.1074/jbc.M703269200

Harris JM, Martin NE, Modi M (2001) Pegylation - a novel process for modifying pharmacokinetics. Clin Pharmacokinet 40:539–551. https://doi.org/10.2165/00003088-200140070-00005

Hashimoto M et al (2009) Effects of docosahexaenoic acid on in vitro amyloid beta peptide 25-35 fibrillation. BBA-Mol Cell Biol Lipids 1791:289–296. https://doi.org/10.1016/j.bbalip.2009.01.012

Haun JB, Yoon TJ, Lee H, Weissleder R (2010) Magnetic nanoparticle biosensors. Wires Nanomed Nanobi 2:291–304. https://doi.org/10.1002/wnan.84

Herscher LL, Cook JA, Pacelli R, Pass HI, Russo A, Mitchell JB (1999) Principles of chemoradiation: theoretical and practical considerations. Oncology-Ny 13:11–22

Hill HD, Mirkin CA (2006) The bio-barcode assay for the detection of protein and nucleic acid targets using DTT-induced ligand exchange. Nat Protoc 1:324–336. https://doi.org/10.1038/nprot.2006.51

Holt KB (2007) Diamond at the nanoscale: applications of diamond nanoparticles from cellular biomarkers to quantum computing. Philos Trans R Soc A 365:2845–2861. https://doi.org/10.1098/rsta.2007.0005

Hostetler MJ et al (1998) Alkanethiolate gold cluster molecules with core diameters from 1.5 to 5.2 nm: core and monolayer properties as a function of core size. Langmuir 14:17–30. https://doi.org/10.1021/la970588w

Hua QX, Weiss MA (2004) Mechanism of insulin fibrillation - the structure of insulin under amyloidogenic conditions resembles a protein-folding intermediate. J Biol Chem 279:21449–21460. https://doi.org/10.1074/jbc.M314141200

Huang RX, Carney RP, Stellacci F, Lau BLT (2013) Protein-nanoparticle interactions: the effects of surface compositional and structural heterogeneity are scale dependent. Nanoscale 5:6928–6935. https://doi.org/10.1039/c3nr02117c

Ikeda K, Okada T, Sawada S, Akiyoshi K, Matsuzaki K (2006) Inhibition of the formation of amyloid beta-protein fibrils using biocompatible nanogels as artificial chaperones. Febs Lett 580:6587–6595. https://doi.org/10.1016/j.febslet.2006.11.009

Iqbal K, Alonso ADC, Gong CX, Khatoon S, Pei JJ, Wang JZ, Grundke-Iqbal I (1998) Mechanisms of neurofibrillary degeneration and the formation of neurofibrillary tangles. J Neural Transm Suppl 53:169–180. https://doi.org/10.1007/978-3-7091-6467-9_15

Iqbal K, Liu F, Gong CX (2016) Tau and neurodegenerative disease: the story so far. Nat Rev Neurol 12. https://doi.org/10.1038/nrneurol.2015.225

Jain RK (2001) Delivery of molecular medicine to solid tumors: lessons from in vivo imaging of gene expression and function. J Control Release 74:7–25. https://doi.org/10.1016/S0168-3659(01)00306-6

Jaishree V, Gupta PD (2012) Nanotechnology: a revolution in cancer diagnosis. Indian J Clin Biochem 27:214–220. https://doi.org/10.1007/s12291-012-0221-z

Jana AK, Sengupta N (2012) Adsorption mechanism and collapse propensities of the full-length, monomeric A beta(1-42) on the surface of a single-walled carbon nanotube: a molecular dynamics simulation study. Biophys J 102:1889–1896. https://doi.org/10.1016/j.bpj.2012.03.036

Ji SR et al (2010) Carbon nanotubes in cancer diagnosis and therapy. BBA-Rev Cancer 1806:29–35. https://doi.org/10.1016/j.bbcan.2010.02.004

Kanazawa T, Akiyama F, Kakizaki S, Takashima Y, Seta Y (2013) Delivery of siRNA to the brain using a combination of nose-to-brain delivery and cell-penetrating peptide-modified nano-micelles. Biomaterials 34:9220–9226. https://doi.org/10.1016/j.biomaterials.2013.08.036

Kang WJ, Chae JR, Cho YL, Lee JD, Kim S (2009) Multiplex imaging of single tumor cells using quantum-dot-conjugated aptamers. Small 5:2519–2522. https://doi.org/10.1002/smll.200900848

Kang B, Mackey MA, El-Sayed MA (2010) Nuclear targeting of gold nanoparticles in cancer cells induces DNA damage, causing cytokinesis arrest and apoptosis. J Am Chem Soc 132:1517. https://doi.org/10.1021/ja9102698

Karnik R, Hong S, Zhang H, Mei Y, Anderson DG, Karp JM, Langer R (2008) Nanomechanical control of cell rolling in two dimensions through surface patterning of receptors. Nano Lett 8:1153–1158. https://doi.org/10.1021/nl073322a

Karve S et al (2012) Revival of the abandoned therapeutic wortmannin by nanoparticle drug delivery. Proc Natl Acad Sci U S A 109:8230–8235. https://doi.org/10.1073/pnas.1120508109

Kheirolomoom A et al (2015) CpG expedites regression of local and systemic tumors when combined with activatable nanodelivery. J Control Release 220:253–264. https://doi.org/10.1016/j.jconrel.2015.10.016

Kim JE, Lee M (2003) Fullerene inhibits beta-amyloid peptide aggregation. Biochem Biophys Res Commun 303:576–579. https://doi.org/10.1016/S0006-291x(03)00393-0
Kircher MF, Mahmood U, King RS, Weissleder R, Josephson L (2003) A multimodal nanoparticle for preoperative magnetic resonance imaging and intraoperative optical brain tumor delineation. Cancer Res 63:8122–8125
Kopp M, Kollenda S, Epple M (2017) Nanoparticle-protein interactions: therapeutic approaches and supramolecular chemistry. Acc Chem Res 50:1383–1390. https://doi.org/10.1021/acs.accounts.7b00051
Lakowicz JR, Gryczynski I, Gryczynski Z, Nowaczyk K, Murphy CJ (2000) Time-resolved spectral observations of cadmium-enriched cadmium sulfide nanoparticles and the effects of DNA oligomer binding. Anal Biochem 280:128–136. https://doi.org/10.1006/abio.2000.4495
Langer K, Balthasar S, Vogel V, Dinauer N, von Briesen H, Schubert D (2003) Optimization of the preparation process for human serum albumin (HSA) nanoparticles. Int J Pharmaceut 257:169–180. https://doi.org/10.1016/S0378-5173(03)00134-0
Lebaron S, Zeltzer LK, Lebaron C, Scott SE, Zeltzer PM (1988) Chemotherapy side-effects in pediatric oncology patients - drugs, age, and sex as risk-factors. Med Pediatr Oncol 16:263–268. https://doi.org/10.1002/mpo.2950160408
Lee YC (1992) Biochemistry of carbohydrate-protein interaction. FASEB J 6:3193–3200
Lee JH et al (2007) Artificially engineered magnetic nanoparticles for ultra-sensitive molecular imaging. Nat Med 13:95–99. https://doi.org/10.1038/nm1467
Lee H, Yoon TJ, Weissleder R (2009) Ultrasensitive detection of bacteria using core-shell nanoparticles and an NMR-filter system. Angew Chem Int Edit 48:5657–5660. https://doi.org/10.1002/anie.200901791
Lilja H, Ulmert D, Vickers AJ (2008) Prostate-specific antigen and prostate cancer: prediction, detection and monitoring. Nat Rev Cancer 8:268–278. https://doi.org/10.1038/nrc2351
Lin CC, Yeh YC, Yang CY, Chen GF, Chen YC, Wu YC, Chen CC (2003) Quantitative analysis of multivalent interactions of carbohydrate-encapsulated gold nanoparticles with concanavalin A. Chem Commun:2920–2921. https://doi.org/10.1039/b308995a
Lin PC, Tseng MC, Su AK, Chen YJ, Lin CC (2007a) Functionalized magnetic nanoparticles for small-molecule isolation, identification, and quantification. Anal Chem 79:3401–3408. https://doi.org/10.1021/ac070195u
Lin S et al (2007b) Quantum dot imaging for embryonic stem cells. BMC Biotechnol 7. https://doi.org/10.1186/1472-6750-7-67
Linse S et al (2007) Nucleation of protein fibrillation by nanoparticles. Proc Natl Acad Sci U S A 104:8691–8696. https://doi.org/10.1073/pnas.0701250104
Liu RT, McAllister C, Lyubchenko Y, Sierks MR (2004) Residues 17-20 and 30-35 of beta-amyloid play critical roles in aggregation. J Neurosci Res 75:162–171. https://doi.org/10.1002/jnr.10859
Lloyd JB (2000) Lysosome membrane permeability: implications for drug delivery. Adv Drug Deliver Rev 41:189–200. https://doi.org/10.1016/S0169-409x(99)00065-4
Loo C, Lowery A, Halas NJ, West J, Drezek R (2005) Immunotargeted nanoshells for integrated cancer imaging and therapy. Nano Lett 5:709–711. https://doi.org/10.1021/nl050127s
Luo S, Ma C, Zhu MQ, Ju WN, Yang Y, Wang X (2020) Application of iron oxide nanoparticles in the diagnosis and treatment of neurodegenerative diseases with emphasis on Alzheimer's disease. Front Cell Neurosci 14. https://doi.org/10.3389/fncel.2020.00021
Ma BBY, Bristow RG, Kim J, Siu LL (2003) Combined-modality treatment of solid tumors using radiotherapy and molecular targeted agents. J Clin Oncol 21:2760–2776. https://doi.org/10.1200/Jco.2003.10.044
Mahmoudi M, Akhavan O, Ghavami M, Rezaee F, Ghiasi SMA (2012) Graphene oxide strongly inhibits amyloid beta fibrillation. Nanoscale 4:7322–7325. https://doi.org/10.1039/c2nr31657a
Mahmoudi M, Quinlan-Pluck F, Monopo MP, Sheibani S, Vali H, Dawson KA, Lynch I (2013) Influence of the physiochemical properties of superparamagnetic iron oxide nanoparticles on

amyloid beta protein fibrillation in solution. ACS Chem Neurosci 4:475–485. https://doi.org/10.1021/cn300196n

Mahtab R, Harden HH, Murphy CJ (2000) Temperature- and salt-dependent binding of long DNA to protein-sized quantum dots: thermodynamics of "inorganic protein"-DNA interactions. J Am Chem Soc 122:14–17. https://doi.org/10.1021/ja9907156

Maltsev AS, Chen J, Levine RL, Bax A (2013) Site-specific interaction between alpha-synuclein and membranes probed by NMR-observed methionine oxidation rates. J Am Chem Soc 135:2943–2946. https://doi.org/10.1021/ja312415q

Mandelkow EM, Mandelkow E (2012) Biochemistry and cell biology of tau protein in neurofibrillary degeneration. Csh Perspect Med 2. https://doi.org/10.1101/cshperspect.a006247

Marrache S, Dhar S (2012) Engineering of blended nanoparticle platform for delivery of mitochondria-acting therapeutics. Proc Natl Acad Sci U S A 109:16288–16293. https://doi.org/10.1073/pnas.1210096109

Marshall KE, Morris KL, Charlton D, O'Reilly N, Lewis L, Walden H, Serpell LC (2011) Hydrophobic, aromatic, and electrostatic interactions play a central role in amyloid fibril formation and stability. Biochemistry-US 50:2061–2071. https://doi.org/10.1021/bi101936c

Maysinger D (2007) Nanoparticles and cells: good companions and doomed partnerships. Org Biomol Chem 5:2335–2342. https://doi.org/10.1039/b704275b

McIntosh CM, Esposito EA, Boal AK, Simard JM, Martin CT, Rotello VM (2001) Inhibition of DNA transcription using cationic mixed monolayer protected gold clusters. J Am Chem Soc 123:7626–7629. https://doi.org/10.1021/ja015556g

Meesaragandla B, Karanth S, Janke U, Delcea M (2020) Biopolymer-coated gold nanoparticles inhibit human insulin amyloid fibrillation. Sci Rep-UK 10. https://doi.org/10.1038/s41598-020-64010-7

Mehdizadeh P et al (2019) Tau folding and cytotoxicity of neuroblastoma cells in the presence of manganese oxide nanoparticles: biophysical, molecular dynamics, cellular, and molecular studies. Int J Biol Macromol 125:674–682. https://doi.org/10.1016/j.ijbiomac.2018.11.191

Michael J, Carroll R, Swift HH, Steiner DF (1987) Studies on the molecular-organization of rat insulin secretory granules. J Biol Chem 262:16531–16535

Miele E, Spinelli GP, Miele E, Tomao F, Tomao S (2009) Albumin-bound formulation of paclitaxel (Abraxane (R) ABI-007) in the treatment of breast cancer. Int J Nanomed 4:99–105

Min YZ et al (2017) Antigen-capturing nanoparticles improve the abscopal effect and cancer immunotherapy. Nat Nanotechnol 12:877. https://doi.org/10.1038/Nnano.2017.113

Mirsadeghi S et al (2015) Protein corona composition of gold nanoparticles/nanorods affects amyloid beta fibrillation process. Nanoscale 7:5004–5013. https://doi.org/10.1039/c4nr06009a

Mishra V, Mahor S, Rawat A, Gupta PN, Dubey P, Khatri K, Vyas SP (2006) Targeted brain delivery of AZT via transferrin anchored pegylated albumin nanoparticles. J Drug Target 14:45–53. https://doi.org/10.1080/10611860600612953

Mohammad-Beigi H et al (2015) Strong interactions with polyethylenimine-coated human serum albumin nanoparticles (PEI-HSA NPs) alter alpha-synuclein conformation and aggregation kinetics. Nanoscale 7:19627–19640. https://doi.org/10.1039/c5nr05663b

Molday RS, Mackenzie D (1982) Immunospecific ferromagnetic iron-dextran reagents for the labeling and magnetic separation of cells. J Immunol Methods 52:353–367. https://doi.org/10.1016/0022-1759(82)90007-2

Nam JM, Thaxton CS, Mirkin CA (2003) Nanoparticle-based bio-bar codes for the ultrasensitive detection of proteins. Science 301:1884–1886. https://doi.org/10.1126/science.1088755

Necula M, Kayed R, Milton S, Glabe CG (2007) Small molecule inhibitors of aggregation indicate that amyloid beta oligomerization and fibrillization pathways are independent and distinct. J Biol Chem 282:10311–10324. https://doi.org/10.1074/jbc.M608207200

Nedumpully-Govindan P et al (2016) Graphene oxide inhibits hIAPP amyloid fibrillation and toxicity in insulin-producing NIT-1 cells. Phys Chem Chem Phys 18:94–100. https://doi.org/10.1039/c5cp05924k

Nie SM, Xing Y, Kim GJ, Simons JW (2007) Nanotechnology applications in cancer. Annu Rev Biomed Eng 9:257–288. https://doi.org/10.1146/annurev.bioeng.9.060906.152025

NIH (2017) Earlier detection and diagnosis. NIH. https://www.cancer.gov/nano/cancer-nanotechnology/detection-diagnosis

Noble W, Hanger DP, Miller CCJ, Lovestone S (2013) The importance of tau phosphorylation for neurodegenerative diseases. Front Neurol 4. https://doi.org/10.3389/fneur.2013.00083

O'Brien EP, Straub JE, Brooks BR, Thirumalai D (2011) Influence of nanoparticle size and shape on oligomer formation of an amyloidogenic peptide. J Phys Chem Lett 2:1171–1177. https://doi.org/10.1021/jz200330k

Olmedo I et al (2008) How changes in the sequence of the peptide CLPFFD-NH2 can modify the conjugation and stability of gold nanoparticles and their affinity for beta-amyloid fibrils. Bioconjugate Chem 19:1154–1163. https://doi.org/10.1021/bc800016y

Otsuka H, Akiyama Y, Nagasaki Y, Kataoka K (2001) Quantitative and reversible lectin-induced association of gold nanoparticles modified with alpha-lactosyl-omega-mercapto-poly(ethylene glycol). J Am Chem Soc 123:8226–8230. https://doi.org/10.1021/ja010437m

Pai AS, Rubinstein I, Onyuksel H (2006) PEGylated phospholipid nanomicelles interact with beta-amyloid((1-42)) and mitigate its beta-sheet formation, aggregation and neurotoxicity in vitro. Peptides 27:2858–2866. https://doi.org/10.1016/j.peptides.2006.04.022

Papahadjopoulos D et al (1991) Sterically stabilized liposomes - improvements in pharmacokinetics and antitumor therapeutic efficacy. Proc Natl Acad Sci U S A 88:11460–11464. https://doi.org/10.1073/pnas.88.24.11460

Patel PJ, Acharya NS, Acharya SR (2013) Development and characterization of glutathione-conjugated albumin nanoparticles for improved brain delivery of hydrophilic fluorescent marker. Drug Deliv 20:143–155. https://doi.org/10.3109/10717544.2013.801050

Peer D, Park EJ, Morishita Y, Carman CV, Shimaoka M (2008) Systemic leukocyte-directed siRNA delivery revealing cyclin D1 as an anti-inflammatory target. Science 319:627–630. https://doi.org/10.1126/science.1149859

Peppercorn J et al (2005) Quality of life among patients with stage II and III breast carcinoma randomized to receive high-dose chemotherapy with autologous bone marrow support or intermediate-dose chemotherapy - results from cancer and leukemia group B 9066. Cancer 104:1580–1589. https://doi.org/10.1002/cncr.21363

Peretz Y, Malishev R, Kolusheva S, Jelinek R (2018) Nanoparticles modulate membrane interactions of human Islet amyloid polypeptide (hIAPP). BBA-Biomembranes 1860:1810–1817. https://doi.org/10.1016/j.bbamem.2018.03.029

Phillips RL, Miranda OR, You CC, Rotello VM, Bunz UHF (2008) Rapid and efficient identification of bacteria using gold-nanoparticle - poly(para-phenyleneethynylene) constructs. Angew Chem Int Edit 47:2590–2594. https://doi.org/10.1002/anie.200703369

Pithadia AS, Bhunia A, Sribalan R, Padmini V, Fierke CA, Ramamoorthy A (2016) Influence of a curcumin derivative on hIAPP aggregation in the absence and presence of lipid membranes. Chem Commun 52:942–945. https://doi.org/10.1039/c5cc07792c

Pouton CW, Wagstaff KM, Roth DM, Moseley GW, Jans DA (2007) Targeted delivery to the nucleus. Adv Drug Deliv Rev 59:698–717. https://doi.org/10.1016/j.addr.2007.06.010

Rabbani G, Khan MJ, Ahmad A, Maskat MY, Khan RH (2014) Effect of copper oxide nanoparticles on the conformation and activity of beta-galactosidase. Colloid Surf B 123:96–105. https://doi.org/10.1016/j.colsurfb.2014.08.035

Raoufi M, Hajipour MJ, Shahri SMK, Schoen I, Linn U, Mahmoudi M (2018) Probing fibronectin conformation on a protein corona layer around nanoparticles. Nanoscale 10:1228–1233. https://doi.org/10.1039/c7nr06970g

Rasoulian B, Kaeidi A, Rezaei M, Hajializadeh Z (2017) Cellular preoxygenation partially attenuates the antitumoral effect of cisplatin despite highly protective effects on renal epithelial cells. Oxid Med Cell Longev 2017. https://doi.org/10.1155/2017/7203758

Saptarshi SR, Duschl A, Lopata AL (2013) Interaction of nanoparticles with proteins: relation to bio-reactivity of the nanoparticle. J Nanobiotechnol 11. https://doi.org/10.1186/1477-3155-11-26
Schellenberger EA, Sosnovik D, Weissleder R, Josephson L (2004) Magneto/optical annexin V, a multimodal protein. Bioconjugate Chem 15:1062–1067. https://doi.org/10.1021/bc049905i
Schroeder JE, Shweky I, Shmeeda H, Banin U, Gabizon A (2007) Folate-mediated tumor cell uptake of quantum dots entrapped in lipid nanoparticles. J Control Release 124:28–34. https://doi.org/10.1016/j.jconrel.2007.08.028
Sergeant N, Wattez A, Delacourte A (1999) Neurofibrillary degeneration in progressive supranuclear palsy and corticobasal degeneration: tau pathologies with exclusively "exon 10" isoforms. J Neurochem 72:1243–1249. https://doi.org/10.1046/j.1471-4159.1999.0721243.x
Shafer-Peltier KE, Haynes CL, Glucksberg MR, Van Duyne RP (2003) Toward a glucose biosensor based on surface-enhanced Raman scattering. J Am Chem Soc 125:588–593. https://doi.org/10.1021/ja028255v
Shanmukh S, Jones L, Driskell J, Zhao YP, Dluhy R, Tripp RA (2006) Rapid and sensitive detection of respiratory virus molecular signatures using a silver nanorod array SERS substrate. Nano Lett 6:2630–2636. https://doi.org/10.1021/nl061666f
Sharma P, Brown S, Walter G, Santra S, Moudgil B (2006) Nanoparticles for bioimaging. Adv Colloid Interfac 123:471–485. https://doi.org/10.1016/j.cis.2006.05.026
Shevtsov MA et al (2015) 70-kDa heat shock protein coated magnetic nanocarriers as a nanovaccine for induction of anti-tumor immune response in experimental glioma. J Control Release 220:329–340. https://doi.org/10.1016/j.jconrel.2015.10.051
Silva AT, Alien N, Ye CM, Verchot J, Moon JH (2010) Conjugated polymer nanoparticles for effective siRNA delivery to tobacco BY-2 protoplasts. BMC Plant Biol 10. https://doi.org/10.1186/1471-2229-10-291
Society AC (2016) Cancer Facts & Figures. American Cancer Society. https://www.cancer.org/research/cancer-facts-statistics/all-cancer-facts-figures/cancer-facts-figures-2016.html
Sonawane SK, Ahmad A, Chinnathambi S (2019) Protein-capped metal nanoparticles inhibit tau aggregation in Alzheimer's disease. ACS Omega 4:12833–12840. https://doi.org/10.1021/acsomega.9b01411
Srinivas PR, Barker P, Srivastava S (2002) Nanotechnology in early detection of cancer. Lab Invest 82:657–662. https://doi.org/10.1038/labinvest.3780460
Sripetchwandee J, Wongjaikam S, Krintratun W, Chattipakorn N, Chattipakorn SC (2016) A combination of an iron chelator with an antioxidant effectively diminishes the dendritic loss, tau-hyperphosphorylation, amyloids-beta accumulation and brain mitochondrial dynamic disruption in rats with chronic iron-overload. Neuroscience 332:191–202. https://doi.org/10.1016/j.neuroscience.2016.07.003
Straubinger RM, Lopez NG, Debs RJ, Hong K, Papahadjopoulos D (1988) Liposome-based therapy of human ovarian-cancer - parameters determining potency of negatively charged and antibody-targeted liposomes. Cancer Res 48:5237–5245
Sun YG, Xia YN (2003) Gold and silver nanoparticles: a class of chromophores with colors tunable in the range from 400 to 750 nm. Analyst 128:686–691. https://doi.org/10.1039/b212437h
Sun Z et al (2010) Aluminum nanoparticles enhance anticancer immune response induced by tumor cell vaccine. Cancer Nanotechnol 1:63–69. https://doi.org/10.1007/s12645-010-0001-5
Tada H, Higuchi H, Wanatabe TM, Ohuchi N (2007) In vivo real-time tracking of single quantum dots conjugated with monoclonal anti-HER2 antibody in tumors of mice. Cancer Res 67:1138–1144. https://doi.org/10.1158/0008-5472.Can-06-1185
Tan WH, Wang KM, He XX, Zhao XJ, Drake T, Wang L, Bagwe RP (2004) Bionanotechnology based on silica nanoparticles. Med Res Rev 24:621–638. https://doi.org/10.1002/med.20003
Toraya-Brown S et al (2014) Local hyperthermia treatment of tumors induces CD8(+) T cell-mediated resistance against distal and secondary tumors. Nanomed-Nanotechnol 10:1273–1285. https://doi.org/10.1016/j.nano.2014.01.011

Tuna M, Knuutila S, Mills GB (2009) Uniparental disomy in cancer. Trends Mol Med 15:120–128. https://doi.org/10.1016/j.molmed.2009.01.005

Tung CH, Mahmood U, Bredow S, Weissleder R (2000) In vivo imaging of proteolytic enzyme activity using a novel molecular reporter. Cancer Res 60:4953–4958

Uversky VN, Eliezer D (2009) Biophysics of Parkinson's disease: structure and aggregation of alpha-synuclein. Curr Protein Pept Sci 10:483–499. https://doi.org/10.2174/138920309789351921

van Rooijen BD, van Leijenhorst-Groener KA, Claessens MMAE, Subramaniam V (2009) Tryptophan fluorescence reveals structural features of alpha-synuclein oligomers. J Mol Biol 394:826–833. https://doi.org/10.1016/j.jmb.2009.10.021

Varshney M, Li YB (2007) Interdigitated array microelectrode based impedance biosensor coupled with magnetic nanoparticle-antibody conjugates for detection of Escherichia coli O157 : H7 in food samples. Biosens Bioelectron 22:2408–2414. https://doi.org/10.1016/j.bios.2006.08.030

Vasir JK, Labhasetwar V (2007) Biodegradable nanoparticles for cytosolic delivery of therapeutics. Adv Drug Deliv Rev 59:718–728. https://doi.org/10.1016/j.addr.2007.06.003

Verma A, Stellacci F (2010) Effect of surface properties on nanoparticle-cell interactions. Small 6:12–21. https://doi.org/10.1002/smll.200901158

Verma A et al (2008) Surface-structure-regulated cell-membrane penetration by monolayer-protected nanoparticles. Nat Mater 7:588–595. https://doi.org/10.1038/nmat2202

von Bergen M, Friedhoff P, Biernat J, Heberle J, Mandelkow EM, Mandelkow E (2000) Assembly of tau protein into Alzheimer paired helical filaments depends on a local sequence motif ((306) VQIVYK(311)) forming beta structure. Proc Natl Acad Sci USA 97:5129–5134. https://doi.org/10.1073/pnas.97.10.5129

von Bergen M, Barghorn S, Biernat J, Mandelkow EM, Mandelkow E (2005) Tau aggregation is driven by a transition from random coil to beta sheet structure. BBA-Mol Basis Dis 1739:158–166. https://doi.org/10.1016/j.bbadis.2004.09.010

Wagner SC, Roskamp M, Pallerla M, Araghi RR, Schlecht S, Koksch B (2010) Nanoparticle-induced folding and fibril formation of coiled-coil-based model peptides. Small 6:1321–1328. https://doi.org/10.1002/smll.200902067

Wang YP, Mandelkow E (2016) Tau in physiology and pathology. Nat Rev Neurosci 17:5–21. https://doi.org/10.1038/nrn.2015.1

Wang M, Thanou M (2010) Targeting nanoparticles to cancer. Pharmacol Res 62:90–99. https://doi.org/10.1016/j.phrs.2010.03.005

Wang EC, Wang AZ (2014) Nanoparticles and their applications in cell and molecular biology. Integr Biol-UK 6:9–26. https://doi.org/10.1039/c3ib40165k

Wang MD, Shin DM, Simons JW, Nie SM (2007) Nanotechnology for targeted cancer therapy. Expert Rev Anticanc 7:833–837. https://doi.org/10.1586/14737140.7.6.833

Wang Y et al (2015) Mesoporous silica nanoparticles in drug delivery and biomedical applications. Nanomed-Nanotechnol 11:313–327. https://doi.org/10.1016/j.nano.2014.09.014

Wang MY, Kakinen A, Pilkington EH, Davis TP, Ke PC (2017) Differential effects of silver and iron oxide nanoparticles on IAPP amyloid aggregation. Biomater Sci-UK 5:485–493. https://doi.org/10.1039/c6bm00764c

Welch MJ, Hawker CJ, Wooley KL (2009) The advantages of nanoparticles for PET. J Nucl Med 50:1743–1746. https://doi.org/10.2967/jnumed.109.061846

Woodle MC, Lasic DD (1992) Sterically stabilized liposomes. Biochim Biophys Acta 1113:171–199. https://doi.org/10.1016/0304-4157(92)90038-C

Worrall JWE, Verma A, Yan HH, Rotello VM (2006) "Cleaning" of nanoparticle inhibitors via proteolysis of adsorbed proteins. Chem Commun:2338–2340. https://doi.org/10.1039/b517421j

Wray S, Saxton M, Anderton BH, Hanger DP (2008) Direct analysis of tau from PSP brain identifies new phosphorylation sites and a major fragment of N-terminally cleaved tau containing four microtubule-binding repeats. J Neurochem 105:2343–2352. https://doi.org/10.1111/j.1471-4159.2008.05321.x

Wu WH et al (2008) TiO2 nanoparticles promote beta-amyloid fibrillation in vitro. Biochem Biophys Res Commun 373:315–318. https://doi.org/10.1016/j.bbrc.2008.06.035

Wu J, Wang C, Sun J, Xue Y (2011) Neurotoxicity of silica nanoparticles: brain localization and dopaminergic neurons damage pathways. ACS Nano 5:4476–4489. https://doi.org/10.1021/nn103530b

Wust P et al (2002) Hyperthermia in combined treatment of cancer. Lancet Oncol 3:487–497. https://doi.org/10.1016/S1470-2045(02)00818-5

Xia N, Hunt TP, Mayers BT, Alsberg E, Whitesides GM, Westervelt RM, Ingber DE (2006) Combined microfluidic-micromagnetic separation of living cells in continuous flow. Biomed Microdevices 8:299–308. https://doi.org/10.1007/s10544-006-0033-0

Xiao LH, Zhao D, Chan WH, Choi MMF, Li HW (2010) Inhibition of beta 1-40 amyloid fibrillation with N-acetyl-L-cysteine capped quantum dots. Biomaterials 31:91–98. https://doi.org/10.1016/j.biomaterials.2009.09.014

Xie HJ, Wu J (2016) Silica nanoparticles induce alpha-synuclein induction and aggregation in PC12-cells. Chem-Biol Interact 258:197–204. https://doi.org/10.1016/j.cbi.2016.09.006

Xu HX, Bjerneld EJ, Kall M, Borjesson L (1999) Spectroscopy of single hemoglobin molecules by surface enhanced Raman scattering. Phys Rev Lett 83:4357–4360. https://doi.org/10.1103/PhysRevLett.83.4357

Xu ZP et al (2008) Subcellular compartment targeting of layered double hydroxide nanoparticles. J Control Release 130:86–94. https://doi.org/10.1016/j.jconrel.2008.05.021

Xu HY et al (2011) Antibody conjugated magnetic iron oxide nanoparticles for cancer cell separation in fresh whole blood. Biomaterials 32:9758–9765. https://doi.org/10.1016/j.biomaterials.2011.08.076

Yamada Y, Harashima H (2008) Mitochondrial drug delivery systems for macromolecule and their therapeutic application to mitochondrial diseases. Adv Drug Deliv Rev 60:1439–1462. https://doi.org/10.1016/j.addr.2008.04.016

Yan B, Thubagere A, Premasiri WR, Ziegler LD, Negro LD, Reinhard BM (2009) Engineered SERS substrates with multiscale signal enhancement: nanoparticle cluster arrays. ACS Nano 3:1190–1202. https://doi.org/10.1021/nn800836f

Yang JA, Johnson BJ, Wu S, Woods WS, George JM, Murphy CJ (2013) Study of wild-type alpha-synuclein binding and orientation on gold nanoparticles. Langmuir 29:4603–4615. https://doi.org/10.1021/la400266u

Yarjanli Z, Ghaedi K, Esmaeili A, Rahgozar S, Zarrabi A (2017) Iron oxide nanoparticles may damage to the neural tissue through iron accumulation, oxidative stress, and protein aggregation. BMC Neurosci 18. https://doi.org/10.1186/s12868-017-0369-9

Yoo SI et al (2011) Inhibition of amyloid peptide fibrillation by inorganic nanoparticles: functional similarities with proteins. Angew Chem Int Edit 50:5110–5115. https://doi.org/10.1002/anie.201007824

Yoshiyama Y et al (2007) Synapse loss and microglial activation precede tangles in a P301S tauopathy mouse model. Neuron 53:337–351. https://doi.org/10.1016/j.neuron.2007.01.010

You CC et al (2007) Detection and identification of proteins using nanoparticle-fluorescent polymer 'chemical nose' sensors. Nat Nanotechnol 2:318–323. https://doi.org/10.1038/nnano.2007.99

You DG et al (2016) ROS-generating TiO2 nanoparticles for non-invasive sonodynamic therapy of cancer. Sci Rep-UK 6. https://doi.org/10.1038/srep23200

Yuan YG, Zhang SM, Hwang JY, Kong IK (2018) Silver nanoparticles potentiates cytotoxicity and apoptotic potential of camptothecin in human cervical cancer cells. Oxid Med Cell Longev 2018. https://doi.org/10.1155/2018/6121328

Zanganeh S et al (2016) Iron oxide nanoparticles inhibit tumour growth by inducing pro-inflammatory macrophage polarization in tumour tissues. Nat Nanotechnol 11:986–994. https://doi.org/10.1038/Nnano.2016.168

Zhang J, Chen YY, Li DF, Cao Y, Wang ZX, Li GX (2016) Colorimetric determination of islet amyloid polypeptide fibrils and their inhibitors using resveratrol functionalized gold nanoparticles. Microchim Acta 183:659–665. https://doi.org/10.1007/s00604-015-1687-1

Bio-nano Interface and Its Potential Application in Alzheimer's Disease

12

Biswajita Pradhan, Bimal Prasad Jit, Sairendri Maharana, Shankar Ramchandani, and Mrutyunjay Jena

Abstract

Bio-nano interface is the boundary where nano scale entities meet biological systems to exert precise biological function. It unlocks new outlooks for more effective diagnostics, tissue renewal, and therapeutic interventions. In organisms, nanoparticles obtain a covering of biomolecules such as lipids, proteins, and polysaccharides from their surroundings, which reduce their surface energy and confer biological uniqueness to particles. Proper maintenance of a healthy life requires efficient and timely monitoring health to detect early onset of diseases. Nano-technological methods are spur development of more sophisticated tools for detecting early stages of diseases such as atherosclerosis, cancer, diabetes, pulmonary diseases, cardiovascular diseases, nephropathy, retinopathy, and Alzheimer's disease (AD). The common neurodegenerative disease is AD that causes memory loss and deterioration of cognitive function. Furthermore, it displays additional neurological symptoms such as delusion, deficiency language, and learning. These symptoms of AD progression gradually ultimately leading to the prostration of individual functions. Therefore, strategies are made to overcome early detection and diagnostics for better management of AD. Developing proper prognosis for AD is one of the most challenging features in the era of modern medicine. In such scenario, nanoparticles (NPs) are emerged as the new tools to overcome the limitation of prognosis for detecting AD. Aggregation of

B. Pradhan · S. Maharana · M. Jena (✉)
Post Graduate Department of Botany, Berhampur University, Bhanja Bihar, Berhampur, Odisha, India

B. P. Jit
Department of Biochemistry, All India Institute of Medical Science, New Delhi, India

S. Ramchandani
Veer Surendra Sai Institute of Medical Science and Research, Sambalpur, Odisha, India

M. Arakha et al. (eds.), *Bio-Nano Interface*,
https://doi.org/10.1007/978-981-16-2516-9_12

amyloid-β peptide in the brain is supposed to be the prime incident in AD pathophysiology. This paper deliberated the role of nanotechnology in premature detection of AD and explored targeting outgrowth in diagnosis and treatment.

Keywords

Alzheimer's disease · Nanoparticle · Nanotechnology · Diagnosis · Amyloid-β peptide

12.1 Introduction

Alzheimer's disease (AD) has attained additional care in last couple of periods due to the rise in dementia cases but has no success in findings of effective treatments. Currently, AD-related death ranks 6th in the world, but still, no positive prevention approach or therapy has been established (Gu et al. 2016). AD is a different kind of dementia that causes memory problems, behavior, and thinking. The AD symptoms generally grow gently but with time the severity of the disease leads to hindrance with everyday responsibilities. AD collects to 60–80% of all types of dementia cases. It is not a usual part of the aging process (Amiri et al. 2013). The highest known risk factor is old age, and the majority of people suffering from AD are above 65 and older.

AD is a heterogeneous, multifaceted disorder categorized by multiple damages of cellular and physiological function caused by epigenetic, genetic, and also ecological factors. In 1906 Alois Alzheimer defined first time the distinctive plaques and neurofibrillary tangles in the brain of 50 years aged older woman with severe cognitive dysfunction (Alzheimer 1907). The cognitive dysfunction or brain fog means loss of intellectual functions such as remembering, thinking, and reasoning leading to interference with daily functions. The incidence and prevalence of AD are increasing day by day. The elderly population was rising worldwide, and now AD becomes one of the foremost worldwide health care problems (Sloane et al. 2002). In the current age, neither accurate diagnostic strategies nor potential therapeutic mediators are existing for the treatment of AD (Nestor et al. 2004; Mortimer et al. 2005). Furthermore, the available therapeutic agents for AD diseases are only to lessen the symptoms (Murman and Colenda 2005).

In recent scenario, nanotechnology has been opening foundation and innovative fields in several aspects of human life, especially the diagnosis and the treatment of diseases, which includes AD. Nanotechnology is progressing in the field of drug discovery, dispersal, molecular detection, and monitoring numerous challenging human diseases, which include neurodegenerative disorders (Jain 2005; Mansoori 2005; Mansoori et al. 2007, 2010; Modi et al. 2010).

The application of nanotechnology towards health is called as Nanomedicine. It exploits the chemical, biological, and physical property of constituents in nanometric scales. Nanomedicine has a possible effect on early prevention and consistent diagnosis and disease treatment. The eventual goal of nano-diagnostics is to aid better detection of disease at a preliminary stage. Nanotechnology offers diagnostic accessories for better specificity, sensitivity, and consistency. The main aim of drug delivery systems is targeting particular cells inside the body. Nanomedicines as a recent technique have been developed to effectually target delivery of drugs to the diseased site, to increase the efficacy of therapeutic intervention.

Additionally, it also provides a better strategy to deliver a new generation of pharmaceutical medicine that could not be effectively delivered by conventional means (Mirkin and Niemeyer 2004; Singh et al. 2008).

12.2 Pathogenesis

AD is a neurological disorder, and worldwide, about 46.8 million of people were suffering from dementia in 2015 and will gradually increase to 131.5 million by 2050 (Cañabate et al. 2017; dos Santos et al. 2018). There are several issues associated with AD, including environmental and genetic factors. There are three genes that are responsible for causing AD. The three single gene mutations associated with the initial onset of AD are the Amyloid precursor protein (APP) on chromosome number 21, Presenilin 1 (PSEN 1) on chromosome number 14, Presenilin 2 (PSEN2) on chromosome number 1. Mutation on the genes resulted in the assembly of abnormal proteins which are linked with AD. Each mutation has a key role in breakdown of APP. The exact function of breakdown protein is not yet fully understood. This breakdown produces hazardous forms of amyloid plaques and creates a hallmark of AD. Development of AD is carried out by two procedures such as extracellular aggregation of beta amyloid (Aβ) and aggregation of intracellular tau protein. The Aβ is the foremost constituent of senile plaques and tau is the constituent of neurofibrillary tangles. The Aβ deposition is primary and particular for AD (Sahni et al. 2011). The pathophysiology of AD is represented in Fig. 12.1.

12.2.1 Amyloid Plaques

The Aβ protein involved in is formed after breakdown of a larger protein called amyloid precursor protein. Aβ 42 is characterized by two extra amino acids and produced over a successive cleavage of amyloid precursor protein (APP) by γ and β secretase (Jarrett and Lansbury Jr 1993). In the brain of Alzheimer's patient, anomalous stages of naturally arising protein clump composed to form plaques which collect among the neurons and disturb functions of the cell. The AD disease progression is based on the idea that alteration in amyloid precursor protein (APP) encourages the Aβ peptides accumulation, whose deposition into senile plaques is followed by the development of neurofibrillary tangles and death of neuronal cells (Grasso and Danani 2020).

In patients suffering from AD, aggregation of the Aβ protein is observed at early stages (Wisniewski and Wegiel 1995). All of the Aβ proteins usually free from

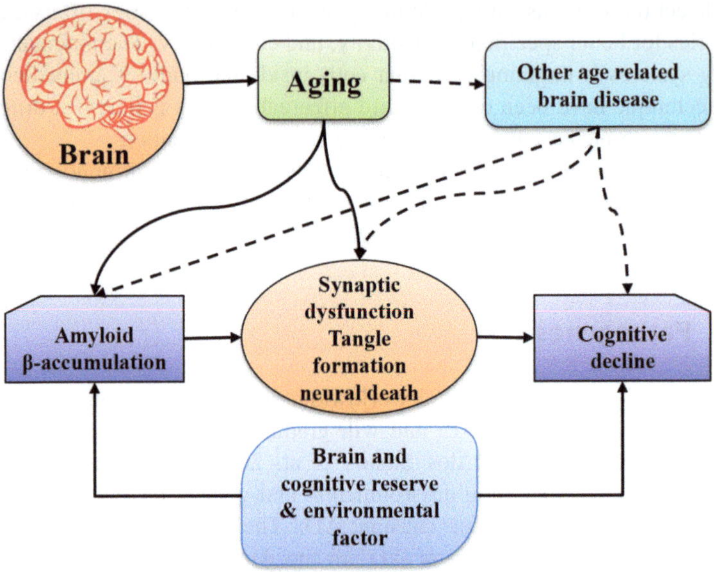

Fig. 12.1 Pathophysiology of Alzheimer's disease

plaques are approximately made up of Aβ 40 and account for 90% of the total, while Aβ 42 accounts for 10–20% of the cell (Tamaoka et al. 1997). Numerous studies report a decline in Aβ 42 concentrations in CSF of AD patients. Moreover, the levels were also found to be lower in AD patients based on the genetic background. For example, patients with ApoE ε4 allele have reduced Aβ 42 concentrations in CSF than individuals without ε4 allele genetic makeup (Tapiola et al. 2000).

12.2.2 Neurofibrillary Tangles

During onset of AD, abnormal accumulation of tau protein is observed in the neurons. In the brain, healthy neurons are sustained by an internal structure called microtubules, which guides the nutrients or the molecules of the cell to the axons and dendrites. However, in AD, abnormal chemical reactions cause tau to separate from microtubules and stick to other tau molecules, forming a thread-like structure that ultimately joins to form tangles inside the neurons. These tangles block the transportation in neurons, eventually affecting the synaptic message between the neurons. Among them, Aβ has become the main focus of neurodegenerative research and many current treatments are targeted to constrain the onset of fibrosis and increase the clearance of Aβ.

AD is commonly divided into premature onset dementia (Presenile) and late onset dementia (Senile). Based on genetic makeup of an individual it is further divided into hereditary and periodic disease (Singh et al. 2008). The pathogenesis of AD is multifactorial. It is not so easy to notice the behavioral change and memory loss caused by AD at an initial stage. Therefore, it is vital to explore early markers to predict the disease onset at an early stage.

12.2.3 Amyloid Precursor Protein (APP)

The characteristic histopathological feature of the AD is accumulation of extracellular amyloid plaques, formed by the deposition of amyloid-β peptides (Aβ) (Chung et al. 1999; Haass and Selkoe 2007; Murphy and LeVine III 2010; Jawhar et al. 2011). Though these plaques contain multiple proteins, their cores are composed of primarily of β amyloid, a 39–42 amino acid proteolytic fragments resulting from the amyloid precursor protein. The APP is a single transmembrane protein with 590–680 amino acids with a long extracellular amino acid terminal domain and approximately 55 amino acid cytoplasmic tail which contains intracellular signals trafficking. Alternatively, splicing of APP mRNA yields eight promising isoforms and out of these three are (695, 751, and 770) predominate in the brain.

12.3 Nanotechnology Used in AD Detection

The following nanoparticles have emerged as the new therapeutic tools for better diagnosis and treatment of AD in patients at an early stage.

12.3.1 Iron Oxide NPs

Due to enormous surface area, magnetic properties, low toxicity, degradability, and biocompatibility, magnetic iron oxide NPs have gained a wide consideration in this field. For the first time Cheng et al. (2015) used super magnetic iron oxide nanoparticle (SPIONs) associated with curcumin covered with poly lactic acid (PLA) and polyethylene glycol (PEG) for detection of amyloid plaques in AD.

12.3.2 Gold NPs

Gold conjugated DNA-NPs have emerged as a new tool for the detection of ADDL (Amyloid beta Derived Diffusible Ligands) at a minute molar concentration (Georganopoulou et al. 2005).

12.3.3 Scanning Tunnelling Microscopy System

The scanning tunnelling microscopy system (STM) is a new current development, which utilizes molecular recognition system for AD detection. This practice is comprised the immobilization of specific antibody fragments on gold (Au) substrate and Au nanoparticles (AuNP). In this technique, addition of the sample solution to the substrate, the nanoparticle-antibody conjugates were developed. These variations among the sample surface and scanning tip current affected (Signal transformation) in the tunnelling. Then the tunnelling current profile was measured based on the pulse frequency peaks and the scanning tip passed over the AuNPs. This technique is very useful for an ultrasensitive detection of Aβ and became more promising even at little concentration (Kang et al. 2009; Lee et al. 2009).

12.3.4 Two Photon Rayleigh Spectroscopy

Recently, AuNPs Rayleigh scattering signals were studied as a converted signal of an immune sensor for tau protein, which acts as the AD biomarkers. In this regard, the synergetic effect of AuNP and anti-tau antibody were used for the detection of tau proteins in sample solution. In the basis of the biomarker transformation indicator was the resultant of deposit of AuNP antibody after adding the tau protein. This technique might be able to detect tau protein at low concentrations (1 pg ml^{-1}) within 35 min. By the use of this technique, researcher claimed to present for first time ultrasensitive and specific nano sensor for detection of tau protein (Neely et al. 2009).

12.4 Nanotechnology in the Treatment of AD

12.4.1 Blood–Brain Barrier (BBB)

Blood–brain barrier (BBB) is the principal blockade to therapeutic drug delivery into the brain. Some of the treatments have affected them to open by generating structural damage to BBB, at which point the BBB has lost the selectivity of drug channels. The carrier system combined with nanotechnology is the utmost proficient treatment strategy for the delivery of drugs into the brain via BBB (Dai et al. 2018).

12.4.2 Nanogels and Fullerene

Nanogels are the most promising drug delivery device because they offer healthier drug stability, encapsulation, and also target based on several factors such as ionic strength, temperature, and pH (Ikeda et al. 2006). The development of colloidal NPs inhibited aggregation of Aβ, thus decreasing its toxicity (Boridy et al. 2009). Additionally, other significant NPs come from carbon family fullerene (C_{60}) and considered as free radical scavenger and antioxidant for AD treatment because of its neuroprotective effect (Dugan et al. 2001).

12.4.3 Diamandoid and Its Derivative

These are the numerous antiviral and antibacterial drugs that are being commercialized and promoted at different progressive stages. Memantine is a diamondoid derivative and has the efficiency to moderate AD. It is a neuroprotective agent possessing the activity against excitotoxicity, excitatory, neurotransmitter glutamate or over initiation of its membrane receptor which causes death of neuronal cell (Lipton 2006; Mansoori et al. 2007).

12.4.4 Curcumin Loaded Nanoparticle

Recently, curcumin has been widely examined in several biological activities, including providing neuroprotective effects (Zaki Ahmad et al. 2014). Many reports suggest that curcumin causes a substantial reduction in Aβ aggregation ratio to neuron toxicity (Ringman et al. 2005). On the other hand, it displays poor aqueous stability and solubility. In addition, it also disposed to photodegradation and oxidation. Nano-capsules are prepared by poly n-butyl cyanoacrylate (PBCA) that have been used for the passage of curcumin across BBB (Anand et al. 2007).

12.4.5 Acetylcholine Loaded (ACh) Nano Carrier

ACh concentration has a straight correlation with loss of memory and has been established as the key symptom for AD. In fact, currently available main AD drug therapy is established on the method to withstand the ACh level by preventing its hydrolysis via enzymes. Drugs employed as an enzyme inhibitor are known to have several side effects. Therefore, the use of neurotransmitter ACh directly as a drug has been suggested as a good approach for controlling AD. On the other hand, its delivery to brain via BBB is interesting. Therefore, nano-technological approach has been projected for delivery of Ach by the use of single-wall carbon nanotubes (SWCNTs) into brain (Yang et al. 2010). Defective cholinergic neuro-transmission is supposed to be the foremost reasons for learning and memory damage in patients of AD (Zaki Ahmad et al. 2014). In the year 2000, FDA approved drug rivastigmine (RT) that is applicable for treatment of AD. However, clinical trials of rivastigmine are still restricted due to poor brain translocation that leads to unsolicited cholinergic properties on peripheral organs and requires regular administration (Gauthier et al. 2010; Wilson et al. 2012). Use of nanotechnology towards the delivery of rivastigmine into brain is a favorable approach to overcome the overhead limitations (Gauthier et al. 2010; Wilson et al. 2012). Rivastigmine (RT) loaded polysorbate 80 coated Poly n-butyl cyanoacrylate nanoparticles (PnBCA-NPs) to enhance the transportation of RT and reduce major side effects (Gauthier et al. 2010; Wilson et al. 2012).

12.4.6 Hormone Loaded Nano carrier

It has been discovered during current inquiries of sex hormones that mainly estrogen and androgens have neuroprotective action against AD such as cytotoxicity, neurotoxicity, and Aβ aggregation. From the reported results, it has been observed that estradiol elevates the progress of cholinergic neurons and extensively decreases the deposition of cerebral Aβ peptide (Amtul et al. 2010).

12.4.7 Polyphenol Drugs Loaded NPs

Newly, epigallocatechin-3-gallate (EGCG) is a polyphenol compound from green tea and possesses the potential therapeutic ability against AD (Rezai-Zadeh et al. 2008). EGCG also abridged Aβ-peptide production separately from the effect of antioxidant. Aβ-peptides are formed after APP proteolysis through a sequence and series of enzymatic activities of β- and γ-secretase. On the other hand, the molecular non-amyloidogenic pathway comprises consecutive APP cleavages via α-secretase (preventing Aβ formation) and γ-secretase, foremost to the growth of non-amyloidogenic fragments. Interruption in the synthesis of these two enzymatic pathways leads to enhanced aggregation of Aβ-peptides to trigger AD (Querfurth and LaFerla 2010).

12.4.8 Zinc Chelators

Clioquinol (Iodochlorohydroxyquin) is a USP drug that possesses chelation property and has the capability to enter BBB. It has shown the possible therapeutic benefit in AD (Ritchie et al. 2003). It is a quinoline derivative and an antibiotic drug possessing a robust chelating affinity for zinc and copper that melts Aβ plaques in vitro and hinders Aβ deposition in AD rats (Ritchie et al. 2003).

12.4.9 Dendrimers

Generally, it is prepared by an iterative sequence of reaction steps. In the last couple of years, such type of NPs has increased attention of biochemical and pharmaceutical investigators. Particularly, the chance to functionalize the peripheral functional groups with desirable ligands is a clever approach to target Aβ peptide for AD (Stiriba et al. 2002).

12.4.10 Antioxidant Nano Carrier

Recently, antioxidant nano carriers in nanotechnology are building blocks, and moreover solid lipid nanoparticles (SLN) have been emphasized to increase the efficacy of antioxidant delivery agents (Bondi et al. 2009; Picone et al. 2009). SLN played the key role of a nano carrier for ferulic acid, a phenolic compound possessing strong antioxidant activity (Bondi et al. 2009). The resultant nano complex holds a lesser colloidal surface and has a highly negative surface charge when dispersed into water. In this regard, the effect of ferulic acid (FA) reduces the production of ROS, decreases cytochrome C release, and controls mitochondrial membrane potential. Collectively, it is the result of increased cell feasibility in the experiments of exposure of neuroblastoma cells to Aβ 42 oligomers. Further, enhanced the drug constancy within biological fluids and making intracellular pointing viable, the SLN nano transporter enhanced the efficiency of this method by reducing the enzymatic degradation of the genetic materials (Nazem and Mansoori 2014).

12.4.11 Gene Nano Carrier

The leading part of gene therapy is the delivery of gene, through which the genetic material is present inside a cell. To protect the cell membrane from biological barrier like cell membrane ionic charge, enzymatic degradation, a carrier should escort the genetic material in the cell (Nikakhtar et al. 2005, 2007; Roy et al. 2008). In AD, mostly viral vectors are used for gene therapy but possess several health issues associated with viral vectors. On the other hand, the use of nanoparticle mediated non-viral gene has significantly developed the efficiency of the method to minimize the enzymatic degradation of genetic materials (Nikakhtar et al. 2005, 2007; Roy et al. 2008). In particular, ceramic polymeric and amino-terminated organically modified silica (ORMOSIL) nanoparticles have displayed some potential in gene transfer to central nervous system (CNS) (Roy et al. 2008). The details of nanotechnology based agents are used in targeting and detection of AD is depicted in (Tables 12.1, 12.2 and Fig. 12.2).

Table 12.1 Nanoparticle used for targeting Alzheimer's disease

Name of bioactive/ drugs	Nano carriers	Remarks	References
Curcumin	Poly (n-butyl) cyanoacrylate (PBCA) nanoparticles	The photostability of curcumin was increased significantly in nanoparticles compared to plain curcumin. ApoE3-PBCA NPs exhibited synergetic activity amyloid induced cytotoxicity along with curcumin	Mulik et al. (2010)
Rivastigmine	PBCA nanoparticle	Intravenous administration of rivastigmine encapsulated polysorbate 80-coated polymeric (T-80) PBCA NPs exhibited 3.8 fold increase in rivastigmine uptake within the brain compartment compared to free rivastigmine	Wilson et al. (2012)
Galanthamine-hydrobromide	Liposome	The efficacy of AChE inhibition of GH was greatly enhanced by intranasal administration compared with its oral administration of galactosamine hydrobromide (Gh sol), bio availability was 3.36 folds higher than those of orally administered GH	Li et al. (2012)
Acetylcholine	Single-wall carbon nanotubes (SWCNTs)	At low dose SWCNTs effectively carried ACh into the lysosome of the neurons and achieve excellent therapeutic effects in AD	Yang et al. (2010)
Estradiol	PLGA nanoparticles	NPs exhibited significantly high brain estradiol levels after 24 h. However the expression of Aβ 42 immuno-reactivity in the hippocampus region of brain	Mittal et al. (2011)
Resveratrol	Polymeric micelles	Protect PC12 cells from Aβ induced damage in a dose-dependent manner by attenuating intracellular oxidative stress and caspase-3 activity without long-term cytotoxicity	Lu et al. (2009)
	Lipid core nano-capsules	Nano-capsules efficiently rescuing the deleterious effect of Aβ1-42	Frozza et al. (2013)

12.5 Limitation of Nanotechnology in the Treatment of AD

It is well known that NPs will be directly enclosed by proteins to make protein crowns after reaching the biological environment (Dai et al. 2018). The influence of protein on Aβ fibrosis has to be assessed. Different diseases may similarly modify the destiny of NPs in the body, so NPs have diverse therapeutic potentials or toxic

Table 12.2 Nanotechnology and neuroprotective agent for treatment of Alzheimer's disease

Nanosytems	Neuroprotective function	Targeted AD for pathology	Study mode	References
Nanogels	Aβ Anti-assembly (incorporate Aβ monomers)	Oligomerization of Aβ	In vitro	Ikeda et al. (2006)
Fullerene (C_{60})	Antioxidant, Aβ anti-assembly/ maintenance of Ca^{2+} homeostasis	Oxidative stress oligomerization of Aβ Ca^{++} influx	In vitro and In vivo (Rats)	Dugan et al. (1996, 1997, 2001), Huang et al. (2000) and Podolski et al. (2007)
Dendrimers	Aβ Anti-assembly	Oligomerization of Aβ cell membrane toxicity	In vitro	Mansoori et al. (2007) and Nikakhtar et al. (2007)
Nanoceria	Antioxidant	Oxidative stress	In vitro	D'Angelo et al. (2009) and Suh et al. (2009)
Gold nanoparticle (AuNP)	Aβ Anti-assembly	Oligomerization of Aβ	In vitro	Kogan et al. (2006)
Diamondoid derivatives	NMDA receptor	Glutamate excitotoxicity	In vivo (FDA approved)	Reisberg et al. (2003)

effects after entering into different patients (Zaki Ahmad et al. 2014; Rahman et al. 2015a, b). These studies propose that in the future research of nanotechnology, a variability of unseen factors in bio-nano interface must be considered. The potential toxicity of Nps is an ongoing issue. In order to resolve these problems, it has been essential to develop multifunctional biocompatible NPs with multiple therapeutics proficiency, for example, a comprehensive switch of Tau protein, inflammatory response, phosphorylation, redox reaction, and improvement of mitochondrial membrane (Zaki Ahmad et al. 2014; Rahman et al. 2015a, b). Advancement of nanotechnology establishes an efficient drug delivery system, and optimistic suggestions have also raised the concern associated with toxicity. The surface morphology and the particle size of nanoparticulate systems are reported to modify the pharmacokinetics and tissue distribution profiling of loaded bioactive/drugs compounds (Zaki Ahmad et al. 2014; Rahman et al. 2015a, b).

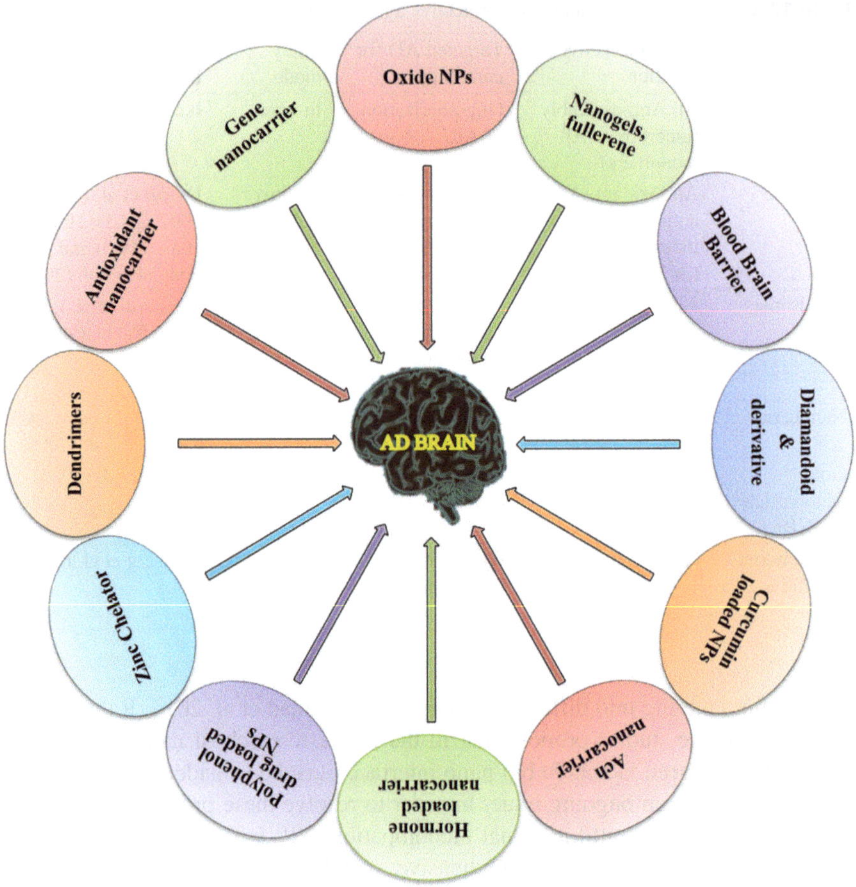

Fig. 12.2 Nanotechnology used in the treatment of Alzheimer's disease

12.6 Conclusion and Future Prospective

Nanotechnology offers an extensive potential for its application in medicine in near future. Since the discovery of degenerative diseases of central nervous system, its diagnosis and treatment have been a colossal medical challenge. In addition, research in the biological field and other fields, particularly the overview of nanotechnology has been promising for researchers. Current research indicates that the external surface of NPs can be adapted its stability towards the drug affinity. In future, we hope to see that the nanotechnology based drug delivery systems can successfully treat more aggressive neurological diseases and overcome drug resistance to diseases.

Conflict of Interests The authors declare that there is no conflict of interest.

Acknowledgments The authors are also thankful to Berhampur University for providing the necessary facilities to carry out the study. The authors also thankful to Mr. Rabindra Nayak for helping to draw the diagram.

References

Alzheimer A (1907) About a peculiar disease of the cerebral cortex. Centralblatt für Nervenheilkunde Psychiatrie 30(1):177–179

Amiri H, Saeidi K, Borhani P, Manafirad A, Ghavami M, Zerbi V (2013) Alzheimer's disease: pathophysiology and applications of magnetic nanoparticles as MRI theranostic agents. ACS Chem Neurosci 4(11):1417–1429

Amtul Z, Wang L, Westaway D, Rozmahel R (2010) Neuroprotective mechanism conferred by 17beta-estradiol on the biochemical basis of Alzheimer's disease. Neuroscience 169 (2):781–786

Anand P, Kunnumakkara AB, Newman RA, Aggarwal BB (2007) Bioavailability of curcumin: problems and promises. Mol Pharm 4(6):807–818

Bondi M, Montana G, Craparo E, Picone P, Capuano G, Carlo M, Giammona G (2009) Ferulic acid-loaded lipid nanostructures as drug delivery systems for Alzheimer's disease: preparation, characterization and cytotoxicity studies. Curr Nanosci 5(1):26–32

Boridy S, Takahashi H, Akiyoshi K, Maysinger D (2009) The binding of pullulan modified cholesteryl nanogels to Aβ oligomers and their suppression of cytotoxicity. Biomaterials 30 (29):5583–5591

Cañabate P, Martínez G, Rosende-Roca M, Moreno M, Preckler S, Valero S, Sotolongo O, Hernández I, Alegret M, Ortega G (2017) Social representation of dementia: an analysis of 5,792 consecutive cases evaluated in a memory clinic. J Alzheimer's Dis 58(4):1099–1108

Cheng Á, Doecke JD, Sharples R, Villemagne VL, Fowler CJ, Rembach A, Martins RN, Rowe CC, Macaulay SL, Masters CL (2015) Prognostic serum miRNA biomarkers associated with Alzheimer's disease shows concordance with neuropsychological and neuroimaging assessment. Mol Psychiatry 20(10):1188–1196

Chung H, Brazil MI, Soe TT, Maxfield FR (1999) Uptake, degradation, and release of fibrillar and soluble forms of Alzheimer's amyloid β-peptide by microglial cells. J Biol Chem 274 (45):32301–32308

D'Angelo B, Santucci S, Benedetti E, Di Loreto S, Phani R, Falone S, Amicarelli F, Ceru MP, Cimini A (2009) Cerium oxide nanoparticles trigger neuronal survival in a human Alzheimer disease model by modulating BDNF pathway. Curr Nanosci 5(2):167–176

Dai C-l, Hu W, Tung YC, Liu F, Gong C-X, Iqbal K (2018) Tau passive immunization blocks seeding and spread of Alzheimer hyperphosphorylated Tau-induced pathology in 3× Tg-AD mice. Alzheimer's Res Ther 10(1):13

dos Santos P, Leide C, Ozela PF, de Fatima de Brito Brito M, Pinheiro AA, Padilha EC, Braga FS, Carlos H, dos Santos CBR, Rosa J (2018) Alzheimer's disease: a review from the pathophysiology to diagnosis, new perspectives for pharmacological treatment. Curr Med Chem 25 (26):3141–3159

Dugan LL, Gabrielsen JK, Shan PY, Lin T-S, Choi DW (1996) Buckminsterfullerenol free radical scavengers reduce excitotoxic and apoptotic death of cultured cortical neurons. Neurobiol Dis 3 (2):129–135

Dugan LL, Turetsky DM, Du C, Lobner D, Wheeler M, Almli CR, Shen CK-F, Luh T-Y, Choi DW, Lin T-S (1997) Carboxyfullerenes as neuroprotective agents. Proc Natl Acad Sci 94 (17):9434–9439

Dugan L, Lovett E, Quick K, Lotharius J, Lin T, O'malley K (2001) Fullerene-based antioxidants and neurodegenerative disorders. Parkinsonism Relat Disord 7(3):243–246

Frozza RL, Bernardi A, Hoppe JB, Meneghetti AB, Matté A, Battastini AM, Pohlmann AR, Guterres SS, Salbego C (2013) Neuroprotective effects of resveratrol against Aβ administration in rats are improved by lipid-core nanocapsules. Mol Neurobiol 47(3):1066–1080

Gauthier S, Cummings J, Ballard C, Brodaty H, Grossberg G, Robert P, Lyketsos C (2010) Management of behavioral problems in Alzheimer's disease. Int Psychogeriatr 22(3):346–372

Georganopoulou DG, Chang L, Nam J-M, Thaxton CS, Mufson EJ, Klein WL, Mirkin CA (2005) Nanoparticle-based detection in cerebral spinal fluid of a soluble pathogenic biomarker for Alzheimer's disease. Proc Natl Acad Sci 102(7):2273–2276

Grasso G, Danani A (2020) Molecular simulations of amyloid beta assemblies. Adv Phys: X 5 (1):1770627

Gu X, Cai Z, Cai M, Liu K, Liu D, Zhang Q, Tan J, Ma Q (2016) Protective effect of paeoniflorin on inflammation and apoptosis in the cerebral cortex of a transgenic mouse model of Alzheimer's disease. Mol Med Rep 13(3):2247–2252

Haass C, Selkoe DJ (2007) Soluble protein oligomers in neurodegeneration: lessons from the Alzheimer's amyloid β-peptide. Nat Rev Mol Cell Biol 8(2):101–112

Huang H-M, Ou H-C, Hsieh S-J, Chiang L-Y (2000) Blockage of amyloid β peptide-induced cytosolic free calcium by fullerenol-1, carboxylate C60 in PC12 cells. Life Sci 66 (16):1525–1533

Ikeda K, Okada T, S-i S, Akiyoshi K, Matsuzaki K (2006) Inhibition of the formation of amyloid β-protein fibrils using biocompatible nanogels as artificial chaperones. FEBS Lett 580 (28–29):6587–6595

Jain KK (2005) The role of nanobiotechnology in drug discovery. Drug Discov Today 10 (21):1435–1442

Jarrett JT, Lansbury PT Jr (1993) Seeding "one-dimensional crystallization" of amyloid: a pathogenic mechanism in Alzheimer's disease and scrapie? Cell 73(6):1055–1058

Jawhar S, Wirths O, Bayer TA (2011) Pyroglutamate amyloid-β (Aβ): a hatchet man in Alzheimer disease. J Biol Chem 286(45):38825–38832

Kang D-Y, Lee J-H, Oh B-K, Choi J-W (2009) Ultra-sensitive immunosensor for β-amyloid (1–42) using scanning tunneling microscopy-based electrical detection. Biosens Bioelectron 24 (5):1431–1436

Kogan MJ, Bastus NG, Amigo R, Grillo-Bosch D, Araya E, Turiel A, Labarta A, Giralt E, Puntes VF (2006) Nanoparticle-mediated local and remote manipulation of protein aggregation. Nano Lett 6(1):110–115

Lee JH, Kang DY, Lee T, Kim SU, Oh BK, Choil JW (2009) Signal enhancement of surface plasmon resonance based immunosensor using gold nanoparticle-antibody complex for beta-amyloid (1-40) detection. J Nanosci Nanotechnol 9(12):7155–7160. https://doi.org/10.1166/jnn.2009.1613

Li W, Zhou Y, Zhao N, Hao B, Wang X, Kong P (2012) Pharmacokinetic behavior and efficiency of acetylcholinesterase inhibition in rat brain after intranasal administration of galanthamine hydrobromide loaded flexible liposomes. Environ Toxicol Pharmacol 34(2):272–279

Lipton SA (2006) Paradigm shift in neuroprotection by NMDA receptor blockade: memantine and beyond. Nat Rev Drug Discov 5(2):160–170

Lu X, Ji C, Xu H, Li X, Ding H, Ye M, Zhu Z, Ding D, Jiang X, Ding X (2009) Resveratrol-loaded polymeric micelles protect cells from Aβ-induced oxidative stress. Int J Pharm 375(1–2):89–96

Mansoori GA (2005) Principles of nanotechnology: molecular-based study of condensed matter in small systems. World Scientific

Mansoori GA, Mohazzabi P, McCormack P, Jabbari S (2007) Nanotechnology in cancer prevention, detection and treatment: bright future lies ahead. World Rev Sci Technol Sustain Dev 4 (2):226

Mansoori GA, Brandenburg KS, Shakeri-Zadeh A (2010) A comparative study of two folate-conjugated gold nanoparticles for cancer nanotechnology applications. Cancers 2(4):1911–1928

Mirkin CA, Niemeyer CM (2004) Nanobiotechnology: concepts, applications, and perspectives. Wiley-VCH

Mittal G, Carswell H, Brett R, Currie S, Kumar MR (2011) Development and evaluation of polymer nanoparticles for oral delivery of estradiol to rat brain in a model of Alzheimer's pathology. J Control Release 150(2):220–228

Modi G, Pillay V, Choonara YE (2010) Advances in the treatment of neurodegenerative disorders employing nanotechnology. Ann N Y Acad Sci 1184(1):154–172

Mortimer JA, Borenstein AR, Gosche KM, Snowdon DA (2005) Very early detection of Alzheimer neuropathology and the role of brain reserve in modifying its clinical expression. J Geriatr Psychiatry Neurol 18(4):218–223

Mulik RS, Mönkkönen J, Juvonen RO, Mahadik KR, Paradkar AR (2010) ApoE3 mediated poly (butyl) cyanoacrylate nanoparticles containing curcumin: study of enhanced activity of curcumin against beta amyloid induced cytotoxicity using in vitro cell culture model. Mol Pharm 7(3):815–825

Murman DL, Colenda CC (2005) The economic impact of neuropsychiatric symptoms in Alzheimer's disease. Pharmacoeconomics 23(3):227–242

Murphy MP, LeVine H III (2010) Alzheimer's disease and the amyloid-β peptide. J Alzheimer's Dis 19(1):311–323

Nazem A, Mansoori G (2014) Nanotechnology building blocks for intervention with Alzheimer's disease pathology: implications in disease modifying strategies. J Bioanal Biomed 6(2):9

Neely A, Perry C, Varisli B, Singh AK, Arbneshi T, Senapati D, Kalluri JR, Ray PC (2009) Ultrasensitive and highly selective detection of Alzheimer's disease biomarker using two-photon Rayleigh scattering properties of gold nanoparticle. ACS Nano 3(9):2834–2840

Nestor PJ, Scheltens P, Hodges JR (2004) Advances in the early detection of Alzheimer's disease. Nat Med 10(7):S34–S41

Nikakhtar A, Nasehzadeh A, Naghibi-Beidokhti H, Mansoori GA (2005) DNA-dendrimer nanocluster electrostatics prediction with the nonlinear Poisson-Boltzmann equation. J Comput Theor Nanosci 2(3):378–384

Nikakhtar A, Nasehzadeh A, Mansoori GA (2007) Formation and stability conditions of DNA-dendrimer nano-clusters. J Comput Theor Nanosci 4(3):521–528

Picone P, Carrotta R, Montana G, Nobile MR, San Biagio PL, Di Carlo M (2009) Aβ oligomers and fibrillar aggregates induce different apoptotic pathways in LAN5 neuroblastoma cell cultures. Biophys J 96(10):4200–4211

Podolski IY, Podlubnaya Z, Kosenko E, Mugantseva E, Makarova E, Marsagishvili L, Shpagina M, Kaminsky YG, Andrievsky G, Klochkov V (2007) Effects of hydrated forms of C60 fullerene on amyloid β-peptide fibrillization in vitro and performance of the cognitive task. J Nanosci Nanotechnol 7(4–5):1479–1485

Querfurth HW, LaFerla FM (2010) Mechanisms of disease. N Engl J Med 362(4):329–344

Rahman M, Akhter S, Ahmad MZ, Ahmad J, Addo RT, Ahmad FJ, Pichon C (2015a) Emerging advances in cancer nanotheranostics with graphene nanocomposites: opportunities and challenges. Nanomedicine 10(15):2405–2422

Rahman M, Zaki Ahmad M, Ahmad J, Firdous J, Jalees Ahmad F, Mushtaq G, Kamal MA, Akhter S (2015b) Role of graphene nano-composites in cancer therapy: theranostic applications, metabolic fate and toxicity issues. Curr Drug Metab 16(5):397–409

Reisberg B, Doody R, Stöffler A, Schmitt F, Ferris S, Möbius HJ (2003) Memantine in moderate-to-severe Alzheimer's disease. N Engl J Med 348(14):1333–1341

Rezai-Zadeh K, Arendash GW, Hou H, Fernandez F, Jensen M, Runfeldt M, Shytle RD, Tan J (2008) Green tea epigallocatechin-3-gallate (EGCG) reduces β-amyloid mediated cognitive impairment and modulates tau pathology in Alzheimer transgenic mice. Brain Res 1214:177–187

Ringman JM, Frautschy SA, Cole GM, Masterman DL, Cummings JL (2005) A potential role of the curry spice curcumin in Alzheimer's disease. Curr Alzheimer Res 2(2):131–136

Ritchie CW, Bush AI, Mackinnon A, Macfarlane S, Mastwyk M, MacGregor L, Kiers L, Cherny R, Li Q-X, Tammer A (2003) Metal-protein attenuation with iodochlorhydroxyquin (clioquinol) targeting Aβ amyloid deposition and toxicity in Alzheimer disease: a pilot phase 2 clinical trial. Arch Neurol 60(12):1685–1691

Roy I, Stachowiak MK, Bergey EJ (2008) Nonviral gene transfection nanoparticles: function and applications in the brain. Nanomed: Nanotechnol Biol Med 4(2):89–97

Sahni JK, Doggui S, Ali J, Baboota S, Dao L, Ramassamy C (2011) Neurotherapeutic applications of nanoparticles in Alzheimer's disease. J Control Release 152(2):208–231

Singh M, Singh S, Prasad S, Gambhir I (2008) Nanotechnology in medicine and antibacterial effect of silver nanoparticles. Digest J Nanomater Biostruct 3(3):115–122

Sloane PD, Zimmerman S, Suchindran C, Reed P, Wang L, Boustani M, Sudha S (2002) The public health impact of Alzheimer's disease, 2000–2050: potential implication of treatment advances. Annu Rev Public Health 23(1):213–231

Stiriba SE, Frey H, Haag R (2002) Dendritic polymers in biomedical applications: from potential to clinical use in diagnostics and therapy. Angewandte Chemie Int Ed 41(8):1329–1334

Suh MK, Kim H, Na DL (2009) Dysphagia in patients with dementia: Alzheimer versus vascular. Alzheimer Dis Assoc Disord 23(2):178–184

Tamaoka A, Sawamura N, Fukushima T, Matsubara E, Shoji M, Hirai S, Furiya Y, Endoh R, Mori H (1997) Amyloid β protein 42 (43) in cerebrospinal fluid of patients with Alzheimer's disease. J Neurol Sci 148(1):41–45

Tapiola T, Pirttilä T, Mikkonen M, Mehta PD, Alafuzoff I, Koivisto K, Soininen H (2000) Three-year follow-up of cerebrospinal fluid tau, β-amyloid 42 and 40 concentrations in Alzheimer's disease. Neurosci Lett 280(2):119–122

Wilson RS, Segawa E, Boyle PA, Anagnos SE, Hizel LP, Bennett DA (2012) The natural history of cognitive decline in Alzheimer's disease. Psychol Aging 27(4):1008

Wisniewski H, Wegiel J (1995) The neuropathology of Alzheimer's disease. Neuroimaging Clin N Am 5(1):45–57

Yang Z, Zhang Y, Yang Y, Sun L, Han D, Li H, Wang C (2010) Pharmacological and toxicological target organelles and safe use of single-walled carbon nanotubes as drug carriers in treating Alzheimer disease. Nanomed: Nanotechnol Biol Med 6(3):427–441

Zaki Ahmad M, Akhter S, Mohsin N, Abdel-Wahab BA, Ahmad J, Husain Warsi M, Rahman M, Mallick N, Jalees Ahmad F (2014) Transformation of curcumin from food additive to multifunctional medicine: nanotechnology bridging the gap. Curr Drug Discov Technol 11 (3):197–213

Potential of Curcumin Nanoparticles in Tuberculosis Management

13

Bhabani Shankar Das, Ashirbad Sarangi, and
Debapriya Bhattacharya

Abstract

Tuberculosis (TB), the deadliest infectious-disease caused by *Mycobacterium tuberculosis* (*M.tb*) created a global-threat to the entire human race since the past few-millennia. The initiation of the antibiotic-era limited the TB infection; however, generation of antibiotic-resistance bacteria causes re-emergence of this infection and makes an alarm to TB-pandemic. Moreover, BCG is the only used vaccine still its inception and its efficacy varies from 0 to 80%. Therefore, the creation of new drugs has become the focus of scientists to combat this deadly-disease. Ethnomedicine was the only therapy at pre-antibiotic-era which in modern-days re-attracted scientists as phytochemical based drug development as it has a huge opportunity. Curcumin is one of the most highlighted phytochemicals found in turmeric of Curcuma plant. From ancient literature, turmeric was found very effective in varieties of infectious and metabolic diseases and in light of modern science, its compound curcumin shows that efficacy. However, due to its less bioavailability initially, scientists were hesitant to use it as a drug molecule but after the emergence of nanotechnology and development of nano-forms of curcumin resolve the bioavailability issue of curcumin and it emerges as a novel drug molecule. Nano-formulation improves its aqueous solubility as well as the guided tissue delivery which will increase the bioavailability of a molecule for a better-transport of drugs and will help in the improvement of TB treatment. In this chapter, we will concentrate on the formulation of different nanocurcumin-particle and its implication in tuberculosis treatment. In

B. S. Das · A. Sarangi · D. Bhattacharya (✉)
Center for Biotechnology, School of Pharmaceutical Sciences, SOA University, Bhubaneswar, Odisha, India
e-mail: debapriyabhattacharya@soa.ac.in

M. Arakha et al. (eds.), *Bio-Nano Interface*,
https://doi.org/10.1007/978-981-16-2516-9_13

brief, we will seek about comprehensive-mechanisms of action of nanocurcumin on therapy and vaccine-development of tuberculosis.

Keywords

Tuberculosis · Curcumin · Limitation · Curcumin nanoparticles · Mechanism of action · Applications

13.1 A Brief Introduction of Tuberculosis

Tuberculosis (TB) ruins the major cause of mortality and socioeconomic crisis for millions of people around the world, even though medical science and therapeutics have made significant progress. This has been plagued by the human race throughout the history of human civilization and prehistory (Bhattacharya et al. 2013). The earliest known molecular evidence for TB was found in an ancient bison fossil (Pleistocene bison) and human remains discovered in an Eastern Mediterranean Neolithic village in 9000 years of age (Rothschild et al. 2001; Hershkovitz et al. 2008). Although, as early as 1689, Dr. Richard Morton identified that a pulmonary type was linked with "*tubercles bacillus*" although it was not diagnosed as a single disease until the 1820s because of its variety of symptoms. And then it was finally named by J.L. Schonlein in 1839 as a "*tuberculosis*" (Lakhtakia 2013). Finally, Robert Koch discovered the *Mycobacterium tuberculosis* (*M.tb*) the causative agent of tuberculosis in 1882 and for this remarkable discovery, he got Nobel prize in 1905 (Gradmann 2001). Tuberculosis which was commonly known as TB and causes serious infectious respiratory disease (Kumar and Robbins 2007). Even, *M.tb* is a strong acid-fast bacillus and having intracellular pathogens to mainly macrophages that have evolved many strategies to escape from the killing of macrophages and it is known as the most effective and continuing disease which is able to live in the host for decades without triggering the infection (Cooper 2009; Rohde et al. 2007; Stewart et al. 2003). After all, TB continues to be a significant issue of global health problems that affects by infecting nearly 10.7 million people's every year and causing the death of 1.3 million peoples annually (WHO 2018). Over 90% of people exposed to TB disease still remains life-long asymptomatically and only ~10% of those people living with *Mycobacterium tuberculosis* are suffering from active TB diseases particularly when they are at immune-compromised condition due to infection with HIV, immune-suppressive therapy, or diabetes-related diseases that create immune deficiency which further leads to reactivation of TB infection (Robertson 1933; Flynn and Chan 2001a, b).

However, success came after antibiotics were discovered in the mid-twentieth century and these antibiotics were incorporated into TB therapy. While antibiotics are worked in many ways, which has been revolutionized in the medical industry and also saved many lives in this revolutionary world. While from this discovery it became a turning point in the evolution of mankind. Even if till this date, we are fully reliant on this use of antibiotics for the treatment of infectious diseases and it also ensures that it is an advanced operation including organ as well as prosthetic transplants which are successful nowadays (Davies and Davies 2010). But regrettably, the use of such wonderful drugs has emerged in the following resistant strains. As a result, multidrug-resistant pathogens and also some "old" pathogens such as *Mycobacterium tuberculosis* also developed resistance towards the new antibiotic of front line anti-tubercular drugs. Moreover, treating with a group of new antibiotics also remains one of the biggest issues among all infectious diseases (Smith 2003). While the region behind all of this thing is that due to long-term treatment with a set of antibiotics which slows down the progression of infectiousness and also for complex immunological responses. In recent years, drug resistance TB like multidrug resistance (MDR) or comprehensive drug resistance (XDR) TB has become the world's largest threat, along with coinfected conditions like TB-HIV or TB-diabetes (Barberis et al. 2017).

However, drug resistance is a tremendous challenge that may require several processes to solve these antibiotic-resistance issues. While the most significant obstacle is that the development of new and effective antimicrobials has been established previously and remaining very few new antibiotic groups have been identified in the last 40 years. So, for that region, a new strategy is required to control bacterial infections as well as also required new drugs for tuberculosis management.

The presence of medicinal plants is a natural gift that has been used since the beginning of human prehistory for the treatments of numerous human diseases. As per the estimates of WHO in developing countries, 80 percent of the population depends on traditional primary health medicinal products. The use of medicinal plants and their phytochemicals gives greater importance to the treatment of many diseases (Ekor 2014).

Turmeric is one of the oldest known medicinal plants which has been used from prehistoric times to cure various disease and disorders in humans. For that being, turmeric is used in various biological sciences and it is scientifically known as curcumin which is a key natural bioactive ingredient of the turmeric root bulbs and it has been also used in Indian continents as the main food ingredient in all kitchens as well as traditional medicines. Moreover, it was reported that curcumin has enormous potential biological assets such as immunomodulatory activity, antimicrobial, antioxidant, anticancer, anti-inflammatory effects as well as cardiovascular effects which indicates that it has excellent anti-inflammatory immunomodulation against infectious diseases and also holds a powerful therapeutic ability to cure the several disease conditions (Gera et al. 2017). However, despite its different activities in varieties of critical diseases, it could not come to the focus as a drug molecule because of its some important disadvantages like that it has low water solubility and low bioavailability in physiological conditions.

Therefore, some modern technology is required to increase the bioavailability of curcumin, then only this miraculous molecule could be used as a drug molecule. The advancement of nanotechnology showed light in this respect. By using nanotechnology development of curcumin nanoparticle has shown the improvement of bioavailability of this molecule, even mono-carbonyl analogs of curcumin inhibit the growth of *M.tb*. In fact, a nanotechnology-based drug delivery system can significantly increase a drug therapeutic index, improve the tolerability of toxic chemotherapeutics, and reduce systemic side effects. It also enables higher stability and had capacity along with immense improvement of drug bioavailability which further leads to a reduction in dose frequency. This review mainly highlights the nano formation of curcumin and its implication in tuberculosis therapy or improvement of BCG vaccine efficacy.

13.2 Epidemiology of Tuberculosis

Several years ago, TB was still considered as one of the chronic infectious diseases in this developing nation. Due to these chronic infections, it remains one of the major respiratory infectious diseases and it has been identified as a global epidemic. Furthermore, these diseases contribute to drug-resistant species which leads to an increase in mortality rates and prolonging the period of hospitalization and creating an immense financial burden to affected individuals, health care facilities, as well as also delaying the objectives of TB managements sustainable growth (Friedman et al. 2016).

Moreover, it was estimated that approximately one-third of the global population is latently infected with TB. That is approximately 1.7 billion people are estimated to be latently infected with TB infection, and about 10% of peoples of them are suffering from active tuberculosis. Globally almost 10 million active TB cases are there and around 1.3 million people have died as a result of this infection each year. WHO also estimated that nearly 484,000 cases of MDR-TB cases occurred in 2018, which included approximately 378,000 cases are infected with MDR-TB and 214,000 deaths cases people died annually. The average ratio of XDR-TB and MDR-TB cases was 6.2%. In addition, India is containing 27%, China 14%, and the Russian Federation 9%, reflecting 50% of the global burden of drug resistance TB (WHO 2018). Looking into its public health importance the Govt. has released a target through the "END TB" strategy. The objective is to reduce the TB mortality rate by 95% and the incidence rate by 90% in 2035 (WHO 2019).

13.3 Nature of *Mycobacterium tuberculosis*

Globally, TB continues to be a massive public health issue. It is established previously that the infection is caused by *Mycobacterium tuberculosis*. The *Mycobacterium* comes under the Actinomycetales class, which consists of many well-known organisms, several of which are related to human-animal diseases (Horsburgh Jr 1991). *M.tb* is an obligatory aerobic and non-motile rod-shaped bacteria that has a dimension of 2–4 μm in length and 0.2–0.5 μm in width and it is highly resistant to alcohol and acid. *M.tb* has a doubling time of 12–24 h under optimal conditions and in clinical cultures, it can take approximately 6–8 weeks. As an obligatory aerobe by nature, it mostly found in the well-aerated and highly vascular region like proximity to apical lobes of the lung. Due to the presence of high mycolic acid present in the cell wall of Mycobacteria, it is less permeable to staining reagents as well as the drugs. The acid-fast nature of bacilli is due to its resistance to up to 25% of conventional acids like sulfuric acid and hydrochloric acid. Hence it is different from other organisms, called Acid-Fast Bacilli (AFB) (Levinson 2020).

13.4 Mode of Transmission and Its Risk Factors

Tuberculosis is an airborne disease that transmits through aerosols from one person to another. During this transmission, people are suffering from active TB which shows some activities such as coughing, sneezing, speaking, and spitting. These infected patient releases small bacilli droplets (0.5–5.0 μm in diameter) which contain the etiological agent of TB, i.e. *M.tb*. An aerosol droplet usually contains up to 40,000 droplets which it can remain suspended in the atmosphere from minutes to hours depending on environmental condition (Cole and Cook 1998).

Patients with extrapulmonary tuberculosis have less chance for transmission of infection as these are paucibacillary by nature. In contrast, pulmonary TB is multibacillary by nature, hence more infectious. People in close contact with active TB infected patients have 22% chances of acquiring the disease and are at high risk (Ahmed and Hasnain 2011).

The most common primary route of entry for tubercle bacilli is via the respiratory route. The bacteria could spread from its prime initial location in the lungs alveoli to other parts of the human organelles like lymph nodes in the case of extrapulmonary TB. Age, gender, nutrition, and low immunity can be affected by tuberculosis. The diseases like HIV-AIDS, Diabetes, malnutrition, prolonged therapy of steroids immunodeficiency disorders are common predisposing factors of Tuberculosis.

The likelihood of transmission of bacterium depends on four factors:

(1) The number of organisms expelled by the TB infected person.
(2) The bacterial load, i.e. concentration of organisms present in the air.
(3) An exposure time of the contaminated air by the healthy person.
(4) Immune condition of the exposed individual.

However, along with this, there has also some other risk factors that affect people from TB susceptibility such as

1. Human immunodeficiency virus (HIV) which causes most of the TB cases.
2. Chronic lung disease that causes silicosis.
3. Tobacco and cigarette smoking.
4. Immuno negotiated patients such as Diabetes mellitus, Cancer, and Leukemia.
5. Close contacts with infected TB individuals including household contacts and health care workers.
6. Social mixing together with overcrowding and malnutrition.
7. Immunosuppressive drugs.

So, these are the factors along with the health system which delay in diagnosis of TB during transmission of the bacilli (Narasimhan et al. 2013).

13.5 Pathogenesis of Tuberculosis

Persons with active pulmonary tuberculosis (TB) serve as the main source of transmission. Though it was generally transmitted through small droplets of TB bacilli, which are particles in size of 1–5 μm in diameter that containing *M. tuberculosis*. Due to its smaller size of TB bacilli, it can remain suspended for several minutes in the air (Frieden et al. 2003; Hill et al. 2004). Whereas, this can happen when somebody was untreated during diagnosis which further leads to an active form of tuberculosis. This microorganism spreads easily through pulmonary infection. After infection bacteria travels through the lungs. Which further bacilli are ingested by alveolar macrophages and then multiply intracellularly by inhibiting the formation of phagolysosomes (Frieden et al. 2003). After that mycobacterium interacts with T lymphocytes cells which further contributes to the creation in the histiocytes cell as well as in epithelioid cells of different macrophages (Chaudhary and Garg 2015). Then T cell triggers stimulations and releases different cytokine as well as interferon-gamma cytokines. Further, γ-interferon cytokines stimulate the infected macrophage cells and release TNF-α, IL-8, IL-12, cytokines as well as also release other proinflammatory cytokines with the macrophage cells. After that different small groups of both epithelioid histiocytes and lymphocytes cells form into granulomas cell which indicates a hallmark of tuberculosis (Jeong and Lee 2008).

Moreover, it has been identified that CD4 + (helper) and CD8 + (cytotoxic) T cells are involved in cell-mediated immune response, both of which work as major importance in the TB defense system. CD4 + T helper cells help to enhance the antibacterial macrophage activity through releases of different cytokines such as interferon-gamma and TNF-α, while CD8 + cells destroy the infected macrophages and also release different types of cytotoxic mediators, such as perforins, granulysin as well as granzymes (Cooper 2009).

If the immune response is strong enough, then the bacteria kills within granuloma, this type of granuloma is called a solid granuloma, ultimately the patient becomes asymptomatic. But if the immune response is weak, then the bacteria replicate within the macrophages and granuloma becomes necrotic and ultimately caseous granuloma formed from where bacteria spill and come out to the airway as an aerosol. It can infect other individuals (Guirado and Schlesinger 2013).

In spite of our improved knowledge still, there is a lack of comprehension of the immune response in *M.tb*. Even if, a different type of immune response is necessarily required for successive vaccine-induced immunity which is till now not fully understood (Horvath et al. 2012).

13.6 Symptoms of Tuberculosis

The TB disease is generally characterized by coughing continuously, cough with blood, consistent fever, chest pain, night sweating, weight loss, and loss of appetite are the symptoms of TB which usually occur when it spread from an infected person through cough, sneeze, spits, or talk (Colledge et al. 2010).

13.7 Susceptibility and Resistance Form of TB and Their Drugs

Tuberculosis is divided into the following two categories, i.e. Latent tuberculosis (LTB) and Active tuberculosis (ATB).

13.7.1 Latent TB

In latent TB, bacteria persist in an inactive state in the body and cause no symptoms. It is also known as inactive TB and it does not affect any other organs of the body and this process may remain many in dormant form many years if bacteria do not have the suitable condition to grow it remains in the lungs. It is usually treated for 9 months with only one prescribed drug (Siddiqui 2020).

13.7.2 Active TB

It is a disease that is initiated by primary infection or else reactivation of latent TB. A TB could be in the form of primary tuberculosis or reactivation of tuberculosis, it totally depends upon the defense mechanism of our immunity system. If the immune system is poor it occurs primary infection. It is mainly found in the lung and also affects the other parts of our body system which include gastrointestinal system, lymphoreticular system, liver, musculoskeletal system, skin, as well as the reproductive system (Siddiqui 2020).

13.7.3 Multidrug-Resistant Tuberculosis (MDR-TB)

It is the active form of tuberculosis which is characterized by resistance to both isoniazid and rifampicin first-line drugs (Prasad et al. 2018). Drug-resistant TB therapy takes nearly about two years to treat but it is so complicated, costly, and harmful to health. While MDR-TB therapy consists of second-line drugs that are fluoroquinolones and an injectable antibiotic that is used for TB treatment. Moreover, several second-line drugs have lethal side effects as well as have debilitating side effects (MINHAS et al. 2020).

13.7.4 Extensively Drug-Resistant Tuberculosis (XDR-TB)

While XDR-TB also emerges a major health risk problem in this present world. It is a brutal form of MDR-TB and this form of MDR-TB is more aggressive and has the strength to resist any fluoroquinolone drugs and one-second line injectable drugs like capreomycin drug, kanamycin, or amikacin that are being used in MDR-TB. Therefore, it poses serious problems for tuberculosis treatment and also significantly raises TB cases annually (Control and Prevention 2006).

13.7.5 Different Antibiotics Regimen Are Used to Treat Antibiotic Sensitive and Drug Resistance Tuberculosis (Table 13.1)

The above scenario indicates the complications of drug resistance TB treatment and its management. Since, conventional drugs have known toxicity and different pharmaceutical challenges, even the recovery rate of different drug-resistant tuberculosis is not up to the mark. Therefore, to address these challenges new anti-tubercular drugs are needed to resolve these cycling processes as well as to improve the compliance of drug resistance issues. So, in the present scenario screening of different natural bioactive molecules is going on to find out some anti-tubercular compound and plants are a potential source of the bioactive compound. After getting these bioactive molecules their efficacy was increased either by modifying the active groups of the molecule or by nano-formulation with advanced nanotechnology which helps to cure this tuberculosis disease.

13.8 Plants Are the Possible Sources to Anti-mycobacterial Agent

Since the beginning of creation, human communities have been in close contact with their environments and used environmental components to acquire food and medicine. The knowledge and use of plants for food and medicine were achieved by trial and error, and human beings were slowly able to fulfill their requirements from their environment. Knowledge regarding medicinal plants has long been passed on slowly and with the development of cultures and the availability of more resources human understanding has steadily been completed from generation to generation. In almost every community, medicinal plants are used as a medical aid. Herbal medicines will help to establish a new age in the medical care system to cope with human diseases in the future by standardizing and improving the health system using active plant-based compounds. Sensitizing the conventional knowledge that medicinal plants play a vital role as a natural plant resource that has been utilized and explored for the treatment in many diseases (Jamshidi-Kia et al. 2018). The WHO estimates that 80% of the global population today depends mostly on conventional medicines, mainly plants, which serve as the main source of health care system (Farnsworth et al. 1985). Nowadays, medicinal plants set to be an important base for the treatment of many

Table 13.1 Different drugs used in different types of tuberculosis

M.tb infection	Drugs	Duration of regimen	Gene	Gene function	Mode of action	Reference
First-line drugs for drug sensitive *M.tb* infection	Isoniazid	6 months	*Kat-G, inhA*	Catalase-peroxidase Enoyl-ACP reductase	It blocks the synthesis of α-mycolic acid	Barclay et al. (1953) and Carlos Palomino and Martin (2013)
	Rifampicin	6 months	*rpoB*	β subunit of RNA polymerase	It blocks sDNA-dependent RNA polymerase	White et al. (1971)
	Ethambutol	First 2 months	*embB*	Arabinosyl transferase	It inhibits biosynthesis of arabinan	Mikusova et al. (1995)
	Pyrazinamide	First 2 months	*pncA*	Nicotinamidase, pyrazinamidase	It kills non-replicating persisters	Zhang et al. (2013)
Second-line drugs for multiple drug-resistant *M.tb* infection	Streptomycin	9–24 months	*rpsL, rrs, gidB*	S12 ribosomal protein, 16A rRNA, 7-methylguanosine methyltransferase	It binds to the 16S rRNA, and interferes with translation proofreading, inhibits the protein synthesis	Moazed and Noller (1987) and Dalton et al. (2016)
	Kanamycin, capreomycin, amikacin, and viomycin	9–24 months	*Rrs, tlyA*	16S rRNA, rRNA methyltransferase	It blocks the synthesis of protein by binding to the four nucleotides of 16S rRNA and the single amino acid of S12 protein and protein synthesis by binding to the 70S ribosomal unit	McClatchy et al. (1977), Krüüner et al. (2003) and Stanley et al. (2010)
	Para-aminosalicylic acid	9–24 months	*thyA, folC*	Thymidylate synthase A	It inhibits the synthesis of folic acid	Rengarajan et al. (2004)
	Ethionamide	9–24 months	*ethA*	Enoyl-ACP reductase	It blocks the biosynthesis of mycolic acid	DeBarber et al. (2000), Vilchèze et al. (2008) and Brossier et al. (2011)

(continued)

Table 13.1 (continued)

M.tb infection	Drugs	Duration of regimen	Gene	Gene function	Mode of action	Reference
	Fluoroquinolones	9–24 months	*gyrA/ gyrB*	DNA gyrase	It blocks the activity of both DNA gyrase enzyme and the topoisomerase IV enzymes	Von Groll et al. (2009)
Third-line drugs for extensively drug-resistant *M. tb* infection	Linezolid	9–24 months	*rplC, rrl*	50S ribosomal protein L3 and 23S RNA	Inhibits protein synthesis	Zhang (2005) and Richter et al. (2007)
	Clofazimine	9–24 months	*rv0678, rv1979c, rv2535c*	Transcriptional regulator, permease involved in aminoacid transport, PepQ putative aminopeptidase	It shows an effect in the external membrane and releases bactericidal levels in reactive oxygen (ROS) species	Yano et al. (2011), Cholo et al. (2012) and Zumla et al. (2013)
	Bedaquiline	9–24 months	*atpE, rv0678*	ATP synthase, subunit F0, Transcriptional regulator	Inhibits ATP synthase	Diacon et al. (2012), Carlos Palomino and Martin (2013) and Chahine et al. (2014)
	Delamanid	9–24 months	*ddn, fgd-1, fbiA, fbiB, fbiC*	Deazaflavin (F420)-dependent nitroreductase, Glucose-6-phosphate dehydrogenase, Protein FbiA, Protein FbiB, Protein FbiC	Inhibits synthesis of mycolic acid	Matsumoto et al. (2006) and Carlos Palomino and Martin (2013)

diseases as well as good sources of development of antimicrobials drugs in the future. It also plays a key role in providing important ecosystem services to mankind. While it has a broad variety of biological resources, bioactive compounds, resins or pigments showed antimicrobial, anti-histamine, and antioxidant properties (Kavitha et al. 2013). It has also reported that phytochemicals of some plant also show a greater efficiency towards tuberculosis such as steroids, alkaloids, phenols, phenolic acids, flavonoids, glycosides, terpenoids, and resins which are used in plant defense mechanisms in response to microbial infections and are often found to be effective as antimicrobial agents (Adebiyi et al. 2015). Plant metabolites are serving as valuable sources of medicines, a targeted drug delivery system, and are considered to be using in treatment for tuberculosis (Singh et al. 2015).

13.9 Curcumin Plays a Key Role in Tuberculosis Disease

Since, time immemorial, people are using herbs and plants as a source to fight against various ailments of infectious diseases. There is a variety of flora and fauna which enrich the Indian subcontinent. From ancient times, it is known that turmeric is one of the earliest known medicinal plants which has been used to treat different diseases in human beings. It is a natural herb that has been used worldwide as a food as well as folk medicine. Moreover, curcumin has been identified and isolated as bioactive compound of turmeric which is an essential curcuminoid of the rhizome of the ethnomedicinal plant *Curcuma longa* (L.) which is known as "Wonder drug of life." The turmeric which is commonly used in India for thousands of years as a key ingredient spice and also as the medicinal herb. It belongs to family *Zingiberaceae*, which is well known for its wide range of pharmacological properties and it is non-toxic (Duvoix et al. 2005; Li et al. 2008).

The structure of curcumin was first identified by Lampe and Milobedeska in 1910 and it is an asymmetric molecule, also known as diferuloylmethane (Milobedeska et al. 1910). The IUPAC nomenclature for curcumin is (1E-6E)-1,7-bis (4-hydroxy-3-methoxy-phenyl)-1,6-heptadiene-3,5-dione, with the chemical formula ($C_{21}H_{20}O_6$) consisting of two aromatic ring compounds that include ortho-methoxy-phenolic OH, groups which are symmetrically connected with the ß-diketone moiety (Fig. 13.1). The molecular weight of curcumin is 368.38 g/mole and its melting temperature was 183 °C (Jovanovic et al. 1999). Curcumin contains 3 main functional groups such as phenolic groups, di keto groups, and ortho-methoxy groups (Sreejayan and Rao 1996).

It was insoluble in water at acidic and neutral pH condition but it was soluble in ethanol, methanol, acetone, and dimethylsulfoxide (Priyadarsini 2009). Curcumin

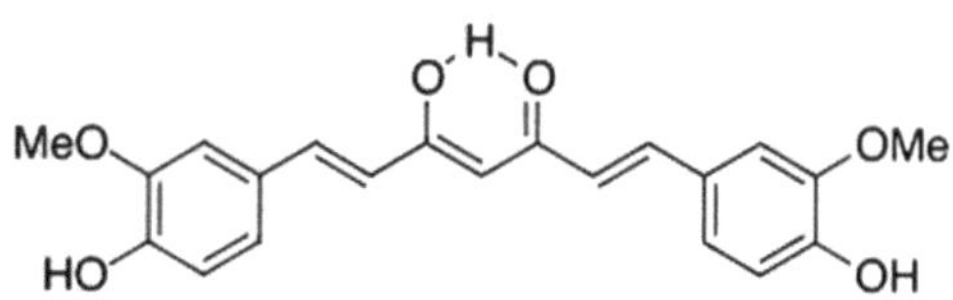

Fig. 13.1 Chemical structure of curcumin (Nelson et al. 2017)

has widespread applications such as antioxidant, anticancer, anti-arthritic, antimicrobial, anti-diabetic, and anti-inflammatory activities (Pizzo et al. 2010; Sugiyama et al. 1996; Lee et al. 2009; Srimal and Dhawan 1973; Aggarwal and Harikumar 2009).

Different authors have found that curcumin can induce apoptosis in M.*tb* infected macrophage which plays an important role in the destruction of tubercle bacilli cells (Li et al. 2014a; Ahmad et al. 2019). In addition, it provides the animals with sterile immunity by restricting the antibiotic-induced AICD (activation-induced cell death) (Tousif et al. 2017). In contrast, curcumin has also inhibitory effects on NF-κB and JAK/STAT pathway which deregulated the inflammatory response and activation of T helper cells (Surh et al. 2001). Regardless of the curcumin's pharmacological activities, it has an excellent immune-modulatory and therapeutic potential in a variety of diseases. However, due to its low solubility in the body fluids, less bioavailability and very short half-life restrict this wonder molecule in the use of the clinical application (Maiti and Dunbar 2018; Anand et al. 2007). Therefore, different techniques were used to increase the aqueous solubility and bioavailability of this molecule. Even varieties of drug delivery methods were used to sustain the release of this molecule so that a continuous presence of the molecule is there in physiological condition.

13.10 Nanoparticle Used as Anti-tubercular Agents

Nanotechnology is the product of its interdisciplinary existence which is considered to be a "central technology of the modern world." Though it is believed as a groundbreaking path with its greater technological evolution which deals with nanometer-scale material management (i.e., one billion times less than a meter) (Rai et al. 2009) which it makes the material lighter, stronger, quicker, smaller in size as well as more durable. Moreover, these materials are also often used in various applications which are implemented as in the form of physical, chemical, and biological structures at the level of nanoscales varying from individual molecules or atoms of a nanomaterial substance (Arakha et al. 2016, 2017; Prince et al. 2016). Even these nanomaterials also have different shapes and sizes such as a spherical, triangular, rod, cylindrical, conical, tubular, hollow core, flat, and spiral which make them distinctive because of their smaller size, i.e. 1 nm to 100 nm. However, this advancement of nanotechnology has brought together technological developments in civilization with a wide range of applications and application of nano-scale structures with new properties for a health remedial intervention. Apart from this nanotechnology has proved that it is efficient in facilitating the detection and distribution of novel conjugated drugs to a particular cell (Singh et al. 2015). Nanoparticles are also having an immense utility in the area of biomedical research because of their appearance, form, and chemical and biologic properties. These nanoparticle structures can be further updated or adjusted according to the goal for improved functionality (Singh et al. 2015). The high surface-to-volume ratios of these nanoparticles show an excellent antibacterial profile which is further improved by the good ligand conjugations that improve the pharmacological and therapeutic efficient penetration, internalization, and effective antimicrobial action of nanosized medicinal products (Banu and Rathod 2013; Yah and Simate 2015). For examples of some nanoparticles, such as gold (Au), silver (Ag), as well as copper (Cu), were synthesized by both chemical and biological pathways, were also effectively used in various *Mycobacterium* species (Rawashdeh and Haik 2009; Li et al. 2014b; Yah and Simate 2015).

The sizes of nanoparticles known as colloidal sub microns (<1 μm) shall be used as drug carriers. For the purposes of therapeutic medicines, it can be integrated covalently into the surface of the particles or in the particle-matrix (Kreuter 2004). Nanoparticles are biocompatible compounds that can be biodegradable, such as polymers which can include as natural materials (for example, gelatin and albumin), synthetic materials (for example, polylactides and polyalkylcyanoacrylates), and stable lipids (SLNR and NLCR) (Müller et al. 1995).

Nanoparticle have some advantage in drug delivery system such as it increases the bioavailability of hydrophobic drugs, increases the drug half-life period, it has fewer side effects, and helps in drug-controlled releases in a systemic distribution, and also helps in to reduce administration dose frequency and dose rate (Santos-Magalhães and Mosqueira 2010; Zhang et al. 2008). In addition, it has benefits in achieving a specific and synergistic effect by controlling the co-delivery of multi-drug combinations with nanoparticle transport systems for further evading the process of drug resistance within microbial populations. Therefore, nano-carriers are highly useful for improving different therapeutics' and save adequate time and energy for creating new drugs.

There are also some advantages in TB systems based on nanoparticles carrier system, for example, site-specific ligands, they are acting as smart particles for site-specific or cell-specific drug delivery as sub-micron (Chellat et al. 2005). This nanoparticle transport system also increases the bioavailability of drug molecules as well as also helps to resolve the resistance problem such as efflux pumps, drug toxicity, drug solubility, and drug retention period (Zazo et al. 2016; Tiuman et al. 2011). In addition, this nano-drug delivery system is a perfect way to provide medicines against mycobacterial intracellular pathogens, since macrophages quickly engulf them by recognizing cell-specific surface ligands (Chellat et al. 2005).

Moreover, these nanoparticles-based drug delivery approach shows that it has a possible solution to resolve the curcumin solubility issues as well as also helps to improve the bioavailability of curcumin molecules (Sun et al. 2012; Jiang et al. 2008; Shutava et al. 2009; Depan 2015). While, in some research studies it has been shown that it is a promising carrier of curcumin delivery using various nanomaterials such as proteins, nanoemulsion, liposomes, solid-lipid nanoparticles, micelles, and polymeric nanoparticles (Deljoo et al. 2019). However, under all nanoparticles-based delivery approaches, polymeric nanoparticles show important nanoparticles which have been used for the delivery of curcumin molecules against different disease (Chirio et al. 2011; Basnet et al. 2012; Dhule et al. 2012). Furthermore, nanoparticles have their own significance in the modern drug delivery system with the continuing growth of nanotechnology. The use of nanoparticles also provides new approaches for improving solubility, stability, and bioavailability as well as pharmacological activity, and holds the capacity to prevent physical or chemical degradation. Therefore, the implementation of nanotechnology with curcumin provides a solution for improved bioavailability as well as therapeutic effectiveness for tuberculosis. Moreover, it could be improved the therapeutics of drug-resistant tuberculosis.

13.11 Synthesis of Curcumin Nanoparticles That Are Used in TB

Various types of nano-carriers were implemented for the synthesis of curcumin nanoparticles such as polymeric, polymeric micelles, solid-lipid, curcumin nanocrystals, nano-liposome-encapsulated, nano-emulsions, curcumin nanosuspension, dendrimers, and cyclodextrin nanoparticles in drug delivery system (Gregory et al. 2017). Whereas, among all of the above-synthesized nanoparticles, polymeric nanoparticles have gained a greater interest among nanomaterials in the preparation of curcumin nanoparticles and it also shows a promising way to synthesize curcumin nanoparticles in a novel way for the tuberculosis drug delivery method. The polymeric nanoparticles can be synthesized with nanoprecipitation approaches (~200 nm in size) developed by Sultan et al. 2017 (Tousif et al. 2017). During this process, the required polymeric solvent (i.e., ethanol) with organic solvent (curcumin herbal medicinal product) was directly dissolved in the polymer solution and then the drug solution was constantly mixed as well as filtered into a fresh test tube. After that, the solution of drug-polymer was subsequently added during the stirring process and then 0.1% citric acid Milli Q water was subsequently added in a dropwise manner under high-pressure homogenizer equipment for a duration of 30 min. Then nanoparticles will start precipitating in the solution during these processes. After that, the polymeric solution was homogenized with 30,000 PSI over ice for 20 cycles in high-pressure homogenizer equipment. Then 0.1% polysorbate 80 solvent was added to the aqueous solution and again it was taken into homogenizer equipment for 1 h. After completing this homogenizer step solution was filtered using nano filter paper and then it was evaporated by using hot air flow technique. Afterward the fine curcumin nanoparticle powder will be produced that can be used in further phases (Tousif et al. 2017). Similarly, in another experiment Priyanka S et al., 2020 showed that rifampicin with curcumin-loaded polyethylene sebacate (PES) was synthesized with a technique of nanoprecipitation using 20 mg of rifampicin drug, 60 mg of curcumin molecules, and 100 mg of PES solvent. Which was further dissolved in a 12.5 mL of tetrahydrofuran (THF) mixture using the bath sonication method. After that, the obtained organic mixture with 1 percent of Tween 80 was added in a 25 mL of distilled water during the stirring condition. Then the following solvent was then mixed at room temperature in a magnetic stirrer plate for 4 h which further helps the solvent to evaporate. Then the solvent nanoparticles were centrifuged for 20 min at 16,350 rcf at 20 °C and then the pellet was washed with distilled water for three times in a periodic centrifugation processes to remove the THF and Tween 80 residues from the mixtures. After completing this step, then the pellet was again dissolved with 10 mL of distilled water and then cryoprotectant was added in a bath sonication process for 10 min. After that solvent again proceeded for 10 min using probe sonication processes. Then the final prescribed lyophilization fined nanoparticle powder will be produced which can be used in further processes (Jahagirdar et al. 2020). Therefore, the above nanoprecipitation technique shows that the polymeric nanoparticles have a strong and durable nanomaterial, which also retains a polymeric micelle in its hydrophobic center that contributes to the main supply of hydrophobic substances to curcumin molecules which further helps to stabilize this curcumin molecule within this polymeric material. After that, it also helps to enhance the curcumin molecules for the drug delivery system. Moreover, the solubility of curcumin has been assessed as the main matrix system which allows the uniform distribution of these polymeric matrixes by integrating curcumin into the micelle and the nanophase (Rai et al. 2017). Well after inserted curcumin with this polymeric nanomaterial, it has shown some advantages that the use of this polymeric nanomaterial is very effective in nature itself and it can also be managed by choosing a polymer length, stabilizing agents, protecting drugs from degradation, reduces the side effects, increases the bioavailability effects, increases the drug concentration at infected sites, and uniformly distributed drug doses for target-specific drug delivery (Fig. 13.2).

Furthermore, these polymer nanoparticles have indicated that it can be a good carrier for curcumin molecules, to enhance the efficacy of targeted cells or tissues. Therefore, these polymeric nanomaterial-based approaches are one of the efficient methods for water-soluble drug encapsulation, solubility, stabilization, and drug-controlled releases of curcumin molecules in tuberculosis therapy.

13.12 Importance and Advantages of Curcumin Nanoparticle in TB

Curcumin being the principal bioactive compound of Curcuma species inhibits human Kv1.3 potassium channels, which leads to a definite inhibition of T-effector memory cell proliferation mainly causing inflammation in Multiple sclerosis and rheumatoid arthritis (Lian et al. 2013). Reports show that curcumin analogs demonstrate *in vitro* anti-mycobacterial activity against drug-resistant strains of *M.tb* (Changtam et al. 2010; Baldwin et al. 2015). Recently, it was found that curcumin molecules induce apoptosis in macrophages that helps in killing of the mycobacterium (Xiyuan et al. 2016; Ahmad et al. 2019). In addition, curcumin potently inhibits hepatotoxicity generated by antibiotic therapies including conventional antibiotic of TB (Tousif et al. 2017; Naik et al. 2011). Considering all these data curcumin can be reflected as an anti-inflammatory immunomodulator molecule having therapeutic potential in many infectious diseases. However, due to its poor intestinal absorption, faster metabolism, and rapid systemic elimination make this molecule very short bioavailability creating a limiting factor for clinical use. To come out from this limitation, nano-formulation of curcumin has been set up. Therefore, in next section focus has been given to the utility of curcumin nanoparticles against tuberculosis treatment.

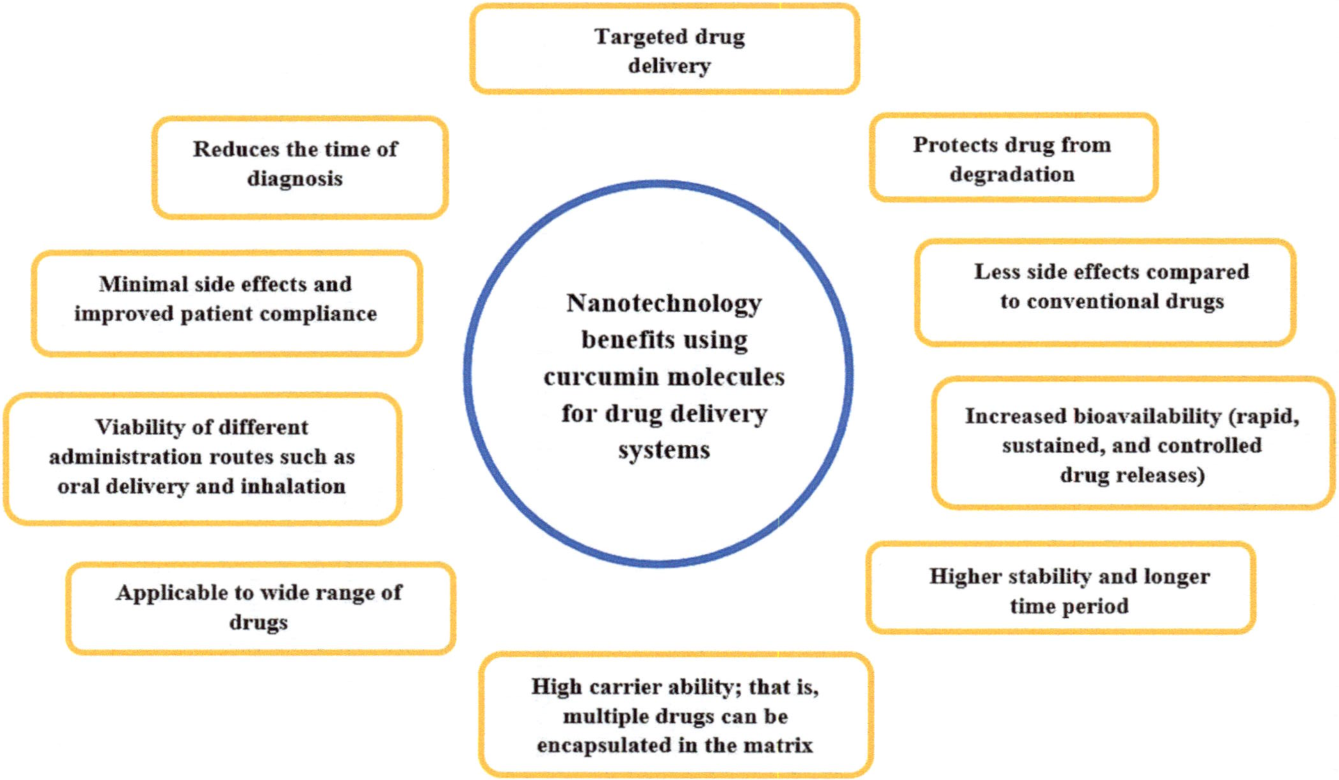

Fig. 13.2 Advantages of nanotechnology-based drug delivery system using curcumin nanoparticle for tuberculosis treatment (Nasiruddin et al. 2017; Singh et al. 2016b)

13.13 Application of Curcumin Nanoparticle

Curcumin is an important natural Phyto-molecule that has a diversified role in the protection of different diseases including cancers, inflammatory diseases, and infectious diseases. The way of action of curcumin is by immune-modulatory, anti-inflammatory, or pharmaceutical blocking of receptor ligands. Moreover, few reports have come out showing the in vitro anti-mycobacterial activity of curcumin (Gupta et al. 2013). Even it has shown the intracellular killing of *M.tb* in macrophages through immune modulation of innate immunity and induction of anti-mycobacterial peptide cathelicidin (Bai et al. 2016; Guo et al. 2013). However, because of its poor bioavailability, it could not be upgraded into a drug molecule. Therefore, to overcome the poor bioavailability and fast metabolism of curcumin in physiological condition varieties of nano-formulation were tried with this molecule for different disease models (Yallapu et al. 2015). Tousif et al. (2017) have formulated the polymeric nanoparticle of curcumin and evaluate the role of this molecule in TB. This nano-formulation of curcumin exhibits a uniform size of ~200 nm as measured by electron microscopy with an extended bioavailability in physiological condition (Tousif et al. 2017). In mouse experiment curcumin nanoparticles, at 100 mg/kg, body weight when injected intraperitoneally to mice, reflected about five times better bioavailability of curcumin at 2 h post-injection with peak plasma concentration (*C*MAX) of 106 ng/mL and a half-life (*T* 1/2) of 13.1 h in comparison to natural curcumin which peaked at 1 h with *C*MAX of 26 ng/mL and *T* 1/2 of 2.65 h.

To decipher the effect of this nano-formulation of curcumin in TB these molecules were co-treated with one of the conventional antibiotic isoniazid in the preclinical experiment and observed that nanocurcumin alone could reduce one log of *M.tb*. However, the animals that receive both nanoparticle and INH together exhibited a drastic clearance of the bacilli in both lung and spleen. Moreover, the nanocurcumin treatment also restored the activation-induced cell death of antigen-specific protective T helper cells that occurred by INH (Tousif et al. 2014). Therefore, the use of non-formulated curcumin-induced sterile immunity to INH treated animals thereby protects the animals from reinfection or further relapse of *M.tb* infection in the murine model of tuberculosis (Tousif et al. 2014). Above all most importantly curcumin nanoparticle reduces the hepatotoxicity and over inflammation of lungs occurred by conventional TB antibiotic INH (Tousif et al. 2014). It is well known that the induction of initial innate immune response is important for the generation of the particular adaptive immune response. To achieve sufficient induction of adaptive response upregulation of costimulatory molecules and increase of MHC expression are very important. Nano-formulated curcumin is found to induce CD80 or CD86 molecules and MHC expression in isolated peritoneal macrophages infected with *M.tb* H37Rv at ex vivo experiment (Ahmad et al. 2019). Even nanocurcumin induces the macrophages to its M1 state and initiates the apoptosis in infected macrophages. It is well known that curcumin blocked the potassium channel Kv1.3 in effector memory T cells (TEM) in rheumatic arthritis that causes reduction of inflammation (Lian et al. 2013). Even in tuberculosis TEM cells that provide short-term memory response also create inflammation (Ahmad et al. 2020). Therefore, nanocurcumin was used with BCG to determine its effect on long-term memory response. It treated intraperitoneally along with BCG to determine its effect on BCG vaccination. Surprisingly it was found that nanocurcumin increases the efficacy of BCG vaccine by clearing the *M.tb* faster and reduces the TEM cells and increases the T central memory (TCM) cells and makes them a critical balance that provides long-term memory responses in both lung and spleen at a murine model of tuberculosis. It is well known that Th1 and Th17 cells are host protective and Th2 and Treg subsets are pathogenic in nature in the tuberculosis model (Ahmad et al. 2020; Bhattacharya et al. 2014a, b; Singh et al. 2016a). In the mouse model of tuberculosis co-treatment of nanocurcumin induces the host protective Th1 and Th17 subsets and reduces Treg in BCG vaccinated animals which protect the animals more potently (Ahmad et al. 2019). Therefore, it was confirmed that nano-formulation of curcumin increases the availability of curcumin in host cells or host physiology, and by that, it showed a diverse effect on a different strategy of tuberculosis treatment.

Very recently another work has come out with polymeric nano-formulation of dual molecule curcumin and most important conventional TB antibiotic rifampicin in intracellular *M.tb* infection (Jahagirdar et al. 2020). This dual drug-containing nanoparticle efficiently enters within the macrophages and releases both the drug molecules 1.5 folds higher than the free drug molecules. Most importantly this nano-formulated curcumin and rifampicin molecule did not exhibit any toxicity in Raw267.4 macrophages. Moreover, it kills the intracellular *M.tb* bacteria almost 100% at minimum dose than any of the free drug molecules or separately nano-formulated drug molecules. Therefore, this strategy showed some light to combat with different drug-resistant TB as in different drug-resistant TB efflux pumps of bacteria throw out the drugs from the bacterial cells so a very high dose of the drugs is recommended. This nano-formulation of curcumin with antibiotic could help in precise drug delivery. Even antibiotic and nanocurcumin have a different mode of action to eradicate the bugs and by using such combined drug could efficiently kill the intracellular *M.tb* as well as induce such host immune response that also restricts the bacteria potently. Even these strategies will be helpful for the shortening of TB antibiotic regimen by efficient drug delivery and induction of host anti TB sterile immunity.

13.14 Conclusion

It is well known that TB is one of the devastating infectious diseases in this present world irrespective of socioeconomic conditions. Perhaps most of the developing countries are now being affected by drug-resistant TB. In order to overcome this situation, a new drug regimen is needed to prevent these drug resistance cases as well as to cure TB disease. Natural products are proposed as a therapeutic alternative to conventional antimicrobial treatment. Among them, curcumin has been a well-known natural product for its effective therapeutic application in the treatment of various diseases since ancient times. Even if curcumin is also referred to as one of the best natural antibiotics because of its host immune-modulatory and novel antimicrobial effects in various infectious diseases, which is now attracting scientists and researchers in this biological field. Well, it is also a good anti-inflammatory agent as well as an excellent immune-modulatory agent that enhances the immunity of anti-mycobacterial along with this it also showed some direct killing effect for tuberculosis. With this remarkable property, it can decrease the period of TB treatment and this thing may bring to a new possibility to reduce the risk for drug-resistant TB variants. In addition, curcumin also induces apoptosis in *M.tb* infected cells thereby minimizing the risk of this disease reactivation. While curcumin has huge healing potential to mitigate this TB disease, but due to low bioavailability and low solubility makes them restrict to use in this medical field. Thus, modern drug delivery methods focused on nanotechnology aim to solve these challenges. The nanoparticle-based techniques can be used to improve curcumin bioavailability as well as solubility, which helped to treat TB. Although curcumin-loaded nanoparticles have several advantages because it can reduce the harmful effects of antibiotic drugs and reduce the risk for disease reactivation as well as TB reinfection in tuberculosis disease. Moreover, the use of curcumin nanoparticles along with antibiotic treatment could reduce the time period of TB therapy as well as produced low side effects with site-specific drug release which could be an alternative therapy to treat tuberculosis disease. Therefore, curcumin nanoparticles signify a novel drug molecule that helps to eradicate this TB and has a huge potentiality to be entered within the TB drug or adjunct therapy of tuberculosis in the near future that can also induce the sterile immunity to host against TB which is important to develop a self-propelled vaccine.

References

Adebiyi A, Ayo R, Bello I, Habila J (2015) Phytochemical screening and anti-TB activity of root extracts of Guiera senegalensis (JF Gmel). Am J Biosci Bioeng 3(6):208–213

Aggarwal BB, Harikumar KB (2009) Potential therapeutic effects of curcumin, the anti-inflammatory agent, against neurodegenerative, cardiovascular, pulmonary, metabolic, autoimmune and neoplastic diseases. Int J Biochem Cell Biol 41(1):40–59

Ahmad S, Bhattacharya D, Kar S, Ranganathan A, Van Kaer L, Das G (2019) Curcumin nanoparticles enhance Mycobacterium bovis BCG vaccine efficacy by modulating host immune responses. Infect Immun 87(11):e00291–e00219

Ahmad S, Bhattacharya D, Gupta N, Rawat V, Tousif S, Van Kaer L, Das G (2020) Clofazimine enhances the efficacy of BCG revaccination via stem cell-like memory T cells. PLoS Pathog 16 (5):e1008356

Ahmed N, Hasnain SE (2011) Molecular epidemiology of tuberculosis in India: moving forward with a systems biology approach. Tuberculosis 91(5):407–413

Anand P, Kunnumakkara AB, Newman RA, Aggarwal BB (2007) Bioavailability of curcumin: problems and promises. Mol Pharm 4(6):807–818

Arakha M, Borah SM, Saleem M, Jha AN, Jha S (2016) Interfacial assembly at silver nanoparticle enhances the antibacterial efficacy of nisin. Free Radic Biol Med 101:434–445

Arakha M, Roy J, Nayak PS, Mallick B, Jha S (2017) Zinc oxide nanoparticle energy band gap reduction triggers the oxidative stress resulting into autophagy-mediated apoptotic cell death. Free Radic Biol Med 110:42–53

Bai X, Oberley-Deegan RE, Bai A, Ovrutsky AR, Kinney WH, Weaver M, Zhang G, Honda JR, Chan ED (2016) Curcumin enhances human macrophage control of Mycobacterium tuberculosis infection. Respirology 21(5):951–957

Baldwin PR, Reeves AZ, Powell KR, Napier RJ, Swimm AI, Sun A, Giesler K, Bommarius B, Shinnick TM, Snyder JP (2015) Monocarbonyl analogs of curcumin inhibit growth of antibiotic sensitive and resistant strains of Mycobacterium tuberculosis. Eur J Med Chem 92:693–699

Banu A, Rathod V (2013) Biosynthesis of monodispersed silver nanoparticles and their activity against Mycobacterium tuberculosis. Int J Biomed Nanosci Nanotechnol 3(1–2):211–220

Barberis I, Bragazzi N, Galluzzo L, Martini M (2017) The history of tuberculosis: from the first historical records to the isolation of Koch's bacillus. J Prev Med Hyg 58(1):E9

Barclay W, Ebert R, Koch-Weser D (1953) Mode of action of isoniazid. Am Rev Tuberc 67 (4):490–496

Basnet P, Hussain H, Tho I, Skalko-Basnet N (2012) Liposomal delivery system enhances anti-inflammatory properties of curcumin. J Pharm Sci 101(2):598–609

Bhattacharya D, Dwivedi VP, Das G (2013) Revisiting immunotherapy in tuberculosis. J Mycobac Dis 4:e123

Bhattacharya D, Dwivedi VP, Kumar S, Reddy MC, Van Kaer L, Moodley P, Das G (2014a) Simultaneous inhibition of T helper 2 and T regulatory cell differentiation by small molecules enhances Bacillus Calmette-Guerin vaccine efficacy against tuberculosis. J Biol Chem 289 (48):33404–33411

Bhattacharya D, Dwivedi VP, Maiga M, Maiga M, Van Kaer L, Bishai WR, Das G (2014b) Small molecule-directed immunotherapy against recurrent infection by Mycobacterium tuberculosis. J Biol Chem 289(23):16508–16515

Brossier F, Veziris N, Truffot-Pernot C, Jarlier V, Sougakoff W (2011) Molecular investigation of resistance to the antituberculous drug ethionamide in multidrug-resistant clinical isolates of Mycobacterium tuberculosis. Antimicrob Agents Chemother 55(1):355–360

Carlos Palomino J, Martin A (2013) Tuberculosis clinical trial update and the current anti-tuberculosis drug portfolio. Curr Med Chem 20(30):3785–3796

Chahine EB, Karaoui LR, Mansour H (2014) Bedaquiline: a novel diarylquinoline for multidrug-resistant tuberculosis. Ann Pharmacother 48(1):107–115

Changtam C, Hongmanee P, Suksamrarn A (2010) Isoxazole analogs of curcuminoids with highly potent multidrug-resistant antimycobacterial activity. Eur J Med Chem 45(10):4446–4457

Chaudhary C, Garg T (2015) Scaffolds: a novel carrier and potential wound healer. Critical Reviews™ in Therapeutic Drug Carrier Systems 32(4)

Chellat F, Merhi Y, Moreau A (2005) Therapeutic potential of nanoparticulate systems for macrophage targeting. Biomaterials 26(35):7260–7275

Chirio D, Gallarate M, Peira E, Battaglia L, Serpe L, Trotta M (2011) Formulation of curcumin-loaded solid lipid nanoparticles produced by fatty acids coacervation technique. J Microencapsul 28(6):537–548

Cholo MC, Steel HC, Fourie PB, Germishuizen WA, Anderson R (2012) Clofazimine: current status and future prospects. J Antimicrob Chemother 67(2):290–298

Cole EC, Cook CE (1998) Characterization of infectious aerosols in health care facilities: an aid to effective engineering controls and preventive strategies. Am J Infect Control 26(4):453–464

Colledge NR, Walker BR, Ralston S, Davidson S (2010) Davidson's principles and practice of medicine. Churchill Livingstone/Elsevier

Control CfD, Prevention (2006) Emergence of Mycobacterium tuberculosis with extensive resistance to second-line drugs--worldwide, 2000-2004. MMWR Morb Mortal Wkly Rep 55 (11):301

Cooper AM (2009) Cell-mediated immune responses in tuberculosis. Annu Rev Immunol 27:393–422

Dalton JP, Uy B, Phummarin N, Copp BR, Denny WA, Swift S, Wiles S (2016) Effect of common and experimental anti-tuberculosis treatments on Mycobacterium tuberculosis growing as biofilms. PeerJ 4:e2717
Davies J, Davies D (2010) Origins and evolution of antibiotic resistance. Microbiol Mol Biol Rev 74(3):417–433
DeBarber AE, Mdluli K, Bosman M, Bekker L-G, Barry CE (2000) Ethionamide activation and sensitivity in multidrug-resistant Mycobacterium tuberculosis. Proc Natl Acad Sci 97 (17):9677–9682
Deljoo S, Rabiee N, Rabiee M (2019) Curcumin-hybrid nanoparticles in drug delivery system. Asian J Nanosci Mater 2(1):66–91
Depan D (2015) Biodegradable polymeric nanocomposites: advances in biomedical applications. CRC Press
Dhule SS, Penfornis P, Frazier T, Walker R, Feldman J, Tan G, He J, Alb A, John V, Pochampally R (2012) Curcumin-loaded γ-cyclodextrin liposomal nanoparticles as delivery vehicles for osteosarcoma. Nanomed: Nanotechnol Biol Med 8(4):440–451
Diacon AH, Dawson R, von Groote-Bidlingmaier F, Symons G, Venter A, Donald PR, van Niekerk C, Everitt D, Winter H, Becker P (2012) 14-day bactericidal activity of PA-824, bedaquiline, pyrazinamide, and moxifloxacin combinations: a randomised trial. Lancet 380 (9846):986–993
Duvoix A, Blasius R, Delhalle S, Schnekenburger M, Morceau F, Henry E, Dicato M, Diederich M (2005) Chemopreventive and therapeutic effects of curcumin. Cancer Lett 223(2):181–190
Ekor M (2014) The growing use of herbal medicines: issues relating to adverse reactions and challenges in monitoring safety. Front Pharmacol 4:177
Farnsworth NR, Akerele O, Bingel AS, Soejarto DD, Guo Z (1985) Medicinal plants in therapy. Bull World Health Organ 63(6):965
Flynn JL, Chan J (2001a) Immunology of tuberculosis. Annu Rev Immunol 19(1):93–129
Flynn JL, Chan J (2001b) MINIREVIEW tuberculosis: latency and reactivation. Infect Immun
Frieden T, Sterling S, Munsiff C, Watt C, Dye C (2003) Tuberculosis. Lancet 362:887–899
Friedman ND, Temkin E, Carmeli Y (2016) The negative impact of antibiotic resistance. Clin Microbiol Infect 22(5):416–422
Gera M, Sharma N, Ghosh M, Huynh DL, Lee SJ, Min T, Kwon T, Jeong DK (2017) Nanoformulations of curcumin: an emerging paradigm for improved remedial application. Oncotarget 8(39):66680
Gradmann C (2001) Robert Koch and the pressures of scientific research: tuberculosis and tuberculin. Med Hist 45(1):1–32
Gregory M, Sarmento BFCC, Martins JAR, Silva CJR, Sárria MP, Gomes A, Coutinho OP, Dias ACP, Franklin G (2017) Curcumin encapsulated into methoxy Poly (Ethylene Glycol) Poly (ε-Caprolactone) nanoparticles increases cellular uptake and neuroprotective effect in glioma cells. Planta Med
Guirado E, Schlesinger L (2013) Modeling the Mycobacterium tuberculosis granuloma–the critical battlefield in host immunity and disease. Front Immunol 4:98
Guo C, Rosoha E, Lowry MB, Borregaard N, Gombart AF (2013) Curcumin induces human cathelicidin antimicrobial peptide gene expression through a vitamin D receptor-independent pathway. J Nutr Biochem 24(5):754–759
Gupta PK, Kulkarni S, Rajan R (2013) Inhibition of intracellular survival of multi drug resistant clinical isolates of Mycobacterium tuberculosis in macrophages by curcumin. Open Antimicrob Agents J 4(1)
Hershkovitz I, Donoghue HD, Minnikin DE, Besra GS, Lee OY, Gernaey AM, Galili E, Eshed V, Greenblatt CL, Lemma E (2008) Detection and molecular characterization of 9000-year-old Mycobacterium tuberculosis from a Neolithic settlement in the Eastern Mediterranean. PLoS One 3(10):e3426
Hill PC, Brookes RH, Fox A, Fielding K, Jeffries DJ, Jackson-Sillah D, Lugos MD, Owiafe PK, Donkor SA, Hammond AS (2004) Large-scale evaluation of enzyme-linked immunospot assay

and skin test for diagnosis of Mycobacterium tuberculosis infection against a gradient of exposure in The Gambia. Clin Infect Dis 38(7):966–973

Horsburgh CR Jr (1991) Mycobacterium avium complex infection in the acquired immunodeficiency syndrome. N Engl J Med 324(19):1332–1338

Horvath C, Shaler C, Jeyanathan M, Zganiacz A, Xing Z (2012) Mechanisms of delayed anti-tuberculosis protection in the lung of parenteral BCG-vaccinated hosts: a critical role of airway luminal T cells. Mucosal Immunol 5(4):420–431

Jahagirdar PS, Gupta PK, Kulkarni SP, Devarajan PV (2020) Intramacrophage delivery of dual drug loaded nanoparticles for effective clearance of Mycobacterium tuberculosis. J Pharm Sci

Jamshidi-Kia F, Lorigooini Z, Amini-Khoei H (2018) Medicinal plants: past history and future perspective. J Herbmed Pharmacol 7(1)

Jeong YJ, Lee KS (2008) Pulmonary tuberculosis: up-to-date imaging and management. Am J Roentgenol 191(3):834–844

Jiang W, Kim BY, Rutka JT, Chan WC (2008) Nanoparticle-mediated cellular response is size-dependent. Nat Nanotechnol 3(3):145–150

Jovanovic SV, Steenken S, Boone CW, Simic MG (1999) H-atom transfer is a preferred antioxidant mechanism of curcumin. J Am Chem Soc 121(41):9677–9681

Kavitha K, Baker S, Rakshith D, Kavitha H, Yashwantha Rao H, Harini B, Satish S (2013) Plants as green source towards synthesis of nanoparticles. Int Res J Biol Sci 2(6):66–76

Kreuter J (2004) Nanoparticles as drug delivery systems. In: Encyclopedia of nanoscience and nanotechnology, vol 7. American Scientific Publishers, pp 161–180

Krüüner A, Jureen P, Levina K, Ghebremichael S, Hoffner S (2003) Discordant resistance to kanamycin and amikacin in drug-resistant Mycobacterium tuberculosis. Antimicrob Agents Chemother 47(9):2971–2973

Kumar V, Robbins SL (2007) Robbins basic pathology. Saunders. Elsevier, Philadelphia, PA

Lakhtakia R (2013) Of Animalcula, Phthisis and Scrofula: historical insights into tuberculosis in the pre-Koch era. Sultan Qaboos Univ Med J 13(4):486

Lee YK, Lee WS, Hwang JT, Kwon DY, Surh YJ, Park OJ (2009) Curcumin exerts antidifferentiation effect through AMPKα-PPAR-γ in 3T3-L1 adipocytes and antiproliferatory effect through AMPKα-COX-2 in cancer cells. J Agric Food Chem 57(1):305–310

Levinson WE (2020) Review of medical microbiology and immunology. McGraw Hill Professional

Li H-L, Liu C, De Couto G, Ouzounian M, Sun M, Wang A-B, Huang Y, He C-W, Shi Y, Chen X (2008) Curcumin prevents and reverses murine cardiac hypertrophy. J Clin Invest 118 (3):879–893

Li M, Wang H, Huang J, Shi G, Wan Y, Wang J, Xi X (2014a) Curcumin inhibits 19-kDa lipoprotein of Mycobacterium tuberculosis induced macrophage apoptosis via regulation of the JNK pathway. Biochem Biophys Res Commun 446(2):626–632

Li X, Robinson SM, Gupta A, Saha K, Jiang Z, Moyano DF, Sahar A, Riley MA, Rotello VM (2014b) Functional gold nanoparticles as potent antimicrobial agents against multi-drug-resistant bacteria. ACS Nano 8(10):10682–10686

Lian YT, Yang XF, Wang ZH, Yang Y, Yang Y, Shu YW, Cheng LX, Liu K (2013) Curcumin serves as a human kv1. 3 blocker to inhibit effector memory T lymphocyte activities. Phytother Res 27(9):1321–1327

Maiti P, Dunbar GL (2018) Use of curcumin, a natural polyphenol for targeting molecular pathways in treating age-related neurodegenerative diseases. Int J Mol Sci 19(6):1637

Matsumoto M, Hashizume H, Tomishige T, Kawasaki M, Tsubouchi H, Sasaki H, Shimokawa Y, Komatsu M (2006) OPC-67683, a nitro-dihydro-imidazooxazole derivative with promising action against tuberculosis in vitro and in mice. PLoS Med 3(11):e466

McClatchy JK, Kanes W, Davidson PT, Moulding TS (1977) Cross-resistance in M. tuberculosis to kanamycin, capreomycin and viomycin. Tubercle 58(1):29–34

Mikusova K, Slayden RA, Besra GS, Brennan PJ (1995) Biogenesis of the mycobacterial cell wall and the site of action of ethambutol. Antimicrob Agents Chemother 39(11):2484–2489

Milobedeska J, Kostanecki V, Lampe V (1910) Structure of curcumin. Ber Dtsch Chem Ges 43:2163–2170
Minhas P, Sharma D, Katual MK (2020) Nano-technological developments in tuberculosis management: an update. UGC Care
Moazed D, Noller HF (1987) Interaction of antibiotics with functional sites in 16S ribosomal RNA. Nature 327(6121):389–394
Müller R, Mehnert W, Lucks J-S, Schwarz C, Zur Mühlen A (1995) Solid lipid nanoparticles (SLN): an alternative colloidal carrier system for controlled drug delivery. Eur J Pharm Biopharm 41(1):62–69
Naik SR, Thakare VN, Patil SR (2011) Protective effect of curcumin on experimentally induced inflammation, hepatotoxicity and cardiotoxicity in rats: evidence of its antioxidant property. Exp Toxicol Pathol 63(5):419–431
Narasimhan P, Wood J, MacIntyre CR, Mathai D (2013) Risk factors for tuberculosis. Pulm Med 2013
Nasiruddin M, Neyaz M, Das S (2017) Nanotechnology-based approach in tuberculosis treatment. Tuberculosis research and treatment 2017
Nelson KM, Dahlin JL, Bisson J, Graham J, Pauli GF, Walters MA (2017) The essential medicinal chemistry of curcumin: miniperspective. J Med Chem 60(5):1620–1637
Pizzo P, Scapin C, Vitadello M, Florean C, Gorza L (2010) Grp94 acts as a mediator of curcumin-induced antioxidant defence in myogenic cells. J Cell Mol Med 14(4):970–981
Prasad R, Gupta N, Banka A (2018) Multidrug-resistant tuberculosis/rifampicin-resistant tuberculosis: principles of management. Lung India 35(1):78
Prince A, Sandhu P, Ror P, Dash E, Sharma S, Arakha M, Jha S, Akhter Y, Saleem M (2016) Lipid-II independent antimicrobial mechanism of nisin depends on its crowding and degree of oligomerization. Sci Rep 6(1):1–15
Priyadarsini KI (2009) Photophysics, photochemistry and photobiology of curcumin: studies from organic solutions, bio-mimetics and living cells. J Photochem Photobiol C: Photochem Rev 10 (2):81–95
Rai M, Yadav A, Gade A (2009) Current trends in phytosynthesis of metal nanoparticles (vol 28, pg 277, 2008). Crit Rev Biotechnol 29(1):78–78
Rai M, Pandit R, Paralikar P, Nagaonkar D, Rehman F, dos Santos CA (2017) Pharmaceutical applications of curcumin-loaded nanoparticles. In: Nanotechnology applied to pharmaceutical technology. Springer, pp 139–154
Rawashdeh R, Haik Y (2009) Antibacterial mechanisms of metallic nanoparticles: a review. Dyn Biochem Process Biotechnol Mol Biol 3(2):12–20
Rengarajan J, Sassetti CM, Naroditskaya V, Sloutsky A, Bloom BR, Rubin EJ (2004) The folate pathway is a target for resistance to the drug para-aminosalicylic acid (PAS) in mycobacteria. Mol Microbiol 53(1):275–282
Richter E, Rüsch-Gerdes S, Hillemann D (2007) First linezolid-resistant clinical isolates of Mycobacterium tuberculosis. Antimicrob Agents Chemother 51(4):1534–1536
Robertson H (1933) The persistence of tuberculous infections. Am J Pathol 9(Suppl):711
Rohde K, Yates RM, Purdy GE, Russell DG (2007) Mycobacterium tuberculosis and the environment within the phagosome. Immunol Rev 219(1):37–54
Rothschild BM, Martin LD, Lev G, Bercovier H, Bar-Gal GK, Greenblatt C, Donoghue H, Spigelman M, Brittain D (2001) Mycobacterium tuberculosis complex DNA from an extinct bison dated 17,000 years before the present. Clin Infect Dis 33(3):305–311
Santos-Magalhães NS, Mosqueira VCF (2010) Nanotechnology applied to the treatment of malaria. Adv Drug Deliv Rev 62(4–5):560–575
Shutava TG, Balkundi SS, Vangala P, Steffan JJ, Bigelow RL, Cardelli JA, O'Neal DP, Lvov YM (2009) Layer-by-layer-coated gelatin nanoparticles as a vehicle for delivery of natural polyphenols. ACS Nano 3(7):1877–1885
Siddiqui T (2020) Active tuberculosis. StatPearls Publishing, Treasure Island (FL)

Singh A, Gopinath K, Sharma P, Bisht D, Sharma P, Singh N, Singh S (2015) Comparative proteomic analysis of sequential isolates of Mycobacterium tuberculosis from a patient with pulmonary tuberculosis turning from drug sensitive to multidrug resistant. Indian J Med Res 141 (1):27

Singh DK, Dwivedi VP, Ranganathan A, Bishai WR, Van Kaer L, Das G (2016a) Blockade of the Kv1. 3 K+ channel enhances BCG vaccine efficacy by expanding central memory T lymphocytes. J Infect Dis 214(9):1456–1464

Singh J, Garg T, Rath G, Goyal AK (2016b) Advances in nanotechnology-based carrier systems for targeted delivery of bioactive drug molecules with special emphasis on immunotherapy in drug resistant tuberculosis–a critical review. Drug Deliv 23(5):1676–1698

Smith I (2003) Mycobacterium tuberculosis pathogenesis and molecular determinants of virulence. Clin Microbiol Rev 16(3):463–496

Sreejayan N, Rao M (1996) Free radical scavenging activity of curcuminoids. Arzneimittelforschung 46(2):169

Srimal R, Dhawan B (1973) Pharmacology of diferuloyl methane (curcumin), a non-steroidal anti-inflammatory agent. J Pharm Pharmacol 25(6):447–452

Stanley RE, Blaha G, Grodzicki RL, Strickler MD, Steitz TA (2010) The structures of the anti-tuberculosis antibiotics viomycin and capreomycin bound to the 70S ribosome. Nat Struct Mol Biol 17(3):289

Stewart GR, Robertson BD, Young DB (2003) Tuberculosis: a problem with persistence. Nat Rev Microbiol 1(2):97–105

Sugiyama Y, Kawakishi S, Osawa T (1996) Involvement of the β-diketone moiety in the antioxidative mechanism of tetrahydrocurcumin. Biochem Pharmacol 52(4):519–525

Sun M, Su X, Ding B, He X, Liu X, Yu A, Lou H, Zhai G (2012) Advances in nanotechnology-based delivery systems for curcumin. Nanomedicine 7(7):1085–1100

Surh Y-J, Chun K-S, Cha H-H, Han SS, Keum Y-S, Park K-K, Lee SS (2001) Molecular mechanisms underlying chemopreventive activities of anti-inflammatory phytochemicals: down-regulation of COX-2 and iNOS through suppression of NF-κB activation. Mutat Res/Fundam Mol Mech Mutagen 480:243–268

Tiuman TS, Santos AO, Ueda-Nakamura T, Dias Filho BP, Nakamura CV (2011) Recent advances in leishmaniasis treatment. Int J Infect Dis 15(8):e525–e532

Tousif S, Singh DK, Ahmad S, Moodley P, Bhattacharyya M, Van Kaer L, Das G (2014) Isoniazid induces apoptosis of activated CD4+ T cells implications for post-therapy tuberculosis reactivation and reinfection. J Biol Chem 289(44):30190–30195

Tousif S, Singh DK, Mukherjee S, Ahmad S, Arya R, Nanda R, Ranganathan A, Bhattacharyya M, Van Kaer L, Kar SK (2017) Nanoparticle-formulated curcumin prevents posttherapeutic disease reactivation and reinfection with Mycobacterium tuberculosis following isoniazid therapy. Front Immunol 8:739

Vilchèze C, Av-Gay Y, Attarian R, Liu Z, Hazbón MH, Colangeli R, Chen B, Liu W, Alland D, Sacchettini JC (2008) Mycothiol biosynthesis is essential for ethionamide susceptibility in Mycobacterium tuberculosis. Mol Microbiol 69(5):1316–1329

Von Groll A, Martin A, Jureen P, Hoffner S, Vandamme P, Portaels F, Palomino JC, da Silva PA (2009) Fluoroquinolone resistance in Mycobacterium tuberculosis and mutations in gyrA and gyrB. Antimicrob Agents Chemother 53(10):4498–4500

White R, Lancini G, Silvestri L (1971) Mechanism of action of rifampin on Mycobacterium smegmatis. J Bacteriol 108(2):737–741

WHO (2018) Global Tuberculosis Report 2018 World Health Organization, Geneva, Switzerland. World Health Organization, Geneva

WHO (2019) Global tuberculosis report 2019. Geneva (Switzerland): World Health Organization; 2019. World Health Organization, Geneva

Xiyuan B, Oberley-Deegan R, An B, Ovrutsky A, Kinney W, Weaver M, Zhang G, Honda J, Edward C (2016) Curcumin enhances human macrophage control ofMycobacterium tuberculosisinfection. Respirology 21(5):951–957

Yah CS, Simate GS (2015) Nanoparticles as potential new generation broad spectrum antimicrobial agents. DARU J Pharm Sci 23(1):43

Yallapu MM, Nagesh PKB, Jaggi M, Chauhan SC (2015) Therapeutic applications of curcumin nanoformulations. AAPS J 17(6):1341–1356

Yano T, Kassovska-Bratinova S, Teh JS, Winkler J, Sullivan K, Isaacs A, Schechter NM, Rubin H (2011) Reduction of clofazimine by mycobacterial type 2 NADH: quinone oxidoreductase a pathway for the generation of bactericidal levels of reactive oxygen species. J Biol Chem 286 (12):10276–10287

Zazo H, Colino CI, Lanao JM (2016) Current applications of nanoparticles in infectious diseases. J Control Release 224:86–102

Zhang Y (2005) The magic bullets and tuberculosis drug targets. Annu Rev Pharmacol Toxicol 45:529–564

Zhang L, Gu F, Chan J, Wang A, Langer R, Farokhzad O (2008) Nanoparticles in medicine: therapeutic applications and developments. Clin Pharmacol Ther 83(5):761–769

Zhang Y, Shi W, Zhang W, Mitchison D (2013) Mechanisms of pyrazinamide action and resistance. Microbiol Spectr 2:1–12. https://doi.org/10.1128/microbiolspec.MGM2-0023-2013

Zumla A, Nahid P, Cole ST (2013) Advances in the development of new tuberculosis drugs and treatment regimens. Nat Rev Drug Discov 12(5):388–404

Application of Nanobiosensor in Health Care Sector

14

Subham Preetam, Lipsa Dash, Suman Sudha Sarangi, Mitali Madhusmita Sahoo, and Arun Kumar Pradhan

Abstract

Nanotechnology is expected to have a revolutionary effect on biology and medicine. It has been able to provide special perspectives to the probe and control a variety of biological along with medical processes that take place at nanometer scales. Nanoparticles offer some advantages like sensing, image enhancement, and delivery agents. These are the approaches for exploiting nanotechnology in medicine as well as in the health care sector. Nanobiosensors are nano-scaled quantifiable devices, those involve nano-conjugated biological particulars as a transducing system to detect the minute quanta of any physical, chemical, or biological analytes. The appearance of nanobiosensors has actually developed the detection of many disorders like cancer, allergic responses, diabetes, and many other diseases and malfunctions related to these diseases. The present review comprehensively scrutinizes various types of cancer detection biosensors fabricated so far along with their technical evaluation regarding the potency and efficiency of selected bioreceptors biotransducers. For the diagnosis as well as monitoring of prostate cancer, the prostate-specific antigen (PSA) has been identified and considered as the most valid clinical tool. By using amperometric immunosensor, prostate cancer can be monitored. For lung cancer detection a new rapid, as well as sensitive technique for screening, is required. With this approach, for the identification of many major disease biomarkers through sensitive, easy as well as rapid applications, biosensors have provided a promising technology as diagnostics tools. In this paper we discuss about lung cancer, prostate cancer, brain cancer, breast cancer, and pancreatic cancer. Also, current

S. Preetam · L. Dash · S. S. Sarangi · M. M. Sahoo · A. K. Pradhan (✉)
Centre for Biotechnology, Siksha O Anusandhan (Deemed to be University), Bhubaneswar, Odisha, India
e-mail: arunpradhan@soa.ac.in

M. Arakha et al. (eds.), *Bio-Nano Interface*,
https://doi.org/10.1007/978-981-16-2516-9_14

disease diagnostics techniques, protein and genetic biomarkers of diseases, as well as emerging biosensor technology for quick identification, are discussed. Moreover, the use of different nanomaterials in various forms as components of biosensors for the detection of various human pathogens has been further explored.

Keywords

Nanobiosensor · Cancer · Biomarkers · Quantum dots · Nanoparticles

14.1 Introduction

In the twenty-first-century humanity has approaches better life habitat by dominating some diseases. The Healthcare sector has been upgraded due to the evolution of many modern approaches (Chamorro-Garcia and Merkoçi 2016). Over the last few years, nanotechnology has altered dramatically healthcare tactics and visualized to have an enormous effect to supply stronger amenities in the healthcare sector (Prasad et al. 2018). The valid and precise information on the required biochemical frameworks is a crucial prerequisite for functional healthcare. Based on this, nanobiosensors are considered to layout possible solutions to the complications caused by the current medical industry. The reason is that these biosensing gadgets give significant advantages, like specificity, small size quick response, and cost (Malhotra and Chaubey 2003). Nanotechnology has increased the scope to modify the area of diagnostics in health, in the field of medicine and many other sectors by altering theoretical outlooks into the practical output. Furthermore, it has a vital role in the progression and innovation which increases the sensitivity and ascribes the applications based on nanosensors and nanobiosensors (Srivastava et al. 2018).

For our better idea, let us take a glance at the development of nanobiosensors in which nanotechnology plays a crucial role. Especially mankind is gifted with nanomaterials by nanotechnology. These nanomaterials are very small in size having one of their dimensions between 1 and 100 nm (Pandit et al. 2016). Due to this unique property of nanomaterials they can readily enter into the human body and by crossing the biological membranes, it can be introduced with the cells, tissues, and organs which cannot be possible by larger-sized particles (Pathak et al. 2007). Nanomaterials have very essential roles in the sensing mechanism associated with the biosensing technology (Pandit et al. 2016). Biosensors are devices that can analyze the conversion or transduction of a biological response into an observable signal. These devices are directly involved in the bioanalytical assay, so these are considered as very essential components of care devices (Shandilya et al. 2019). Through the latest signal transduction mechanics, by the utilization of nanomaterials, the performance and sensitivity of biosensors are upgraded (Sagadevan and Periasamy 2014). As the name suggests, nanobiosensors are the combination of nanotechnology and biosensors. Generally, nanobiosensors are the sensors developed by nanomaterials (Pandit et al. 2016). Biosensors based on nanoparticles are extremely attractive. The reason is that they can be simply synthesized in large quantities by applying standard chemical methods, and do not need improved fabrication techniques. As they have a very tiny size, they can provide extremely high surface areas (Sagadevan and Periasamy 2014). Moreover, these nanobiosensors are nano-scaled quantifiable devices that involve nano-conjugated biological particulars as a transducing system to detect the minute quanta of any physical, chemical, or biological analytes. For conveying the appropriate data in the form of signals, the fabricates use electrochemical, optical, thermometric, magnetic, piezoelectric, and micro mechanical techniques (Shandilya et al. 2019). Due to many advantages, the nanobiosensors are more attractive than the traditional biosensors. Some nanoparticles have an enzymatic property which plays an important role in the achievement of a very good level of sensitivity in a well-developed biosensor (Prasad et al. 2018).

The definition and representation of the idea of the effectiveness of nanobiosensors leave no space for their applications as they are extremely versatile and multifunctional, so many and possibly unlimited. The appearance of biosensors has actually developed the detection of many disorders like cancer, allergic responses, diabetes, and many other diseases and malfunctions related to these diseases (Chamorro-Garcia and Merkoçi 2016). From a clinical perspective, there are countless clinical approaches that are mainly being enabled applying biosensors on a daily basis. Detection of glucose in diabetic patients, detection of HIV-AIDS, the detection of bacterial infections in the urinary tract along with the cancer diagnosis are the major clinical applications of nanobiosensor (Chamorro-Garcia and Merkoçi 2016). In the environment, the presence of pathogens can be detected by using nanotechnology (Pandit et al. 2016). The nanobiosensors have reached LOD in the picomolar as well as femtomolar range that is giving new approaches in the initial stage diagnosis of autoimmune diseases in vitro and in vivo (Ghorbani et al. 2019). DNA modified by Gold nanoparticle has been utilized in improving a microcantilever-based DNA biosensor which is used to identify DNA even at a very lower concentration by a hybridization reaction process. Nanotechnology has increased the customization of oncology for modified diagnostics and therapeutics of each according to the demand. Several genomic tests have been methodically reviewed that could be utilized in screening, identification, and treatment of target cancer affected part in an individual's body (Prasad et al. 2018). When the cells are treated with a drug or attacked by a biological pathogen, the reactions of the cells can be analyzed by several other applications of nanobiosensors. This has major involvement ranging from drug treatment improvement to national safety, protection of the environment, and a greater understanding of molecular biology at a systems level (Pathak et al. 2007). Research on Nanobiosensor concentrates on improving new technologies that have the capacity to create remarkable contributions in the areas of detection of human and animal disease markers, encouraging diagnostic compound detection and examination, nano-and biomaterials characterization, and biocatalyst development.

14.2 Use of NanoBiosensor for Detection of Cancer

In the rise of the Modern science in Nanotechnology, Biosensor has played a crucial role in disease diagnosis. In that field nanobiosensor has a major impact on the detection of cancer in the primary stages.

14.2.1 Prostate Cancer

Cancer in the prostate is always a significant reason for mortality in the male populace. As the widely recognized tumor in men reproductive system, the subsequent driving reason for malignant growth passings among guys in the US. The prostate carcinoma (CAP) is a lethal threat and a significant reason for death in men populace matured somewhere in the range of 55 and 80 years. Prostate explicit antigen PSA might be a composition of glycoprotein with 93% peptide and seven sugar molecules and an isoelectric pH of 6.9, which might present inside the human serum in the form of f-PSA with modern (PSA-ACT) structures. Ordinarily total PSA (t-PSA) level is over 10 ng/ml (Sarkar et al. 2002). A significant utilization of PSA testing to screen the patients determined to have CAP which was demonstrated to be very valuable serologic test in arranging and checking CAP particularly in the early discovery of repetitive illness (Blijenberg et al. 1996). PCa happens when a harmful tumor is incited by the transformation of DNA. As we probably are aware the most normally lethal sort of disease in matured males 50–80 years old (Fritz 2009). Early detection of prostate cancer (PCa) is basic for the anticipation of metastasis and early treatment; another technique for manufacture for GOssDNA (graphene oxide/ssDNA) based biosensors by at the same time utilizing dual-antibody which was modified PLLA NPs (poly-L-lactide nanoparticles) for signal intensification, VEGF (vascular endothelial development factor), and PSA (prostate-specific antigen) identification. On the other hand, VEGF is a basic biomarker for disease development and is regularly overexpressed in tumors. The convergences of PSA higher than 4.0 ng/ml are viewed as strange. Hence, the mix of PSA and VEGF location should be useful in the exact early analysis of PCa (Pan et al. 2017). Another touchy, barely surface plasmon resonance (SPR) and few-mode fiber (FMF) biosensor with the sandwich measure for the location of PSA (Jang et al. 2009). The proposed FMF SPR biosensor was an incredible capacity for the constant examination of resistance among biomolecules and consequently upsides of high affectability and label-free recognition. The benefits of SPR biosensors incorporate include high sensitivity, real-time detection, and label-free detection. Upsides of the side-polished FMF SPR biosensors with sandwich test incorporate superfluity (Maksim and Andrey 2006), high affectability, basic creation, and name free identification.

For now, the nanoparticle (NP) mark/immunochromatographic electrochemical biosensor (IEB) is used for quick with the touchy discovery of prostate-specific antigen in the human serum (Lin et al. 2008). The high affectability is because of the NP-produced signal intensification and natural high affectability of the electrochemical techniques. The most solid tumor marker was the serum prostate-specific antigen for distinguishing PCa in the principal stage (Stephan et al. 2006). We have built up a straightforward and financially savvy method for recognizable proof and checking of prostatic adenocarcinoma utilizing amperometric immunosensor (Sarkar et al. 2002). The strategy for immunoassay which is an amperometric screen-printed cathode for the location of malignancy initiator be utilized for another age of assays for clinical labs and might adjusted for screening gadgets for doctors. The technique for distinguishing proof and observing of prostate malignancy is amperometric immunosensor.

14.2.2 Lungs Cancer

Nowadays, the lungs cancer shows the best death rate around the world (1.3 million in a year) with significance with other regular malignancies as prostate, colon, and breast cancer (Altintas and Tothill 2013). Cancer in the lungs is that the uncontrolled development of irregular cells in one or both lungs. Anomalous cells degrade their typical usefulness and quit forming and separating into normal tissue. There is an earnest need to create quick screening techniques for checking the measure of immunogenic tumor-related antigen (TAAs) in samples used for clinical treatment and malignancy immunological research. r-Enolase (ENO1), a p48 particle, is cosmopolitan in an extreme type of tissues, though γ-enolase (ENO2) and -enolase (ENO3) are covered in neuron and muscle tissues, respectively. We have found out the practicality of utilizing nanogold marks on a dispensable SPCE in an immunoassay that might be delicately identified utilizing after the oxidative arrival of Au (III) particles. We presume that this single-use, dispensable SPCE-base (screen-printed carbon cathodes) electrochemical immunosensors in additional improvement into useful clinical cancer in the lungs determination frameworks (Ho et al. 2010). Epidermal protein receptor (EGFR) exon 19 transformation status could be a significant forecast list for tyrosine kinase inhibitors (TKIs) treatment (Xu et al. 2016). A better specific sandwich-type electrochemical biosensor that distinguishes in-frame deletions of exon 19 of EGFR in genuine examples of people with non-little cell lung cancer in the lungs. Roughly 80% of all cellular breakdown in the lungs cases are non-small cell lungs cancer (NSCLC), that incorporates helpless visualization in middle endurance of one year when treated with ordinary chemotherapy (Schiller et al. 2002). These days, tyrosine kinase inhibitors (TKIs) as first-line treatment for particular metastatic during this work, as DNA electrochemical biosensor has intended to discover EGFR exon 19 statuses of cancer in lungs cancer patients. The system that adjusts the erasure area in target deliberately per diverse hybridization stage is prepared to support a sandwich-type DNA biosensor. Discoveries may have suggestions for the arranging of biosensors in better execution. Carcinoembryonic antigen (CEA) is one in everything about preeminent examined protein markers for cancer in the lungs (Homola 2003). There are numerous clinical and examination put together an application with respect to CEA (Sakao et al. 2004). Carcinoembryonic antigen is created during a lethal turn of events and furthermore, its creation ends before human birth. The protein mostly inside the blood of normal individuals aside from cigarette-smoker; nonetheless, its amount shows a major increment in certain conditions including in the lung cancer. Its level is explained to a cellular breakdown in the lungs inside which CEA focus is raised fundamentally in malignant growth cases than in healthy people. At the time the concentration of CEA is an increase of over 3 ng/ml, cancer in the lungs may start (Sakao et al. 2004).

So neuron-specific enolase (NSE) has another nanobiosensor in which neuron-specific enolase could be a catalyst of glycolytic which present in grown-up neurons on grounds that the most bountiful assortment of the glycolytic compounds. NSE shows calcium level to play out its capacities-dependent manner and wants magnesium cofactor for catalysis. When the amount of NSE is over the 9 ng/ml lung cancer occurs (Barlési et al. 2004). It is a valuable marker to dissect lungs cancer disease. The optical biosensors include evanescent wave fiber optic biosensors, optrode-based fiber optic biosensors, surface plasmon resonance biosensors (SPR), interferometric biosensors, the resonant mirror optical biosensor, and the resonant mirror optical biosensor (Homola 2003). Numerous commercially accessible stages are supported by fluorescence detection systems for better recognition.

14.2.3 Breast Cancer

These days, breast cancer is the commonest ladies' disease around the world. Improvement of early recognition procedures for breast malignancy is essentially significant for improving endurance rates. For the early location of cancer diseases, coursing miRNAs are developing as new dependable biomarkers (Mostafa et al. 2016). At that point, a nanobiosensor indicated high specificity as well as was prepared for the forceful segregation among reciprocal target miRNA, single-, three-base mismatch and non-complementary miRNA. Close by the exceptional affectability and selectivity, this nanobiosensor had an incredible capacity, reproducibility, and indicated a decent reaction brighten the genuine example investigation with plasma cloud, clinically be utilized in the primary discovery of the breast cancer growth, by direct location of the plasma miR-155 which is a genuine clinical example. This way the quick, precise, and sensitive determination of hydrogen peroxide (H_2O_2) is of incredible hugeness inside the physiological, obsessive as well as natural fields that have proposed a delicate and particular amperometric biosensor for the recognition of extracellular H_2O_2 delivered from human breast cancer cells with the help of a sequence-specific peptide, the proposed biosensor can distinguish H_2O_2 in an extremely wide direct range from 1.0 107 M to 1.0 104 M with a few location cutoffs of 3.0 108 M (Zhao et al. 2013). This work is expected to style for the essential time a catalyst based biosensor that is equipped for identifying miR21, a sort of microRNA that is known to be upregulated in breast malignancy tissues, from cell lysate of absolute RNA (Mihi et al. 2009).

A delicate and recyclable crystal stone microbalance (QCM) biosensor was created utilizing chitosan (CS) and folic acid (FA), producing forms that are mostly perceived by MCF-7 cancer cells over-communicated folic acid receptors (Hao et al. 2009). Surface plasmon resonance (SPR) bio transducers give an opportunity of constant detection of bimolecular communication in an exceptionally sensitive, quick as well as label-free way. It has been utilized in the discovery of point changes in genes related to breast cancer and it has significant application in breast cancer biosensing. Biosensors have changed the dread of breast cancer growth determination offering fast, basic, and worth powerful courses. For low recognition limits, glycoproteins are the first focuses for breast cancer finding among CTC (circulatory tumor cell) due to the low wealth of last ones (Mittal et al. 2017).

In that way, protein biomarkers have been vigorously explored with their parts in the identification, evaluation, and checking of forceful kinds of breast cancer. During this work, we portray a total unique, mark-free methodology for identifying the HER2 extra-cell space breast cancer malignancy biomarker in human serum tests utilizing the opto-fluidic ring resonator (OFRR). The OFRR consolidates microfluidics and optical ring resonator detecting innovation to accomplish rapid label-free detection in a little and minimal effort stage (Capobianco et al. 2008). HER2 proteins were spiked in serum at different fixations, during this examination. Results show that the OFRR identifies HER2 at restoratively significant focuses in serum beginning from 13 to 100 ng/ml in 30 min. The whole identification methodology can be fundamentally improved since it performs label-free detection (Suter et al. 2008).

14.2.4 Brain Cancer

The central nervous system (CNS) tumors have a significant purpose behind horribleness and mortality around the world. Most of them, gliomas are preeminent normal and deadly essential brain tumors in humans (Agarwal et al. 2011). Accordingly, precisely imaging glioma cells pressing for primary conclusion along with powerful treatment in brain glioma. Ongoing examination shows that fluorescence imaging methods are wont to recognize the tumors for its various benefits, for affectability, a high difference (Kobayashi et al. 2009). A shiny new assortment of carbon dabs (CD-Asp) with focusing on work toward brain malignancy glioma was combined by means of a straightforward pyrolysis route. As-arranged CD-Asp displays focusing on C6 glioma cells without assistance extra focusing on particles. After 15 min of tail vein infusion in vivo fluorescence pictures indicated high contrast biodistribution of CD-Asp, more heavy fluorescent sign was identified inside the glioma location than in typical cerebrum, however, L-glutamic corrosive (CD-Glu) do not have any or less selectivity in glioma. Thusly, CD-Asp used as a fluorescence imaging and focusing on a specialist for non-intrusive glioma determination (Zheng et al. 2015).

The improvement of sensitive and specific biological sensors for the checking of disease biomarkers is a significant inspiration for nanoscience research. The restricted surface plasmon resonance (LSPR) nanosensor has been shown to be a decent stage for the quantitative recognition of biological and chemical species

(Haes et al. 2004). Human brain extracts and humor (CSF) tests from control and ailing patients display definitely various reactions, which demonstrate the presence of raised ADDL (amyloid-â inferred diffusible ligands). That is the reason it identifies little changes to brain cancerous cells. The LSPR nanosensor was a solid device for considering the oligomerization of low concentrations of amyloid antecedents. Given this achievement, the use of LSPR innovation additionally holds guarantee joined of the least difficult identification procedures for the screening of oligomerization blocking drugs (Amanda et al. 2005). They show that the LSPR nanobiosensor will have the option to consider human samples and should help inside the comprehension of the instrument and conclusion of Alzheimer's illness. Electrochemical biosensor is viewed as a better technique than formaldehyde detection method due to its constant estimations, high affectability, and selectivity (Herschkovitz et al. 2000).

The development of an electrochemical biosensor for the detection of formaldehyde in solution, supported the coupling of the enzyme formaldehyde dehydrogenase and a carbon nanotubes (CNT)-modified screen-printed electrode (SPE) (Lilach et al. 2010). We introduced a biosensor in the location of formaldehyde upheld coupling the selectivity of the compound FDH (formaldehyde dehydrogenase) in a profoundly proficient NADH identification framework utilizing a (carbon nanotubes) CNT-adjusted SPE (Vianello et al. 2007). Biosensor distinguished formaldehyde delivered from human brain cancer cells U251 accordingly formaldehyde-releasing prodrug therapy for anticancer. The detection limit of direct formaldehyde addition within the presence of cells (10 μM) was over in clear solution, probably because of the cells physically blocking formaldehyde access to the electrode surface, we report for the primary time that our in-house developed Localized Surface Plasmon Resonance biosensor with self-assembly gold nanoislands (SAM-AuNIs). Abhimanyu et al. (2017) would be able to detect and distinguish exosomes from MVs isolated from A-549 cells, SH-SY5Y cells, bodily fluid, and urine from a carcinoma mouse model. Tamada et al. (2007) delivered a discernable reaction to the exposed bare LSPR (Local surface plasmon resonance) biosensor without the function, proposing a particular biophysical communication among exosomes and MVs along with SAM AuNIs. That type of sensor achieves the restriction of the location to 0.194μg/ml, hence the direct powerful range covers 0.194–100μg/ml (Ng et al. 2017). This disclosure not just uncovers extraordinary knowledge into the particular layer of tumor-derived exosomes and MVs, yet in addition encourages the occasion of novel LSPR biosensors indirect recognition with the segregation of heterogeneous EVs.

14.2.5 Pancreatic Cancer

Pancreatic cancer forms in the cell of the pancreas. This sickness is more common in male than in female and mostly occurs after the age of 65 while before age 40 are uncommon. Basically, by the method of surgery pancreatic cancer can be completely cured if it is detected in an early stage. And if the cancer has spread to the whole body, it will not be a cure. Chemotherapy may help to control but it will not cure the cancer. So, it is more important to detect cancer at an early stage.

As the research study proceeds in the nanotechnology sphere, in the present decades, we can detect the pancreatic cancer using nanobiosensor. So, nanobiosensors have the ability to observe minor biomarkers in a little sample and also upgrade picture quality and successfully pancreatic cancer is detected at an early stage. To spot cancers at the primary stage in its clinical trial has the power to decrease the rate of morbidity and mortality. Currently, for the spotting and diagnosis of pancreatic cancer, some nanobiosensors have been utilized; these are nanoparticles, nanowires, nanotubes, and nanocantilevers. As research progressed in the nanotechnology field, Nanoparticles play an important role in detecting cancer. Successfully some distinct leveling ligands have been used in surface functionalization for pointing of nanoparticles with (Kumagai et al. 2009) urokinase-plasminogen-activator-receptor (uPAR) that acts as a surface receptor, which further enlarged the expression on pancreatic tumor cells and stromal cells that cover the cancer lump and folate (Yang et al. 2009). To detect CA19-9 cancer biomarkers, ZnO QDs (Quantum dots) utilized as a signaling agent in sandwich immunoassays with the help of both square-wave-stripping-voltammetry and photoluminescence detection procedures (Gu et al. 2011). According to Zaman et al. for pancreatic cancer detection, CEACAM6 which could be utilized as a biomarker that may be targeted by a dominant functionalization of Quantum dots with a single domain antibody, it could be used as a biomarker. He also said that it is far better to use single-domain-antibodies for targeting cancer because these antibodies are more stable in nature and economically very cost-effective than the conventional method of antibody attachment (Zaman et al. 2011). But QDs is very harmful, i.e. highly toxic for human. The utilization of some other nanoparticles like gold and silver nanoparticles has also been done for the identification and diagnosis of pancreatic cancer. Covalently attached F19 human monoclonal antibodies used to gold nanoparticles are much more effective at labeling pancreatic carcinoma tissue and it is possible to detect cancerous parts under darkfield microscopy (Gdowski et al. 2014).

A narrow film sensing unit on interdigitated gold-electrodes joining polyethylenimine and carbon nanotubes in a layer-by-layer method (Ariga et al. 2014; Rydzek et al. 2015) onto which antibodies anti-CA19-9 are adsorbed with a secondary layer of N-hydroxysuccinimide and 1-ethyl-3-(3-dimethylaminopropyl) carbodiimide solution. The pancreatic cancer biomarker CA19-9 is detected in a buffer with an edge of detection of 0.35 U/Ml by the use of obstacular spectroscopy. This high sensitivity permitted for the distinction between samples of blood serum from patients with distinct probabilities to develop pancreatic cancer.

Till now, the medical blood trials have focused on CA19-9 (carbohydrate antigen 19-9) (Morris Stiff and Taylor 2012), CEA (carcinoembryonic antigen) (Ozkan et al. 2003), and recently K-ras gene mutations for the recognition of pancreatic cancer in the preliminary stage (Kinugasa et al. 2015). The furthermost advanced pancreatic cancer test until the age is CancerSEEK13 (Cohen et al. 2018). Here protease and arginase are used as pancreatic biomarkers. Because in genomic and proteomic tests:

a genetic test shows the capability of disease growth, but does not accurately when a tumor really arises. That transition can be detected by protease/arginase assays exactly. Matrix metalloproteinases (MMPs), serine-proteases and cysteine-proteases and also arginase have an important job in malignant progression and immune (dis)regulation in cancer. But using this it can be possible to recognize premature tumors with the help of liquid biopsies. And the major point is that tumor-promoting-proteases perform as a part of an extensive multidirectional network of proteolytic interactions. Gene Expression Analysis (Choi and Kendziorski 2009) is a clear outlook to determine the proteases that are overexpressed in solid tumors, i.e. pancreatic cancer. Here, for the detection of arginase and protease, fluorescence nanobiosensors are used as ultrasensitive (sub-femtomolar), which consisting of water-dispersible Fe/Fe3O4 core/shell nanoparticles and two tethered-fluorescent dyes, i.e. TCPP (Tetrakis(4-carboxyphenyl)porphyrin) and cyanine 5.5. By gene expression analysis it is possible to recognize an enzymatic snap for the finding of pancreatic adenocarcinomas in serum, consisting of arginase, matrix metalloproteinases -1, -3, and -9, cathepsins -B and -E, urokinase plasminogen activator, and neutrophil elastase (Madumali et al. 2018).

MicroRNAs have also been successfully verified to be the reliable biomarker for the initial identification of pancreatic cancer in the blood samples. In the serum of pancreatic carcinoma, the high expression level of microRNA-21, microRNA-155, microRNA-196a, and microRNA-210 can also be found (Zeng et al. 2017). A DSN (duplex-specific nuclease)-assisted dual-SERS (Surface-enhanced Raman spectroscopy) biosensor is developed for microRNA-10b in exosome and residual plasma of blood samples detection depending on the DTNB which can be used as a bio-weapon for pancreatic detection in presence of target microRNA.

14.3 Use of Nanobiosensor for Detection of Pathogenic Bacteria

14.3.1 Detection of *Escherichia coli*

Escherichia coli (*E. coli*) is a microorganism which is usually found in the small intestinal region of humans as a part of the human microbiome. Most of the strains of *E. coli* are non-pathogenic, except for a few that are able to cause diseases such as bloody diarrhea, septicemia, and other gut and urinary tract ailments (Kaper et al. 2004). The successful detection of the bacteria through nanosensors has been achieved by the detection of *E. coli* specific protein or genes in several strains of *E. coli* such as *E. coli* μX, *E. coli* B, and *E. coli* 0157:H7. Cheng et al. have been able to detect the presence of viable but non-culturable (VNBC) cells of *E. coli* by monitoring the fluctuation in the faradaic current caused by the obstruction in the orifices of the platinum electrode altered with nanoporous aluminum oxide (Cheng et al. 2011). Likewise, gold nanoparticles (AuNPs) have been utilized in various forms such as electrodes (Minnaei et al. 2016) and Au-Ag nanorods (Sun et al. 2015) in order to aid in the sensitive acquisition of the bacteria-specific probe. Similarly, nanofibers composed of nitrocellulose by electrospinning have been employed by Luo et al. as capture pads to detect the pathogen through lateral flow immunoassay (LFA) (Luo et al. 2010). Titanium dioxide nanoparticles too have been exploited in several studies as thin films for their biocompatible nature and stability under extreme environments for the sensing of *E. coli* 0157:H7 (Mura et al. 2012; Nadzirah et al. 2020). Likewise, the strains of *E. coli,* namely *E. coli* B and *E. coli* μX have been discerned through their binding to specific T4 phages in conjugation with thin films of silver nanoparticles (AgNPs). This method of detection has been able to sense about a single bacterium in an analyte sample of 10μl (Srivastava et al. 2015). Furthermore, AgNPs in conjugation with the piezoelectric quartz crystal sensors have been able to quantify as low as 1 CFU/ml of *E. coli* in 800 ml of water (Sun et al. 2009). Magnetic nanoparticles (MNPs) have found use in the colorimetric detection of the bacteria in which the presence of *E. coli* specific proteases results in the change in color of the sensory platform. This is a result of degradation of the MNP conjugated peptides mounted on the platform by the proteases, which reveals the gold-layered sensor surface underneath (Suaifan et al. 2017). MNPs-probes are also well capable of directing the selective separation and concentration of the probes after successful binding to the specific *E. coli* antigen. The antigen is then be detected by measuring the impedance charge between electrodes of the biosensor (Varshney and Li 2007). This property of MNPs has also been exploited coupled with other detection methods for increasing the overall sensitivity and effectiveness of the nanobiosensor (Luo et al. 2010; Hossain et al. 2012; Banerjee et al. 2016).

14.3.2 Detection of *Salmonella*

Salmonella is a genus of bacteria which are known as the causative organism of salmonellosis (Woodward et al. 1997). Most of the studies carried out for the detection of salmonella have utilized *S. typhimurium*, though *S. choleraesuis* and *S. enteritidis* have been studied for the purpose as well. The nanobiosensors devised to detect salmonella have mostly employed recognition molecules such as anti-salmonella antibodies (Moongkarndi et al. 2011; Salam et al. 2013; Kim et al. 2015; Xia et al. 2016; Viter et al. 2017), salmonella specific aptamers (Labib et al. 2012; Yuan et al. 2014), or probe nucleic acids (Song et al. 2014; Gokce et al. 2011) to achieve specificity and selectivity for the bacteria. Some of the studies have utilized the conjugation of nanoparticles either directly to the pathogen or its probe to ultimately label the bacteria for its enhanced detection (Yuan et al. 2014; Kim et al. 2015). Yet in some other studies, the nanoparticles have been used as the base surfaces in form of nanotubes, nanopores, or nanofibers (Song et al. 2014; Viter et al. 2017; Gokce et al. 2011) that provide increased surface area, thus improving the overall detection efficiency of salmonella. Nanoparticles have further been used in conjunct with magnetic properties for increased sensitivity of detection of salmonella present in solutions in low concentrations (Xia et al. 2016). Furthermore, in another study by Salam et al. the nanoparticles have aided as amplifiers in mass detection of salmonella (Salam et al. 2013). The final detection of bacteria can be carried out by quantifying the change in fluorescence (Song et al. 2014; Kim et al. 2015; Viter et al. 2017), difference in surface charge transfer (Gokce et al. 2011), deposition of nanoparticles at the site of the pathogen by virtue of redox reactions (Yuan et al. 2014), LFA (Moongkarndi et al. 2011; Xia et al. 2016), or quartz crystal microbalance assay (Salam et al. 2013). Simultaneous detection of two strains of salmonella, *S. typhimurium* and *S. enteritidis* has been achieved through LFA, with lower detection limits of 10^4 CFU/ml and 10^6 CFU/ml, respectively (Moongkarndi et al. 2011). Furthermore, in the recent studies independently carried out by Yuan et al. (2014) and Ranjbar et al. (2018) to detect *S. typhimurium*, the detection limits of the nanosensors were as low as 7 CFU/ml and 1 CFU/ml, respectively.

14.3.3 Detection of *Mycobacterium tuberculosis*

Mycobacterium tuberculosis (*M. tb*) is a rod-shaped bacteria responsible for being the causative agent of tuberculosis (Dannenberg 1982). Various proteins, genes, and the genomic DNA of *M. tb* have been targeted for its acquisition through nanotechnology, among which the IS6110 gene has been employed the most due to its abundant repetitions in the *M. tb* genome that enhances the sensitivity of detection of the pathogen (Barreda-García et al. 2015; Karaballi et al. 2015; Prabowo et al. 2016; Bai et al. 2019; Rabti et al. 2020). AuNPs have been extensively employed for the detection of the *M. tb* where the signal transduction has been mostly performed using colorimetry (Bernacka-Wojcik et al. 2013; Kaewphinit et al. 2013; Tsai et al. 2017), electrochemical detection (Thiruppathiraja et al. 2011; Torres-Chavolla and Alocilja 2011), or detection of surface plasmon resonance of the AuNPs (Xiang et al. 2015; Prabowo et al. 2016; Matsishin et al. 2017). Similarly nanoparticles of different forms of carbon have been utilized in sensors for the electrochemical detection of *M. tb* (Miodek et al. 2015; Mukherjee et al. 2015; Chen et al. 2017). Furthermore, in several studies the nanohybrids of graphene and gold have been able to detect *M. tb* DNA sequence among which the study carried out by Mogha et al. (2018) has reported to have obtained an LOD of as low as 0.1 fM of DNA (Liu et al. 2014; Mogha et al. 2018; Perumal et al. 2018). Besides, graphene nanoparticle in conjugation with nanofibers of polyaniline has been able to determine well around 7.853×10^{-7} M of *M. tb* DNA through FTIR spectroscopy (Mohamad et al. 2017). Yet in another study by Mulpur et al. (2015) usage of a nanocomposite thin film of silver and C_{60} coupled to surface plasmon coupled emission technique has allowed an LOD of 20 Mtb/mm^2.

14.3.4 Detection of *Staphylococcus aureus*

S. aureus is a gram-positive bacterium that belongs to the family Micrococcaceae. Although it has been reported to colonize a small percent of human population (Wertheim et al. 2005), *S. aureus* is known to be responsible for nosocomial infections, toxic shock syndrome, bacteremia, and endocarditis like ailments

(Lowy 1998). A good amount of studies have reported the utilization of AuNPs for the colorimetric detection of *S. aureus*. The aggregation of the modified AuNPs triggered by specific binding to the pathogen-specific antigen has been colorimetrically determined to confirm the presence of *S. aureus* (Shahbazi et al. 2018). Liu et al. have used cysteamine enhanced AuNPs conjugated to pVIII fusion protein of *S. aureus* as a probe to obtain an LOD of 19 CFU/ml of bacteria (Liu et al. 2016). Furthermore in one of the studies performed by Li et al. (2011), the change in color intensity has been converted to quantifiable electrical signals with the aid of a devised portable transducer. Yet in another study by Wang et al., the use of AuNPs has been replaced with that of silver-layered MNPs. These modified nanoparticles upon conjugation with 5,5-dithiobis-(2-nitrobenzoic acid) (DTNB) form DTNB-labeled inside-and-outside plasmonic NPs (DioPNPs), which in turn are able to detect up to 10 CFU/ml of *S. aureus* (Wang et al. 2015). Aptamer conjugated silica nanoparticles are also in frequent use for the detection of *S. aureus* in form of nanokeepers or fluorescence labeled nanoparticles (Shangguan et al. 2015; Borsa et al. 2016). Furthermore, a study by He et al. has focused on flow cytometric detection of *S. aureus* cells by using Fluorescent silica nanoparticles (FSiNPs), where they have used Syber green dye to achieve dual color detection for minimizing the false positive chances (He et al. 2014). There have also been several studies on carbon nanoparticles as biosensor for *S. aureus* detection, such as one study by Zelada-Guillén et al. where the potentiometric detection of unlabeled *S. aureus* cells was feasible by relying on the pathogen-specific binding of the single-walled carbon nanotubes-aptamer conjugates (Zelada-Guillén et al. 2012). Similarly, Zuo et al. (2013) have been able to utilize graphene oxide nanoparticles to detect *S. aureus* cells within a short response duration of 10 min.

14.3.5 Detection of *Pseudomonas aeruginosa*

Pseudomonas aeruginosa (*P. aeruginosa*) is an opportunistic pathogen of human which is known to cause numerous infections including cystic fibrosis and chronic obstructive pulmonary disease (Ohman and Chakrabarty 1981; Murphy et al. 2008; De Bentzmann and Plésiat 2011). Multiple studies have exploited MNPs for the detection of *P. aeruginosa* owing to their ability to effectively separate and concentrate the target molecule via using external magnetic field. In one study by Tang et al. (2013), MNPs have been utilized both as the adsorbing surface for the analyte nucleic acid and labeled probes for the final detection of *P. aeruginosa* specific gyrB gene in the sample through the application of chemiluminescence. Similarly, in a more recent study carried out by. Zhong et al. (2018) the MNPs have been recruited for the fluorescent detection of *P. aeruginosa,* where partially fluorescence conjugated complementary DNA fragments are bound to *P. aeruginosa* aptamers modified with MNPs. The presence of *P. aeruginosa* in the analyte replaces the fluorescent complementary DNA with the *P. aeruginosa* specific DNA. The resultant decrease in the fluorescence acts as the indicator of the presence of *P. aeruginosa* in the sample. Likewise, chitosan nanoparticles (NC) have been employed for the detection of *P. aeruginosa* by virtue of their ability to improve electrochemical conduction. As per the study carried out by Sarabaegi and Roushani (2019) layering of NC on glassy carbon electrode has aided in the improving the functionality of *P. aeruginosa* specific aptamers which has led to the sensitive detection of *P. aeruginosa* specific DNA sequences via measuring the charge impedance. This approach was able to achieve an LOD of as low as 3 CFU/ml. Similarly, fluorescent organic nanoparticles synthesized from a pyrimidine derivative have been used by Kaur et al. (2015) in their study to detect *P. aeruginosa*, where the presence of the bacteria has been correlated with an increase in the fluorescence activity. AuNPs have also contributed to the effective sensing of *P. aeruginosa* through colorimetric detection of the aggregate state of the nanoparticles. The selective binding of AuNP-aptamer probes to the ecfX gene of *P. aeruginosa* avoided the aggregation of AuNPs, which was evident from the color of the reaction solution as bluish-gray to colorless. In contrast, the red color of the solution indicated the accumulation of the AuNPs in the absence of *P. aeruginosa* (Manajit et al. 2018). Likewise, the presence or absence of nanozyme property of AuNPs to act as a peroxidase has been able to indicate the presence or absence of *P. aeruginosa* in the samples, respectively (Das et al. 2019). Furthermore, the use of electrochemical active molecule as a self-assembled monolayer on redox-active AuNP electrodes has been able to detect up to 10 CFU/ml of *P. aeruginosa* in spiked plasma (Lee et al. 2019).

14.4 Conclusion and Future Aspects

Clinically, the cancers can be detected at an early stage with the aid of nanobiosensors through which the chances of recovery of patients can be improved. Presently, numerous cancers are detected when they have already metastasized throughout the body. In the present decade, nanobiotechnology provides an alternative platform to detect an exact and particular biological analyte by basically translating a biological entity into an electrical signal which can be sensed and examined. In cancer identification and monitoring, the application of nanobiosensors holds huge potential. The detection and diagnosis of several types of cancer such as prostate, breast, pancreatic, lung, and brain cancer have utilized the nanobiosensor systems like nanoparticles, nanowires, nanotubes, and nanocantilevers. To determine the effectiveness of anti-cancerous chemotherapy agents at several target sites and for the detection of emerging cancer biomarkers, the biosensors can be designed. Nanobiosensor technology has the ability to give quick and exact detection, clear creditable pictures of cancer progression. Numerous nanobiosensors have also been studied to detect pathogenic microorganisms in various biological and non-biological samples. Strains of *E. coli* such as *E. coli* B and *E. coli* μX have been identified through their binding to specific T4 phages in conjugation with thin films of silver nanoparticles (AgNPs) and platinum electrode altered with nanoporous aluminum oxide, while gold nanoparticles (AuNPs) have been utilized in detection of bacteria such as *E. coli* 0157:H7, *M. tb*, and *S. aureus*. Magnetic nanoparticles (MNPs) have contributed to the colorimetric detection of the E. coli bacteria and detection of *Pseudomonas aeruginosa* upon being subjected to external magnetic field. Proper quantification of the signals so obtained has facilitated accurate determination of the pathogenic concentration present in the sample analyte.

Nanosized ingredients combined into existing clinical diagnostic, detection systems and biosensors have confirmed upgraded and improved specificity as well as sensitivity as compared to the traditional methods of screening for invasive pathogenic species and tumor cells. Through further study on nanobiosensors, it is possible to discover novel methodologies with higher effectiveness that might pave way in the future to a more efficient cure for pathogenic infections and a deadly disease like cancer.

References

Abhimanyu T et al (2017) Direct detection of two different tumor-derived extracellular vesicles by SAM-AuNIs LSPR biosensor. Biosens Bioelectron 94:400–407

Agarwal S, Sane R, Oberoi R, Ohlfest JR, Elmquist WF (2011) Delivery of molecularly targeted therapy to malignant glioma, a disease of the whole brain. Expert Rev Mol Med 13:e17

Altintas Z, Tothill I (2013) Biomarkers and biosensors for the early diagnosis of lung cancer. Sens Actuators B: Chem 188:988–998

Amanda J et al (2005) Detection of a biomarker for Alzheimer's disease from synthetic and clinical samples using a nanoscale optical biosensor. J Am Chem Soc 127:2264–2271

Ariga K, Yamauchi Y, Rydzek G, Ji Q, Yonamine, Wu KCW, Hill JP (2014) Layer-by-layer nanoarchitectonics: invention, innovation, and evolution. Chem Lett 43:36–68

Bai L et al (2019) Ultrasensitive electrochemical detection of Mycobacterium tuberculosis IS6110 fragment using gold nanoparticles decorated fullerene nanoparticles/nitrogen-doped graphene nanosheet as signal tags. Analytica Chimica Acta 1080:75–83

Banerjee T et al (2016) Multiparametric magneto-fluorescent nanosensors for the ultrasensitive detection of Escherichia coli O157: H7. ACS Infect Dis 2:667–673

Barlési F, Gimenez C, Torre JP, Doddoli C, Mancini J, Greillier L, Roux F, Kleisbauer JP (2004) Prognostic value of combination of Cyfra 21-1, CEA and NSE in patients with advanced non-small cell lung cancer. Respir Med 98:357–362

Barreda-García S et al (2015) Attomolar quantitation of Mycobacterium tuberculosis by asymmetric helicase-dependent isothermal DNA-amplification and electrochemical detection. Biosens Bioelectron 68:122–128

Bernacka-Wojcik I et al (2013) Bio-microfluidic platform for gold nanoprobe based DNA detection—application to Mycobacterium tuberculosis. Biosens Bioelectron 48:87–93

Blijenberg BG, Kranse R, Eman I, Schroder FH (1996) Some analytical considerations on the measurement of prostate-specific antigen. Eur J Clin Chem Clin Biochem 34:817–821

Borsa BA et al (2016) Staphylococcus aureus detection in blood samples by silica nanoparticle-oligonucleotides conjugates. Biosens Bioelectron 86:27–32

Capobianco JA, Shih WY, Yuan Q, Adams GP, Shih W (2008) Label-free, all-electrical, in situ human epidermal growth receptor 2 detection. Rev Sci Instrum 79:076101

Chamorro-Garcia A, Merkoçi A (2016) Nanobiosensors in diagnostics. Nanobiomedicine 3: 1849543516663574

Chen Y et al (2017) A polyaniline-reduced graphene oxide nanocomposite as a redox nanoprobe in a voltammetric DNA biosensor for Mycobacterium tuberculosis. Microchim Acta 184:1801–1808

Cheng MS et al (2011) Membrane-based electrochemical nanobiosensor for Escherichia coli detection and analysis of cells viability. Environ Sci Technol 45:6453–6459

Choi Y, Kendziorski C (2009) Statistical methods for gene set co-expression analysis. Bioinformatics 25:2780–2786

Cohen JD, Li L, Wang Y, Thoburn C, Afsari B, Danilova L, Douville C, Javed AA, Wong F, Mattox A, Hruban RH, Wolfgang CL, Goggins MG, Dal Molin M, Wang TL, Roden R, Klein AP, Ptak J, Dobbyn L, Schaefer J, Silliman C, Diaz L, Tomasetti C, Kinzler KW, Vogelstein B, Lennon AM, Papadopoulos N (2018) Detection and localization of surgically resectable cancers with a multianalyte blood test. Science 359:926–930

Dannenberg AM Jr (1982) Pathogenesis of pulmonary tuberculosis. Am Rev Respir Dis 125:25–30

Das R et al (2019) Aptamer-mediated colorimetric and electrochemical detection of Pseudomonas aeruginosa utilizing peroxidase-mimic activity of gold NanoZyme. Anal Bioanal Chem 411: 1229–1238

De Bentzmann S, Plésiat P (2011) The Pseudomonas aeruginosa opportunistic pathogen and human infections. Environ Microbiol 13:1655–1665

Fritz H (2009) Schröder, screening and prostate-cancer mortality in a randomized European study. N Engl J Med 360:1320–1328

Gdowski A, Ranjan AP, Mukerjee A, Vishwanatha JK (2014) Nanobiosensors: role in cancer detection and diagnosis. In: Infectious diseases and nanomedicine I. Springer, New Delhi, pp 33–58

Ghorbani F, Abbaszadeh H, Mehdizadeh A, Ebrahimi-Warkiani M, Rashidi MR, Yousefi M (2019) Biosensors and nanobiosensors for rapid detection of autoimmune diseases: a review. Microchim Acta 186:838

Gokce ZG et al (2011) Impedimetric DNA biosensor based on polyurethane/poly (m-anthranilic acid) nanofibers. Sens Actuators B: Chem 254:719–726

Gu B, Xu C, Yang C, Liu S, Wang M (2011) ZnO quantum dot labeled immunosensor for carbohydrate antigen 19–9. Biosens Bioelectron 26:2720–2723

Haes AJ, Hall WP, Chang L, Klein WL, Van Duyne RP (2004) A localized surface plasmon resonance biosensor: first steps toward an assay for Alzheimer's disease. Nano Lett 4:1029–1034

Hao R, Wang D, Zuo G, Wei H, Yang R, Zhang Z, Zhou Y (2009) Rapid detection of Bacillus anthracis using monoclonal antibody functionalized QCM sensor. Biosens Bioelectron 24: 1330–1335

He X et al (2014) Aptamer-fluorescent silica nanoparticles bioconjugates based dual-color flow cytometry for specific detection of Staphylococcus aureus. J Biomed Nanotechnol 10:1359–1368

Herschkovitz Y, Eshkenazi I, Campbell CE, Rishpon J (2000) An electrochemical biosensor for formaldehyde. J Electroanal Chem 491:182–187

Ho JA, Chang HC, Shih NY, Wu LC, Chang YF, Chen CC, Chou C (2010) Diagnostic detection of human lung cancer-associated antigen using a gold nanoparticle-based electrochemical immunosensor. Anal Chem 82:5944–5950

Homola J (2003) Present and future of surface plasmon resonance biosensors. Anal Bioanal Chem 377:528–539

Hossain SZ et al (2012) Multiplexed paper test strip for quantitative bacterial detection. Anal Bioanal Chem 403:1567–1576

Jang HS et al (2009) Optical fiber SPR biosensor with sandwich assay for the detection of prostate specific antigen. Opt Commun 282:2827–2830

Kaewphinit T, Santiwatanakul S, Chansiri K (2013) Colorimetric DNA based biosensor combined with loop-mediated isothermal amplification for detection of mycobacterium tuberculosis by using gold nanoprobe aggregation. Sens Transducers 149:123

Kaper JB, Nataro JP, Mobley HL (2004) Pathogenic Escherichia coli. Nat Rev Microbiol 2:123–140

Karaballi R et al (2015) Development of an electrochemical surface-enhanced Raman spectroscopy (EC-SERS) aptasensor for direct detection of DNA hybridization. Phys Chem Chem Phys 17: 21356–21363

Kaur G et al (2015) Pyrimidine-based functional fluorescent organic nanoparticle probe for detection of Pseudomonas aeruginosa. Org Biomol Chem 13:4673–4679

Kim G et al (2015) A microfluidic nano-biosensor for the detection of pathogenic Salmonella. Biosens Bioelectron 67:243–247

Kinugasa H, Nouso K, Miyahara K, Morimoto Y, Dohi C, Tsutsumi K, Kato H, Matsubara T, Okada H, Yamamoto K (2015) Detection of K-ras gene mutation by liquid biopsy in patients with pancreatic cancer. Cancer 121:2271–2280

Kobayashi H, Ogawa M, Alford R, Choyke PL, Urano Y (2009) New strategies for fluorescent probe design in medical diagnostic imaging. Chem Rev 110:2620–2640

Kumagai M, Kano MR, Morishita Y, Ota M, Imai Y, Nishiyama N et al (2009) Enhanced magnetic resonance imaging of experimental pancreatic tumor in vivo by block copolymercoated magnetite nanoparticles with TGF-beta inhibitor. J Control Release 140:306–311

Labib M et al (2012) Aptamer-based viability impedimetric sensor for bacteria. Anal Chem 84: 8966–8969

Lee C-W et al (2019) Ultra-sensitive electrochemical detection of bacteremia enabled by redox-active gold nanoparticles (raGNPs) in a nano-sieving microfluidic system (NS-MFS). Biosens Bioelectron 133:215–222

Li C-z et al (2011) Paper based point-of-care testing disc for multiplex whole cell bacteria analysis. Biosens Bioelectron 26:4342–4348

Lilach B et al (2010) Carbon nanotubes based electrochemical biosensor for detection of formaldehyde released from a cancer cell line treated with formaldehyde-releasing anticancer prodrugs. Bioelectrochemistry 77:94–99

Lin Y-Y et al (2008) A nanoparticle label/immunochromatographic electrochemical biosensor for rapid and sensitive detection of prostate-specific antigen. Biosens Bioelectron 23:1659–1665

Liu C et al (2014) An electrochemical DNA biosensor for the detection of Mycobacterium tuberculosis, based on signal amplification of graphene and a gold nanoparticle–polyaniline nanocomposite. Analyst 139:5460–5465

Liu P et al (2016) Gold nanoprobe functionalized with specific fusion protein selection from phage display and its application in rapid, selective and sensitive colorimetric biosensing of Staphylococcus aureus. Biosens Bioelectron 82:195–203

Lowy FD (1998) Staphylococcus aureus infections. N Engl J Med 339:520–532

Luo Y et al (2010) Surface functionalization of electrospun nanofibers for detecting E. coli O157: H7 and BVDV cells in a direct-charge transfer biosensor. Biosens Bioelectron 26:1612–1617

Madumali K et al (2018) Early detection of pancreatic cancers in liquid biopsies by ultrasensitive fluorescence nanobiosensors. Nanomed: Nanotechnol Biol Med 14:1823–1832

Maksim S, Andrey K (2006) Plasmon excitation by the Gaussian-like core mode of a photonic crystal waveguide. Opt Express 14:8419–8424

Malhotra BD, Chaubey A (2003) Biosensors for clinical diagnostics industry. Sens Actuators B: Chem 91:117–127

Manajit O et al (2018) Development of uracil-DNA-glycosylase-supplemented loop-mediated isothermal amplification coupled with nanogold probe (UDG-LAMP-AuNP) for specific detection of Pseudomonas aeruginosa. Mol Med Rep 17:5734–5743

Matsishin M et al (2017) Selective amplification of SPR biosensor signal for recognition of rpob gene fragments by use of gold nanoparticles modified by thiolated DNA. Nanoscale Res Lett 12: 1–6

Mihi Y et al (2009) Effects of bisphenol A on breast cancer and its risk factors. Arch Toxicol 83: 281–285

Minnaei ME et al (2016) Label-free, PCR-free DNA hybridization detection of Escherichia coli O157: H7 based on electrochemical nanobiosensor. Electroanalysis 28:2582–2589

Miodek A et al (2015) E-DNA sensor of Mycobacterium tuberculosis based on electrochemical assembly of nanomaterials (MWCNTs/PPy/PAMAM). Anal Chem 87:9257–9264

Mittal S et al (2017) Biosensors for breast cancer diagnosis: a review of bioreceptors, biotransducers and signal amplification strategies. Biosens Bioelectron 88:217–231

Mogha NK et al (2018) Reduced graphene oxide nanoribbon immobilized gold nanoparticle based electrochemical DNA biosensor for the detection of Mycobacterium tuberculosis. J Mater Chem B 6:5181–5187

Mohamad FS et al (2017) Synthesis and characterization of polyaniline/graphene composite nanofiber and its application as an electrochemical DNA biosensor for the detection of mycobacterium tuberculosis. Sensors 17:2789

Moongkarndi P, Rodpai E, Kanarat S (2011) Evaluation of an immunochromatographic assay for rapid detection of Salmonella entericaserovars Typhimurium and Enteritidis. J Vet Diagn Invest 23:797–801

Morris Stiff G, Taylor MA (2012) Cal9-9 and pancreatic cancer: is it really that good? J Gastrointest Oncol 3:88–89

Mostafa A et al (2016) An electrochemical nanobiosensor for plasma miRNA-155, based on graphene oxide and gold nanorod, for early detection of breast cancer. Biosens Bioelectron 77:99–106

Mukherjee MD et al (2015) Facile synthesis of 2-dimensional transparent graphene flakes for nucleic acid detection. Sens Actuators B: Chem 210:281–289

Mulpur P et al (2015) Flexible Ag–C 60 nano-biosensors based on surface plasmon coupled emission for clinical and forensic applications. Phys Chem Chem Phys 17:25049–25054

Mura S et al (2012) FTIR nanobiosensors for Escherichia coli detection. Beilstein J Nanotechnol 3: 485–492

Murphy TF et al (2008) Pseudomonas aeruginosa in chronic obstructive pulmonary disease. Am J Respir Crit Care Med 177:853–860

Nadzirah S et al (2020) Titanium dioxide–mediated resistive nanobiosensor for E. coli O157: H7. Microchim Acta 187:1–9

Ng SP, Qiu G, Ding N, Lu X, Wu CML (2017) Biosens Bioelectron 89:468–476

Ohman DE, Chakrabarty A (1981) Genetic mapping of chromosomal determinants for the production of the exopolysaccharide alginate in a Pseudomonas aeruginosa cystic fibrosis isolate. Infect Immun 33:142–148

Ozkan H, Kaya M, Cengiz A (2003) Comparison of tumor marker CA 242 with CA 19-9 and carcinoembryonic antigen (CEA) in pancreatic cancer. Hepatogastroenterology 50:1669–1674

Pan et al (2017) Electrochemical biosensor to simultaneously detect VEGF and PSA for early prostate cancer diagnosis based on graphene oxide/ssDNA/ PLLA nanoparticles. Biosens Bioelectron 89:598–605

Pandit S, Dasgupta D, Dewan N, Prince A (2016) Nanotechnology based biosensors and its application. Pharma Innov 5:18

Pathak P, Katiyar VK, Giri S (2007) Cancer research-nanoparticles, nanobiosensors and their use in cancer research. J Nanotechnol 3:1–4

Perumal V et al (2018) Gold nanorod embedded novel 3D graphene nanocomposite for selective bio-capture in rapid detection of Mycobacterium tuberculosis. Biosens Bioelectron 116:116–122

Prabowo BA et al (2016) Graphene-based portable SPR sensor for the detection of Mycobacterium tuberculosis DNA strain. Procedia Eng 168:541–545

Prasad M, Lambe UP, Brar B, Shah I, Manimegalai J, Ranjan K, Iqbal HM (2018) Nanotherapeutics: an insight into healthcare and multi-dimensional applications in medical sector of the modern world. Biomed Pharmacother 97:1521–1537

Rabti A, Raouafi A, Raouafi N (2020) DNA markers and nano-biosensing approaches for tuberculosis diagnosis. In: Nanotechnology based approaches for tuberculosis treatment, pp 207–230

Ranjbar S, Shahrokhian S, Nurmohammadi F (2018) Nanoporous gold as a suitable substrate for preparation of a new sensitive electrochemical aptasensor for detection of Salmonella typhimurium. Sens Actuators B: Chem 255:1536–1544

Rydzek G, Ji Q, Li M, Schaaf P, Hill JP, Boulmedais F, Ariga K (2015) Electrochemical nanoarchitectonics and layer-by-layer assembly: from basics to future. Nano Today 10:138–167

Sagadevan S, Periasamy M (2014) Recent trends in nanobiosensors and their applications-a review. Rev Adv Mater Sci 36:62–69

Sakao Y, Tomimitsu S, Takeda Y, Natsuaki M, Itoh T (2004) Carcinoembryonic antigen as a predictive factor for postoperative tumor relapse in early-stage lung adenocarcinoma. Eur J Cardio-thoracic Surg 25:520–522

Salam F, Uludag Y, Tothill IE (2013) Real-time and sensitive detection of Salmonella Typhimurium using an automated quartz crystal microbalance (QCM) instrument with nanoparticles amplification. Talanta 115:761–767

Sarabaegi M, Roushani M (2019) A nano-sized chitosan particle based electrochemical aptasensor for sensitive detection of P. aeruginosa. Anal Methods 11:5591–5597

Sarkar P et al (2002) Amperometric biosensors for detection of the prostate cancer marker (PSA). Int J Pharm 238:1–9

Schiller et al (2002) Comparison of four chemotherapy regimens for advanced non-small-cell lung cancer. N Engl J Med 346:92–98

Shahbazi R et al (2018) Highly selective and sensitive detection of Staphylococcus aureus with gold nanoparticle-based core-shell nano biosensor. Mol Cell Probes 41:8–13

Shandilya R, Bhargava A, Bunkar N, Tiwari R, Goryacheva IY, Mishra PK (2019) Nanobiosensors: point-of-care approaches for cancer diagnostics. Biosens Bioelectron 130:147–165

Shangguan J et al (2015) A combination of positive dielectrophoresis driven on-line enrichment and aptamer-fluorescent silica nanoparticle label for rapid and sensitive detection of Staphylococcus aureus. Analyst 140:4489–4497

Song Y et al (2014) Nicking enzyme-assisted biosensor for Salmonella enteritidis detection based on fluorescence resonance energy transfer. Biosens Bioelectron 55:400–404

Srivastava SK et al (2015) Highly sensitive and specific detection of E. coli by a SERS nanobiosensor chip utilizing metallic nanosculptured thin films. Analyst 140:3201–3209

Srivastava AK, Dev A, Karmakar S (2018) Nanosensors and nanobiosensors in food and agriculture. Environ Chem Lett 16:161–182

Stephan C, Klaas M, Muller C, Schnorr D, Loening S, Jung K (2006) Interchangeability of measurements of total and free prostate-specific antigen in serum with 5 frequently used assay combinations: an update. Clin Chem 52:59–64

Suaifan GA, Alhogail S, Zourob M (2017) Paper-based magnetic nanoparticle-peptide probe for rapid and quantitative colorimetric detection of Escherichia coli O157: H7. Biosens Bioelectron 92:702–708

Sun H et al (2009) Nano-silver-modified PQC/DNA biosensor for detecting E. coli in environmental water. Biosens Bioelectron 24:1405–1410

Sun J et al (2015) DNA biosensor-based on fluorescence detection of E. coli O157: H7 by Au@ Ag nanorods. Biosens Bioelectron 70:239–245

Suter JD, Sun Y, Howard JD, Viator J, Fan X (2008) PDMS embedded opto-fluidic microring resonator lasers. Opt Express 16:10248–10253

Tamada K, Nakamura F, Ito M, Li X, Baba A (2007) SPR-based DNA detection with metal nanoparticles. Plasmonics 2:185–191

Tang Y et al (2013) Highly sensitive and rapid detection of Pseudomonas aeruginosa based on magnetic enrichment and magnetic separation. Theranostics 3:85

Thiruppathiraja C et al (2011) Specific detection of Mycobacterium sp. genomic DNA using dual labeled gold nanoparticle based electrochemical biosensor. Anal Biochem 417:73–79

Torres-Chavolla E, Alocilja EC (2011) Nanoparticle based DNA biosensor for tuberculosis detection using thermophilic helicase-dependent isothermal amplification. Biosens Bioelectron 26: 4614–4618

Tsai TT et al (2017) Diagnosis of tuberculosis using colorimetric gold nanoparticles on a paper-based analytical device. ACS Sens 2:1345–1354

Varshney M, Li Y (2007) Interdigitated array microelectrode based impedance biosensor coupled with magnetic nanoparticle–antibody conjugates for detection of Escherichia coli O157: H7 in food samples. Biosens Bioelectron 22:2408–2414

Vianello F, Boscolo-Chio R, Signorini S, Rigo A (2007) On-line detection of atmospheric formaldehyde by a conductometric biosensor. Biosens Bioelectron 22:920–925

Viter R et al (2017) Toward development of optical biosensors based on photoluminescence of TiO2 nanoparticles for the detection of Salmonella. Sens Actuators B: Chem 252:95–102

Wang J et al (2015) Magnetically assisted surface-enhanced Raman spectroscopy for the detection of Staphylococcus aureus based on aptamer recognition. ACS Appl Mater Interfaces 7:20919–20929

Wertheim HF et al (2005) The role of nasal carriage in Staphylococcus aureus infections. Lancet Infect Dis 5:751–762

Woodward DL, Khakhria R, Johnson WM (1997) Human salmonellosis associated with exotic pets. J Clin Microbiol 35:2786–2790

Xia S et al (2016) Developing a novel immunochromatographic test strip with gold magnetic bifunctional nanobeads (GMBN) for efficient detection of Salmonella choleraesuis in milk. Food Control 59:507–512

Xiang Y et al (2015) Real-time monitoring of mycobacterium genomic DNA with target-primed rolling circle amplification by a Au nanoparticle-embedded SPR biosensor. Biosens Bioelectron 66:512–519

Xu X-W et al (2016) Detection EGFR exon 19 status of lung cancer patients by DNA electrochemical biosensor. Biosens Bioelectron 80:411–417

Yang L, Mao H, Cao Z, Wang YA, Peng X, Wang X et al (2009) Molecular imaging of pancreatic cancer in an animal model using targeted multifunctional nanoparticles. Gastroenterology 136: 1514–1525

Yuan J et al (2014) A visual detection method for Salmonella typhimurium based on aptamer recognition and nanogold labeling. Food Control 37:188–192

Zaman M, Baral TN, Jakubek ZJ, Zhang J, Wu X, Lai E et al (2011) Single-domain antibody bioconjugated near-IR quantum dots for targeted cellular imaging of pancreatic cancer. J Nanosci Nanotechnol 11:3757–3763

Zelada-Guillén GA et al (2012) Label-free detection of Staphylococcus aureus in skin using real-time potentiometric biosensors based on carbon nanotubes and aptamers. Biosens Bioelectron 31:226–232

Zeng D et al (2017) DNA tetrahedral nanostructure-based electrochemical miRNA biosensor for simultaneous detection of multiple miRNAs in pancreatic carcinoma. ACS Appl Mater Interfaces 9:24118–24125

Zhao J et al (2013) An amperometric biosensor for the detection of hydrogen peroxide released from human breast cancer cells. Biosens Bioelectron 41:815–819

Zheng M et al (2015) Self-targeting fluorescent carbon dots for diagnosis of brain cancer cells. ACS Nano 9:11455–11461

Zhong Z et al (2018) Selective capture and sensitive fluorometric determination of Pseudomonas aeruginosa by using aptamer modified magnetic nanoparticles. Microchim Acta 185:377

Zuo P et al (2013) A PDMS/paper/glass hybrid microfluidic biochip integrated with aptamer-functionalized graphene oxide nano-biosensors for one-step multiplexed pathogen detection. Lab Chip 13:3921–3928

15 Bioactive Nanoparticles: A Next Generation Smart Nanomaterials for Pollution Abatement and Ecological Sustainability

Showkat Mir, Nirius Jenan Ekka, Binata Nayak, and Iswar Baitharu

Abstract

The level of environmental pollution is increasing rapidly with increased urbanization and rapid industrialization across the globe. To abate pollution, there is utmost necessity to develop technology that can monitor, detect and clean contaminants from the air, water and soil with higher efficiency. Recently nanotechnology has emerged as a highly effective and reliable technique that offers a wide range of capabilities to improve the quality of existing environment. Due to its large surface area, the nanoparticles adsorb large amount of pollutants at a much faster rate. Nanomaterials can reach to inaccessible areas making in-situ remediation of pollutants effective. Coating of nanomaterials with various ligands provides opportunities to develop sensor with high selectivity and specificity toward pollutants. However, nanomaterials used for pollution abatement can itself cause environmental pollution. There are limited studies exploring the fate of nanomaterials after their end use. Nanotoxicological studies conducted so far indicate damaging impact of nanomaterials in ecological functioning and maintenance of ecosystem integrity. Bioactive nanoparticles on the other hand are biodegradable, have shorter life span and minimal negative impact on environment. Although application of bioactive nanomaterials in environmental pollution abatement is in its infancy, it is gradually gaining wider acceptance in pollution management because of its promising potential.

S. Mir · N. J. Ekka · B. Nayak
School of Life Sciences, Sambalpur University, Burla, Odisha, India

I. Baitharu (✉)
P.G. Department of Environmental Sciences, Sambalpur University, Jyoti Vihar, Burla, Odisha, India
e-mail: iswarbaitharu@suniv.ac.in

M. Arakha et al. (eds.), *Bio-Nano Interface*,
https://doi.org/10.1007/978-981-16-2516-9_15

Keywords

Bioactive nanomaterials · Pollution abatement · Ecological functioning

15.1 Introduction

Ever increasing industrialization and urbanization has filled the world with number of toxic chemicals such as carbon monoxide (CO), chlorofluorocarbons (CFCs), heavy metals, hydrocarbons, nitrogen oxides, organic compounds, sulfur dioxide and particulates. Water bodies are polluted with fertilizers, herbicides, pesticides and by-products of number of industries across the globe. Contaminants are mostly found mixed in the air, water and soil (Ang et al. 2005). Environmental pollution is one of the most challenging problem, the modern world is confronting now. As reported by The World Health Organization, every year 270,000 children loss their life due to lack of clean water and acute air pollution. In low income countries having poor technology and resources to deal with emission of VOCs such as polycyclic Aromatic Hydrocarbons (PAHs), people are getting exposed to number of carcinogenic/mutagenic contaminants in air above their acceptable level causing increased mortality from cancer in the community (Montero-Montoya et al. 2018).

Conventional methods such as adsorption, biological remediation, chemical oxidation, ion exchange, chemical precipitation, electrochemical treatment, membrane filtration, reverse osmosis, coagulation, extraction and irradiation are used primarily for remediating environmental pollutions (Ouyang et al. 2019). However, these methods have their own disadvantages such as the use of large amounts of reagents for precipitation methods, high operational costs for ion exchange methods, low adsorption capacity and selectivity for adsorption methods. In this context, nanotechnology offers a wide range of capabilities and technologies to improve the quality of existing environment by enhancing the performance of pollution control technologies and reducing their cost of operation. Though the implication of nanomaterials in pollution remediation is still in its infancy, people across the globe have begun to recognise the promising potential of nanotechnology. Because of the nano scale size of the nanomaterials and high surface area to volume ratio, it can be used for detection of sensitive environmental pollutants (Willner and Vikesland 2018).

As an alternative to conventional materials, nanomaterials can provide new opportunities to cope with these challenges. Nanomaterials can be used for the preparation of excellent adsorbents, catalysts, and sensors due to their unique properties, including large surface area to volume ratio, high reactivity, reduced size, strong sorption, active surface sites, specific interaction with contaminants, etc. With the use of nanomaterials, polymer nanocomposites can be obtained with improved sorption, removal and filtration properties when they are used in environmental applications. In environmental remediation applications, iron oxides, titanium dioxide, cadmium sulphide, etc. are used as nanomaterials and PVP, PMMA, CMC, PANI, PHB, alginate, etc. are used as polymers to prepare nanocomposites

(Guerra et al. 2018). These materials can be blended using mainly two methods as direct compounding and in situ synthesis. The existing technologies used for pollution remediation though are effective but the cost of their implementation in different sectors is prohibitively high. It is realized now that there is a dare necessity for development of technology capable of monitoring, detecting and cleaning the contaminants from the air, water and soil in ecofriendly, cost effective and sustainable manner.

The number of new nanomaterials with specific desired characteristics is now designed, synthesized and produced in large-scale industrially. Currently, it has been observed that the conventional engineering environmental nanomaterials (EENM) lose their functionalities when subjected to changing environmental condition making them non-functional (Chang et al. 2018). To overcome such problems, smart environmental bio-nanomaterials (SEBN) having capacity to self adjust under changing environmental conditions are now been designed and synthesized. When the concentration of pollutants is above the MPL, state-of-the-art nanomaterials could be used, such as bio-inspired self-healing nanomaterials without external stimuli, not energy input, among others. However, there is also growing awareness of the need to understand and characterize the properties of ENMs as they change from the time of synthesis to their final state during application and possible release in the environment (Karakoti et al. 2012). Nanomaterials generally do not retain the same properties from their point of synthesis to their state of application and both the particle processing and storage histories often are poorly documented.

15.2 Nanomaterials as Environmental Pollutants

In the last one decade, the application of nanomaterials in the field of medicines, cosmetics, electronic devices and number of consumer products has seen a spectacular rise resulting in the generation of emerging class of environmental pollutants. As it is not known with certainty how nanomaterials after their end use behave in the environmental compartments like air, water and soil, appropriateness of the existing regulations for chemical environmental pollutants is debatable (Gupta and Xie 2018). Significant progress in understanding the roles of various factors influencing the fate and transport of nanomaterials in environmental compartments have been made in the last few years. Researches demonstrate that the nanomaterials behave differently in aquatic environment as a dissolved chemicals compared to their solid or colloidal counterpart. However, there are not enough investigations made to understand the structure and activity relation of different classes of nanomaterials in different condition. How one nanomaterial interacts with other nanomaterial present in the same environmental compartment with respect to their toxicity on organism living within is still an unanswered question requiring further investigations. There is necessity of establishing predictive models for nanomaterials in environment to suitably access and manage the risk associated with exposure to various nanomaterials singly or in combination.

Nanomaterials are known to undergo significant settling under normal gravitational condition in various environmental media. Analytical techniques show that they also exhibit reduced diffusivity compared to other dissolved species. In the absence of gravitational and inertial impaction, air/water, air/soil and water/soil inter medium transport of nanomaterials is known to be governed by diffusive processes. Eco-toxicity by nanoparticles is a major concern in present scenario and gains the interest of scientists and researchers. These nano-materials may drift-off from the manufactured site to alternate places like water bodies and agricultural fields through physical and natural processes. Pollution by nanomaterial's in air, water and soil cause adverse effects on terrestrial and aquatic eco-system. It also cause severe damage to micro and macro habitat through dermal contact, inhalation or by pore penetrations (Hoet et al. 2004). The industrial waste approximately of nano range is common nano-sized material which causes a major exploitation of physcio-chemical properties of living creature and existence of nature. The nature of nanoparticles depends upon techniques used during synthesis procedure, chemical and physical synthesized nano material is highly toxic as compared to biologically synthesized material (Das et al. 2017). It has been observed that atmospheric nanomaterials can have residence time of about twenty days and their aggregate may not cause any effect on human respiratory system. Similarly exposure assessment to nanomaterials in water may not always bring identical result as nanomaterials particle stabilize in aquatic system (Ray et al. 2009). Nevertheless there is very few research to establish the environmental pollution by nanomaterials barring a handful of modeling studies that have investigated ENM release to the environment. The major source of release of nanomaterials to environmental compartments is mostly sewage sludge, wastewater, and waste incineration of products containing ENM. However, there is lack of proper evidence on release of ENM during their production and application in various fields. No quantitative information is available till now linking occupational exposure and emission flow of ENM into environment.

15.2.1 Monitoring of Nanowastes in Environment

The traditional methodologies used for quantification of nanomaterials during their manufacturing cannot be applied adequately while analysing nanomaterials in environmental sample. There are no available standard methods for analysis of nanomaterials in environmental sample especially their fate, transport and toxicological effect on living being. The existing methods for detection and quantification of nanomaterials cannot differentiate the fraction of the nanomaterials which are generated and released to environment through natural processes from the manufactured nanomaterials that are released to the environmental compartment because of their application in different sectors by human being (Laux et al. 2018). The traditional analytical techniques proved inadequate for analysing the physicochemical forms of ENMs. The effects of the surrounding medium on most of the nanomaterials in environmental compartment that influence their properties make analysis inaccurate and difficult to understand because of the artefact effect (Lin et al.

2014). Now it is high time to develop appropriate technique for extraction, cleanup, separation, and sample storage inducing minimal artefacts, increase sensitivity and add specificity of analytical techniques.

15.3 Bio-nanomaterials as a Degradable Smart Option in Pollution Abatement

Last few years has seen rapid growth in the use of nanomaterials in different fields. Today, because of widespread applications of nanomaterials, several thousand tons of nanomaterials are manufactured daily which enters to environmental compartments after their end use (Buzea et al. 2007). However, most of these nanomaterials are non-biodegradable, toxic and are conservative in nature and hence tend to accumulate in environment (Sharifi et al. 2012) Unlike metallic and carbon based nanomaterials, bio-nanomaterials have shorter life, biodegradable in nature and easy to prepare (Mishra et al. 2018). Therefore bio-nanomaterials can be used as a safer alternative to the existing nanomaterials based technologies without compromising with their performance.

15.3.1 Bioactive Nanoparticles for Water Pollution

There are several techniques popularly used at present to remediate water pollution in different sectors. Technology like Electro coagulation, though effective in removing fluoride and other contaminants from water, sometime leaching of the coagulants into the treated water during treatment process make the water toxic (Kabdaşlı et al. 2012; Tetteh and Rathilal 2020). Moreover, it leads to the generation of toxic sludge which are very difficult to manage. Ion exchange techniques used for water treatment is highly expensive and its sensitivity to *p*H make it difficult to operate. The most effective water treatment technique at present is membrane based processes which are highly efficient in removing contaminants. But its drawback is that it is prohibitively costly and energy intensive and sometime causes ionic imbalance in drinking water. In addition, it releases huge volume of water with higher concentration of initial contaminants. On the other hand, adsorption based processes are less efficient in removal of contaminants, lack reusability and require high energy input (Guillossou et al. 2020). The effort to overcome the disadvantages of existing water treatment techniques is underway across the globe. Scientists believe that nnanotechnology has the potential to address all of these drawbacks and can be a magical instrument to deal with the current and upcoming water crisis. Focussed researches are being undertaken all over the world to develop novel nanomaterials with high affinity, capacity, and selectivity toward contaminants (Table 15.1).

Current research intends to enhance the removal efficiency of bio-based materials used in water treatments processes by impregnating them with various nanomaterials. Activated bio char impregnated with metal oxide nanomaterials like Fe_2O_4, Al_2O_3, and ZrO_2 have been reported to be highly effective as an

Table 15.1 List of the selected nanomaterials used popularly for pollution abatement

Nanomaterial	Type of pollutant removed	Origin of removed pollutant material	References
Graphene oxide(GO)	Removes cationic, anionic, and/or amphiphilic and detection of chlorophenol	Water	Scalese et al. (2016) and Yi et al. (2019)
Chitosan and alginate coated with silver nanoparticles	Highly efficient in removal of fluoride and chromium	Water	Kumar et al. (2017)
Bioactive carbon nanotubes CNT	Helps in removal of chlorophenol and excessive fluoride, NO_X, SO_2 and CO_2	Water and air	Khin et al. (2012)
Nanocellulose	Removes heavy metals and commercial dyes	Water	Suman et al. (2015)
Bioactive silver nanoparticles	The Ag nanoparticles removes hazardous dyes like Safranine O, Methyl red, Methyl orange and Methylene blue	Industrial waste management	Jyoti and Singh (2016)
Bioactive graphene, fullerene, and nanotubes	Acts as an adsorbent for polyaromatic hydrocarbons PAHs	Soil and water	Wang et al. (2014)
Bioactive TiO_2	Integrated with air pollution devices also have a potential to remove benzene, toluene, ethylbenzene and xylene (BTEX)	Air and water	Patel et al. (2020)
Bioactive ZnO	Acts as a photocatalyst and remediate chlorocatechol	Air	El Golli et al. (2021)
Single walled (SWNTs) and multi walled nano-tubes MWNTs	Acts as a sensor for NO_2 and NH_3	Air	Pandhi et al. (2020)
Zero-valent (nZVI)	This may highly remove Cu^{2+}, chlorinated hydrocarbons, $CrO2^{-2}$ and $NO3^{-}$	Water and soil	Zhao et al. (2016)
Manganese oxide conjugated with gold nanoparticles	Japanese technology which are used to efficiently remove VOCs, nitrogen and sulfur oxides at room temperature	Air	Sinha and Suzuki (2007)
SeNps	It helps in removal of bacteria (A. faecalis SeO_3) also acts as an friendly candidate for bioremediation	Soil and water	Sakr et al. (2018)

adsorbent in removing contaminants from water compared to bio char alone (Pathak et al. 2003). Several studies reported enhanced removal efficiency of cellulose by trapping MnO_2 nanoparticles in the cellulose matrix (Maliyekkal et al. 2010). Impregnation of aluminium ions with activated carbon fibers and carbon nanofibers have been reported to elevate their fluoride removal efficiency manifold (Gupta et al. 2009). Doping of ferric oxide nanoparticles to carbon nanotubes improved the pollutant removal from water with higher efficiency. Magnetic alginate beads had very high defluoridation capacities (Gao et al. 2014). This material is not leachable and due to high magnetic sensitivity, the materials can be separated easily from the aqueous solution. The major advantages of these adsorbents are that their effectiveness remains intact over a wide range of *p*Hs. However, the major drawback of such adsorbents is that there is possibility of release of nanomaterials to treated water and their synthesis is energy intensive in nature.

Recently multipurpose bionanomaterials-based scaffold are designed to deal with water pollution effectively. Alginate and chitosan are mixed to prepare the scaffold and aluminium ion is added to it. The surface of the scaffold is coated with silver nanoparticles. The scaffold effectively removed fluoride, chromium and several reactive dyes mixed with water with remarkable efficiency. Regeneration of the scaffold after use depends upon whether the pollutants in water form covalent linkage with scaffold or not. Pollutants forming covalent bonds with scaffold make it difficult to regenerate and hence limit its reusability. For Cr (VI) and dye, the scaffold can be regenerated easily but fluoride form covalent bond with scaffold making it difficult to regenerate (Kumar et al. 2017).

Cleansing of polluted ground water can be achieved easily using nanotechnology because application of nanoparticles into the underground source of water is cheaper compared to pumping water for treatment. Nanosized fibre can be used as elctrode for efficient deionisation of water with very little investment and energy input (NRC 2006). In a recent study, starch-based nanocomposites (starch/SnO2) were used for removal of Hg2+ from an aqueous medium, and the removal percentage reached up to 97% (Thines et al. 2017). Nanofibre filter which has ability to filter out virus from drinking water has already been developed (Gupta et al. 2007). Ion exchange resins made up of bionanomaterials with nanosized pore on the surface to trap and exchange ions have already developed and in use for separation, purification and decontamination of water (Zhang and Liu 2020). These resins are widely used for softening of hard water. They also remove toxic heavy metals from water and replace them with sodium and potassium.

15.3.2 Bio-nanomaterials as a Degradable Smart Option in Air Pollution

Around 4.2 million people lost their life and 103.1 million people became disable because of exposure to particulate matter $PM_{2.5}$ world wide in 2015 as per WHO report (Oberdorster et al. 2007). Pollutants like volatile organic compounds are the most common contaminant in both indoor and outdoor air. VOCs like Polycyclic aromatic hydrocarbons (PAHs) which are released to air from combustion processes are highly toxic, mutagenic and carcinogenic in nature. Therefore, appropriate technical advancement must be achieved rapidly and efficiently to reduce these contaminants in air to a safer level and in a cost-effective manner (Gómez et al.

2012). Applications of nanocatalyst that can speed up the chemical reactions transforming harmful pollutants into harmless gases are gaining popularity in pollution control. Currently nanofiber catalyst made of manganese oxide is used popularly to remove VOCs from industrial smokestacks (Kuleyin 2007). Nanostructured membranes having pore size small enough to separate methane or carbon dioxides are used near the automobile exhaust. Researchers are now developing carbon nanotubes (CNT) to trap GHGs emitted from mining and thermal power industries. These nanotubes can trap GHGs much faster than any other conventional methods. Therefore, CNTs can purify large volume of air effictively within small time. Studies conducted in Japan in 2006 demonstrated a technique to convert soot filtered out of diesel fuel emissions into single walled CNT filter through laser vaporisation process (Uchida et al. 2006). This technique resulted in no waste because essentially the waste generated out of filtration process became the filter.

Wide range of toxic chemical pollutants are removed from air and water using polyamide nanofilter fabricated with MgO, TiO2 and other oxides (Ibrahim et al. 2016). The decontamination of toxic nerve gases such as paragon from air has been achieved with higher efficiency using electrospun nanofibres. Environmental remediation using polymer-supported metal and metal oxides mostly silver, iron, aluminum, titanium and magnesium are more widely used because of their high reactivity and ability catalyze degradation of air pollutants (Yang et al. 2019). Semiconductor metal oxides such as titanium dioxide (TiO2), zinc oxide (ZnO), tin dioxide (SnO2), and copper oxide (CuO), are promising materials for photocatalytic degradation of pollutants (Uddin et al. 2020).

15.4 Challenges in Synthesis of Bioactive Nanomaterials

Recent research has shown that green method for synthesis of nanomaterials is highly effective and low cost in nature. It has been observed that severe stress on environment is exerted when nanomaterials are synthesized using physical and chemical methods due to release of toxic metabolite in larger quantities. However, Bio based synthesis of nanomaterials are easy and less troublesome where metal salts is synthesized with desired plant extract and the process get completed within minutes to hours at room temperature. Bio nanomaterials of silver and gold metals synthesized using green synthesis process has been reported to be more stable and secure compared to their metallic counterpart. Large scale production of bio-nanomaterials using green synthesis protocol can be easily achieved and are cost effective (Singh et al. 2018). Nanomaterials synthesized using conventional methods where large number of toxic and harmful chemicals is used which make the composition of nanomaterials synthesized uncertian and could pose greater risk on human health and environment. Green methods of synthesis are significantly attractive because of their potential to reduce the toxicity of NPs. Accordingly, the use of vitamins, amino acids, plants extracts is being greatly popularized nowadays (Gour and Jain 2019).

In the recent decades, several research activities are being undertaken to prepare bio-composites by blending/reinforcing bio-nanomaterials in a wide variety of polymer matrices (Rossi et al. 2014; Uddin et al. 2012). The applications of nano-composites based on entirely renewable polymerase versatile (Yang et al. 2015). Unlike other nanomaterials, bionanomaterials can be synthesized easily using plant and animal resources (Mir et al. 2017). The synthesis of nano crystallites involves mechanical stirring and acid hydrolysis process Acid hydrolysis removes lower order regions. The resulting water insoluble high crystalline regime further gets converted to stable suspension via mechanical shearing (Rossi et al. 2014). There are significant challenges which need to be addressed are as follows:

(i) It is difficult to develop effective separation route suitable for extraction of nano-reinforcements from renewable resources,
(ii) It is hard to Find proper compatibility between nano-reinforcement and the polymer matrix and
(iii) It is essential to process the bio-nanocompsites using suitable techniques

Further commercialisation of bionanomaterials require the entire process to be cost effective and less dependent on high energy supplementation. Though at present nano-reinforcement in the polymer matrix is the widely used method for improving the properties bioolymers, the techniques itself is in the developmental phase and is not full proof. Bionanofiller have several advantages over number of commercially available nanofillers such as nanoclay, inorganicfillers, activated carbon, graphene and carbon nanotubes. Bionanofiller are renewable and biodegradable while nanofillers are not. Preparation of PVA composite films is done using solution casting method. The thermal stability and crystallisation propoerties of chitosan matrix remain unaffected when chitin is incorporated in the chitosan matrix (Grząbka-Zasadzińska et al. 2017). The tensile strength of the nanocomposite has been observed to increase with increase in chitin whiskers.

15.4.1 Green Synthesis from Enzymes and Vitamins

Enzymes because of their well defined structure and purity are best option for utilization in green synthesis of nanomaterials. Silver bionanomaterials are synthesized by combining them with enzymes involved in growth process of organisms. For synthesis of bimetallic nanomaterials Fe/Pd particles, enzymes were inserted into the lipid bilayer through electrostatic interactions (Gao et al. 2008). Extracellular amylase is utilized to generate Au NPs. It has been reported that large size silver nanomaterials were obtained when reduced with decreasing amount of beet juice. To get bimetallic nanomaterials, Green tea extracts is suitably used which produces nanomaterials Fe/Pd due to the fact that the extracts of green tea can act as both reductive as well as capping agent (Krishna et al. 2012). Use of natural agents in the field of nanosynthesis is big advancement toward green nanotechnology Vitamin B_2 has been used as reducing and capping agent in synthesis of Ag and palladium nanosphere and nanowires. Chitosan is popularly used as stabilizing agent because of its ability to bind strongly with metal ions along with Ascorbic acid as capping and reducing agent (Lu et al. 2010).

15.4.2 Green Synthesis Using Bacteria, Yeasts, Algae, Fungi and Actinomycetes

Psychrophilic bacteria *Pseudomonas antarctica,* , *Pseudomonas meridiana, Arthrobacter kerguelensis, Arthrobacter gangotriensis, Pseudomonas proteolytica, Bacillus indicus* and *Bacillus cecembensis* have been demonstrated to synthesize AgNPs which are highly stable and smaller in size (Shivaji et al. 2011). Silver nanocrystals of different compositions were successfully synthesized by *Pseudomonas stutzeri* AG259 (Klaus et al. 1999). Number of different species of yeast and fungi are used for synthesis of bionanomatterials. *F. oxysporum* metabolically transform Silver nitrate into well dispersed Ag oxide nanoparticles (Salem et al. 2015). *Alternaria alternata* are popularly used for synthesis of nano-platinum with partcile size varying from 2 to 30 nm (Sarkar et al. 2012). The fungus *Trichoderma viride* synthesize silver nanoparticles from silver nitrate through extracellular biosynthesis process (Elamawi et al. 2018). Very stable silver nanoparticles with size range of 5–15 nm have been reported to be synthesized using *Fusarium oxysporum* (Ahmad et al. 2003). Cyanobacterial Species like *L. majuscule, S. subsalsa, R. hieroglyphics, C. vulgaris, C. prolifera, P. pavonica, S. Platensis* and *S. fluitans* can be used as cost effective materials for bio recovery of metal and green synthesis of metal gold NPs (Bakir et al. 2018; Uma Suganya et al. 2015).

15.4.3 Green Synthesis Using Plants and Phytochemicals

The pomegranate was found to have the ability to produce more uniform size and shape NPs of Au and Ag in the range of 20–500 nm. *F. herba* isolate was used to reduce the platinum compound, the closeness of hydrogen and carbonyl in polyphenolic compound mainly goes about as fixing expert for metal particles (Dobrucka 2016). Formation of NPs could be completed in salt solution within short duration of time depending on the nature of plant extracts; the main reason being the concentration of the extracts, metal salt, pH and contact. It has been discovered that decrease of $AgNO_3$ to AgNPs by dihydroquercetin, quercetin and rutin prompted the development of an intensive surface plasmon resonance (SPR) band, which suggests reduction of this constituent (Veisi et al. 2018). Kou and Varma reported a simple, green and fast (complete within 5 min) approach for the construction of Ag NPs by MW irradiation using beet juice as a reducing reagent.

15.5 Future Prospective of Bioactive Nanomaterials in Pollution Abatement

The pollution of environments today has reached to its peak causing serious harm to the living beings on the earth. Toxic gases emitted from manufacturing industries and several sectors are now reaching to levels that directly deteriorate the quality of air. Relentless and unsustainable uses of numerous hazardous chemicals in agricultural sector are making the water bodies unsuitable for human use and are damaging the soil microbes that maintain the soil health. Increased cases of oil spillage from industries, warships, and leakage during transportation are major concerns that pollute the environment (Mohamed 2017). Under such condition, nanotechnology can prove itself a boon in pollution remediation. Nanomaterials can be used to develop an excellent environmental pollution cleaning system because of its unique structural characteristics. However the main concern in synthesis of nanomaterials requires the use of reagents, volatile organic and inorganic chemicals that are toxic in nature (Palit 2017). Although nanotechnology water treatment is far better and more efficient than conventional water treatment systems, the environmental fate of used nanomaterials, their transport in environmental compartments and their interaction with living beings are not well understood.

In the last few years, efforts to replace the toxic chemicals used in preparation of nanomaterials by safer and biologically conducive materials are underway. Regulatory bodies have already put stern regulations that ensure reduction of use of toxic chemicals during synthesis of nanomaterials so that their levels remain within safe limits in the environmental compartments. The greener route of synthesis of nanomaterials could be one of the options to get rid of these hazardous chemicals used in synthesis of nanomaterials. The biogenic method involves natural substances contain in plants, bacteria, algae, fungi, yeast, actinomycetes that act as reducing, capping and stabilizing agents for nanomaterials. Hence the biogenic method is both ecologically sustainable and economically viable. The manufacture of metallic nanomaterials using natural vitamins, polyphenols, proteins, amino acids and natural surfactants is slowly becoming more appropriate (Dhillon et al. 2012). As bionanomaterials have potential to clean and prevent environmental pollution and greener route of their production do not involve any toxic chemicals, nanotechnology involving bionanomaterials could be the next generation technology dealing with pollution abatement in near future.

15.6 Conclusion

Demand for clean and safe environment is growing worldwide with increased incidences of pollution related health issues in the community and development of awareness in public. Recent reports shows that there is a strong association between environmental chemicals and heavy metals pollutants in air, water and soil and rapid increase in the cases of cancer, lungs and kidney related diseases across the globe. Rapid progress in the nanotechnology-based pollution abatement techniques has generated a ray of hope for creating a world free of pollutants. Advancement of research and innovations in the field of nanotechnology could lead to the development of highly effective environmental pollution treatment techniques in the near future. Treatment of environmental pollution using bioactive nanomaterials not only ensure clean environment but also has the potential to overcome the hurdles of high cost and technical capacity for present and future generations. Large scale production of nanomaterials should only be done using greener route. Biogenic production of nanomaterials should be scaled up, commercialized and proper evaluation of their toxicity to environment and health should be performed in natural environmental condition in order to develop a clean and pollution free environment.

References

Ahmad A, Mukherjee P, Senapati S, Mandal D, Kahn MI, Kumar R et al (2003) Extracellular biosynthesis of silver nanoparticles using the fungus *Fusarium oxysporum*. Colloids Surf B Biointerfaces 28:313–318

Ang EL, Zhao H, Obbard JP (2005) Recent advances in the bioremediation of persistent organic pollutants via biomolecular engineering. Enzyme Microb Technol 37(5):487–496

Bakir EM, Younis NS, Mohamed ME et al (2018) Cyanobacteria as nanogold factories: chemical and anti-myocardial infarction properties of gold nanoparticles synthesized by *Lyngbya majuscula*. Mar Drugs 16:217

Buzea C, Pacheco II, Robbie K (2007) Nanomaterials and nanoparticles: sources and toxicity. Biointerphases 2(4):MR17–MR71

Chang J, Zhang L, Wang P (2018) Intelligent environmental nanomaterials. Environ Sci Nano 5 (4):811–836

Das RK, Pachapur VL, Lonappan L, Naghdi M, Pulicharla R, Maiti S, Cledon M, Dalila LMA, Sarma SJ, Brar SK (2017) Biological synthesis of metallic nanoparticles: plants, animals and microbial aspects. Nanotechnol Environ Eng 2(1):18

Dhillon GS, Brar SK, Kaur S, Verma M (2012) Green approach for nanoparticle biosynthesis by fungi: current trends and applications. Crit Rev Biotechnol 32:49

Dobrucka R (2016) Biofabrication of platinum nanoparticles using *Fumariae herba* extract and their catalytic properties. Saudi J Biol Sci 26:31–37

El Golli A, Fendrich M, Bazzanella N, Dridi C, Miotello A, Orlandi M (2021) Wastewater remediation with ZnO photocatalysts: green synthesis and solar concentration as an economically and environmentally viable route to application. J Environ Manage 286:112226

Elamawi RM, Al-Harbi RE, Hendi AA (2018) Biosynthesis and characterization of silver nanoparticles using Trichoderma longibrachiatum and their effect on phytopathogenic fungi. Egypt J Biol Pest Co 28(1):28

Gao S, Shi Y, Zhang S, Jiang K, Yang S, Li Z, Takayama-Muromachi E (2008) Biopolymer-assisted green synthesis of iron oxide nanoparticles and their magnetic properties. J Phys Chem C 112:10398–10401

Gao C, X-Y Y, Luo T, Jia Y, Sun B, Liu J-H, Huang X-J (2014) Millimeter-sized Mg–Al- LDH nanoflake impregnated magnetic alginate beads (LDH-n-MABs): a novel bio-based sorbent for the removal of fluoride in water. J Mater Chem A 2:2119

Gómez M, Murcia MD, Dams R, Christofi N, Gómez E, Gómez JL (2012) Removal efficiency and toxicity reduction of 4-chlorophenol with physical, chemical and biochemical methods. Environ Technol 33(7–9):1055–1064

Gour A, Jain NK (2019) Advances in green synthesis of nanoparticles. Artif Cell Nanomed B 47 (1):844–885

Grząbka-Zasadzińska A, Amietszajew T, Borysiak S (2017) Thermal and mechanical properties of chitosan nanocomposites with cellulose modified in ionic liquids. J Therm Anal Calorim 130 (1):143–154
Guerra FD, Attia MF, Whitehead DC, Alexis F (2018) Nanotechnology for environmental remediation: materials and applications. Molecules 23(7):1760
Guillossou R, Roux JL, Mailler R, Pereira-Derome CS, Varrault G, Bressy A, Vulliet E, Morlay C, Nauleau F, Rocher V et al (2020) Influence of dissolved organic matter on the removal of 12 organic micropollutants from wastewater effluent by powdered activated carbon adsorption. Water Res 172:115487
Gupta R, Xie H (2018) Nanoparticles in daily life: applications, toxicity and regulations. J Environ Pathol Tox 37(3)
Gupta R, Kumar AN, Bandhu S, Gupta S (2007) Skeletal fluorosis mimicking seronegative arthritis. Scand J Rheumatol 36(2):154–155
Gupta AK, Deva D, Sharma A, Verma N (2009) Adsorptive removal of fluoride by micronanohierarchalweb of activated carbon fibers. Ind Eng Chem Res 48:9697–9707
Hoet PH, Brüske-Hohlfeld I, Salata OV (2004) Nanoparticles–known and unknown health risks. J Nanobiotechnol 2(1):12
Ibrahim RK, Hayyan M, AlSaadi MA, Hayyan A, Ibrahim S (2016) Environmental application of nanotechnology: air, soil, and water. Environ Sci Pollut Res 23(14):13754–13788
Jyoti K, Singh A (2016) Green synthesis of nanostructured silver particles and their catalytic application in dye degradation. Biotechnol Genet Eng 14(2):311–317
Kabdaşlı I, Arslan-Alaton I, Ölmez-Hancı T, Tünay O (2012) Electrocoagulation applications for industrial wastewaters: a critical review. Environ Technol Rev 1(1):2–45
Karakoti AS, Munusamy P, Hostetler K, Kodali V, Kuchibhatla S, Orr G et al (2012) Preparation and characterization challenges to understanding environmental and biological impacts of ceria nanoparticles. Surf Interface Anal 44(8):882–889
Khin MM, Nair AS, Babu VJ, Murugan R, Ramakrishna S (2012) A review on nanomaterials for environmental remediation. Eng Environ Sci 5(8):8075–8109
Klaus T, Joerger R, Olsson E, Granqvist CG (1999) Silver-based crystalline nanoparticles, microbially fabricated. Proc Natl Acad Sci U S A 96:13611–13614
Krishna R, Titus E, Krishna R, Bardhan N, Bahadur D, Gracio J (2012) Wet-chemical green synthesis of L-lysine amino acid stabilized biocompatible iron-oxide magnetic nanoparticles. J Nanosci Nanotechnol 12:6645–6651
Kuleyin A (2007) Removal of phenol and 4-chlorophenol by surfactant-modified natural zeolite. J Hazard Mater 144(1–2):307–315
Kumar A, Paul P, Nataraj SK (2017) Bionanomaterial scaffolds for effective removal of fluoride, chromium, and dye. ACS Sustain Chem Eng 5(1):895–903
Laux P, Tentschert J, Riebeling C, Braeuning A, Creutzenberg O, Epp A et al (2018) Nanomaterials: certain aspects of application, risk assessment and risk communication. Arch Toxicol 92(1):121–141
Lin PC, Lin S, Wang PC, Sridhar R (2014) Techniques for physicochemical characterization of nanomaterials. Biotechnol Adv 32(4):711–726
Lu W, Shen Y, Xie A, Zhang W (2010) Green synthesis and characterization of superparamagnetic Fe_3O_4 nanoparticles. J Magn Magn Mater 322:1828–1833
Maliyekkal M, Lisha KP, Pradeep T (2010) A novel cellulose–manganese oxide hybridmaterial by in situ soft chemical synthesis and its application for the removal of Pb (II) from water. J Hazard Mater 181:986–995
Mir SA, Acharya P, Nayak B (2017) Nanoparticle from biological source: their role in medicine and drug delivery. Int J Creat Res Thoug 6:310–318
Mishra RK, Ha SK, Verma K, Tiwari SK (2018) Recent progress in selected bio-nanomaterials and their engineering applications: an overview. J Sci-Adv Mater Dev 3(3):263–288
Mohamed EF (2017) Nanotechnology: future of environmental air pollution control. Scanning 6(2)

Montero-Montoya R, López-Vargas R, Arellano-Aguilar O (2018) Volatile organic compounds in air: sources, distribution, exposure and associated illnesses in children. Ann Glob Health 84 (2):225

National Research Council (NRC), Fluoride in drinking water: a scientific review of EPA's standards, 2006. Washington, DC

Oberdorster G, Oberdorster E, Oberdorster J (2007) Concepts of nanoparticle dose metric and response metric. Environ Health Perspect 115(6):A290–A290

Ouyang W, Chen T, Shi Y, Tong L, Chen Y, Wang W et al (2019) Physico-chemical processes. Water Environ Res 91(10):1350–1377

Palit S (2017) Nanomaterials for industrial wastewater treatment and water purification. In: Handbook of ecomaterials, pp 1–41

Pandhi T, Chandnani A, Subbaraman H, Estrada D (2020) A review of inkjet printed graphene and carbon nanotubes based gas sensors. Sensors 20(19):5642

Patel HK, Kalaria RK, Khimani MR (2020) Nanotechnology: a promising tool for bioremediation. In: Removal of toxic pollutants through microbiological and tertiary treatment, pp 515–547

Pathak A, Panda AB, Tarafdar A, Pramanik P (2003) Synthesis of nano-sized metal oxide powders and their application in separation technology. J Ind Chem Soc 80:289–296

Ray PC, Yu H, Fu PP (2009) Toxicity and environmental risks of nanomaterials: challenges and future needs. J Environ Sci Heal C 27(1):1–35

Rossi M, Cubadda F, Dini L, Terranova ML, Aureli F, Sorbo A, Passeri D (2014) Scientific basis of nanotechnology, implications for the food sector and future trends. Trends Food Sci Technol 40:127–148

Sakr TM, Korany M, Katti KV (2018) Selenium nanomaterials in biomedicine—an overview of new opportunities in nanomedicine of selenium. J Drug Deliv Sci Technol 46:223–233

Salem W, Leitner DR, Zingl FG, Schratter G, Prassl R, Goessler W, Reidl J, Schild S (2015) Antibacterial activity of silver and zinc nanoparticles against Vibrio cholerae and enterotoxic Escherichia coli. Int J Med Microbiol 305:85–95

Sarkar J, Ray S, Chattopadhyay D, Laskar A, Acharya K (2012) Mycogenesis of gold nanoparticles using a phytopathogen *Alternaria alternate*. Bioprocess Biosyst Eng 35:637–643

Scalese S, Nicotera I, D'Angelo D, Filice S, Libertino S, Simari C et al (2016) Cationic and anionic azo-dye removal from water by sulfonated graphene oxide nanosheets in Nafion membranes. New J Chem 40(4):3654–3663

Sharifi S, Behzadi S, Laurent S, Forrest ML, Stroeve P, Mahmoudi M (2012) Toxicity of nanomaterials. Chem Soc Rev 41(6):2323–2343

Shivaji S, Madhu S, Singh S (2011) Extracellular synthesis of antibacterial silver nanoparticles using psychrophilic bacteria. Process Biochem 46:1800–1807

Singh J, Dutta T, Kim KH, Rawat M, Samddar P, Kumar P (2018) 'Green' synthesis of metals and their oxide nanoparticles: applications for environmental remediation. J Nanobiotechnol 16 (1):84

Sinha AK, Suzuki K (2007) Novel mesoporous chromium oxide for VOCs elimination. Appl Catal B: Environ 70(1–4):417–422

Suman, Kardam A, Gera M, Jain VK (2015) A novel reusable nanocomposite for complete removal of dyes, heavy metals and microbial load from water based on nanocellulose and silver nano-embedded pebbles. Environ Technol 36(6):706–714

Tetteh EK, Rathilal S (2020) Evaluation of different polymeric coagulants for the treatment of oil refinery wastewater. Cogent Eng 7(1):1785756

Thines KR, Abdullah EC, Mubarak NM (2017) Effect of process parameters for production of microporous magnetic biochar derived from agriculture waste biomass. Micropor Mesopor Mater 253:29–39

Uchida T, Ohashi O, Kawamoto H, Yoshimura H, Kobayashi KI, Tanimura M et al (2006) Synthesis of single-wall carbon nanotubes from diesel soot. Jpn J Appl Phys 45(10R):8027

Uddin AJ, Araki J, Fujie M, Sembo S, Gotoh Y (2012) Interfacial interaction and mechanical properties of chitin whisker-poly(vinyl alcohol) gel-spun nanocomposite fibers. Polym Int 61:1010–1015

Uddin MT, Hoque ME, Bhoumick MC (2020) Facile one-pot synthesis of heterostructure SnO 2/ZnO photocatalyst for enhanced photocatalytic degradation of organic dye. RSC Adv 10 (40):23554–23565

Uma Suganya KS, Govindaraju K, Ganesh Kumar V et al (2015) Blue green alga mediated synthesis of gold nanoparticles and its antibacterial efficacy against Gram positive organisms. Mater Sci Eng C 47:351–356

Veisi H, Azizi S, Mohammadi P (2018) Green synthesis of the silver nanoparticles mediated by *Thymbra spicata* extract and its application as a heterogeneous and recyclable nanocatalyst for catalytic reduction of a variety of dyes in water. J Clean Prod 170:1536–1543

Wang X, Liu B, Lu Q, Qu Q (2014) Graphene-based materials: fabrication and application for adsorption in analytical chemistry. J Chromatogr A 1362:1–15

Willner MR, Vikesland PJ (2018) Nanomaterial enabled sensors for environmental contaminants. J Nanobiotechnol 16(1):1–16

Yang C, Wei H, Guan L, Guo J, Wang Y, Yan X et al (2015) Polymer nanocomposites for energy storage, energy saving, and anticorrosion. J Mater Chem 3(29):14929–14941

Yang J, Hou B, Wang J, Tian B, Bi J, Wang N et al (2019) Nanomaterials for the removal of heavy metals from wastewater. Nanomaterials (Basel) 9(3):424

Yi W, Li Z, Dong C, Li HW, Li J (2019) Electrochemical detection of chloramphenicol using palladium nanoparticles decorated reduced graphene oxide. Microchem J 148:774–783

Zhang X, Liu Y (2020) Nanomaterials for radioactive wastewater decontamination. Environ Sci Nano 7(4):1008–1040

Zhao X, Liu W, Cai Z, Han B, Qian T, Zhao D (2016) An overview of preparation and applications of stabilized zero-valent iron nanoparticles for soil and groundwater remediation. Water Res 100:245–266

Smart Nanomaterials for Bioimaging Applications: An Overview

16

Kunal Biswas, Avik Sett, Debashis De, Jaya Bandyopadhyay, and Yugal Kishore Mohanta

Abstract

Nanomaterials are the classes of nanomaterials which basically fall in the range of 1–100 nm dimensions and have achieved its ubiquitous role in the domain of bioelectronics to biomedical area. The applications in the biomedical domain comprise bioimaging, targeted drug delivery, anti-cancerous activity, etc., which have attracted much attention in the recent times. Due to the specific unique activity of nano-scaled materials such as its higher surface to volume ratio, higher surface energy, more surface reactivity, etc., such materials find applications in the field of biological domain to electronical area. The electron confinement activity to differentially capped nano-scaled materials have been used extensively now-a-days in the bioimaging and applications. Due to the higher quantum yield and more stability in the biological systems, nanomaterials have been a material of choice for an augmented biological activity as compared to their bulk counterparts. Such nanomaterials due to their biocompatible dimensions having

K. Biswas · J. Bandyopadhyay
Department of Biotechnology, Maulana Abul Kalam Azad University of Technology, West Bengal, Haringhata, West Bengal, India

A. Sett
Department of Electronics and Electrical Communication Engineering, IIT Kharagpur, Kharagpur, West Bengal, India

D. De
Department of Computer Science and Engineering, Maulana Abul Kalam Azad University of Technology, West Bengal, Haringhata, West Bengal, India

Y. K. Mohanta (✉)
Biotechnology and Omics Laboratory, Natural and Medical Sciences Research Center University of Nizwa, Nizwa, Sultanate of Oman

Department of Botany, Maharaja Sriram Chandra Bhanj Deo University, Baripada, Odisha, India

M. Arakha et al. (eds.), *Bio-Nano Interface*,
https://doi.org/10.1007/978-981-16-2516-9_16

resemblance to biological systems, enable for an effective drug delivery carrier molecule as compared to other drug carrier systems. Moreover, such materials bear unique photo-physical and luminescence properties which are harnessed for different bioimaging applications, showing robust imaging behaviour as compared to the bulk systems. Finally, targeted drug delivery (TDD) means has been achieved in a better and refined manner using such nano-scaled materials owing to their unique specificity and sensitivity attributes in the biological interface. Understanding the comprehensive role of nano-scaled materials in diverse biomedical area specifically in myriad imaging techniques makes nanomaterials of different types a preferred choice in bioimaging applications making a revolutionary measure in the biomedical field of research and applications.

Keywords

Nanomaterials · Biomedical · Bioimaging · Surface area · Luminescence

16.1 Introduction

With the introduction of nano-scaled targeted drug delivery systems, the efficiency of many conventional pharmaceutical therapies can be significantly improved. It helps to avoid unnecessary drug loss because nanocarriers show higher loading efficiency, thereby providing protection from the harsh surrounding. It helps to minimise unwanted side effects because it possesses higher specificity, biocompatibility and biodegradability behaviour (Naeem et al. 2020).

Nanotechnology is used for multiple purposes like titanium dioxide (Tio2) or zinc oxide (ZnO) which absorbs UV radiation and is thus used in sunscreen (Petrick and Ibadurrohman 2020), silver (AgNPs) which is used in antimicrobial activity and is used in soap formulations, surgical implants, etc. (Gunputh et al. 2020). It is also reported that Gold (AuNPs) nanoparticles were also used in different bioimaging applications and diagnosis (Mousavi et al. 2020). Nanoparticle formation and distribution has diverse bioimaging varieties thereby finding advantages in early diagnosis and treatment of diseases (Viana et al. 2020). Upon using such nano-scaled bioimaging materials cellular localisation, interaction dynamic changes of a variety of bioactive substances can be observed (Placido et al. 2019). It can be observed that environmental nanoparticle from atmospheric, terrestrial, aquatic regions is obtained and synthesised biogenically (Órdenes-Aenishanslins et al. 2020) for different bioimaging applications. Nanomaterials of different types are basically synthesised using bottom-up method and top-down technology. Bottom-up technology refers to the process where the smaller subunits are added to form larger complex materials whereas in case of top-down technique, the larger chunk of precursor material is broken down into smaller subunits or monomers for different applications. Usually bottom-up technique is mostly preferred as compared to the top-down principles because of the sensitivity approach involved in the formation of the polymeric molecules upon gradual addition of smaller contributing monomeric forms resulting into sensitive applications in diverse fields mostly in bioimaging, tagging agents, etc. (Li et al. 2019).

In the recent times, nanomaterials both chemically synthesised and biogenically synthesised are used in diverse applications ranging from toxic contaminant chelation and removal mechanism to dreaded disease treatment. In case of environmental detection and removal of heavy metals from the polluted sources like contaminated water bodies, industrial belts, etc., several tailor made nano-scaled materials were employed by several groups for the successful detection and removal of the pollutants like iron oxide nanoparticles (FeONPs) for heavy metal removal from water bodies, etc. (Samrot et al. 2019). Similarly, several groups have also fabricated pristine as well as decorated/functionalised nanomaterials for the effective detection mechanism of the heavy metal ions like Cadmium, Arsenic, etc. (Shi et al. 2019). Also, in the biomedical field of research, different classes of nanomaterials like Gold nanoparticles, etc., are used extensively these days for effective cancer diagnosis and treatment (Sayyadi et al. 2019) for bearing the sensitivity and selectivity properties of such nano-scaled materials thereby achieving a targeted drug delivery response in the cancer treatment process. In this chapter, specifically the bioimaging nature and mechanism of different classes of nanomaterials has been studied besides different bio-activity of nanomaterials in the current scenario to uncover and present the imaging potentialities of such nanomaterials in the biomedical field of research which can aid in understanding and formulating nano-based future imaging agents which can outcompete the conventional manner of imaging of the bio systems.

16.2 Inorganic Nanocarriers in Bioimaging and Drug Delivery

Inorganic nanomaterials find widespread applications in the field of biochemistry, biotechnology and biomedical domains (Yang et al. 2019). Such applications include Gold nanoparticles, 2D-graphene, carbon nanotubes, quantum dots, mesoporous silica nanoparticles, quantum dots, magnetic nanoparticles, multifunctional composite nanoparticles, etc., which have become a promising choice in diverse biomedical research fields (Yogesh et al. 2020). In bioimaging, inorganic nanoparticles have shown great potential and diverse applications and are beneficial for targeted drug delivery and cancer therapies (Caires et al. 2020).

16.2.1 Carbon Nanotube as Imaging Agents

For a plethora of biomedical applications, carbon nanotubes (CNTs) have been used extensively in the current scenario (Yadav et al. 2020). Carbon nanotubes are also useful for delivery vehicles for carrying drugs, proteins, DNA, imaging agents and other materials (Prajapati et al. 2020). New CNT based contrast agent (CA) synthesis can be used for X-ray computed tomography (CT) imaging (Deshmukh et al. 2020). The CA is a hybrid material derived by using ultrashort single-walled carbon nanotubes (20–80 nm long, US-tubes) & Bi (iii) oxo-salicylate clusters with four

Bi (iii) ions per clusters (Bi4C). Over iodine, bismuth element was chosen that is conventional element used for CT Cas in the clinic because of its high X-ray attenuation capability and its low toxicity, due to this bismuth is a more promising element for new CT CA design.

16.2.2 Colloidal Gold Nanoparticles as Imaging Agents

Colloidal gold nanoparticles can be used for the biomedicine and drug delivery purposes (Masse et al. 2019). By preparation of the monodispersed GNPs (Gold Nanoparticles) by using the citrate reduction method, it helps to find out the importance of colloidal gold (Biterge-Süt and Canpolat 2019). For a variety of purposes like contrast agents, drugs, in vivo delivery of genes, colloidal gold nanoparticle is of choice by several researchers. It is because of its easy synthesis, large surface area, it has become proposed DDSs (Drug Delivery Systems) in different biomedical fields (Daraee et al. 2016). It can create DDSs with the help of smart polymers that help to release their payload to respond to external stimuli. It can also be used as photothermal agents in photothermal therapy because of bearing the property of higher molar absorption coefficient (Takeuchi et al. 2017). GNPs lie in the visible to near infra-red region. Also, biological applications of GNPs range from biomolecular sensing to therapeutic interventions due to its intrinsic surface plasmon resonance behaviour (Weissman et al. 2016). The Schiffrin–Brust biphasic method has been the most acceptable choice owing to the requirement of lesser reagents and solutions for the synthesis of GNPs, which was developed in the year 1994. Owing to the lack of different types of surface capping ligands and the presence of available chemical modifications over the surface of the GNPs structure, the same has been in use in diverse applications from nanotechnology to biomedicinal domains (Shimoni and Valenzuela 2017). Different steps are been employed for the production of water-dispersive GNPs with functionalised moieties for applications in different fields. Several groups have introduced various ligand exchange reactions on alkane thiol protected GNPs via careful surface modification (Lopez-Chaves et al. 2018). It is also possible to create gold nanostructures with active targeting capabilities and would be really useful for the detection of overexpressed cancerous cells using the mechanism of target-receptor modalities.

Making the recognition specific in nature (Wolfe et al. 2016) and their cellular toxicity has indeed been examined by several research groups as for the cytotoxicity of GNPs (Jia et al. 2017). The requirement of distinction between the toxicity of the GNP core and the exterior ligands is of paramount importance. It is found that upon carboxylate terminal capping using alkylthiolate-GNPs, the toxicity of the GNPs seems to be reduced, whereas the cationic GNPs are seem to be non-toxic in nature (Bhawawet et al. 2018). It has been reported that Human leukaemia cells (K562), upon treatment with functionalised GNPs with glucose, citrate, cysteine, biotin exhibit no significant toxicity in the cell lines as comparted to the salt solutions of Gold chloride (HuCL4), which has a tremendous toxicity of 90% inhibition when treated to same cell lines (Martínez-Torres et al. 2019). For the cancer treatment, most of the work on drug delivery of GNPs has been studied thereby limiting the exposure towards the healthy tissue through both passive and active targeting, where the concentration of drug can be increased at the tumour site resulting in return a robust targeted anti-cancerous activity as compared to the conventional way of cancerous treatment (Penon et al. 2017).

16.2.3 Mesoporous Silica Nanoparticles

Another important group of inorganic delivery system is colloidal mesoporous silica nanoparticles (MSNPs). Owing to their unique controlled morphologies, significant mesostructures with tremendous biocompatibility with a flexibility of effective functionalisation approach, such mesoporous silica nanoparticles find effective applications in the biological domains (Manzano and Vallet-Regí 2020). It can be also made hydrophilic by decorating them with the abundant silanol groups on the surface of MSNPs. Using various means, the controlled holding/release of cargo molecules has also been achieved by several groups. The mesoporous materials pore capacity and large internal surface area enable the effective entrapment of the therapeutic cargo molecules inside the porous structure of the material leading to the efficient therapeutic efficiency towards the desired target moieties (Siminzar et al. 2020). To improve the efficacy of the delivery of various hydrophobic anticancer drugs within the bloodstream the MSNPs, it takes advantages of the larger intrinsic pore capacity which make the payload drug solubilised. It is because of the effectiveness of such drugs encapsulation, the efficacy of the drug release and the kinetics is achieved more robustly than other delivery means. On both the exterior and the interior surfaces, the modification of MSNPs can be achieved. In order to improve the nanocarrier drug delivery mechanism and provide a range of functionalities, modification of SNPs is more acceptable. MSNPs in their mesoporous structure with a high surface area to volume ratio contribute to the extensive functionalization via covalent or electrostatic interactions, where there the augmented rate of organic molecules is placed in the silanol groups of the outside surface of the material structures. The side effects of the drug delivery approach could be reduced and the increasing in the specificity of the drug delivery modalities could be harnessed using such versatile MSNPs surface properties as potential drug delivery vector platform.

Functionalisation of the internal structures of the MSNPs intrinsic structure could also augment the rate of drug delivery approach using such nano-scaled mesoporous structure for potentially viable molecules like proteins, nucleic acids for myriad therapeutic areas along with the underlying exploration of effective cytotoxicity regime of the material towards the biological systems (Downing and Jain 2020). With both healthy cells and cancer cell lines cellular uptake phenomenon of MSNPs and their good biocompatibility properties were confirmed. Surface charge, effective shape and overall dimension of the nanomaterials determine the effective cellular uptake behaviour and their associated cytotoxicity when introduced into the biological. It has been reported that non-modified MSNPs in the concentration up to 100 mg/ml, dimensions ~100 nm exhibits no significant cytotoxicity has been observed which is not significant for the effective therapeutic efficacy. The effective therapeutic activity of the drug delivery mechanism is only been achieved by the utilisation of the lowest drug molecule when confronted in the healthy tissues with the compromised side effects in the healthy cells, which is especially important in the cancer therapy regime. Depend on chemically modifying MSNPs with targeted moieties most strategies used for cell targeting (Choi et al. 2020). Some of the moieties includes the proteins, FA, antibodies, peptides molecules, etc. For cancer cell targeting based on PEI-functionalised and FA- conjugated MSNPs developed a selective nanoparticulate system. Using FA as the targeting ligand the PEI-MSNP hybrid nanoparticles are non-toxic and can be specifically endocytosed. In order to gain a tremendous cancer treatment effectivity, Folate receptors play a significant role in the treatment of the disease and the higher rate of material internalisation is achieved when diagnosis of the tumour cells in the cancer therapy management is concerned (Song et al. 2020).

16.2.4 Quantum Dots in Imaging

The engineering of multifunctional nanodevices Quantum dots (QDs), semiconductor nanoparticles with unique photo-physical properties have become one of the dominant classes of imaging probes as well as universal platforms (Shukla et al. 2020). The exceptional photostability along with the one photon-multiphoton absorption behaviour with the size dependent tunable absorption and emission attributes is achieved using the QDs (Quantum Dots). The bleeding nature during the multiplexed bioimaging process could be achieved using such QDs due to the narrower photoluminescence band phenomenon of such material of zero electronical dimensions. At the single molecular level, such QDs play a crucial role in achieving the durability and the sensitivity along with the brightest and stable photoluminescence properties associated with them, making the bioimaging approach using QDs a significant choice in the current imaging research (Wang et al. 2020b). The narrow photoluminescence bands of QDs are beneficial in advancing device technology and in the area of biotechnology (Arshad et al. 2019). The effective biological applications of the QDs are bestowed in their structure due to the intrinsic optical properties in their structure which make the mediation of the organometallic routes along with the ligand exchange phenomenon associated with the QDs treatment modalities making them an ideal choice in the bioimaging regime (He et al. 2020). Generally, highly fluorescent QDs can be prepared for both exchanging hydrophobic QDs from organic to aqueous phase and introducing functional groups for bioconjugation. The effective stability against the hydrolysis and the biochemical reactions for the QDs could be achieved by capping the QD surface using the thiol groups, which enables the QDs a best choice in achieving the enhanced biocompatibility phenomenon when introduced into the biological systems (Shi et al. 2020). Coating or conjugation of polymers onto QD surface is another method with a tri-block amphiphilic copolymer successfully over-coated CdSe/ZnS QDs which protect QDs against hydrolysis and enzymatic degradation. Previous study have shown that entrapping the QDs inside the silica shells significantly improves the coating of a silica layer from sodium silicate initially by conjugating a layer of 3-mercaptopropyl trimethoxysilane (MPS) on the surface of citrate-stabilised CdS QDs for sensing gene and drug delivery, and cellular and biomolecular imaging (Rathee et al. 2020). Recently, bioconjugated QDs have become regular parts of biology by using bioconjugated QDs, which can be classified into nonspecific and targeted formulation (Díaz-González et al. 2020). QDs are also conjugated with bio-entities like proteins, peptides, nucleic acids, liposomes, etc., making such QDs used in the direct and indirect labelling of the extracellular proteins and subcellular organelles for an effective bioimaging process. CdSe/ZnS QDs coated bind to human epithelial kidney (HEK) cells to a greater extent than to mouse fibroblast cells (NIH3T3) with the cytoplasmic location of rhodamine dextran. The nuclear localisation signal (NLS) conjugated QDs have been shown by several groups by co-localisation of QDs and MitoTracker Red (Paesano et al. 2020). Similarly, mitochondrial localisation signal (MLS)-conjugated QDs were observed around mitochondria, thereby making the application of mitochondria in diverse biomedicine and diagnosis (Jiang et al. 2019). The specific labelling of cellular tissue for in vivo applications would result in higher resolution and sensitivity cell, where the main applications of QDs lie in the multimodal imaging probes, biological applications, which is used by using a mixture of magnetic nanoparticles, quantum dots and an amphiphilic polymer, followed by functionalization of the bead surface with folic acid (Feng et al. 2020).

It is further noted that multicolour Quantum Dots (QDs) capped magnetic nanorings are 4 times larger than the superparamagnetic iron oxides, which find applications in diverse fields of applications in imaging, etc. (Yue et al. 2019). The major obstacle which comes in the pathway of QDs for bioimaging lies in the fact that several QDs release heavy metal ions such as Cd+ while treating the cells with CdSe quantum dots in bioimaging (Yadav et al. 2019). Such research challenges are the major impediments while dealing with the quantum dot based in vivo bioimaging applications. The understanding of the basic degradation phenomenon and the accumulation attributes of the inorganic QDs along with the association of the intrinsic toxicity of the material remains a serious concern while researching bioimaging using QDs (Garmanchuk et al. 2019).

16.3 Multifunctional Composite Nanoparticles

16.3.1 Graphene

Graphene is a 2-dimensional carbonaceous sheet, which is having several unique physico-chemical properties like higher mechanical strength, robust conductivity, larger surface area (2630 m^2 per gram), etc. (Thamer et al. 2020). Owing to such higher level of enhanced physico-chemical attributes, Graphene has been used in diverse applications like Quantum communications, nanoelectronic applications, energy research, etc. (Stanford et al. 2020). Besides electronic and energy applications, the Graphene Oxide sheets and its different allied forms are extensively been used in biological applications like anti-bacterial activities, anti-cancerous activities, bioimaging, etc. (Biswas et al. 2018; Yogesh et al. 2020).

The higher surface to volume ratio of Single layered Graphene Oxide sheets enables it to be used in myriad applications like drug delivery and other physico-chemical applications in biomedical fields (Liang et al. 2020). Using Graphene, anti-cancerous activities using NIR based photothermal ablation mechanism have been achieved by several researchers (Wang et al. 2020a). It has been also reported that decoration of several inorganic nanoparticles like Silver nanoparticles, Gold nanoparticles over the Graphene surface, resulted into the synthesis of Graphene based composite systems, which bears augmented optical and superb magnetic properties, finding applications in diverse imaging applications (Griep et al. 2020). The toxicity of Graphene based materials has been studied by several researchers in different biological systems both in vitro and in vivo. The mechanism of bio-distribution and the associated toxicity in the biological systems is mainly regulated by the particulate size of the sheets and the surface chemistry properties. The effect of functionalisation of the Graphene sheets plays a significant role in reducing the toxicity of the nano-scaled Graphene sheets when treated in the biological models like mice at a particular dosage concentration (Chen et al. 2020).

Several techniques are available for the synthesis of Graphene Oxide sheets using chemical vapour deposition (CVD), chemical route of synthesis, etc. Using chemical route of synthesis, Hummers method and its different modification protocol has been used like Improved Hummers method, Modified Hummers method, etc. (Kudus et al. 2020). The interplay of reducing agents and oxidising agents plays a critical role in the synthesis of Graphene from Graphite powder using such methods. The purity of Graphene production has been achieved in large scale at the industrial level using scale-up technologies yielding ultrapure graphene powder of 99.99% purity. Based upon such advantageous properties of Graphene, zero dimensional Graphene Quantum dots (GQDs) (Kang et al. 2020) were also been employed in the different biomedical applications bearing enhanced shelf life and net quantum yield as compared to the conventional tagging agents like rhodamine, etc., which has a compromised shelf life and quantum yield phenomenon.

16.3.2 Magnetic Nanoparticles

Due to their superparamagnetic properties, tunable size and other biological functionalities magnetic nanoparticles (MNPs) such as Fe3O4 magnetite and y-Fe2O3 maghemite are particularly appealing (Krans et al. 2020). MNPs exhibit superparamagnetism at room temperature, when the particle size is smaller than the single domain limit. Owing to the unique magnetic properties of the nanoparticles, such low scaled nanomaterials having intrinsic magnetism have been employed for the applications in the fields of cancer therapy, biomedical treatment, gene delivery, etc. (Alphandéry 2020).

It is because of the presence of higher proportions of iron as an element, MNPs are used safely with huge prospects of lower rate of immuno rejection and a higher rate of clearance from the body system (Fopase and Pandey 2020). It has been reported previously that Fe per kg determines the overall toxicity in the biological system. A study reported that at a higher concentration of 60 mg Fe per kg, it exhibits a Fe toxicity in the biological systems. In comparison to the bulk Fe content, the proportion of Fe in coatings as microemulsions, nanoparticulate encapsulations exhibits a lesser value or negligible toxicity in the biological systems (Sosa-Acosta et al. 2020). Also, the average particulate size of the nanoparticles and its hydrodynamic size determines the net toxicity in the systems. The study enables to understand that the positive attributes of the magnetism and the nano-scaled materials properties make the magnetic nanoparticles (MNPs) an ideal candidate in the applications of different biological imaging applications (El-Sherbiny et al. 2020). Co-precipitation of iron salts and thermal decomposition of organometallic compounds determine 90% of the available reports, which find applications in diverse biomedical applications.

Modification of an organic shell surrounding the magnetic core is the first step for the preparation of targeted/therapeutic MNPs. With chemically reactive groups available this reaction would yield a water-soluble biocompatible product. The best methodology of stripping off the organic molecular capping of nanoparticles is the ligand exchange technique (Reaz et al. 2020). Different technique for the modification of the nanoparticles is the functionalisation of the nanoparticles using terminal carboxylic groups. Several reports have been highlighted that decoration of the nanoparticles with the specific targeted molecules or specific drugs for achieving targeted drug delivery (Ahmad et al. 2020). In that regard, various synthetic materials like FA (Folate receptors), RGD peptides (directed against avb 3 integrin), etc., have been employed for the purpose of model targeting moieties (Aisida et al. 2020). MRI stands for the magnetic resonance imaging which is based upon the resonance between the radio frequency emanated from the bodily system and the external source (Le Page et al. 2020). The major drawbacks of the instrumentation lies in the drawbacks associated like it has minimised sensitivity as well as insufficient spatial or temporal resolution, although MRI currently can be used for combining two or more imaging modalities. The disadvantages of MRI have been made by MRI/PET, MRI/CT and triple-modality imaging. In order to realise the high resolution, higher sensitivity and excellent soft-tissue contrast, the combination of MRI and CT imaging is highly desirable for MRI/optical properties from CT and MRI (Lally et al. 2019). Few research groups have explained that Fe3O4–TaOx core–shell NPs can provide complementary information by CT (computed tomography), and the internal tumour microenvironment with newly formed blood vessels in the tumours can be clearly imaged using MRI (Khmara et al. 2019). Such imaging technique shows the hypoxic and oxygenated regions, which can be evaluated for both optical and MR imaging and is highly desirable to understand and develop NIR dyes and specific contrast agents to determine quantitatively the long-term bio-distribution and tumour localisation both in the presence and absence of the external magnetic fields when injected in the xenograft breast tumour cases in mice

(Mohd Tamsir et al. 2019). The optical means of the MNPs make them an ideal candidate for the soft-tissue and superior spatial applications (Chaves et al. 2019) with the mediation in the clinical oncology applications (Basini et al. 2019). Moreover, MRI/PET bimodal imaging has great potential impact of voxel-based MRI-guided PVE correction in functional FDG-PET brain imaging. Different imaging modalities have been designed and discovered which find applications in diagnosis, soft-tissue contrast and different clinical studies involving cells and tissues of the biological systems. Also, the interstitial hyperthermia and thermoablation based method based upon magnetic field-induced excitation of biocompatible superparamagnetic nanoparticles has been designed (Shakil et al. 2019). Previous reports have shown that a new technique based upon thermotherapy means using the magnetic nanoparticles, in which there is an involvement of the external alternating magnetic field exposed to the targeted cells or tissues like cancerous tumours, where upon subjection of the combination of magnetic nanoparticles and magnetic field, the targeted area of interest gets ablated upon increase in the temperature at around 42–43 °C (Kandasamy et al. 2019). Different functionalised MNPs have been used in varied rat tumour model studies. Using the histopathological examinations of the brain and tumour, it is noticed that aminosilane and dextran-coated superparamagnetic iron oxide nanoparticles (SPIONS) have exhibited potential good rate of thermotherapy (Alphandéry 2019). As compared with the dextran-coated particles, thermotherapy with the aminosilane coated nanoparticles exhibits a 4.5 fold prolongation in the survival process. The effectiveness of treatment was determined using the properties of average size distribution profile, magnetic anisotropic nature as well as heat efficiency properties (Yalcin 2019). Further, few research groups have demonstrated that SPIONS have a good properties to exhibit a higher rate of magnetic hyperthermia in various media and is acting a good platform for attaching other target molecules. Togetherly, in combination with the other magnetic based chemotherapy and radiation therapy, such nano-based magnetic therapeutic modalities can act as a suitable and a prospectus means for the treatment of cancer and allied therapy (El-Sherbiny et al. 2020).

16.3.3 Layered Double Hydroxides

The class of nanosystems could be denoted by the formula [M2 + 1 × M3 + x(OH)2] (An)x/n − mH2O, in which M^{2+} and M^{3+} cations are located in the brucite-like layers and Anis the charge-balancing interlayer anion (Barik 2019). Layered double hydroxides (LDHs) are the class of molecular therapeutic modalities where 2D layered structure and allied are used in the different gene and drug delivery mechanisms in various biological domains (Sanjay et al. 2019). By virtue of the versatility in chemical composition as well as the stability and biocompatibility of LDH materials in drug/gene delivery and biomedicine, LDH materials show the following advantages which make LDHs an ideal drug nanocarrier system because of its superior biocompatibility and low cytotoxicity behaviour (Yan et al. 2019).

Due to the intrinsic layered structure of the LDHs, the payload of interest like proteins, peptides, drug molecules, antibodies can be safely loaded into the layered structure of the LDHs, resulting into an effective treatment regime outcompeting the conventional drug delivery systems which is further associated with the disadvantages like ligand exchange reaction, surface modification requirement, etc., which makes the LDHs based drug/gene delivery mechanism a suitably controlled approach (Yazdani et al. 2019).

16.4 Graphene Based Bioimaging

16.4.1 Role in Fluorescence Imaging in Biological Tissues

It is a probe imaging technique which is a non-invasive approach (Wang et al. 2019b) used for the specific binding to Raji B cells, which is sensed using InGaAs detector upon subjection to laser excitation of 658 nm wavelength. The first report was performed for the B-cell specific antibody rituxan (anti –CD 20) which is further conjugated to pegylated nGO(nGO-PEG-Rituxan). Owing to the extraordinary intrinsic photoluminescence properties of the nano Graphene Oxide (nGO) sheets, such behaviour of the material is exploited in diverse in vivo applications (Lu et al. 2020).

Fluorescence imaging of nGO exhibits a lower quantum yield both in vitro and in vivo imaging. In order to image the tumour xenografted mice, some research group first demonstrated the NIR fluorescent dye (Cy 7 conjugated nGO-PEG-Cy7). Also, the tagged graphene oxide sheets with the Cystein molecules enabled a good rate of cell permeation and tumour accumulation after the post injection at 24 h time period, indicating a promising tagging agent in bioimaging applications (Zaboli et al. 2020).

16.4.2 Implication in the Two-Photon Fluorescence Imaging

In the basic biomedical diagnostics and research, laser imaging depth analyses have been done owing to the lower Rayleigh scattering and lower tissue absorption of the NIR light. Such result caused into reduced photobleaching and phototoxicity as compared to single-photon Fluorescence imaging (Hu et al. 2020). Because of the minor autofluorescence background, more detailed analysis of various cellular/subcellular activities in the deep location of biological samples is achieved. It can obtain a high reflux of excitation photons as compared to one-photon excitation wavelength using simple continuous-wave lasers, two-photon nonlinear excitation usually uses a nonlinear femtosecond laser (Li et al. 2020a).

The appropriate wavelength range used for the deep-sited organ imaging and tissues, two-photon excitation is employed which lies in 700–1350 nm. The field of two-photon fluorescence imaging (TPFI) in context to Graphene sheet has been used rigorously by many researchers. Also, few research groups have also reported the first instance of transferrin functionalised GO-PEG employing the two-photon luminescence behaviour as a non-bleaching optical probe for the three dimensional TPFI and cancer microsurgery therapeutic mechanism (Gong et al. 2019).

For the cellular and deep-tissue imaging purpose, several nanoprobes were designed by several research groups. Among the probes, nitrogen-doped GQDs (N-GQDs) bearing an average size of ~3 nm have been employed as efficient two-photon fluorescent probes (Kuo et al. 2020). N-GQDs were synthesised by wet-chemical approach by several research groups. Some research groups have shown that by a facile one-pot solvothermal method for the synthesis of N-GQDs (Fu et al. 2019b). The method employed by taking dimethylformamide as solvent and nitrogen sources, the nitrogen was successfully doped to GQDs enabling the extraction of smaller sp^2 domains from the large GO sheet, which shows chemical structure of N-GQDs with dimethylamine binding to GOs under 800 nm femtosecond pulse laser excitation, at 400 μm. The fluorescence imaging is significantly differed as compared with the Two-photon microscopy of the GQDs at 1800 μm. In the one-photon fluorescence imaging (OPFI) (Liu et al. 2019a) technique, for in vivo investigation of biostructures in the 800–1500 μm region, TPFI using N-GQDs as fluorescent probe is particularly suitably accepted (Singh et al. 2019).

16.4.3 Effect in the Radionuclide Based Bioimaging

Radionuclide bioimaging comprises the photo bleaching or the fluorescence quenching phenomenon along with the light absorption nature scattering of tissues with an auto fluorescence background (Day et al. 2020). Optical imaging cannot provide quantitative results and sometimes may be interfered at the in vivo applications in a quantitative manner with excellent sensitivity ~10–12 mol/l and nearly no penetration depth limit (Wu et al. 2019). Radiolabelling method would be able to accurately track the labelled substances (Knight et al. 2019). PET and SPECTe radionuclide-based imaging mainly comprises over a nominally low background signal PET and SPECT images, which are acquired and required little signal amplification in PET/SPECT imaging (Edem et al. 2019). Graphene based nanomaterials are promising nano-platforms which play an important role in such imaging modalities. It is reported that upon decorating iodine atoms onto the available defects and the surface edges of the GO sheets, few groups have demonstrated a method to label nGO–PEG with ^{125}I (Farzin et al. 2019). Using ^{64}Cu labelled nGO–PEG, a research group has explored the in vivo active tumour targeting behaviour by conjugating nGO–PEG with an antibody, also the in vivo PET imaging results confirmed the active tumour targeting cases (Ge et al. 2020).

16.4.4 Significance in Magnetic Resonance Imaging

It is observed that employing MRI, the anatomy and the different functions of the tissues are best monitored quantitatively as compared to other examining techniques

(Sprooten et al. 2019). It is a non-invasiveness mechanism without ionising radiation, which has been extensively employed. Compared with the optical and radionuclide imaging for improving, the T1 or T2 contrast in the observable water pool, it is observed that the brighter images of the T1 agents usually employed Gadolinium based composites (Clough et al. 2019), whereas for the T2 agents it employed for the composites which comprise iron oxide nanoparticles (Heckman et al. 2019). Gadolinium (Gd) and manganese (Mn) are generally toxic that are ions of paramagnetic in nature (García-Hevia et al. 2019). The intercalation of such toxic ions within the layers of Graphene layers minimises the toxicity response of the ions by coordinating with the available functional groups and moieties of the graphene architecture (Si et al. 2020). Such molecular coordinated architectured system enables to understand and enable to apply in the enhanced MRI relaxivity ($r1 = 70 \pm 6\ mM^{-1}\ s^{-1}$ and $r2 = 108 \pm 9\ mM^{-1}\ s^{-1}$) phenomenon, which forms the 16 and 21 times greater than the current clinically available Gd 3^+ based T1 agents which are chelated using $Gd3^+$ ions with the carboxyphenylated graphene nanoribbons (GNRs) (Li et al. 2020b).

16.4.5 Effect in the Photoacoustic Imaging

Imaging technique which comprises the conversion of short pulsed electromagnetic energy (non-ionising laser pulses) into heat production is categorised into the class of imaging technique known as *Photoacoustic Imaging,* which is based upon the photoacoustic effect (PA effect) (Liu et al. 2019b). Such observations resulted into the acoustic emission attributes because of the transient thermoeleastic expansion phenomenon (Fu et al. 2019a). In other words, the lower range electromagnetic waves have a potentiality to penetrate deeper into the tissues and cells as compared to the shorter wavelength. The radio frequency waves have the property of lower scattering nature in the biological specimens which seems to be more favourable in the measurement in the biological systems (Hariri et al. 2019). The mechanism of lower scattering phenomenon could be associated with the localised thermoelastic expansion, which in return produces wide-band acoustic waves. The sound waves from the biological specimens produce a deep-tissue imaging which is recorded with an interplay of ultra sound transducers (Moore et al. 2019). The significance of Photo Acoustic Imaging lies with the fact that PAI gives an optical absorption contrast agent along with the resolution at the ultrasound magnitude level. Such response of the optical contrasting nature of the Graphene sheet is because of the fact that Graphene bears sp2 domain architecture. In the larger view, among the graphene family nanomaterials, RGO sheets are potential PA contrast agents, which can absorb NIR light more efficiently than the native Gos (Jun et al. 2019). Such differing nature of GO based platform for the contrast agent lies in the fact that GO are hydrophilic whereas RGO sheets upon further reduction from GO sheets with an interplay of reducing agents, resulted into poor water solubility and more hydrophobicity nature. In one of the previous study, in order to obtain lesser oxygenated Graphene, one research group hase produced microwave assisted lower oxygen content Graphene layer which has the potentiality to generate photoacoustic signals with the NIR excitation (Li et al. 2020c).

16.4.6 Multimodal Imaging Applications

Biomedical applications has gained popularity, the idea of using multiple imaging modalities in conjunction for bringing the significance of multiple imaging modalities at one place thereby harnessing the valuable properties of different imaging techniques in a comprehensive manner (Rouffiac et al. 2020). The mechanism of the multimodal imaging applications comprises the comprehensive integration of different detectability parameters at one place for detection process of the different dosages in the body system (Wang et al. 2019a). Graphene owing to its unique physico-chemical properties and enhanced multifunctional chemistry, the material has been find application in diverse multimodal imaging contrast agents (Chawda et al. 2019). The structural configuration of such multimodal contrast agents was exploited based on its RGO-IONP for triple modal Fluorescence, Photoacoustic (PA) and MR (Magnetic Resonance) imaging, respectively (Qian et al. 2019).

16.5 Conclusion

The revolution of the nanomaterials breakthrough has resulted into the widespread application in different domains of scientific arenas. Owing to the associated demerits of the available tagging agents and imaging applications in engineering and biomedical fields, the need of a more robust imaging agents with higher efficiency has become the need of the hour. The present study predominantly highlights the myriad types of the available nano-scaled materials and their different unique mechanism methodology in varieties of biological systems, which are employed in different bioimaging applications. The study focusses the significance of inorganic and organic based nanomaterials which are used in biomedical fields, which due to their unique physico-chemical properties and un-conventional electrochemical properties replace the already available conventional tagging agents in biomedical domain. The study in the nut shell opens up the new horizon in un-covering the different varieties of nanomaterials currently utilised in different types of imaging aspects from biomedical to bioengineering domain. The significance of the underlying electronical and other electromechanical, electrical properties of selected nano-scaled materials and their derivatives are the next choice in the intricate bioimaging applications, which this study will showcase for employing in different spheres of science and technological applications.

Acknowledgement Author KB thanks TEQIP III, MAKAUT, WB for providing research fellowship during the work.

References

Ahmad W, Khan AU, Shams S, Qin L, Yuan Q, Ahmad A et al (2020) Eco-benign approach to synthesize spherical iron oxide nanoparticles: a new insight in photocatalytic and biomedical applications. J Photochem Photobiol B Biol 205:111821

Aisida SO, Akpa PA, Ahmad I, Zhao TK, Maaza M, Ezema FI (2020) Bio-inspired encapsulation and functionalization of iron oxide nanoparticles for biomedical applications. Eur Polym J 122:109371

Alphandéry E (2019) Biodistribution and targeting properties of iron oxide nanoparticles for treatments of cancer and iron anemia disease. Nanotoxicology 13:573–596

Alphandéry E (2020) Iron oxide nanoparticles for therapeutic applications. Drug Discov Today 25:141–149

Arshad A, Akram R, Iqbal S, Batool F, Iqbal B, Khalid B et al (2019) Aqueous synthesis of tunable fluorescent, semiconductor CuInS2 quantum dots for bioimaging. Arab J Chem 12:4840–4847

Barik S (2019) Layered double hydroxide-decorated hydrogel for biomedical applications. In: Nanostructured polymer composites for biomedical applications. Elsevier, pp 367–383

Basini M, Guerrini A, Cobianchi M, Orsini F, Bettega D, Avolio M et al (2019) Tailoring the magnetic core of organic-coated iron oxides nanoparticles to influence their contrast efficiency for magnetic resonance imaging. J Alloys Compd 770:58–66

Bhawawet N, Essner JB, Atwood JL, Baker GA (2018) On the non-innocence of the imidazolium cation in a rapid microwave synthesis of oleylamine-capped gold nanoparticles in an ionic liquid. Chem Commun 54:7523–7526

Biswas K, De D, Bandyopadhyay J, Sen P (2018) Differential antibacterial response exhibited by graphene nanosheets toward gram-positive bacterium Staphylococcus aureus. IET Nanobiotechnol 12:733–740

Biterge-Süt B, Canpolat E (2019) Evaluation of gold nanoparticles in terms of their use in biomedical applications. Turkish J Agric Sci Technol 7:1694–1698

Caires AJ, Mansur AA, Carvalho IC, Carvalho SM, Mansur HS (2020) Green synthesis of ZnS quantum dot/biopolymer photoluminescent nanoprobes for bioimaging brain cancer cells. Mater Chem Phys 244:122716

Chaves NL, Amorim DA, Lopes CAP, Estrela-Lopis I, Böttner J, de Souza AR et al (2019) Comparison of the effect of rhodium citrate-associated iron oxide nanoparticles on metastatic and non-metastatic breast cancer cells. Cancer Nanotechnol 10:7

Chawda N, Basu M, Majumdar D, Poddar R, Mahapatra SK, Banerjee I (2019) Engineering of gadolinium-decorated graphene oxide nanosheets for multimodal bioimaging and drug delivery. ACS Omega 4:12470–12479

Chen L, Li J, Chen Z, Gu Z, Yan L, Zhao F et al (2020) Toxicological evaluation of graphene-family nanomaterials. J Nanosci Nanotechnol 20:1993–2006

Choi E, Lim DK, Kim S (2020) Hydrolytic surface erosion of mesoporous silica nanoparticles for efficient intracellular delivery of cytochrome c. J Colloid Interface Sci 560:416–425

Clough TJ, Jiang L, Wong KL, Long NJ (2019) Ligand design strategies to increase stability of gadolinium-based magnetic resonance imaging contrast agents. Nat Commun 10:1–14

Daraee H, Eatemadi A, Abbasi E, Fekri Aval S, Kouhi M, Akbarzadeh A (2016) Application of gold nanoparticles in biomedical and drug delivery. Artif Cells Nanomed Biotechnol 44:410–422

Day AH, Domarkas J, Nigam S, Renard I, Cawthorne C, Burke BP et al (2020) Towards dual SPECT/optical bioimaging with a mitochondrial targeting, 99m Tc (i) radiolabelled 1, 8-naphthalimide conjugate. Dalt Trans 49:511–523

Deshmukh MA, Jeon JY, Ha TJ (2020) Carbon nanotubes: an effective platform for biomedical electronics. Biosens Bioelectron 150:111919

Díaz-González M, de la Escosura-Muñiz A, Fernandez-Argüelles MT, Alonso FJG, Costa-Fernandez JM (2020) Quantum dot bioconjugates for diagnostic applications. Top Curr Chem 378:1–44

Downing MA, Jain PK (2020) Mesoporous silica nanoparticles: synthesis, properties, and biomedical applications. In: Nanoparticles for biomedical applications. Elsevier, pp 267–281

Edem PE, Sinnes JP, Pektor S, Bausbacher N, Rossin R, Yazdani A et al (2019) Evaluation of the inverse electron demand Diels-Alder reaction in rats using a scandium-44-labelled tetrazine for pretargeted PET imaging. EJNMMI Res 9:1–8

El-Sherbiny IM, El-Sayed M, Reda A (2020) Superparamagnetic iron oxide nanoparticles (SPIONs) as multifunctional cancer theranostics. In: Magnetic nanoheterostructures. Springer, Cham, pp 223–241

Farzin L, Sheibani S, Moassesi ME, Shamsipur M (2019) An overview of nanoscale radionuclides and radiolabeled nanomaterials commonly used for nuclear molecular imaging and therapeutic functions. J Biomed Mater Res Part A 107:251–285

Feng S, Pan J, Li C, Zheng Y (2020) Folic acid-conjugated nitrogen-doped graphene quantum dots as a fluorescent diagnostic material for MCF-7 cells. Nanotechnology 31:135701

Fopase R, Pandey LM (2020) Iron oxide based magnetic nanomaterials for biomedical applications. Magnetochem Mater Appl 66:276

Fu Q, Zhu R, Song J, Yang H, Chen X (2019a) Photoacoustic imaging: contrast agents and their biomedical applications. Adv Mater 31:1805875

Fu Y, Gao G, Zhi J (2019b) Electrochemical synthesis of multicolor fluorescent N-doped graphene quantum dots as a ferric ion sensor and their application in bioimaging. J Mater Chem B 7:1494–1502

García-Hevia L, Bañobre-López M, Gallo J (2019) Recent progress on manganese-based nanostructures as responsive MRI contrast agents. Chem Eur J 25:431–441

Garmanchuk LV, Borovaya MN, Nehelia AO, Inomistova M, Khranovska NM, Tolstanova GM et al (2019) CdS quantum dots obtained by "green" synthesis: comparative analysis of toxicity and effects on the proliferative and adhesive activity of human cells. Cytol Genet 53:132–142

Ge J, Zhang Q, Zeng J, Gu Z, Gao M (2020) Radiolabeling nanomaterials for multimodality imaging: new insights into nuclear medicine and cancer diagnosis. Biomaterials 228:119553

Gong P, Sun L, Wang F, Liu X, Yan Z, Wang M et al (2019) Highly fluorescent N-doped carbon dots with two-photon emission for ultrasensitive detection of tumor marker and visual monitor anticancer drug loading and delivery. Chem Eng J 356:994–1002

Griep MH, Demaree JD, Cole DP, Henry TC, Karna SP (2020) Protein-mediated synthesis of Au nanocluster decorated reduced graphene oxide: a multifunctional hybrid nano-bio platform. Plasmonics:1–7

Gunputh UF, Le H, Lawton K, Besinis A, Tredwin C, Handy RD (2020) Antibacterial properties of silver nanoparticles grown in situ and anchored to titanium dioxide nanotubes on titanium implant against Staphylococcus aureus. Nanotoxicology 14:97–110

Hariri A, Zhao E, Jeevarathinam AS, Lemaster J, Zhang J, Jokerst JV (2019) Molecular imaging of oxidative stress using an LED-based photoacoustic imaging system. Sci Rep 9:1–10

He C, Ruan F, Jiang S, Zeng J, Yin H, Liu R et al (2020) Black phosphorus quantum dots cause nephrotoxicity in organoids, mice, and human cells. Small 2001371

Heckman AJ, Agarwal A, Sacco KA, Hiroto BT, Guru PK (2019) An ounce of caution: superparamagnetic iron oxide nanoparticle based magnetic resonance imaging contrast–associated anaphylaxis. Ann Allergy Asthma Immunol 122:210–211

Hu L, Zhang G, Liu Y, Guo T, Shao L, Ying L (2020) Efficient dendrimers based on naphthalene indenofluorene for two-photon fluorescent imaging in living cells and tissues. J Mater Chem C 8:2160–2170

Jia YP, Ma BY, Wei XW, Qian ZY (2017) The in vitro and in vivo toxicity of gold nanoparticles. Chin Chem Lett 28:691–702

Jiang XL, Liu JH, Que YT, Que YM, Hu PP, Huang CZ et al (2019) Multifunctional single-layered graphene quantum dots used for diagnosis of mitochondrial malfunction-related diseases. ACS Biomater Sci Eng 6:1727–1734

Jun SW, Manivasagan P, Kwon J, Mondal S, Ly CD, Lee J et al (2019) Folic acid–conjugated chitosan-functionalized graphene oxide for highly efficient photoacoustic imaging-guided tumor-targeted photothermal therapy. Int J Biol Macromol

Kandasamy G, Sudame A, Maity D, Soni S, Sushmita K, Veerapu NS et al (2019) Multifunctional magnetic-polymeric nanoparticles based ferrofluids for multi-modal in vitro cancer treatment using thermotherapy and chemotherapy. J Mol Liq 293:111549

Kang S, Kim KM, Son Y, Mhin S, Ryu JH, Shim KB et al (2020) Publisher correction: graphene oxide quantum dots derived from coal for bioimaging: facile and green approach. Sci Rep 10

Khmara I, Strbak O, Zavisova V, Koneracka M, Kubovcikova M, Antal I, Kavecansky V et al (2019) Chitosan-stabilized iron oxide nanoparticles for magnetic resonance imaging. J Magn Magn Mater 474:319–325

Knight JC, Mosley MJ, Kersemans V, Dias GM, Allen PD, Smart S et al (2019) Dual-isotope imaging allows in vivo immunohistochemistry using radiolabelled antibodies in tumours. Nucl Med Biol 70:14–22

Krans NA, Weber JL, Van den Bosch W, Zečević J, de Jongh PE, de Jong KP (2020) Influence of promotion on the growth of anchored colloidal iron oxide nanoparticles during synthesis gas conversion. ACS Catal 10:1913–1922

Kudus MHA, Zakaria MR, Akil HM, Ullah F, Javed F (2020) Oxidation of graphene via a simplified Hummers' method for graphene-diamine colloid production. J King Saud Univ 32:910–913

Kuo WS, Chang CY, Huang KS, Liu JC, Shao YT, Yang CH et al (2020) Amino-functionalized nitrogen-doped graphene-quantum-dot-based nanomaterials with nitrogen and amino-functionalized group content dependence for highly efficient two-photon bioimaging. Int J Mol Sci 21:2939

Lally PJ, Montaldo P, Oliveira V, Soe A, Swamy R, Bassett P et al (2019) Magnetic resonance spectroscopy assessment of brain injury after moderate hypothermia in neonatal encephalopathy: a prospective multicentre cohort study. Lancet Neurol 18:35–45

Le Page LM, Guglielmetti C, Taglang C, Chaumeil MM (2020) Imaging brain metabolism using hyperpolarized 13C magnetic resonance spectroscopy. Trends Neurosci

Li L, Guo Z, Wang S, Li D, Hou X, Wang F et al (2019) Facile synthesis of MoS 2 quantum dots as fluorescent probes for sensing of hydroquinone and bioimaging. Anal Methods 11:3307–3313

Li W, Liu Z, Fang B, Jin M, Tian Y (2020a) Two-photon fluorescent Zn2+ probe for ratiometric imaging and biosensing of Zn2+ in living cells and larval zebrafish. Biosens Bioelectron 148:111666

Li Y, Dong H, Tao Q, Ye C, Yu M, Li J et al (2020b) Enhancing the magnetic relaxivity of MRI contrast agents via the localized superacid microenvironment of graphene quantum dots. Biomaterials:120056

Li Z, Zhu X, Wu T, Han X, Xing F, Wang G (2020c) Broadband and ultrasensitive graphene-based mechanical wave detector with nanosecond response used for biological photoacoustic imaging. ACS Appl Mater Interfaces 12:17268–17275

Liang S, Wang B, Li X, Chu R, Yu H, Zhou S et al (2020) In vivo pharmacokinetics, transfer and clearance study of graphene oxide by La/Ce dual elemental labelling method. NanoImpact 17:100213

Liu L, Zhang Y, Zhou J, Yang J, Zhong C, Zhang Y et al (2019a) Design of a quinazolinone-based environment-sensitive fluorescent dye: solvatochromic fluorescence and application for one-photon and two-photon bioimaging. Dye Pigment 165:58–64

Liu Y, Bhattarai P, Dai Z, Chen X (2019b) Photothermal therapy and photoacoustic imaging via nanotheranostics in fighting cancer. Chem Soc Rev 48:2053–2108

Lopez-Chaves C, Soto-Alvaredo J, Montes-Bayon M, Bettmer J, Llopis J, Sanchez-Gonzalez C (2018) Gold nanoparticles: distribution, bioaccumulation and toxicity. In vitro and in vivo studies. Nanomed Nanotechnol Biol Med 14:1–12

Lu Z, Zhai X, Yi R, Li Z, Zhang R, Wei Q et al (2020) Photoluminescence emission during photoreduction of graphene oxide sheets as investigated with single-molecule microscopy. J Phys Chem C 124:7914–7921

Manzano M, Vallet-Regí M (2020) Mesoporous silica nanoparticles for drug delivery. 30:1902634

Martínez-Torres AC, Lorenzo-Anota HY, García-Juárez MG, Zarate-Triviño DG, Rodríguez-Padilla C (2019) Chitosan gold nanoparticles induce different ROS-dependent cell death modalities in leukemic cells. Int J Nanomed 14:7173

Masse F, Desjardins P, Ouellette M, Couture C, Omar MM, Pernet V et al (2019) Synthesis of ultrastable gold nanoparticles as a new drug delivery system. Molecules 24:2929

Mohd Tamsir N, Mohd Esa N, Shafie NH, Hussein MZ, Hamzah H, Abdullah MA (2019) The acute effects of oral administration of phytic acid-chitosan-magnetic iron oxide nanoparticles in mice. Int J Mol Sci 20:4114

Moore MJ, Hysi E, Fadhel MN, El-Rass S, Xiao Y, Wen XY et al (2019) Photoacoustic F-mode imaging for scale specific contrast in biological systems. Commun Phys 2:1–10

Mousavi SM, Zarei M, Hashemi SA, Ramakrishna S, Chiang WH, Lai CW et al (2020) Gold nanostars-diagnosis, bioimaging and biomedical applications. Drug Metab Rev:1–20

Naeem M, Awan UA, Subhan F, Cao J, Hlaing SP, Lee J et al (2020) Advances in colon-targeted nano-drug delivery systems: challenges and solutions. Arch Pharm Res:1–17

Órdenes-Aenishanslins N, Anziani-Ostuni G, Monrás JP, Tello A, Bravo D, Toro-Ascuy D et al (2020) Bacterial synthesis of ternary CdSAg quantum dots through cation exchange: tuning the composition and properties of biological nanoparticles for bioimaging and photovoltaic applications. Microorganisms 8:631

Paesano L, Marmiroli M, Bianchi MG, White JC, Bussolati O, Zappettini A et al (2020) Differences in toxicity, mitochondrial function and miRNome in human cells exposed in vitro to Cd as CdS quantum dots or ionic Cd. J Hazard Mater 122430

Penon O, Marín MJ, Russell DA, Pérez-García L (2017) Water soluble, multifunctional antibody-porphyrin gold nanoparticles for targeted photodynamic therapy. J Colloid Interface Sci 496:100–110

Petrick J, Ibadurrohman M (2020) Development of chitosan/TiO2 nanocomposite for multifunctional sunscreen application. In: IOP conference series: materials science and engineering. IOP Publishing, p 012030

Placido J, Bustamante-Lopez S, Meissner KE, Kelly DE, Kelly SL (2019) Comparative study of the characteristics and fluorescent properties of three different biochar derived carbonaceous nanomaterials for bioimaging and heavy metal ions sensing. Fuel Process Technol 196:106163

Prajapati SK, Malaiya A, Kesharwani P, Soni D, Jain A (2020) Biomedical applications and toxicities of carbon nanotubes. Drug Chem Toxicol:1–16

Qian R, Maiti D, Zhong J, Xiong S, Zhou H, Zhu R et al (2019) Multifunctional nano-graphene based nanocomposites for multimodal imaging guided combined radioisotope therapy and chemotherapy. Carbon N Y 149:55–62

Rathee N, Khurana K, Jaggi N (2020) Toxicology studies of CdSe QDs and Ag composites with CdSe for in-vitro bio imaging applications. In: AIP conference proceedings. AIP Publishing LLC, p 020064

Reaz M, Haque A, Cornelison DM, Wanekaya A, Delong R, Ghosh K (2020) Magneto-luminescent Zinc/Iron oxide core-shell nanoparticles with tunable magnetic properties. Phys E Low-dimensional Syst Nanostruct 114090

Rouffiac V, Roux KSL, Salomé-Desnoulez S, Leguerney I, Ginefri JC, Sébrié C et al (2020) Multimodal imaging for tumour characterization from micro-to macroscopic level using a newly developed dorsal chamber designed for long-term follow-up. J Biophotonics 13:e201900217

Samrot AV, Sahithya CS, Selvarani J, Pachiyappan S (2019) Surface-engineered super-paramagnetic iron oxide nanoparticles for chromium removal. Int J Nanomed 14:8105

Sanjay MR, Siengchin S, Pruncu CI, Jawaid M, Kumar TSM, Rajini N (2019) Biomedical applications of polymer/layered double hydroxide bionanocomposites. In: Nanostructured polymer composites for biomedical applications. Elsevier, pp 315–322

Sayyadi K, Rahdar A, Esmaili N, Sayyadi J (2019) Application of gold nanoparticles in cancer diagnosis and therapy: a mini-review. Adv Nanochem 1:47–51

Shakil MS, Hasan M, Sarker SR (2019) Iron oxide nanoparticles for breast cancer theranostics. Curr Drug Metab 20:446–456

Shi H, Zhu F, Zhou X, Li H, Yang F, Zhang X et al (2019) Large scale fabrication of disposable carbon cloth electrochemical sensors for simultaneous determination of heavy metal ion. J Electroanal Chem 840:328–337

Shi K, Xu X, Li H, Xie H, Chen X, Zhan Y (2020) Biosynthesized quantum dots as improved biocompatible tools for biomedical applications. Curr Med Chem

Shimoni O, Valenzuela SM (2017) Gold nanoparticles with organic linkers for applications in biomedicine. In: Nanotechnology in biology and medicine: methods, devices, and applications. CRC Press/Taylor & Francis Group, LLC

Shukla D, Das M, Kasade D, Pandey M, Dubey AK, Yadav SK et al (2020) Sandalwood-derived carbon quantum dots as bioimaging tools to investigate the toxicological effects of malachite green in model organisms. Chemosphere 248:125998

Si X, Wu K, Si Y, Yousaf B (2020) Mechanistic insights into the reactive radicals-assisted degradation of sulfamethoxazole via calcium peroxide activation by manganese-incorporated iron oxide–graphene nanocomposite: formation of radicals and degradation pathway. Chem Eng J 384:123360

Siminzar P, Omidi Y, Golchin A, Aghanejad A, Barar J (2020) Targeted delivery of doxorubicin by magnetic mesoporous silica nanoparticles armed with mucin-1 aptamer. J Drug Target 28:92–101

Singh H, Sreedharan S, Tiwari K, Green NH, Smythe C, Pramanik SK et al (2019) Two photon excitable graphene quantum dots for structured illumination microscopy and imaging applications: lysosome specificity and tissue-dependent imaging. Chem Commun 55:521–524

Song Y, Zhou B, Du X, Wang Y, Zhang J, Ai Y et al (2020) Folic acid (FA)-conjugated mesoporous silica nanoparticles combined with MRP-1 siRNA improves the suppressive effects of myricetin on non-small cell lung cancer (NSCLC). Biomed Pharmacother 125:109561

Sosa-Acosta JR, Iriarte-Mesa C, Ortega GA, Díaz-García AM (2020) DNA–iron oxide nanoparticles conjugates: functional magnetic nanoplatforms in biomedical applications. Top Curr Chem 378:1–29

Sprooten E, O'Halloran R, Dinse J, Lee WH, Moser DA, Doucet GE et al (2019) Depth-dependent intracortical myelin organization in the living human brain determined by in vivo ultra-high field magnetic resonance imaging. Neuroimage 185:27–34

Stanford MG, Zhang C, Fowlkes JD, Hoffman A, Ivanov IN, Rack PD et al (2020) High-resolution laser-induced graphene. Flexible electronics beyond the visible limit. ACS Appl Mater Interfaces 12:10902–10907

Takeuchi I, Nobata S, Oiri N, Tomoda K, Makino K (2017) Biodistribution and excretion of colloidal gold nanoparticles after intravenous injection: effects of particle size. Biomed Mater Eng 28:315–323

Thamer S, Al-Tamimi BH, Farid SB (2020) Preparation of graphene nano-sheets from graphite flakes via milling-ultrasonication promoted process. Mater Today Proc 20:579–582

Viana B, Richard C, Castaing V, Glais E, Pellerin M, Liu J et al (2020) NIR-persistent luminescence nanoparticles for bioimaging, principle and perspectives. In: Benayas A, Hemmer E, Hong G, Jaque D (eds) Near infrared-emitting nanoparticles for biomedical applications. Springer International Publishing, pp 163–197

Wang C, Fan W, Zhang Z, Wen Y, Xiong L, Chen X (2019a) Advanced nanotechnology leading the way to multimodal imaging-guided precision surgical therapy. Adv Mater 31:1904329

Wang F, Chen B, Yan B, Yin Y, Hu L, Liang Y et al (2019b) Scattered light imaging enables real-time monitoring of label-free nanoparticles and fluorescent biomolecules in live cells. J Am Chem Soc 141:14043–14047

Wang C, Wang X, Chen Y, Fang Z (2020a) In-vitro photothermal therapy using plant extract polyphenols functionalized graphene sheets for treatment of lung cancer. J Photochem Photobiol B Biol 204:111587
Wang Z, Liu Z, Gu B, Gao B, Wang T, Zheng X et al (2020b) Ultraviolet light-driven controllable doping of graphene quantum dots with tunable emission wavelength for fluorescence bio-imaging. Mater Lett 266:127468
Weissman MR, Winger KT, Ghiassian S, Gobbo P, Workentin MS (2016) Insights on the application of the retro Michael-type addition on maleimide-functionalized gold nanoparticles in biology and nanomedicine. Bioconjug Chem 27:586–593
Wolfe T, Chatterjee D, Lee J, Grant JD, Bhattarai S, Tailor RC et al (2016) Corrigendum to: "Targeted gold nanoparticles enhance sensitization of prostate tumors to megavoltage radiation therapy in vivo" [Nanomedicine: Nanotechnology, Biology, and Medicine 11 (2015) 1277–1283]. Nanomed Nanotechnol Biol Med 12:851–852
Wu Y, Ali MR, Chen K, Fang N, El-Sayed MA (2019) Gold nanoparticles in biological optical imaging. Nano Today 24:120–140
Yadav AN, Kumar R, Jaiswal RK, Singh AK, Kumar P, Singh K (2019) Surface modification of CdS quantum dots: an effective approach for improving biocompatibility. Mater Res Express 6:055002
Yadav N, Tyagi M, Wadhwa S, Mathur A, Narang J (2020) Few biomedical applications of carbon nanotubes. In: Nanoarchitectures built with carbon nanotub. magn. nanoparticles, vol 630, p 347
Yalcin S (2019) Dextran-coated iron oxide nanoparticle for delivery of miR-29a to breast cancer cell line. Pharm Dev Technol 24:1032–1037
Yan L, Gonca S, Zhu G, Zhang W, Chen X (2019) Layered double hydroxide nanostructures and nanocomposites for biomedical applications. J Mater Chem B 7:5583–5601
Yang Y, Wang L, Wan B, Gu Y (2019) Optical active nanomaterials for bioimaging and targeted therapy. Front Bioeng Biotechnol 7:320
Yazdani P, Mansouri E, Eyvazi S, Yousefi V, Kahroba H, Hejazi MS et al (2019) Layered double hydroxide nanoparticles as an appealing nanoparticle in gene/plasmid and drug delivery system in C2C12 myoblast cells. Artif Cells Nanomed Biotechnol 47:436–442
Yogesh GK, Shuaib EP, Roopmani P, Gumpu MB, Krishnan UM, Sastikumar D (2020) Synthesis, characterization and bioimaging application of laser-ablated graphene-oxide nanoparticles (nGOs). Diam Relat Mater 104:107733
Yue L, Li H, Liu Q, Guo D, Chen J, Sun Q et al (2019) Manganese-doped carbon quantum dots for fluorometric and magnetic resonance (dual mode) bioimaging and biosensing. Microchim Acta 186:315
Zaboli M, Raissi H, Moghaddam NR, Farzad F (2020) Probing the adsorption and release mechanisms of cytarabine anticancer drug on/from dopamine functionalized graphene oxide as a highly efficient drug delivery system. J Mol Liq 112458

Biology of Earthworm in the World of Nanomaterials: New Room, Challenges, and Future Perspectives

17

Jayanta Kumar Sahu, Rajendra Kumar Behera, Iswar Baitharu, and Prajna Paramita Naik

Abstract

The environmental basin is continuously getting exposed to the nanomaterials (NMs) with the advancement of the nanotechnological world. Though the eco-toxicological effect of nanoparticles (NPs) is currently widely investigated, we have only a handful of knowledge regarding their effects on the biology of naturally available scavenger like an earthworm. Moreover, our current understanding in the context of the possible influence of nanomaterial on earthworm and their effectiveness in performing the remediation process and nullification of toxic forms is scarce. Again, as the earthworms represent soil invertebrates, they are always focussed on and occupy a crucial trophic level. Many nanotoxicologists and ecotoxicologists have focussed their study on earthworms to find out how and why nanoparticles exert their toxic effects and how to remediate nanotoxicity. Studies showed that nanomaterials exert a minor effect on the survival and growth pattern of adult earthworms, but NPs exposure reduces the reproductive activity of earthworms. However, it is interesting to mention that earthworm has the unique ability to biodegrade or bioaccumulate or biotransform NPs and released them as a component of soil.

J. K. Sahu
School of Life Sciences, Sambalpur University, Burla, Sambalpur, Odisha, India

Department of Biology, Odisha Adarsha Vidyalaya Kursud, Balangir, Odisha, India

R. K. Behera
School of Life Sciences, Sambalpur University, Burla, Sambalpur, Odisha, India

I. Baitharu
P.G. Department of Environmental Sciences, Sambalpur University, Jyoti Vihar, Burla, Odisha, India

P. P. Naik (✉)
P.G. Department of Zoology, Vikram Deb (Auto) College, Jeypore, Odisha, India

M. Arakha et al. (eds.), *Bio-Nano Interface*,
https://doi.org/10.1007/978-981-16-2516-9_17

Keywords

Earthworm · Nanotoxicity · Remediation · Coelomocyte

17.1 Introduction

Nanotechnology has emerged as a widely beneficial field through the production of various nanomaterials (NMs) or devices that have specific characteristics for specific functions (McKee and Filser 2016). Nanoparticles (NPs) are of tiny size that is in nanometre range, which makes them suitable for extensive use in commercial products. For example, zinc oxide and titanium dioxide NPs are used in certain food products and cosmetics. Moreover, silver NPs are used in disinfectants, food packaging, household appliances, and clothing materials. Till date, various application-oriented works have been explored; however, only a few investigations have been carried out on aspects of general and occupational health hazards and safety measures (McNeil 2005). In this connection, a new field of study called nanotoxicology emerged that deals with the effect of nanomaterials or nanoparticles on the environment and human health. Studies have already shown the toxic effects of nanoparticles as cytotoxic, pro-apoptotic, carcinogenic, and cell growth suppressor (Donaldson et al. 2004). Recently, there has been an increased global concern over the role of the toxic effects of NPs and leading to the framing of management strategies to control NP uses and release. OECD (Organisation for Economic Co-operation and Development) founded the Working Party on Manufactured Nanomaterials (WPMN) and Working Party on Nanotechnology (WPN). The use of NMs has been restricted by the U.S. Environmental Protection Agency (USEPA) by following the Toxic Substances Control Act. Again, the regulation program, REACH (Registration, Evaluation, Authorisation and Restriction of Chemicals), is followed by the European Union, which gives directions to handle the manufacture and import of NPs. Furthermore, the use of silver NPs and carbon nanotubes (CNTs) in electronic goods is restricted in the European Union by abiding the guidelines on the use of Ag-NPs and carbon nanotubes in electronic goods by following the restriction of the use of certain hazardous substances (RoHS) in electrical and electronic equipment, 2002/95/EC (Kwak and An 2015). Various ecotoxicologists have studied the adverse effect of NPs on a range of aquatic and terrestrial organisms (Dahms et al. 2011; Kokalj et al. 2018; Kwak and An 2015). The primary cause of NP contamination in soil environment is through the usage to land during wastewater treatment or via accidental spillage (Batley et al. 2013). Such contamination by NMs/NPs is a matter of great worry to terrestrial ecosystems. In particular, the effects of NMs on the terrestrial earthworm species are presently gaining a lot of attention (Hu et al. 2010; Kwak et al. 2014). Moreover, the remediation ability of earthworm about the NM is not extensively investigated. As recommended by OECD and USEPA, the soil-dwelling invertebrates like earthworms are generally used for soil toxicity assessments (ISO 2012; Kwak and An 2015; OECD 1984). Substantial studies regarding the ecotoxicity of various nanoparticles have been reported in terms of bioaccumulation, growth, mortality, and reproduction rate of earthworms. Again, recent reports suggest the role of earthworm as nanoscavengers, which can be potentially employed for remediation of NMs in the contaminated soil.

17.2 Biology of Earthworms

The earthworms are simple, segmented, and cylindrical organisms with a pointed head and slightly flattened posterior. They vary significantly in size. Some species may reach up to 1 m in length (e.g. *Drawida nilamburensis* and *D. grandis*), whereas some species *are* even less than 20 mm long (e.g. *Microscolex phosphoreus*, *Dichogester saliens,* and *Bimastos parvus*) (Bano et al. 1987). The world's most giant known earthworm is about 7 m long *Microchaetus microchaetus* which is found in South Africa. Earthworm occurs in very hydrophilic environments. However, there are species that can thrive under the snow. Many earthworms are reported to be arboreal that can survive in the accumulated detritus in the axils of trees like bamboo, palm, and banana. They lack skeletal structures like bones or cartilage. The bristles like setae on the soft body help in movement and locomotion as they lack true legs. Moreover, they do not have any appendages, but hooks like chaetae help them to attach with substratum. Earthworms are coelomate, and coelomic fluid fills the cavity between the dermal layer and internal organs. The pressure exerted by the coelomic fluid against the dermal layer contributes to form its shape (Edwards and Bohlen 1996; Laverack 2013; Ravindran et al. 2008). The coelomic fluid plays a vital role in various physiological processes like transport of respiratory gases, transportation of nutrients and metabolites, and shipping of signalling molecules. Coelomic fluid exhibits a specific composition with characteristic cell types that help in the physiological processes like immune response, wound healing, and coagulation. Moreover, the coelomic fluid possesses free coelomocytes. Through the dorsal pore every body cavity segment of earthworm face the environment. These dorsal pores enable the microorganisms to enter the coelomic cavity (Yadav 2017b). Coelomocytes maintain the homeostatic regulation, immune reactions, and body regeneration. Coelomocytes play a critical role in the process of phagocytosis, coagulation of coelomic fluid, inflammatory processes, innate immunity, and graft rejection. These cells are the key contributors of the humoral immune response which release mediators like lysozyme, agglutinin, phenoloxidases, and peroxidase. It also secretes antimicrobial factors such as lysenin, eiseniapore, fetidin, and coelomic cytolytic factor. Moreover, chloragocytes are important regulators of various physiological processes through these signalling molecules (Hanley et al. 2009). Coelomocytes have features like macrophages (phagocytosis) and natural killer cells (lytic reactions) (Cooper 2002; Cooper and Kleinschmidt 1995; Porchet-Hennere et al. 1992). Coelomocytes are classified into three types: (1) acidophils, (2) basophils, and (3) chloragocytes (chloragogen cells or eleocytes) (Valembois et al. 1992). Coelomocytes possess a type of granules called chloragosomes that protects against foreign substances (Adamowicz 2005).

Though earthworms are hermaphrodites, they require another earthworm for the mating process. The clitellum is a distinctive epidermal ring-shaped area present in sexually mature worms. The clitellum is glandular in nature and is responsible for the formation of cocoons. Generally, from one end of the cocoon, two or more baby worms hatch. At the age of 6 weeks, the cocoon production starts till the end of six months. In temperate worms, the incubation period of a cocoon ranges between 3 and 30 weeks, whereas 1 and 8 weeks in tropical worms. The new born worms are whitish to almost transparent in colour. They are roughly 1.27–2.54 cm long. It takes 4–6 weeks by the worms to become sexually mature. The process of cocoon production and reproduction can occur throughout the year. Earthworms are highly nutritious as their body contains high quality lysine-rich protein (65%), fats (14%), carbohydrates (14%), and ash (3%) (Graff 1981; Zhenjun et al. 1997). A wide variety of microbial decomposers inhabits the gut of earthworm, which includes the algae, bacteria, fungi, actinomycetes, and protozoa. Sometimes nematodes are also seen throughout the length of the gut of earthworm. Moreover, according to report, more than 50 species of bacteria could be seen in the gut of earthworm (Doube 1998) that are similar to the bacteria those reside in surrounding soil or organic matter they feed (Edwards et al. 1985). Earthworms intake soil for which they have ‘chemoreceptor’ that help in searching for food. The digestive system of earthworm contains pharynx, oesophagus gizzard, and intestine. Basically, the intestine has two parts, i.e. anterior intestine that secretes digestive enzymes and posterior intestine that absorbs nutrients. Earthworms are best in the ingestion and mixing of different soil components and surface or subsurface castings production. However, the pattern of surface or subsurface cast released cast might vary among species. For example, *Pontoscolex corethrurus* and *Polypheretima elongata* deposit thick and sticky mounds, whereas *Lampito mauritii* releases granular casts on the soil surface (Edwards and Bohlen 1996; Laverack 2013).

17.2.1 Earthworm: The Golden Decomposer

Vermicomposting is a widely used biotechnological process of composting where particular earthworm species are exploited to augment the process of bioconversion of waste into better compost. It is mesophilic in nature where microorganisms and earthworms used are active at 10–32 °C of temperature. This process is generally quicker than the regular composting process. The subsequent earthworm castings are also called as worm manure. Worm manure is abundant in plant growth regulators, microbial activity, and pest repellent characteristics. In other words, earthworms are biotransformers that are proficient in converting garbage into ‘gold’, out of a sort of biological alchemy (Adhikary 2012; Dominguez et al. 1997). Earthworm forms a sphere of influence within the soil, which is called the drilosphere system. It encompasses the casts of surface and below ground earthworm, burrow systems, gut of earthworm, and physical, chemical, microbiological, and biological interactions when the earthworm comes in contact with soil environment (Brown et al. 2000; Edwards 2004). Recently, composting and vermicomposting are being combinedly used to stabilize soil substrates for improving soil fertility (Saranraj and Stella 2012; Tognetti et al. 2007). Composting is a better way for the removal of toxic materials and waste sanitization. Furthermore, the subsequent vermicomposting process decreases the particle size of the compost that increases nutrient availability in soil. The vermiconversion of wastes from industries, urban areas, and agro-industries into biofertilizers is now-a-days practised (Das et al. 2016; Elvira et al. 1996; Sangwan et al. 2008; Suthar 2006) as varieties of organic wastes ingested by earthworms are egested as vermicast which is much porous, fine, and microbially active product of humification and decomposition (Dominguez et al. 1997; Ndegwa and Thompson 2001). During vermicomposting, bio-oxidation and stabilization of organic materials occur by the collective activities of microorganisms and earthworms. Initially, the microorganisms play a crucial role in the degradation of the organic matter biochemically. However, earthworms also play a pivotal role in the bio-composting process as they condition, aerate and fragment the substrates for further microbial activity (Garg et al. 2006; Lazcano et al. 2008). Earthworms are natural mechanical blenders. They alter the physico-chemical status of soil by commuting the organic matter and by reducing the C:N ratio that gradually facilitate the generation of increased surface area for microorganisms to act upon for subsequent decomposition process (Dominguez et al. 1997; Gajalakshmi and Abbasi 2004). Vermicompost is a mixture of organic material, worm castings, humus, cocoons, living earthworms, and other organisms. Various organic wastes, including agricultural waste, food wastes, city waste, animal dung, forestry wastes, and leaf litter, are regular substrates of the vermicomposting process (Domínguez et al. 2010; Gajalakshmi and Abbasi 2004; Hand 1988; Logsdon 1994). However, various wastes of industrial origin, including paper pulp, distillery wastes, and guar gum, are also subjected to vermicomposting these days (Sundaravadivel and Ismail 1995; Suthar 2006, 2007). However, to date, the process of vermicomposting has not been carried out in full-fledged industrial scale (Dominguez et al. 1997) as it is a mesophilic process where it is not possible to remove the pathogens (Monroy et al. 2008). Various epigeic forms of earthworms are reported to accelerate the decomposition of waste (Suthar 2009) and produce better quality manure than traditional earthworm species. Before going into the process of vermicomposting, one must have a proper understanding of feeding habits, biomass production potential, and reproduction of particular earthworm species (Domfnguez 2004; Sharma et al. 2005). Since different agro-climatic conditions and soil types drastically alter earthworm diversity, it must be ensured that the selected species of earthworm for the vermicomposting suit the particular climatic condition of the region where the process is being operated (Najar and Khan 2014; Sharma and Poonam 2014).

17.3 Exposure and Bioaccumulation of Nanoparticles in Earthworm

The release of NPs to the soil microenvironment further opens the arena to study the interactions of NPs with biochar and soil. It is also a need of the hour to study the influence of NP availability and toxicity to soil biota. Moreover, the quantification of

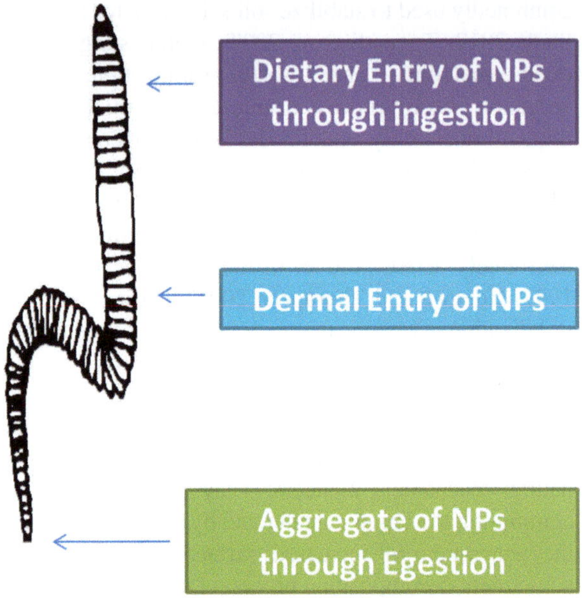

Fig. 17.1 Schematic diagram of routes of exposure of nanoparticle

the uptake and elimination of NMs by various biotas is critical in evaluating its potential in environmental risk. Several studies are reported to explain the bioaccumulation, depuration, and excretion of NPs in body of earthworms. Various techniques are being utilized to study the bioaccumulation of various NPs including inductively coupled plasma (Coleman et al. 2010; Hu et al. 2010; Schlich et al. 2013; Shoults-Wilson et al. 2011a, b, c; Stewart et al. 2013; Unrine et al. 2010a, b), gamma spectrometry (Coutris et al. 2012; Oughton et al. 2008), X-ray micro-spectroscopy (Unrine et al. 2010a), micro X-ray fluorescence spectrometry (Lapied et al. 2011), liquid scintillation counting (Lapied et al. 2010; Petersen et al. 2008, 2009, 2011), flame atomic absorption spectrophotometry (Heggelund et al. 2014; Li et al. 2011), and microwave (Li et al. 2013). It observed that dermal uptake of metal ions of NMs occurs in soft-bodied invertebrate species like earthworms. The process of bioaccumulation and biomagnification of metal-based NPs has been poorly addressed to date. Bioaccumulation of Cu-NPs (Unrine et al. 2010b), Au-NPs (Coutris et al. 2012; Oughton et al. 2008) ZnO-NPs (Hu et al. 2010), Al-NPs (Coleman et al. 2010), carbon-based NPs (Li et al. 2010, 2013; Petersen et al. 2011), Co-NPs (Coutris et al. 2012; Oughton et al. 2008), Ti-NPs (Hu et al. 2010; Lapied et al. 2011) has been studied. Usually, some of the metal-based NPs partly dissolve to form aggregates and agglomerates and accumulate in the environment. As per the study, the bioaccumulation Ti-NPs may occur as aggregates are formed in water and soil solution that can be readily internalized. Moreover, the interaction of NPs with organic molecules (e.g. lipids and carbohydrates) may lead to bioaccumulation and biomagnifications (Baun et al. 2008; Hu et al. 2010; Unrine et al. 2008). NPs can be introduced to earthworms by various routes of exposure (Fig. 17.1). One of them is the direct dermal uptake via the skin. However, the ingestion of contaminated food is the major route of exposure (French et al. 2009). According to reports, earthworms can store molecules such as phosphate, calcium, and sulphur in its chloragogenous tissue (Makama et al. 2015; Van Den Brink et al. 2019). A toxicokinetics investigation with earthworm *Lumbricus rubellus* indicated that Ag-NMs might have an exclusive pathway for the uptake, detoxification, and excretion through the gut wall, chloragogenous tissue, and nephridia (Diez-Ortiz et al. 2015). In Sandy loam with characteristics like low organic content and lower pH, there occurs a greater Ag availability. For instance, *E. fetida* accumulated considerable higher Ag content upon Ag-NPs exposure (Shoults-Wilson et al. 2011c). Moreover, earthworms showed increased avoidance of Ag-NPs in natural sandy loam as compared to artificial soil (Shoults-Wilson et al. 2011a). The extended X-ray absorption fine structure spectroscopy analysis demonstrated that accumulation of Ag-NPs in earthworms upon exposure in soil indicates that both dissolved and non-dissolved particles from NMs contribute to Ag accumulation (Heckmann et al. 2011; Shoults-Wilson et al. 2011a, c). Furthermore, accumulation of aluminium is more in the earthworms upon exposure to NPs as compared to micro-sized particles (Wang et al. 2009).The accumulation of NPs is governed by the particle size and exposure concentration. Again, 20 nm and 55 nm particle size have no effect on accumulation of Au-NP in earthworms. However, most of the Au was internalized from the 20-nm exposure at lower exposure concentration. Likewise, at the highest exposure concentrations like the 55-nm particles exposure, causes higher body burdens (Unrine et al. 2010a). Interestingly, it was found that 65 mg/kg dry soil concentration of Cu-NPs exposure in artificial soil had no effects on the gene expression, growth, survival, and reproduction of earthworms (Unrine et al. 2010b). It is intriguing to note that the Ti-NP bioaccumulation did not occur in *Lumbricus terrestris* when exposed via water, soil, and food. The NPs were prevented from passing the tissue barrier, which resulted in lesser bioaccumulation (Lapied et al. 2011). Moreover, Zn-NPs exposure causes its internalization in body wall tissues of earthworms in the less active form, as evidenced by the backscattered electron imaging with associated EDX element maps (Hooper et al. 2011). It was found that Co-NPs have higher bioaccumulation ability than Ag-NPs in earthworms. Moreover, Zn-NPs have greater retention capacity and toxicity to earthworms than Ti-NPs. Zn and Ti-NPs both are reported to be accumulated in earthworm tissues. However, Zn-NPs shows greater accumulation than Ti-NPs in earthworm tissues (Hu et al. 2010). As soon as the NP enters the biological system, it comes in contact with biomolecules that instantaneously coat the surface of NP and form the protein corona, i.e. the biological identity of NPs. In this form, they can be selectively transported via exocytosis and endocytosis into and out of cells. To be internalized, the NPs bind with the receptor protein of vesicles through the specific chemical groups on NPs and ultimately get metabolized. This may lead to the generation of oxidative stress and trigger the stress-dependent cellular signalling pathways, direct impairment of subcellular organelles, and chromatin condensation, DNA condensation and fragmentation, changes in the pattern of gene expression, inflammatory response, and apoptosis. However, the size of NPs determines the uptake efficiency, internalization pathway selection, intracellular localization, and related cytotoxicity

(Yadav 2017b). The process of entry of NPs into the cell occurs via two endocytosis processes. The first is a phagocytic mechanism that is associated with 0.1–10μm NP size. Phagocytosis is triggered when the ingested NPs comes in with particular receptors present on the cells. Moreover, it is hypothesized that NP can use receptors that are used for the large proteins uptake (Simon-Deckers et al. 2008). The second mechanism of endocytosis involves the receptor-mediated endocytosis. The size of the NP determines which pathway will be followed for the endocytosis. Accordingly, it can be of three types: clathrin-mediated endocytosis (NP $<$ 200 nm), caveolae-mediated endocytosis (NP $=$ 200 nm–1μm), and macropinocytosis (NP $>$ 1μm) (Rejman et al. 2004).

17.4 Toxicity of Nanoparticles to Earthworm

The nanotoxicity in earthworms can be studied in various organizational levels like molecular, cellular, tissue, individual organism, and population level (Fig. 17.2).

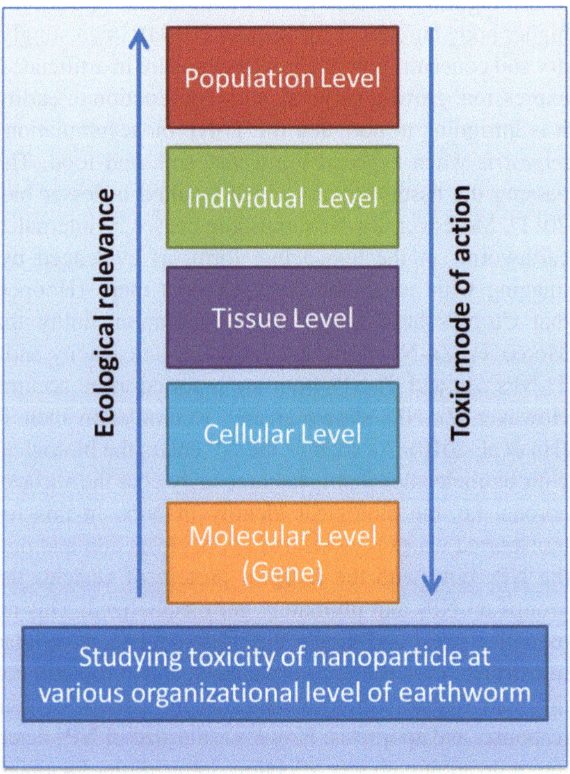

Fig. 17.2 Schematic illustration of nanotoxicity study at various organizational levels

17.4.1 Toxicity at Gene Level/Molecular Level (Genotoxicity)

The expression profile of gene, proteins, and metabolites delivers a quick and sensitive strategy to explore the effect of NPs exposure in earthworms (Hayashi et al. 2013; Hu et al. 2010, 2013; Tsyusko et al. 2012; Unrine et al. 2010a, b; Van der Ploeg et al. 2011; Whitfield Åslund et al. 2012). For any kind of environmental stress, DNA is the principal target (Frenzilli et al. 2001; Nigro et al. 2002). Cells accumulate NPs and then damage the DNA after coming in contact through various modes which further triggers the reactive oxygen species (ROS) mediated cellular signalling processes, activates inflammation, and finally aggravates cell cycle arrest or cell death (Asha Rani et al. 2009; Correia et al. 2017). Ag-NPs were reported to instigate genotoxic effects, as there is increased micronuclei (MN) content in coelomycetes of *A. caliginosa* earthworms (Khalil 2016). Moreover, Si-NPs were reported to aggregate on the cell membrane of eleocytes that damage DNA through the generation of ROS by facilitating the entry of silica (Di Marzio et al. 2018; Hayashi and Engelmann 2013). Moreover, in *E. Hortensis*, earthworms, Co-NPs exposure led to aggravated DNA damage, chromosomal aberrations, and cytokinesis failure (Ciğerci et al. 2016). A statistically significant increase in DNA and chromosomal damage was observed after Fe-NPs exposure along with the elevation of micronuclei and binucleate cells (Ciğerci et al. 2018). Again, a number of studies have been conducted to investigate the effect of NPs on the gene level by studying the expression profile of different antioxidant enzymes owing to the fact that NPs exert their toxic effect by inducing the ROS level. Ag-NPs exposure affected the catalase (CAT) and heat shock protein 70 (HSP70) gene expression in earthworm (Tsyusko et al. 2012). Similarly, *in Eisenia fetida*, the transcriptional levels of CAT, HSP70 GST, and superoxide dismutase (SOD) were slightly induced after Ag-NPs exposure in a dose-dependent manner. However, the enzyme glutathione-S-transferase (GST) transcript level increased significantly with no upregulation in the metallothionein (MT) transcript level (Choi and Park 2015). In another study, however, it led to upregulation of MT expression in coelomocytes of *Eisenia fetida* earthworms (Patricia et al. 2017). Unlikely, it observed that Au-NPs exposure did not enhance metallothionein expression in earthworms even after its application of 60 mg/kg soil for 56 days (Unrine et al. 2010a). But, metallothionein expression was significantly altered against Cu-NPs (20–40 nm) exposure at concentrations of 20 and 65 mg/kg soil for 28 days (Unrine et al. 2010a). The Cu-NPs could not alter the gene expression profile of CAT, HSP60, HSP70, superoxide dismutase (SOD), ubiquitin (UBQ) (Unrine et al. 2010b). Again, C60 was reported not to affect glutathione-S-transferase (GST) and CAT activity. But, C60 could potentially reduce the cytokine-like protein coelomic cytolytic factor-1 (CCF-1) and HSP70 gene expression (Van der Ploeg et al. 2011). In another in vitro study, exposure of C60 to coelomocytes from *Lumbricus rubellus* showed no reduction in cellular viability but a decline in the CCF-1 gene expression, indicating immune suppression or nano-immunotoxicity (van der Ploeg et al. 2014b). Interestingly, the multi-walled carbon nanotubes (MWCNTs) exposure triggered the augmentation of glutathione content and subsequent DNA damage but not the glutathione reductase activity (Van

Der Ploeg et al. 2013). Furthermore, the application of MWCNTs absorbed in nonylphenol caused severe DNA damage as compared to either MWCNT or nonylphenol alone (Hu et al. 2013). Both CNTs triggered the oxidative damage by producing ROS, which subsequently stimulated the lipid peroxidation with malondialdehyde (MDA) generation, but at high NPs stress inhibition SOD, and CAT activity occurred (Yang et al. 2017). According to Hu et al., as reported in 2013 there is also inhibition of cellulase enzyme in response to Zn-NPs and Ti-NPs exposure at 5000 mg/kg concentrations for seven days in the soil. However, the SOD activity did not get altered with the same exposure condition (Hu et al. 2010). Moreover, exposure of Ag-NPs decreased the cellulase activity in the *A. caliginosa* earthworms gut (Khalil 2016). Ti- NPs treatment brought about an escalation of MDA and CAT activities. The exposure to Zn-NPs led to increased CAT and MDA activities. In an intriguing time-course profiling report on *Eisenia fetida,* it was mentioned that at day 1, Ag-NPs treatment caused the upregulation of oxidative stress related genes and enzymes, whereas at day 2, there was suppression of marker genes of energy metabolism. However, on day 14, Ag-NPs exposure resulted in a temporal pattern shift to immune genes followed by metabolic upregulation on day 7 (Hayashi et al. 2013). Moreover, Cu-NPs treatment in *Enchytraeus albidus* led to differences in transcripts level of energy metabolism genes, including monosaccharide transporting NADH dehydrogenase subunit 1, ATPase, and cytochrome c (Gomes et al. 2012). Again, certain amino acids like leucine, isoleucine, valine, and phenylalanine, the sugars glucose, and maltose, and the nucleoside inosine level increased after C60 NPs treatment in *Eisenia fetida* (Lankadurai et al. 2015).

17.4.2 Toxicity at the Cellular Level (Cytotoxicity)

Lapied et al., in a study with *Lumbricus terrestris* found that when the earthworms were exposed to Ag-NPs, it induced a remarkable rise in apoptosis in earthworm (Lapied et al. 2010). Again Hu et al. found severe damage to mitochondria in the gut cells of earthworm upon Zn-NPs and Ti-NPs exposure for seven days in the soil at 5000 mg/kg concentration (Hu et al. 2010). Moreover, the treatment of 50 mg/L MWCNTs in earthworm led to a decrease in the mitochondrial membrane potential (MMP) due to the lipid peroxidation of the membrane (Yang et al. 2017). In specific tissues like cuticle, intestinal epithelium, and chloragogenous tissue of earthworms, TiO2 nanocomposites exposure was shown to promote apoptosis. However, the treatment of nanocomposite for 2–8 weeks of concentration within 0 to 100 mg/kg did not cause any apoptosis (Lapied et al. 2011). Nanoparticles are also reported to damage the cytological membranes. In nanotoxicology studies, the lactate dehydrogenase (LDH) is a sensitive index of the loss of cell membrane integrity. MWCNTs were shown to trigger more severe damage of cytomembrane as extrapolated from the LDH activity (Yang et al. 2017). In another report, the Ag-NP mediated trophic indexes and cytotoxicity were evaluated in coelomocytes of earthworm. The amoebocytes of earthworms eliminate foreign material through the process of phagocytosis and encapsulation, whereas chloragocytes or eleocytes synthesize and secret cytolytic components that stimulate degradation of non-self-material (Bilej et al. 2000). Similarly, in vivo exposure of Ag-NPs demonstrated the decreased viability of coelomocytes in both dose- and time-dependent manner. Ag-NPs entered through the dorsal pores of the earthworm's integument and impacted the cellular membrane permeability of coelomocytes stimulating the cytotoxic effects (Garcia-Velasco et al. 2016, 2017; Irizar et al. 2015; McShane et al. 2012). Again, the citrate coated Ag-NPs induced lysosomal cytotoxicity in *Eisenia andrei* coelomocytes (Kwak et al. 2014). Moreover, Ag-NPs are reported to augment the relative number of eleocytes and increase the mortality of amoebocytes (Bilej et al. 2000). Moreover, exposure of Ag-NP led to an increase in the frequency of MN and binucleate (BN) cells in coelomocytes. It is found that Ag-NPs can cause chromosomal aberrations and failure of cytokinesis in *Aporrectodea caliginosa* earthworm (Zhan 2012). Ag-NP treatment was discovered to enhance the granulocytes proliferation on day 2 in earthworm *Aporrectodea caliginosa* (Bate 2015).

17.4.3 Toxicity at the Tissue Level

Reports indicate that nanoparticles-exposed earthworms show erosion of the epithelium. Ce-NPs treatment led to erosion of the epithelium of *Eisenia fetida*. Furthermore, the earthworms displayed loss of circular muscle architecture, mucocyte proliferation, and fibrosis upon Ce-NPs treatment (Lahive et al. 2014; Spurgeon and Svendsen 2014). *Lumbricus rubellus* earthworms, when exposed to Ag-NPs demonstrated similar results (van der Ploeg et al. 2014a). Magnetite NPs could also induce gut disintegration in earthworms. Again, the exposure to NPs could promote the deposition of lipofuscin in the circular muscle of earthworms (Samrot et al. 2017). Another study on *Eisenia fetida* earthworm upon Ag-NPs exposure induced the histopathological alteration like cuticle disruption. Moreover, it was found that mucocytes were very conspicuous, and their secretion caused complete covering of earthworm's body (Garcia-Velasco et al. 2016). NP treatment is also reported to drive mild hypoplasia of mucous cells in the epidermis of earthworms (Tatsi et al. 2018). Moreover, Ce-NPs exposure encouraged the formation of diffuse architecture of the fatty parenchyma in clitellum (Lahive et al. 2014). Again, exposure of 50 mg/kg of earthworm biomass of fullerenes ended up in the loss of 16% of the epidermal cuticle fibres of the worms (Pakarinen et al. 2011).

17.4.4 Toxicity at the Individual Level

In many cases, NPs in natural soils are shown not to induce significant mortality of earthworms. However, the absence of lethality does not indicate that there will be no damage in earthworms. Interestingly, there was a change in colour of earthworm body from light brown to dark brown/black when exposed to magnetic nanoparticles though a process of melanization involved basically in the defence mechanism. TLR (toll-like receptors) and pattern-recognition receptors (PRRs) present on the surface of the coelomic cell help in the internalization of nanoparticle, leading to the activation of coelomic cytolytic factor (CCF) which further activates the prophenoloxidase (ProPO) cascade to produce melanin (Yadav 2017b). Moreover, the Zn- and Cu-NP treatment causes a reduction in the weight of earthworm and cocoon production (Alahdadi and Behboudi 2015). Moreover, there were substantial decline in the earthworms biomass in response to Ag-NPs exposure (Khalil 2016). Ag-NPs were also reported to exert a negative influence in the reproduction rate of earthworms (Schlich et al. 2013). In another study, Ag-NPs exposure did not alter the fertility of adults but the developmental stage of earthworm, i.e. cocoons and the survival rate of juveniles were affected (Makama et al. 2016; Schlich et al. 2013). Similarly, the reproduction rate of *E. andrei* decreased significantly by Zn-NPs. A substantial decrease in reproductive potential of *E. fetida* (i.e. by 72%) was also reported post-Zn-NPs exposure based on the number of juveniles produced per cocoon (Alves et al. 2019). Moreover, Zn-NPs were reported to decrease the cocoon production, hatching as well as reproduction rate in a concentration, temperature and exposure period dependent manner (Alves et al. 2019; Chouhan and Tripathi 2020). Nano-sized zero-valent iron (nZVI) in *Eisenia fetida* and *Lumbricus rubellus* caused avoidance, weight changes, and mortality (El-Temsah and Joner 2012; Yirsaw et al. 2016). Cu-NP too indicated a lower reproductive output and higher avoidance (Amorim and Scott-Fordsmand 2012). Earthworms could sense exposure to environmental nanoparticles and alter their behaviour. According to a study *Eisenia fetida* consistently avoids soils with Ag-NPs silver (Shoults-Wilson et al. 2011a). Similarly, earthworms were also reported to avoid Ti-NPs at a concentration between 1000 and 5000 mg per kilogram of soil (McShane et al. 2012).

17.4.5 Toxicity at the Population Level

The consequence of exposure of nanoparticle at the population level might be a more appropriate approach to investigate than the individual level making it better predictors of nano-toxicological hazards to earthworms under field conditions. Moreover, the assessment of exposure of NP on the individual level could be subsequently incorporated into effects on the rate of growth of earthworm population and their stage distribution, i.e. population development as composition and number of individuals. In a study, C_{60} exposure in *Lumbricus rubellus* triggered its effect at the population level. In this study, data collected at individual level showed a decline in production of cocoon, and growth rate of juvenile along with a higher rate of the mortality of juveniles. Upon extrapolation at population level, it indicated a lower growth rate in association with a stage distribution shift towards juvenile population. This suggests that juvenile stage represents a more sensitive target for long-term effects of NPs on earthworms populations (Van der Ploeg et al. 2011). Again population modelling study demonstrated a reduction in population growth rates after Ag-NP treatments. Further, no population growth was observed at high Ag-NP concentration as a result of increased juvenile mortality (van der Ploeg et al. 2014a).

Similarly, Zn-NPs exposure is involved with the depletion of the earthworm population over the time (García-Gómez et al. 2014).

17.5 Earthworm as Nanoscavenger

By virtue of their biological and physico-chemical actions, earthworms are potential candidate for bioremediation approaches to stimulate biodegradation of organic contaminants. Earthworms posess an exclusive quality of biotransforming the chemical contaminants into a lesser toxic form that remain bio-accumulated in their tissues through a process call bioremediation (Hickman and Reid 2008; Yadav 2017a). Earlier earthworms were demonstrated to accelerate and disperse microorganisms that degrade organic contaminant and to slow down the binding of organic contaminants with soil and release soil-bound contaminants for further degradation (Ceccanti et al. 2006; Hickman and Reid 2008). Similarly, when exposed to contaminants like NPs, they absorb it through their moist body wall as well as through the process of ingestion by mouth and movement of food through the gut. In the present advancing era of nanotechnology, there is continuous release of NMs, engineered NMs, as well as their derivatives to the environment. However, there is limited understanding regarding the remediation of toxic NMs through natural scavengers, otherwise called as nanoscavengers (Gupta and Yadav 2014; Prabha 2015). The coelomocytes of earthworm have the ability to uptake and accumulate NPs. Therefore, they are being highly investigated in the context of nanoscavenging activity.

17.5.1 Coelomocytes in Nanoscavenging Activity

Uptake of NPs by the coelomocytes occurs through the clathrin-mediated endocytosis, caveolae-mediated phagocytosis, and micro-pinocytosis. Moreover, scavenger receptor class A, a pattern-recognition receptor present on the surface of amebocytes is reported to be crucial for phagocytosis. Again, NPs are recognized by the macrophage receptor (MACRO) of amebocytes for phagocytic clearance. The secretion of cytokine is significantly affected by nanoparticles. This suggests that during exocytosis of NPs, there occurs an indirect communication of coelomocytes in coelomic fluid (Gupta et al. 2014). The phagocytic coelomocytes do possess the ability to uptake and release out NPs in the scavenging aggregates from/into the soil systemand they could be a suitable candidate for the process of nanoremediation (Fig. 17.3). A recent study in *Eisenia fetida* reported the aggregation of Zn-NPs in the coelomic fluid and tissues. According to the report, the cellular uptake of Zn-NPs occurred in the chloragocytes of *Eisenia fetida* when they are exposed to ZnO-NPs of 100 and 50 nm sizes. It is demonstrated that coelomocytes when treated with a dose lower than 3 mg/L do not show substantial DNA damage suggesting its ability in up taking the Zn-NPs and transforming them into microparticles in soil system (Gupta et al. 2014). Again, in a similar kind of study in *E. fetida,* Au-NPs

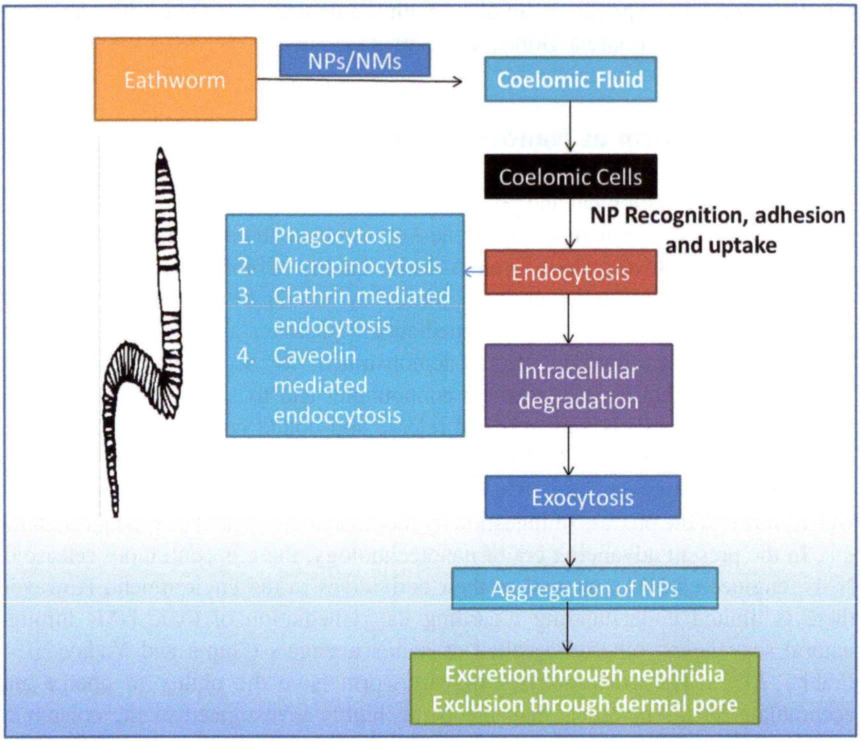

Fig. 17.3 Schematic of representation of role of coelomic cell in NP bioconversion

aggregation was found in gut tissues Unrine et al. (2010a, b). As many NPs have a mild effect on the survival ability, mortality, or the reproductive potential of earthworm, it could be exploited in nanoremediation (van der Ploeg et al. 2014b). Moreover, uptake and accumulation of C60 were reported in the in vitro study of coelomocytes extracted from *Lumbricus rubellus* which showed no decrease in their cellular viability (van der Ploeg et al. 2014b). Furthermore, in *Eisenia andrei* earthworm, microscopic and dissolution studies displayed the aggregation of citrate coated Ag-NPs in coelomic fluid (Kwak et al. 2014). In a study with Ti-NPs, its uptake was found to be performed mainly through the endocytosis process. Moreover, the transmission electron microscopy (TEM) showed that intracellular accumulation of Ti-NP aggregates occurs in the amoebocytes mostly which is irrespective of duration of exposure. Moreover, study found the presence of large and small NP aggregates in the granulations (lysosomes) or entrapped in large vesicles (phagosomes) after 12 and 24 h of NPs exposure (Bigorgne et al. 2012). Recently, a study with Ag-NPs showed the biotransformation of Ag indicating the promising nanoscavenging activities of earthworm. X-ray micro-computed tomography generated 3D images of intact *E. fetida* earthworm demonstrated the preferential accumulation of Ag in the coelomocytes, nephridial epithelium, and chloragogen tissue. This bioaccumulation of Ag by *E. fetida* in the internal tissue indicates a relationship between silver and the thiol-rich proteins like metallothioneins that mostly occur in these tissues and are involved in the sequestration and detoxification of metals (Courtois et al. 2020). The NP aggregation in coelomic fluid is associated with the glycoproteins and glycans (high-mannose-type, N-linked glycans, mucin-type O-linked glycans), and elevated amounts of pentose-containing oligomers because of the interactions of electric charges allowing the NPs to bond together culminating in a zero point of charge. Hence, size and density of aggregates of NP in the coelomic fluid of earthworm is the outcome of interactions between glycoproteins/glycans and NP surfaces (Yadav 2017b).

17.6 Conclusion and Future Perspective

As there is an increasing trend in the release of NPs into the environment, studying the ecotoxicological effects of nanoparticles is the need of the hour. The effect of NPs at various levels of biological endpoints is actively studied, viz. molecular level (differential gene expression, enzyme activity, etc.), cellular level (cytotoxicity), tissue level (histo-toxicity), individual organism level (i.e. bioaccumulation, growth, survival, reproduction, and behaviour), and population level. In various reports, the mortality rate of earthworms does not seem to be adversely influenced by NPs exposure. Rather, it causes a reduction in the reproductive potential of earthworms. As they exert the toxic effect on genes, the earthworms of next generation may get affected. Therefore, genotoxicity and reproductive toxicity of NPs to earthworms must be taken care of. Again, there is a very minimal study on the exploitation of earthworms in the field of nanoparticle remediation. The phagocyte population of coelomocytes present in the coelomic fluid of earthworm is a susceptible target of NMs. Coelomocytes, especially chloragocytes are involved in NP recognition for cellular uptake as well as their accumulation due to sub- and inter-cellular making the earthworm a potential nanoscavenger.

References

Adamowicz A (2005) Morphology and ultrastructure of the earthworm *Dendrobaena veneta* (Lumbricidae) coelomocytes. Tissue Cell 37(2):125–133. https://doi.org/10.1016/j.tice.2004.11.002

Adhikary S (2012) Vermicompost, the story of organic gold: a review. Agric Sci 3(7):905–917. https://doi.org/10.4236/as.2012.37110

Alahdadi I, Behboudi F (2015) The effects of CuO and ZnO nanoparticles on survival, reproduction, absorption, overweight, and accumulation in *Eisenia fetida* earthworm tissues in two substrates. Int J Environ Res 9(1):35–42. https://doi.org/10.22059/ijer.2015.871

Alves ML, Oliveira Filho LCID, Nogueira P, Ogliari AJ, Fiori MA, Baretta D, Baretta CRDM (2019) Influence of ZnO nanoparticles and a non-nano ZnO on survival and reproduction of earthworm and springtail in tropical natural soil. Rev Bras Ciênc Solo. https://doi.org/10.1590/18069657rbcs20180133

Amorim MJB, Scott-Fordsmand JJ (2012) Toxicity of copper nanoparticles and CuCl2 salt to *Enchytraeus albidus* worms: survival, reproduction and avoidance responses. Environ pollut 164:164–168. https://doi.org/10.1016/j.envpol.2012.01.015

Asha Rani P, Low Kah Mun G, Hande MP, Valiyaveettil S (2009) Cytotoxicity and genotoxicity of silver nanoparticles in human cells. ACS Nano 3(2):279–290. https://doi.org/10.1021/nn800596w

Bano K, Kale R, Gajanan G (1987) Culturing of earthworm *Eudrilus eugeniae* for cast production and assessment of worm cast as bio fertilizer. J Soil Biol Ecol 7:98–104

Bate AE (2015) Speciation of and the effects of silver nanoparticles on the earthworm *Aporrectodea caliginosa*.. https://researcharchive.lincoln.ac.nz/handle/10182/6765. Accessed 15 May 2020

Batley GE, Kirby JK, McLaughlin MJ (2013) Fate and risks of nanomaterials in aquatic and terrestrial environments. Acc Chem Res 46(3):854–862. https://doi.org/10.1021/ar2003368

Baun A, Hartmann NB, Grieger K, Kusk KO (2008) Ecotoxicity of engineered nanoparticles to aquatic invertebrates: a brief review and recommendations for future toxicity testing. Ecotoxicology 17(5):387–395. https://doi.org/10.1007/s10646-008-0208-y

Bigorgne E, Foucaud L, Caillet C, Giambérini L, Nahmani J, Thomas F, Rodius F (2012) Cellular and molecular responses of *E. fetida* cœlomocytes exposed to TiO2 nanoparticles. J Nanopart Res 14:959. https://doi.org/10.1007/s11051-012-0959-5

Bilej M, De Baetselier P, Beschin A (2000) Antimicrobial defense of the earthworm. Folia Microbiol (Praha) 45:283. https://doi.org/10.1007/BF02817549

Brown GG, Barois I, Lavelle P (2000) Regulation of soil organic matter dynamics and microbial activity in the drilosphere and the role of interactions with other edaphic functional domains. Eur J Soil Biol 36(3–4):177–198. https://doi.org/10.1016/S1164-5563(00)01062-1

Ceccanti B, Masciandaro G, Garcia C, Macci C, Doni S (2006) Soil bioremediation: combination of earthworms and compost for the ecological remediation of a hydrocarbon polluted soil. Water Air Soil Pollut 177:383–397. https://doi.org/10.1007/s11270-006-9180-4

Choi JS, Park J-W (2015) Molecular characterization and toxicological effects of citrate-coated silver nanoparticles in a terrestrial invertebrate, the earthworm (*Eisenia fetida*). Mol Cell Toxicol 11:423–431. https://doi.org/10.1007/s13273-015-0045-z

Chouhan N, Tripathi G (2020) A comparative study of zinc oxide nanotoxicity on reproductive potential of an earthworm in natural and artificial substrates. Int J Nanosci. https://doi.org/10.1142/S0219581X19500303

Ciğerci İH, Ali MM, Kaygısız ŞY, Liman R (2016) Genotoxicity assessment of cobalt chloride in *Eisenia hortensis* earthworms coelomocytes by comet assay and micronucleus test. Chemosphere 144:754–757. https://doi.org/10.1016/j.chemosphere.2015.09.053

Ciğerci İH, Ali MM, Kaygısız ŞY, Kaya B, Liman R (2018) Genotoxic assessment of different sizes of iron oxide nanoparticles and ionic iron in earthworm (*Eisenia hortensis*) coelomocytes by comet assay and micronucleus test. Bull Environ Contam Toxicol 101(1):105–109. https://doi.org/10.1007/s00128-018-2364-y

Coleman JG, Johnson DR, Stanley JK, Bednar AJ, Weiss CA Jr, Boyd RE, Steevens JA (2010) Assessing the fate and effects of nano aluminum oxide in the terrestrial earthworm, *Eisenia fetida*. Environ Toxicol Chem 29(7):1575–1580. https://doi.org/10.1002/etc.196

Cooper EL (2002) A new model for analyzing antimicrobial peptides with biomedical applications, vol 343 of NATO science series, I: life and behavioural sciences. IOS Press, pp 175–178

Cooper RG, Kleinschmidt EJ (1995) Benchmarking the firm's critical success factors in new product development. J Prod Innovat Manag 12:374–391. https://doi.org/10.1016/0737-6782(95)00059-3

Correia B, Lourenco J, Marques S, Nogueira V, Gavina A, da Graça Rasteiro M, Antunes F, Mendo S, Pereira R (2017) Oxidative stress and genotoxicity of an organic and an inorganic nanomaterial to *Eisenia andrei*: SDS/DDAB nano-vesicles and titanium silicon oxide. Ecotoxicol Environ Saf 140:198–205. https://doi.org/10.1016/j.ecoenv.2017.02.035

Courtois P, Rorat A, Lemiere S, Levard C, Chaurand P, Grobelak A, Lors C, Vandenbulcke F (2020) Accumulation, speciation and localization of silver nanoparticles in the earthworm *Eisenia fetida*. Environ Sci Pollut Res Int. https://doi.org/10.1007/s11356-020-08548-z
Coutris C, Hertel-Aas T, Lapied E, Joner EJ, Oughton DH (2012) Bioavailability of cobalt and silver nanoparticles to the earthworm *Eisenia fetida*. Nanotoxicology 6:186–195. https://doi.org/10.3109/17435390.2011.569094
Dahms H-U, Hagiwara A, Lee J-S (2011) Ecotoxicology, ecophysiology, and mechanistic studies with rotifers. Aquat Toxicol 101:1–12. https://doi.org/10.1016/j.aquatox.2010.09.006
Das D, Bhattacharyya P, Ghosh B, Banik P (2016) Bioconversion and biodynamics of *Eisenia foetida* in different organic wastes through microbially enriched vermiconversion technologies. Ecol Eng 86:154–161. https://doi.org/10.1016/j.ecoleng.2015.11.012
Di Marzio W, Curieses S, Scodeller P, Alberdi J, Sáenz M (2018) Cyto and genotoxicity of positive and negative coated silica nanoparticles on celomocytes of earthworms *Eisenia fetida* (Oligochaeta, Annelida). Adv Environ Stud 2(1):74–81. https://doi.org/10.36959/742/205
Diez-Ortiz M, Lahive E, Kille P, Powell K, Morgan AJ, Jurkschat K, Van Gestel CA, Mosselmans JFW, Svendsen C, Spurgeon DJ (2015) Uptake routes and toxico-kinetics of silver nanoparticles and silver ions in the earthworm *Lumbricus rubellus*. Environ Toxicol Chem 34:2263–2270. https://doi.org/10.1002/etc.3036
Domfnguez J (2004) State-of-the-art and new perspectives on vermicomposting research. In: Edwards CA (ed) Earthworm ecology, 2nd edn. CRC Press, pp 401–424
Dominguez J, Edwards C, Subler S (1997) A comparison of vermicomposting and composting. BioCycle 38:57–59
Domínguez J, Aira M, G6mez-Brand6n M (2010) Vermicomposting: earthworms enhance the work of microbes. In: Insam H, Franke-Whittle I, Goberna M (eds) Microbes at work. Springer, Berlin, Heidelberg, pp 93–114. https://doi.org/10.1007/978-3-642-04043-6_5
Donaldson K, Stone V, Tran C, Kreyling W, Borm PJ (2004) Nanotoxicology. Occup Environ Med 61:727–728. https://doi.org/10.1136/oem.2004.013243
Doube BM (1998) Life in a complex community: functional interactions between earthworms, organic matter, microorganisms, and plants. Earthworm Ecol:179–211
Edwards CA (2004) The importance of earthworms as key representatives of the soil fauna. In: Edwards CA (ed) Earthworm ecology, 2nd edn. CRC Press, pp 3–11. https://doi.org/10.1201/9781420039719
Edwards CA, Bohlen PJ (1996) Biology and ecology of earthworms, 3rd edn. Springer Science & Business Media, pp 11–12
Edwards C, Burrows I, Fletcher K, Jones B (1985) The use of earthworms for composting farm wastes. In composting of agricultural and other wastes. JKR Gasser, Elsevier, Amsterdam
El-Temsah YS, Joner EJ (2012) Ecotoxicological effects on earthworms of fresh and aged nano-sized zero-valent iron (nZVI) in soil. Chemosphere 89:76–82. https://doi.org/10.1016/j.chemosphere.2012.04.020
Elvira C, Goicoechea M, Sampedro L, Mato S, Nogales R (1996) Bioconversion of solid paper-pulp mill sludge by earthworms. Bioresour Technol 57:173–177. https://doi.org/10.1016/0960-8524(96)00065-X
French RA, Jacobson AR, Kim B, Isley SL, Penn RL, Baveye PC (2009) Influence of ionic strength, pH, and cation valence on aggregation kinetics of titanium dioxide nanoparticles. Environ Sci Technol 43:1354–1359. https://doi.org/10.1021/es802628n
Frenzilli G, Nigro M, Scarcelli V, Gorbi S, Regoli F (2001) DNA integrity and total oxyradical scavenging capacity in the Mediterranean mussel, *Mytilus galloprovincialis*: a field study in a highly eutrophicated coastal lagoon. Aquat Toxicol 53:19–32. https://doi.org/10.1016/s0166-445x(00)00159-4
Gajalakshmi S, Abbasi S (2004) Earthworms and vermicomposting. Ind J Biotechnol 3:486–494
García-Gómez C, Babin M, Obrador A, Álvarez J, Fernández M (2014) Toxicity of ZnO nanoparticles, ZnO bulk, and ZnCl 2 on earthworms in a spiked natural soil and toxicological

effects of leachates on aquatic organisms. Arch Environ Contam Toxicol 67:465–473. https://doi.org/10.1007/s00244-014-0025-7

Garcia-Velasco N, Gandariasbeitia M, Irizar A, Soto M (2016) Uptake route and resulting toxicity of silver nanoparticles in *Eisenia fetida* earthworm exposed through Standard OECD Tests. Ecotoxicology 25:1543–1555. https://doi.org/10.1007/s10646-016-1710-2

Garcia-Velasco N, Peña-Cearra A, Bilbao E, Zaldibar B, Soto M (2017) Integrative assessment of the effects produced by Ag nanoparticles at different levels of biological complexity in *Eisenia fetida* maintained in two standard soils (OECD and LUFA 2.3). Chemosphere 181:747–758. https://doi.org/10.1016/j.chemosphere.2017.04.143

Garg P, Gupta A, Satya S (2006) Vermicomposting of different types of waste using *Eisenia foetida*: a comparative study. Bioresour Technol 97:391–395. https://doi.org/10.1016/j.biortech.2005.03.009

Gomes SI, Novais SC, Scott-Fordsmand JJ, De Coen W, Soares AM, Amorim MJ (2012) Effect of Cu-nanoparticles versus Cu-salt in *Enchytraeus albidus* (Oligochaeta): differential gene expression through microarray analysis. Comp Biochem Phys C Toxicol Pharmacol 155(2):219–227. https://doi.org/10.1016/j.cbpc.2011.08.008

Graff O (1981) Preliminary experiments by vermicomposting of different waste materials using *Eudrilus eugeniae*. In: Proceedings of a workshop on the role of earthworms in the stabilization of organic residues, Kalamazoo, Michigan, April 9–12, 1980, pp 179–191

Gupta S, Yadav S (2014) Bioaccumulation of ZnO-NPs in Earthworm *Eisenia fetida* (Savigny). J Bioremed Biodeg 5:250. https://doi.org/10.4172/2155-6199.1000250

Gupta S, Kushwah T, Yadav S (2014) Earthworm coelomocytes as nanoscavenger of ZnO NPs. Nanoscale Res Lett 9(1):259. https://doi.org/10.1186/1556-276X-9-259

Hand P (1988) Earthworm biotechnology (vermicomposting). In: Greenshields R (ed) Resources and applications of biotechnology. Palgrave Macmillan, London, pp 49–58. https://doi.org/10.1007/978-1-349-09574-2_5

Hanley C, Thurber A, Hanna C, Punnoose A, Zhang J, Wingett DG (2009) The influences of cell type and ZnO nanoparticle size on immune cell cytotoxicity and cytokine induction. Nanoscale Res Lett 4(12):1409–1420. https://doi.org/10.1007/s11671-009-9413-8

Hayashi Y, Engelmann P (2013) Earthworm's immunity in the nanomaterial world: new room, future challenges. Invert Surviv J 10:69–76

Hayashi Y, Heckmann L-H, Simonsen V, Scott-Fordsmand JJ (2013) Time-course profiling of molecular stress responses to silver nanoparticles in the earthworm *Eisenia fetida*. Ecotoxicol Environ Saf 98:219–226. https://doi.org/10.1016/j.ecoenv.2013.08.017

Heckmann L-H, Hovgaard MB, Sutherland DS, Autrup H, Besenbacher F, Scott-Fordsmand JJ (2011) Limit-test toxicity screening of selected inorganic nanoparticles to the earthworm *Eisenia fetida*. Ecotoxicology 20(1):226–233. https://doi.org/10.1007/s10646-010-0574-0

Heggelund LR, Diez-Ortiz M, Lofts S, Lahive E, Jurkschat K, Wojnarowicz J, Cedergreen N, Spurgeon D, Svendsen C (2014) Soil pH effects on the comparative toxicity of dissolved zinc, non-nano and nano ZnO to the earthworm *Eisenia fetida*. Nanotoxicology 8(5):559–572. https://doi.org/10.3109/17435390.2013.809808

Hickman ZA, Reid BJ (2008) Earthworm assisted bioremediation of organic contaminants. Environ Int 34(7):1072–1081. https://doi.org/10.1016/j.envint.2008.02.013

Hooper HL, Jurkschat K, Morgan AJ, Bailey J, Lawlor AJ, Spurgeon DJ, Svendsen C (2011) Comparative chronic toxicity of nanoparticulate and ionic zinc to the earthworm *Eisenia veneta* in a soil matrix. Environ int 37(6):1111–1117. https://doi.org/10.1016/j.envint.2011.02.019

Hu C, Li M, Cui Y, Li D, Chen J, Yang L (2010) Toxicological effects of TiO2 and ZnO nanoparticles in soil on earthworm *Eisenia fetida*. Soil Biol Biochem 42(4):586–591. https://doi.org/10.1016/j.soilbio.2009.12.007

Hu C, Cai Y, Wang W, Cui Y, Li M (2013) Toxicological effects of multi-walled carbon nanotubes adsorbed with nonylphenol on earthworm *Eisenia fetida*. Environ Sci Process Impacts 15(11):2125–2130. https://doi.org/10.1039/c3em00376k

Irizar A, Rivas C, García-Velasco N, de Cerio FG, Etxebarria J, Marigómez I, Soto M (2015) Establishment of toxicity thresholds in subpopulations of coelomocytes (amoebocytes vs. eleocytes) of *Eisenia fetida* exposed in vitro to a variety of metals: implications for biomarker measurements. Ecotoxicology 24(5):1004–1013. https://doi.org/10.1007/s10646-015-1441-9

ISO (2012) ISO 11268–2. Soil quality-Effects of pollutants on earthworms-Part 2. Determination of effects on reproduction of *Eisenia fetida/Eisenia andrei*. Geneva, International Organization for Standardization Switzerland. Accessed 14 May 2020

Khalil AM (2016) Physiological and genotoxic responses of the earthworm *Aporrectodea caliginosa* exposed to sub-lethal concentrations of AgNPs. J Basic Appl Zool 74:8–15. https://doi.org/10.1016/j.jobaz.2015.12.004

Kokalj AJ, Drobne D, Novak S (2018) Nanotoxicity evaluation using experimental animals: a detailed review. In: Kumar V et al (eds) Nanotoxicology toxicity evaluation, risk assessment and management. CRC Press, pp 389–418

Kwak JI, An Y-J (2015) Ecotoxicological effects of nanomaterials on earthworms: a review. Hum Ecol Risk Assess 21:1566–1575. https://doi.org/10.1080/10807039.2014.960302

Kwak JI, Lee WM, Kim SW, An YJ (2014) Interaction of citrate-coated silver nanoparticles with earthworm coelomic fluid and related cytotoxicity in *Eisenia andrei*. J Appl Toxicol 34(11):1145–1154. https://doi.org/10.1002/jat.2993

Lahive E, Jurkschat K, Shaw BJ, Handy RD, Spurgeon DJ, Svendsen C (2014) Toxicity of cerium oxide nanoparticles to the earthworm *Eisenia fetida*: subtle effects. Environ Chem 11(3):268–278. https://doi.org/10.1071/EN14028

Lankadurai BP, Nagato EG, Simpson AJ, Simpson MJ (2015) Analysis of *Eisenia fetida* earthworm responses to sub-lethal C60 nanoparticle exposure using 1H-NMR based metabolomics. Ecotoxicol Environ Saf 120:48–58. https://doi.org/10.1016/j.ecoenv.2015.05.020

Lapied E, Moudilou E, Exbrayat J-M, Oughton DH, Joner EJ (2010) Silver nanoparticle exposure causes apoptotic response in the earthworm *Lumbricus terrestris* (Oligochaeta). Nanomedicine (Lond) 5(6):975–984. https://doi.org/10.2217/nnm.10.58

Lapied E, Nahmani JY, Moudilou E, Chaurand P, Labille J, Rose J, Exbrayat J-M, Oughton DH, Joner EJ (2011) Ecotoxicological effects of an aged TiO2 nanocomposite measured as apoptosis in the anecic earthworm *Lumbricus terrestris* after exposure through water, food and soil. Environ Int 37(6):1105–1110. https://doi.org/10.1016/j.envint.2011.01.009

Laverack MS (2013) The physiology of earthworms: international series of monographs on pure and applied biology: zoology. Elsevier

Lazcano C, Gómez-Brandón M, Domínguez J (2008) Comparison of the effectiveness of composting and vermicomposting for the biological stabilization of cattle manure. Chemosphere 72:1013–1019. https://doi.org/10.1016/j.chemosphere.2008.04.016

Li D, Fortner JD, Johnson DR, Chen C, Li Q, Alvarez PJ (2010) Bioaccumulation of 14C60 by the earthworm *Eisenia fetida*. Environ Sci Technol 44(23):9170–9175. https://doi.org/10.1021/es1024405

Li L-Z, Zhou D-M, Peijnenburg WJ, van Gestel CA, Jin S-Y, Wang Y-J, Wang P (2011) Toxicity of zinc oxide nanoparticles in the earthworm, *Eisenia fetida* and subcellular fractionation of Zn. Environ Int 37(6):1098–1104. https://doi.org/10.1016/j.envint.2011.01.008

Li S, Irin F, Atore FO, Green MJ, Cañas-Carrell JE (2013) Determination of multi-walled carbon nanotube bioaccumulation in earthworms measured by a microwave-based detection technique. Sci Tot Environ 445–446:9–13. https://doi.org/10.1016/j.scitotenv.2012.12.037

Logsdon G (1994) Worldwide progress in vermicomposting: earthworms and composting. BioCycle 35(10):63–65

Makama S, Peters R, Undas A, van den Brink NW (2015) A novel method for the quantification, characterisation and speciation of silver nanoparticles in earthworms exposed in soil. Environ Chem 12(6):643–651. https://doi.org/10.1071/EN15006

Makama S, Piella J, Undas A, Dimmers WJ, Peters R, Puntes VF, van den Brink NW (2016) Properties of silver nanoparticles influencing their uptake in and toxicity to the earthworm

Lumbricus rubellus following exposure in soil. Environ Pollut 218:870–878. https://doi.org/10.1016/j.envpol.2016.08.016

McKee MS, Filser J (2016) Impacts of metal-based engineered nanomaterials on soil communities. Environ Sci Nano 3:506–533. https://doi.org/10.1039/c6en00007j

McNeil SE (2005) Nanotechnology for the biologist. J Leukoc Biol 78(3):585–594. https://doi.org/10.1189/jlb.0205074

McShane H, Sarrazin M, Whalen JK, Hendershot WH, Sunahara GI (2012) Reproductive and behavioral responses of earthworms exposed to nano-sized titanium dioxide in soil. Environ Toxicol Chem 31(1):184–193. https://doi.org/10.1002/etc.714

Monroy F, Aira M, Domínguez J (2008) Changes in density of nematodes, protozoa and total coliforms after transit through the gut of four epigeic earthworms (Oligochaeta). Appl Soil Ecol 39(2):127–132. https://doi.org/10.1016/j.apsoil.2007.11.011

Najar IA, Khan AB (2014) Factors affecting distribution of earthworms in Kashmir Valley: a multivariate statistical approach. Proc Zool Soc 67:126–135. https://doi.org/10.1007/s12595-013-0081-4

Ndegwa P, Thompson S (2001) Integrating composting and vermicomposting in the treatment and bioconversion of biosolids. Bioresour Technol 76:107–112. https://doi.org/10.1016/s0960-8524(00)00104-8

Nigro M, Frenzilli G, Scarcelli V, Gorbi S, Regoli F (2002) Induction of DNA strand breakage and apoptosis in the eel Anguilla anguilla. Mar Environ Res 54(3–5):517–520. https://doi.org/10.1016/s0141-1136(02)00178-2

OECD ATT (1984) OECD guideline for testing of chemicals, Earthworm Acute Toxicity Tests. www.oecd.org/chemicalsafety/risk-assessment/1948293.pdf

Oughton DH, Hertel-Aas T, Pellicer E, Mendoza E, Joner EJ (2008) Neutron activation of engineered nanoparticles as a tool for tracing their environmental fate and uptake in organisms. Environ Toxicol Chem 27(9):1883–1887. https://doi.org/10.1897/07-578.1

Pakarinen K, Petersen E, Leppänen M, Akkanen J, Kukkonen J (2011) Adverse effects of fullerenes (nC60) spiked to sediments on *Lumbriculus variegatus* (Oligochaeta). Environ Pollut 159(12):3750–3756. https://doi.org/10.1016/j.envpol.2011.07.014

Patricia CS, Nerea G-V, Erik U, Elena SM, Eider B, Manu S (2017) Responses to silver nanoparticles and silver nitrate in a battery of biomarkers measured in coelomocytes and in target tissues of *Eisenia fetida* earthworms. Ecotoxicol Environ Saf 141:57–63. https://doi.org/10.1016/j.ecoenv.2017.03.008

Petersen EJ, Huang Q, Weber J, Walter J (2008) Bioaccumulation of radio-labeled carbon nanotubes by *Eisenia foetida*. Environ Sci Technol 42(8):3090–3095. https://doi.org/10.1021/es071366f

Petersen EJ, Pinto RA, Landrum PF, Weber J, Walter J (2009) Influence of carbon nanotubes on pyrene bioaccumulation from contaminated soils by earthworms. Environ Sci Technol 43(11):4181–4187. https://doi.org/10.1021/es803023a

Petersen EJ, Pinto RA, Zhang L, Huang Q, Landrum PF, Weber WJ Jr (2011) Effects of polyethyleneimine-mediated functionalization of multi-walled carbon nanotubes on earthworm bioaccumulation and sorption by soils. Environ Sci Technol 45(8):3718–3724. https://doi.org/10.1021/es103004r

Porchet-Hennere E, Dugimont T, Fischer A (1992) Natural killer cells in a lower invertebrate, *Nereis diversicolor*. Eur J Cell Biol 58(1):99–107

Prabha ML (2015) Bioremediation of heavy metal contaminated soil by *Eisenia fetida* and the characterization of synthesized nanoparticles. J Adv Biol Basic Res 1:01

Ravindran B, Dinesh S, Kennedy LJ, Sekaran G (2008) Vermicomposting of solid waste generated from leather industries using epigeic earthworm *Eisenia foetida*. Appl Biochem Biotechnol 151 (2-3):480–488. https://doi.org/10.1007/s12010-008-8222-3

Rejman J, Oberle V, Zuhorn IS, Hoekstra D (2004) Size-dependent internalization of particles via the pathways of clathrin-and caveolae-mediated endocytosis. Biochem J 377:159–169. https://doi.org/10.1042/BJ20031253

Samrot AV, Justin C, Padmanaban S, Burman U (2017) A study on the effect of chemically synthesized magnetite nanoparticles on earthworm: *Eudrilus eugeniae*. Appl Nanosci 7:17–23. https://doi.org/10.1007/s13204-016-0542-y

Sangwan P, Kaushik C, Garg V (2008) Vermiconversion of industrial sludge for recycling the nutrients. Bioresour Technol 99:8699–8704. https://doi.org/10.1016/j.biortech.2008.04.022

Saranraj P, Stella D (2012) Vermicomposting and its importance in improvement of soil nutrients and agricultural crops. Novus Nat Sci Res 1:14–23

Schlich K, Klawonn T, Terytze K, Hund-Rinke K (2013) Effects of silver nanoparticles and silver nitrate in the earthworm reproduction test. Environ Toxicol Chem 32:181–188. https://doi.org/10.1002/etc.2030

Sharma R, Poonam B (2014) Earthworm diversity in trans-gangetic habitats of Haryana, India. Res J Agric For Sci 2(2):1–7

Sharma S, Pradhan K, Satya S, Vasudevan P (2005) Potentiality of earthworms for waste management and in other uses–a review. J Am Sci 1:4–16

Shoults-Wilson W, Zhurbich OI, McNear DH, Tsyusko OV, Bertsch PM, Unrine JM (2011a) Evidence for avoidance of Ag nanoparticles by earthworms (*Eisenia fetida*). Ecotoxicology 20(2):385–396. https://doi.org/10.1007/s10646-010-0590-0

Shoults-Wilson WA, Reinsch BC, Tsyusko OV, Bertsch PM, Lowry GV, Unrine JM (2011b) Effect of silver nanoparticle surface coating on bioaccumulation and reproductive toxicity in earthworms (*Eisenia fetida*). Nanotoxicology 5(3):432–444. https://doi.org/10.3109/17435390.2010.537382

Shoults-Wilson WA, Reinsch BC, Tsyusko OV, Bertsch PM, Lowry GV, Unrine JM (2011c) Role of particle size and soil type in toxicity of silver nanoparticles to earthworms. Soil Sci Soc Am J 75:365–377. https://doi.org/10.2136/sssaj2010.0127nps

Simon-Deckers A, Gouget B, Mayne-L'Hermite M, Herlin-Boime N, Reynaud C, Carriere M (2008) In vitro investigation of oxide nanoparticle and carbon nanotube toxicity and intracellular accumulation in A549 human pneumocytes. Toxicology 253:137–146. https://doi.org/10.1016/j.tox.2008.09.007

Spurgeon DJ, Svendsen C (2014) Toxicity of cerium oxide nanoparticles to the earthworm. Environ Chem 11(3):268–278. https://doi.org/10.1071/EN14028

Stewart DT, Noguera-Oviedo K, Lee V, Banerjee S, Watson DF, Aga DS (2013) Quantum dots exhibit less bioaccumulation than free cadmium and selenium in the earthworm *Eisenia andrei*. Environ Toxicol Chem 32:1288–1294. https://doi.org/10.1002/etc.2182

Sundaravadivel S, Ismail S (1995) Efficacy of a biological filter unit in the treatment of distillery effluents. J Ecotoxicol Environ Monit 5:125–129

Suthar S (2006) Potential utilization of guar gum industrial waste in vermicompost production. Bioresour Technol 97:2474–2477. https://doi.org/10.1016/j.biortech.2005.10.018

Suthar S (2007) Production of vermifertilizer from guar gum industrial wastes by using composting earthworm *Perionyx sansibaricus* (Perrier). Environmentalist 27:329–335. https://doi.org/10.1007/s10669-007-9032-9

Suthar S (2009) Vermistabilization of municipal sewage sludge amended with sugarcane trash using epigeic *Eisenia fetida* (Oligochaeta). J Hazard Mater 163:199–206. https://doi.org/10.1016/j.jhazmat.2008.06.106

Tatsi K, Shaw BJ, Hutchinson TH, Handy RD (2018) Copper accumulation and toxicity in earthworms exposed to CuO nanomaterials: effects of particle coating and soil ageing. Ecotoxicol Environ Saf 166:462–473. https://doi.org/10.1016/j.ecoenv.2018.09.054

Tognetti C, Mazzarino M, Laos F (2007) Improving the quality of municipal organic waste compost. Bioresour Technol 98:1067–1076. https://doi.org/10.1016/j.biortech.2006.04.025

Tsyusko OV, Hardas SS, Shoults-Wilson WA, Starnes CP, Joice G, Butterfield DA, Unrine JM (2012) Short-term molecular-level effects of silver nanoparticle exposure on the earthworm, *Eisenia fetida*. Environ Pollut 171:249–255. https://doi.org/10.1016/j.envpol.2012.08.003

Unrine J, Bertsch P, Hunyadi S (2008) Bioavailability, trophic transfer, and toxicity of manufactured metal and metal oxide nanoparticles in terrestrial environments. In: Grassian

VH (ed) Nanoscience and nanotechnology: environmental and health impacts, pp 345–366. https://doi.org/10.1002/9780470396612.ch14
Unrine JM, Hunyadi SE, Tsyusko OV, Rao W, Shoults-Wilson WA, Bertsch PM (2010a) Evidence for bioavailability of Au nanoparticles from soil and biodistribution within earthworms (*Eisenia fetida*). Environ Sci Technol 44:8308–8313. https://doi.org/10.1021/es101885w
Unrine JM, Tsyusko OV, Hunyadi SE, Judy JD, Bertsch PM (2010b) Effects of particle size on chemical speciation and bioavailability of copper to earthworms (*Eisenia fetida*) exposed to copper nanoparticles. J Environ Qual 39:1942–1953. https://doi.org/10.2134/jeq2009.0387
Valembois P, Lassegues M, Roch P (1992) Formation of brown bodies in the coelomic cavity of the earthworm *Eisenia fetida andrei* and attendant changes in shape and adhesive capacity of constitutive cells. Dev Comp Immunol 16:95–101. https://doi.org/10.1016/0145-305x(92)90010-a
Van Den Brink NW, Kokalj AJ, Silva PV, Lahive E, Norrfors K, Baccaro M, Khodaparast Z, Loureiro S, Drobne D, Cornelis G (2019) Tools and rules for modelling uptake and bioaccumulation of nanomaterials in invertebrate organisms. Environ Sci Nano 6:1985–2001. https://doi.org/10.1039/C8EN01122B
Van der Ploeg M, Baveco J, Van der Hout A, Bakker R, Rietjens I, Van den Brink N (2011) Effects of C60 nanoparticle exposure on earthworms (*Lumbricus rubellus*) and implications for population dynamics. Environ Pollut 159(1):198–203. https://doi.org/10.1016/j.envpol.2010.09.003
Van Der Ploeg MJ, Handy RD, Heckmann L-H, Van Der Hout A, Van Den Brink NW (2013) C60 exposure induced tissue damage and gene expression alterations in the earthworm *Lumbricus rubellus*. Nanotoxicology 7(4):432–440. https://doi.org/10.3109/17435390.2012.668569
van der Ploeg MJ, Handy RD, Waalewijn-Kool PL, van den Berg JH, Herrera Rivera ZE, Bovenschen J, Molleman B, Baveco JM, Tromp P, Peters RJ (2014a) Effects of silver nanoparticles (NM-300K) on *Lumbricus rubellus* earthworms and particle characterization in relevant test matrices including soil. Environ Toxicol Chem 33:743–752. https://doi.org/10.1002/etc.2487
van der Ploeg MJ, van den Berg JH, Bhattacharjee S, de Haan LH, Ershov DS, Fokkink RG, Zuilhof H, Rietjens IM, van den Brink NW (2014b) In vitro nanoparticle toxicity to rat alveolar cells and coelomocytes from the earthworm *Lumbricus rubellus*. Nanotoxicology 8(1):28–37. https://doi.org/10.3109/17435390.2012.744857
Wang H, Wick RL, Xing B (2009) Toxicity of nanoparticulate and bulk ZnO, Al2O3 and TiO2 to the nematode *Caenorhabditis elegans*. Environ Pollut 157(4):1171–1177. https://doi.org/10.1016/j.envpol.2008.11.004
Whitfield Åslund ML, McShane H, Simpson MJ, Simpson AJ, Whalen JK, Hendershot WH, Sunahara GI (2012) Earthworm sub-lethal responses to titanium dioxide nanomaterial in soil detected by 1H NMR metabolomics. Environ Sci Technol 46(2):1111–1118. https://doi.org/10.1021/es202327k
Yadav S (2017a) Contribution of earthworm to bioremediation as a living machine: bioremediation. In: Bhakta JN (ed) Handbook of research on inventive bioremediation techniques. IGI Global, pp 324–340. https://doi.org/10.4018/978-1-5225-2325-3.ch014
Yadav S (2017b) Potentiality of earthworms as bioremediating agent for nanoparticles. In: Ghorbanpour M, Manika K, Varma A (eds) Nanoscience and plant–soil systems. Soil biology, vol 48. Springer, Cham, pp 259–278. https://doi.org/10.1007/978-3-319-46835-8_8
Yang Y, Xiao Y, Li M, Ji F, Hu C, Cui Y (2017) Evaluation of complex toxicity of canbon nanotubes and sodium pentachlorophenol based on earthworm coelomocytes test. PLoS One 12(1):e0170092. https://doi.org/10.1371/journal.pone.0170092
Yirsaw BD, Mayilswami S, Megharaj M, Chen Z, Naidu R (2016) Effect of zero valent iron nanoparticles to *Eisenia fetida* in three soil types. Environ Sci Pollut Res Int 23(10):9822–9831. https://doi.org/10.1007/s11356-016-6193-4
Zhan Y (2012) Effects of silver nanoparticles on bacteria and earthworms. Lincoln University. https://researcharchive.lincoln.ac.nz/handle/10182/5276. Accessed 12 May 2020
Zhenjun S, Xianchun L, Lihui S, Chunyang S (1997) Earthworm as a potential protein resource. Ecol Food Nutr 36:221–236. https://doi.org/10.1080/03670244.1997.9991517

Bioethanol Production from Agricultural Wastes with the Aid of Nanotechnology

18

Manisha Mahapatra and Arun Kumar Pradhan

Abstract

World's energy source is dependent on fossil fuels, which are produced from a natural slow geographical process. But these fossil fuels are non-renewable energy sources. Due to rapid pace of urbanization and industrialization the demand of energy has highly increased. Therefore renewable energy source is needed. Biofuels are renewable energy sources, which are eco-friendly and are produced from biomass. Bioethanol, biodiesel, and HVO (hydrotreated vegetable oil) are types of biofuels. Most biofuels which are produced from sugar, corn or any other agricultural crops are called conventional bio fuels or 1st generation biofuels. To meet the global demand of biofuels, agricultural waste were taken into account as raw materials for bioethanol production, hence called as 2nd generation biofuels. The use of lignocellulosic substances is cost effective, present abundantly, easy to access, renewable. Bioethanol gives economic strategic benefit and are known to be clean and safe fuel. Methods such as pretreatment, enzymatic treatment, fermentation, and distillation are followed. In this treatment method nanotechnology proofs itself a promising technology, by reducing product costs and lowering the harmful environmental impact. Different nanocatalyst, nanomaterials, nanoparticles, nanocomposites are used for production according to the requirement. This bioethanol can be said as fuels for future and globally competitive nature can be felt. As these fuels reduces the carbon dioxide emission, reduces cancer causing gasoline compounds, reduces small particulate matter, less greenhouse gas emission and are compatible with current vehicles, so it can be used in the place of or replacement of other conventional fuels.

M. Mahapatra · A. K. Pradhan (✉)
Center for Biotechnology, Siksha 'O' Anusandhan (Deemed to be University), Bhubaneswar, Odisha, India
e-mail: arunpradhan@soa.ac.in

M. Arakha et al. (eds.), *Bio-Nano Interface*,
https://doi.org/10.1007/978-981-16-2516-9_18

Bioethanol has immense potential in reducing air pollution. Therefore its gives better future without harming the environment.

Keywords

Bioethanol · Agricultural wastes · Nanotechnology · Biofuel · Renewable energy

18.1 Introduction

In today's world everyone's lives are dependent on various fossil energy sources, such as natural gases, oils, coal, petroleum, etc. These energy sources are very much helpful in generation of electricity, production of fuels, and other goods. Everyday world's energy demand is increasing which is continuously fulfilled by continuous and excessive consumption of fossil fuels, coal, etc. This continuous consumption has led to not only depletion of oil/fuels/energy but also elevated the level of environmental pollution. The problem of global warming which everyone is facing is due to emission of gases like carbon dioxide, carbon monoxide, etc. from burning of fossil fuels.

To overcome these problems, now scientists have been working to develop eco-friendly energy sources, which will not affect our precious environment. Biofuels are organic, renewable, energy source which plays crucial role in reducing emission of carbon dioxide.

Bioethanol, biobutanol are alcoholic fuels which are produced by the process of fermentation of sugars, whereas biodiesels are produced from algae, animal fat, vegetable oils, etc. These biofuels are believed to be promising replacement for non-renewable fossil fuels, such as coal, petroleum, etc., which are mainly derived from living organisms. Biofules can be liquid like bioethanol or biobutanol, solid like woods, charcoal, agricultural wastes, dried manure, etc. and gases like hydrogen and methane. Bioethanol is considered as clean and organic. Bioethanol is mainly produced from biomass like sugarcane, rice, wheat, sorghum, corn, potato skin, and agricultural wastes. All these rice, corn, sorghum, etc. are considered as conventional crops due to their valuable as food, hence agricultural wastes are more convenient and better choice to use.

Agricultural wastes are lignocellulosic substances which are abundant in amount, easily available, cost effective, and renewable. Therefore this feedstock has been given very much attention towards the production of bioethanol. Agricultural wastes are pretreated, physical, chemical, and enzymatic methods are used to produce bioethanol. But apart from this, nanotechnology can also be used to produce bioethanol, which also helps in improving the quality of product.

Nanotechnology is a broad field of science which deals with matter at their atomic and molecular level which is called as nanoparticles/nanomaterials. Richard Feynman, in 1965 won the Nobel Prize for the discovery of nanoparticles. This discovery is boon to scientific society by enabling us to do a wide range of research and applications. Nanoparticles or nanomaterials have a wide range of applications in clinical research, disease diagnosis, molecular nanotechnology, research purpose, discovery of chemicals, and many more. Nanoparticles/nanomaterials can be of metals, non-metals, etc., according to need.

Nanomaterials also play an important role in making bioethanol and are very effective. This chapter gives a brief insight about bioethanol, its production by use of nanomaterials and the need of bioethanol production from agro-based waste products.

18.2 Types of Agricultural Wastes Used for Bioethanol

Biofuels like bioethanol, biobutanol, or biodiesels are mainly produced from biomasses. The agriwastes/agricultural wastes are unwanted materials which are produced from agricultural activities during the production of fruits, vegetables, crops, poultry, meat, dairy products. These agricultural wastes are the biomass or raw materials used for the production of bioethanol. These agricultural wastes are classified into crop residues, food wastes, livestock wastes, and fruit and vegetables wastes, agro industries waste (Pattanaik et al. 2009).

The agro industries like food industries, sugar manufacturing industries, slaughterhouse industries include fruit peels, vegetable peels, fruit pomace, starch residue, molasses, sugarcane baggases, edible oils, egg skin, chicken skin, animal fats, etc. The livestock wastes include wastewater solid or liquid manure, etc. Whereas the crop wastes include dry leafs, straws, seed pods, strovers dry bark, dry roots, wood, etc. The three major crop wastes are rice straws, wheat straws, and corn stover. The four major agro-based wastes used for bioethanol production are sugarcane bagasses, rice straw, corn straw, and wheat straw (Fig. 18.1) (Sarkar et al. 2012).

These are cheap, easily available, easily accessible, renewable, organic, and cost effective, therefore most suitable for bioethanol production. These agricultural wastes are mainly utilized as domestic fuel, animal feeder, etc., but wheat and corn straws are the least to be utilized as animal feeder (Kim and Dale 2004).

Looking carefully into the chemical composition of crop agricultural wastes, it is found that they mainly consist of cellulose, hemicelluloses, and lignin. Therefore they are called as lignocellulosic material. Lignocelluloses are complex carbohydrates and hence the 3 major components plays important role in maintaining the structure of cell wall. Cellulose is crystalline and hemicelluloses is branched in structure whereas lignin is hydrophobic which binds or link them together (Peiji et al. 1997). Asia and America are the largest or major producers of agro-based wastes, because rice, wheat, and corn are considered as staple food.

18.3 Nanotechnology in the Field of Bioethanol

The interest in bioethanol as energy fuel has increased because it helps to reduce the emission of various harmful gases into environment and ultimately helps in reducing the pollution. The major impact of burning of fossil fuel is pollution, increase in

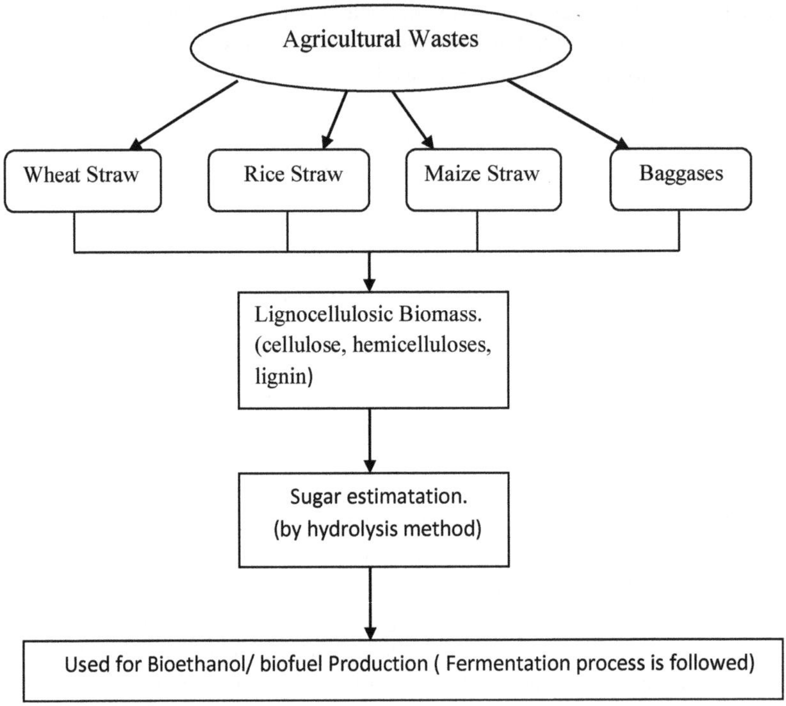

Fig. 18.1 Flow chart of common agricultural wastes used for bioethanol production

greenhouse gases and global warming. Raw materials like rice straw, sorghum, corns are used as food products, feedstuff as well as for bioethanol production. These are food items therefore their cost rises time to time which can increase the cost during bioethanol production (Endo 2009). Therefore non-food products are used, as discussed previously that is agricultural waste. For economically, large production some modern technologies are required, which are more efficient than the conventional materials and methods (Farrell et al. 2006). It is experimentally demonstrated that the lignocelluloses and cellulose are much more effective reducing the carbon gases emission hence good source of biofuels. The general process for bioethanol production involves fermentation of sugars obtained from the raw materials. It includes physical treatment, chemical treatment, biological treatment, enzymatic or acidic hydrolysis, and fermentation and distillation process (Sarkar et al. 2012). Whereas the new technology, that is nanotechnology deals with nanoparticles and it is very efficient in production of bioethanol. Nanoparticles or nanomaterials are suitable, renewable and enhances the production process. The most crucial part is to link the nanoparticle with the different processes of bioethanol production to get the better quality, quantity products. There are various kinds of nanoparticles used such as iron oxides, zinc oxides, nickel cobaltite, and many more (Kushwaha et al. 2018).

Nanotechnology has spread in almost all branches of science from physics, chemistry to most importantly life sciences. Major applications of nanotechnology includes cosmetic industries, treatment of disease, disease diagnosis, fabric industries, for research purpose, coating industries, drug discovery, industrial dyes, nanotoxicology and also in biofuel production (Shah et al. 2010). Use of nanomaterials in the bioenergy field helps in long-term large production, environmental care, and eco-friendly product. Use of these nanoparticles in pretreatment process as well as the fermentation process not only increases the production but also increases the rate of reaction (Chaturvedi et al. 2012). The choosing of type of nanoparticle is also important to get the good end results.

The two major ways involved in bioethanol production are –

- Nanomaterials in the processing of raw materials.
- Nanomaterials in the bioethanol fermentation.

18.3.1 Nanomaterials in the Processing of Raw Materials

Raw material processing is an important part before the fermentation process. This nanomaterials/nanoparticles used for this processes are helpful in bioethanol as well as biobutanol production. This nanaomaterial linked preprocessing of biomasses reduces labor, cost and improves the quality and quantity (Kushwaha et al. 2018). Different kinds of nanoparticles can be used like nanofibres, nanorods, nanowires, nanosheets, nonaclusters like metal and metal oxides (Surendhiran and Sirajunnisa 2014). These tiny materials are used as catalyst to improve the biochemical reaction process and after the reaction these materials are reusable (Kim and Lee 2016). The nanoparticles/nanomaterials are linked with enzymes along with a crosslinker which is called as Nanobiocatalyst/nanobiomaterials. Pretreatment is costly and unavoidable process and challenging affair. These raw materials of agriculture waste consist of lignin, cellulose, and hemicelluloses and are tightly linked together because of lignin hydrophobicity. The preprocessing or pretreatment of raw materials deals with the separation of solubility of different biomass, so the resulted product can undergo for further chemical or biological process to produce bioethanol/biofue (Demirbaş 2005; Sarkar et al. 2012). During this process the crystalline structure of cellulose is broken so amorphous cellulose increases which are suitable for acid hydrolysis and enzymatic attack (Sánchez and Cardona 2008). Following the enzymatic method or acid hydrolysis method monomeric sugars are increased, which can be further processes to yield bioethanol (Mosier et al. 2005). The nanomaterials used in this processes generally penetrate through the cell wall of biomass to release carbohydrates in the form of sugars. Mostly the metal nanoparticles are used and regarded as more effective. Silver nanoparticles were used (AgNP) for the biomass of *chlorella vulgaris* for about 40 min in 100 rpm which resulted in release of carbohydrates and lipids by Abdul Razack et al. (2016). Srivstava N. used NiCo2O4 particles in *Aspergillus fumigates* to produce cellulase enzyme and observed the thermal stability (Srivastava 2014). Wang et al. (2012) used acid functionalized magnetic nanoparticles for the preprocessing of wheat straws. Srivastav et al. used FeSo4 nanoparticles and showed they increase the enzyme immobilization

(Srivastava et al. 2015, 2016). Graphene oxide nanobed was used in *rhizopus oryzae* (Hermanová et al. 2015). Grapheme based nanoparticles also were used in enzyme immobilization and ethanol oxidation (Zhang et al. 2010, Gokhale et al. 2020, Kakaei et al. 2016).The whole purpose of experimenting with these different kind of nanoparticles enables them to be more effective and efficient.

18.3.2 Nanotechnology in the Bioethanol Fermentation

During the production of biofuel, fermentation process is followed and nanoparticles act as the catalyst to influence the biochemical reactions. The metal nanoparticles like nickel, cobalt, iron, manganese, silica, etc. are widely used. From the year 2000 to 2017 mostly used nanoparticles are CoFe2O4.SiO2CH3 nanoparticles, CoFe2O4. SiO2 nanoparticles, Mg-Al hydrotalcite, Fe3O4 nanoparticles, SrO/SiO2, magnetic nanoparticles, carbon nanotubes, nanoparticle sheets, and many more. Carbon nanotubes and magnetic nanoparticles are more advanced and more efficient. They are light weight, have thermal stability, more tensile strength, and mechanical stability (Chaturvedi et al. 2012; Rai et al. 2018). Pan et al. in 2007 showed the hydrogen and carbon monoxide reaction which resulted in ethanol production (Pan et al. 2007). Santos et al. in 2016 used carbon electrode containing the graphite oxide along with copper nanoparticles (Santosh et al. 2016). Lin et al. (2016) synthesized 2D zinc oxide nanosheets which are used along with Ag nanoparticles for the detection of bioethanol detection.

Once the raw material pretreatment is done those pretreatment products are further processed to convert them into bioethanol. The pretreatment product mainly consists of simpler carbohydrates. These are either subjected to acid hydrolysis or enzymatic activity to convert into simpler carbohydrates (sugars) and acids. The sugar part is only used for fermentation hence acid sugar separation is required. They are generally separated by chromatographic separation technique but this is not much effective method. To get pure sugar molecules nanofilters are used. In the nanofilter membrane the acid passes through the membrane but the sugar molecules are rejected by the nanofilter membrane (Fig. 18.2). Most available nanofilter mambranes are GEsepa, NF-90, NF-200, NF-45, SU-600, SU-210, etc. (Niwa et al. 2014). It can be enzyme linked nanofilter or two stage NF assembly. Ultrafiltration and osmosis are unique properties of nanofilter membrane. Lyu (2015) used hydrothermal liquefacation process to recover huge amount of chemicals and sugars.

Now the nanofiltered separated sugar molecules can be utilized for the fermentation process to produce bioethanol. Depending on the need batch, feed batch or continuous fermentation are preferred for large scale production (Chandel et al. 2013). In some cases the final outcome of fermentation process also undergoes distillation or separation process to get the desired product.

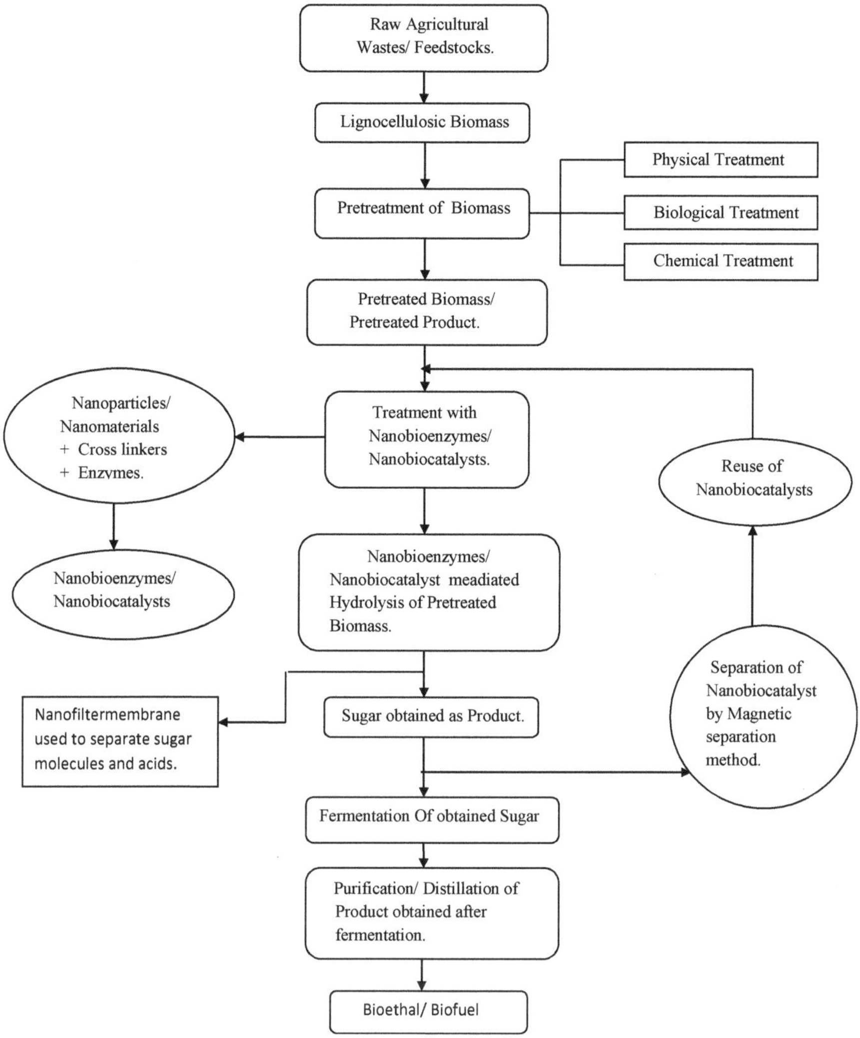

Fig. 18.2 Brief diagrammatic representation of use of nanomaterials in bioethanol production process

18.4 Conclusion and Future Aspects

Bioethanol is green fuel and lignocelluloses biomass (agricultural wastes) is one of the major attractive resources for bioethanol production. Worldwide conventional sources are used, which are not sufficiently present. Hence these agricultural wastes are much more better option for production because they are easily available, cost effective, easily accessible, less labor, and less energy required. These materials does not have food value, hence utilizing them for biofuel or bioethanol production gives a good opportunity for large quantity production due to abundance of material is present. Using nanotechnology for this purpose is a good discovery and boon to society. Most impressive property of the nanomaterials is that it can be renewed again and again utilized for next processes. This chapter explains about the importance of nanoparticles/nanomaterials in eco-friendly bioethanol production. Wide varieties of nanoparticles are used and they are concluded as renewable and reusable which makes it more potent. Nanomaterials are capable of acting upon our desired target and help to convert biomass into liquid bioethanol. In future advancement of this process is possible with respect to advancement of fermentation process. The type and synthesis of nanopartmaterial also determine the product quality after production. For large scale production of these nanoparticles/nanomaterial, enzyme immobilization technique is needed. This immobilization technique also increases their stability, availability and reusability.

The use of nanomaterials yet to be discovered in the other intermediate processes during bioethanol production, to give rise convenient and economically better product. Future challenges involve lots of questions as well as new thoughts, which requires discussion for the possibilities. The future aspects and challenges which require attention are—to make nanoparticles commercial, development of nanoparticles which are more versatile, working on different combinations of nanoparticles to use as catalyst, for detection nanomaterial sensors can be developed so the intermediate processes can be tracked, bionanoparticles synthesis which can be easily utilized during production process ay many more.

There is a huge application of nanotechnology and a lot to discover yet, there problems to overcome and discovery something better and new. Every day the use of nanoparticle and its abilities are discovered and hence this discovery contribute toward eco-friendly bioethanol production.

References

Abdul Razack S, Duraiarasan S, Mani V (2016) Biosynthesis of silver nanoparticle and its application in cell wall disruption to release carbohydrate and lipid from C. vulgaris for biofuel production. Biotechnol Rep 11:70–76

Chandel AK et al (2013) Economics and environmental impact of bioethanol production technologies; an appraisal. Afr J Environ Econ Manag 1(5):126–143

Chaturvedi S, Dave P, Shah N (2012) Applications of nano-catalyst in new era. J Saudi Chem Soc 16(3):307–325

Demirbaş A (2005) Bioethanol from cellulosic materials: a renewable motor fuel from biomass. Energy Sources 27(4):327–337

Endo T (2009) Bioethanol production from woods with the aid of nanotechnology. Synthesiology 2 (4):310–320

Farrell A et al (2006) Ethanol can contribute to energy and environmental goals. Science 311 (5760):506–508

Gokhale A, Lu J, Lee I (2020) Immobilization of cellulase on magnetoresponsive graphene nano-supports. J Mol Catal B: Enzym 90:76–86

Hermanová et al (2015) Graphene oxide immobilized enzymes show high thermal and solvent stability. Nanoscale 7(13):5852–5858
Kakaei K et al (2016) Fabrication of Pt–CeO2 nanoparticles supported sulfonated reduced graphene oxide as an efficient electrocatalyst for ethanol oxidation. Int J Hydrogen Energy 41 (6):3861–3869
Kim S, Dale BE (2004) Global potential bioethanol production from wasted crops and crop residues. Biomass Bioenergy 26(4):361–375
Kim Y, Lee H (2016) Use of magnetic nanoparticles to enhance bioethanol production in syngas fermentation. Bioresource Technol 204:139–144
Kushwaha D, Upadhyay SN, Mishra PK (2018) Nanotechnology in bioethanol/biobutanol production. In: Green nanotechnology for biofuel production. Springer International Publishing A G
Lin L et al (2016) Enhancing ethanol detection by heterostructural silver nanoparticles decorated polycrystalline zinc oxide nanosheets. Ceram Int 42(2):3138–3144
Lyu H (2015) Two-stage nanofiltration process for high-value chemical production from hydrolysates of lignocellulosic biomass through hydrothermal liquefaction. Sep Purif Technol 147:276–283
Mosier N et al (2005) Features of promising technologies for pretreatment of lignocellulosic biomass. Bioresource Technol 96(6):673–686
Niwa M et al (2014) Method for producing sugar solution and method for producing ethanol. Patent number EP 2816124 AI
Pan X et al (2007) Enhanced ethanol production inside carbon-nanotube reactors containing catalytic particles. Nat Mater 6(7):507–511
Pattanaik L et al (2009) Biofuels from agricultural wastes. In: Second and third generation of feedstocks, pp 103–142. https://doi.org/10.1016/B978-0-12-815162-4.00005-7
Peiji G et al (1997) Screening microbial strain for improving the nutritional value of wheat and corn straws as animal feed. Enzyme Microb Technol 20(8):581–584
Rai M et al (2018) Emerging role of nanobiocatalysts in hydrolysis of lignocellulosic biomass leading to sustainable bioethanol production. Catal Rev 61(1):1–26
Sánchez Ó, Cardona C (2008) Trends in biotechnological production of fuel ethanol from different feedstocks. Bioresource Technol 99(13):5270–5295
Santosh FCU et al (2016) Electrochemical determination of total reducing sugars from bioethanol production of lignicellulose cellulose degradating enzymes in the fungus Trametes versicolor. Fuel 163:112–121
Sarkar N et al (2012) Bioethanol production from agricultural waste: an overview. In: Renewal energy, vol 37. Elsevier, West Bengal, India, pp 17–27
Shah V et al (2010) Influence of iron and copper nanoparticle powder on the production of lignocellulose degrading enzymes in the fungus Trametes versicolor. J Hazard Mater 178 (1–3):1141–1145
Srivastava N (2014) Effect of nickel–cobaltite nanoparticles on production and thermostability of cellulases from newly isolated thermotolerant Aspergillus fumigatus NS (Class: Eurotiomycetes). Appl Biochem Biotechnol 174(3):1092–1103
Srivastava N et al (2015) Improved production of reducing sugars from rice straw using crude cellulase activated with Fe3O4/Alginate nanocomposite. Bioresource Technol 183:262–266
Srivastava N et al (2016) Application of ZnO nanoparticles for improving the thermal and pH stability of crude cellulase obtained from Aspergillus fumigatus AA001. Front Microbiol 7
Surendhiran D, Sirajunnisa AR (2014) Nanosilver fabrication mediated by expolysaccharides from Pseudomonas Fluorescens. Am J Pharma Tech Res 4:727–742
Wang D et al (2012) Acid-functionalized nanoparticles for pretreatment of wheat straw. J Biomater Nanobiotechnol 03(03):342–352
Zhang J et al (2010) Graphene oxide as a matrix for enzyme immobilization. Langmuir 26 (9):6083–6085

Nanotechnology for Sustainable Bioenergy Production 19

Yengkhom Disco Singh, Manasa Kumar Panda, Mrinal Kumar Sarma, Rina Ningthoujam, Punuri Jayasekhar Babu, Mohinikanti Das, and Jyoshnamayee Patra

Abstract

Fossil fuel is going to be exhausted in the near future and the emission products of these fuels have been causing serious environmental pollution; therefore, the world is looking for alternate fuel that can be a replacement fuel for fossil fuel. In this context, nanotechnology as multidisciplinary approach can play a major role in improving biofuel production. Nanoparticles like silver, gold and other metallic nanoparticles have been reported to have potential effect in biofuel production. This fuel is more eco-friendly and renewable energy sources and

Y. D. Singh (✉)
Biomaterials and Bioprocessing Research Laboratory, Department of Post Harvest Technology, College of Horticulture and Forestry, Central Agricultural University, Pasighat, Arunachal Pradesh, India

M. K. Panda · J. Patra
Environment and Sustainability Department, CSIR-Institute of Minerals and Materials Technology, Bhubaneswar, Odisha, India

M. K. Sarma
Advance Biofuel Division, The Energy and Resources Institute, New Delhi, India

R. Ningthoujam
Department of Vegetable Science, College of Horticulture and Forestry, Central Agricultural University, Pasighat, Arunachal Pradesh, India

P. J. Babu
Biomaterials and Bioengineering Research Laboratory, Department of Biotechnology, Pachhunga University College, Mizoram University (A Central University), Aizawl, Mizoram, India

M. Das
Department of Botany, College of Basic Science and Humanities, Odisha University of Agriculture and Technology, Bhubaneswar, Odisha, India

M. Arakha et al. (eds.), *Bio-Nano Interface*,
https://doi.org/10.1007/978-981-16-2516-9_19

hence getting more attention in energy research. The unique properties of nanoparticles can be harnessed as carrier to immobilize enzymes in biodiesel and bioethanol production. In this chapter, use of nanoparticles in biofuel production has been discussed elaborately by highlighting the current literature.

Keywords

Nanotechnology · Biofuel · Bioenergy · Sustainable development · Greenhouse gases

19.1 Introduction

Nanotechnology is a technology of science that refers to study and application of small things less than 100 nm in size which have the ability to invent new materials. This novel technology can be employed in fields of physics, chemistry, biology, biotechnology, engineering and many other related subjects of science. Since the particle size is nano, they have physio-chemical properties and can be applied widely in many products (Panda et al. 2020a, b; Gogoi et al. 2015; Babu et al. 2013a). They have higher surface volume ratio and unique quality of solubility, toxicity, strength, magnetism, diffusivity, optics, colour and thermodynamics. The utilization of this technology has extended in different sectors including agriculture, health diagnosis, sports industry, clothing, cosmetics, space elevators, food industry and for faster processing in mobile phone and computers. It also has potential effects in bringing a cleaner energy, better energy storage and treatment of water. The anti-microbial properties of nanomaterial help in preserving, processing, packaging and storage of agriculture and horticulture crops (Baranwal et al. 2018). Silver based nanoparticles are harmful for microorganisms. They are used for coating in medicine, steel coatings, textile fabrics, water treatments, dental work, wound and burns treatment, surgery applications and sunscreen lotions (Durán et al. 2007). Essential nutrient supplements like lycopene, beta carotene, vitamins, anti-oxidant and phytosterols are incorporated in the functional food with help of nanotechnology (Momin et al. 2013). Recently, it is observed that it has potential impacts upon sustainable bioenergy production and its demand has been increasing rapidly. Bioenergy can be defined as renewable form of energy which is derived from natural and biological sources. Renewable energy includes by-products of plants and animals, solar, wind, geothermal, residue of forest and agricultural waste. Bioenergy can be divided into three types based on their derivation. They are solid bioenergy (wood, straw, husks), liquid bioenergy (plant oils, animal fats) and gas bioenergy (methane, carbon dioxide). The advent of nanotechnology for sustainable bioenergy production makes wasteland zone into a useful and potential bioenergy resource. The contamination of our environment is increasing quickly due to burning of fossil fuels and emission of greenhouse gases where it resulted in change of climate (Cherubini 2010; Huber et al. 2006). The exploration of this technology helps in balancing the ecological balance and makes a clean environment. Rai et al. (2016a) reported that nano-based materials like carbon nanotubes, magnetic and metal oxide nanoparticles can be used in the production of sustainable bioenergy. The applications of nanotechnology have the ability to increase the efficiency of water, solar and biogas energies (Hussein 2015). Bioenergy can be in different forms such as biomass, biogas, algae biofuels, solid biofuels, biodiesel and bioethanol. Biomasses are generated from waste of plants and animals for energy production like electricity, heat or may be in any other useful products. They are commonly extracted from crop waste of wheat, straw, sugarcane, vegetables, corn cobs, woods, switchgrass, animal waste (manures) and human waste. The conversion of biomass to bioenergy can be done by process of thermal, chemical, biochemical and electrochemical (Singh et al. 2017; Singh and Satapathy 2018). The presence of molecular size particles and inorganic impurities may cause great effect in the process of conversion. Biogas, biofuel and biodiesel are derived naturally from decomposition of organic waste under anaerobic condition. Organic matters like food scraps, sewage and animal waste break down and undergo fermentation in the absence of oxygen which leads to produced methane and carbon dioxide. Due to the presence of methane, by undergoing several processes it can be used as flame. The conversion of organic waste into other form of energy makes pollution free surrounding and prevents the formation of toxic gases. The application of nanotechnology improves biofuel production and utilized both in scientific and engineering solutions to make an eco-friendly energy resource (Serrano et al. 2009; Ramsurn and Gupta 2013). Biofuel can also be produced from vegetable oils and algae by using various methods. Bioethanol, another form of bioenergy is produced from juice of sugarcane and waste of grains by fermentation process. But it can also be produced by chemical process of reacting ethylene with steam and bioethanol is considered as alternative way to substitute petrol. The production of alcohol can be done from materials after fermentation which is mainly composed of cellulose and hemicellulose, polymeric structure of carbohydrates and lignin (Antunes et al. 2014). The replacement of other fuel by bioethanol has many advantages as it is biodegradable and causes less pollution to the environment. Use of bioethanol reduces the production of carbon monoxide from vehicles and makes a clear and clean air quality. It can also be blended with petroleum to produce a much more efficient fuel. The production of bioethanol is enhanced with cooperation of nano-based catalysts or by immobilizing the enzymes. The approach of this new technology in production of sustainable bioenergy plays a very important role with huge benefit over conventional methods. Different nano-based materials or particles use in production of biofuel, bioethanol and biodiesel directly or indirectly have also been reported. The utilization of this promising tool will bring a change in global energy matrix. Hence, it is believed that the scope of producing renewable form of energy will expand in near future and create solutions for world's energy problems.

Nanoparticle is used for water purification as it has anti-bacterial properties (Stoimenov et al. 2002). They are used for monitoring and removal of metals, cyanide, organics, algae, virus, bacteria and parasites. Nanomaterial is also used in the treatment of water by applying processes like adsorption, sensing, detection, photocatalyst, membrane process and control of bacterial growth (Qu et al. 2013). It has distinct characteristics of high surface area ratio.

19.2 Nanomaterials and Its Properties

The word nanotechnology makes reference to construction, fabrication and production of layout or design by managing the figure and dimensions at nanoscale level (Sarsar et al. 2013). Nanotechnology has a wide area in the current research field for the synthesis of nanoparticles as well as material sciences (Jahn 1999). It also exhibits various applications in many scientific fields including life science, optics, electronics, engineering, environmental science, physics, biotechnology, microbiology and medical sciences (Slawson et al. 1992).

The particles of size range in between 1 and 100 nm produced through nanotechnology process by controlling the shape and size at its nanometre level are known as nanoparticles or nanomaterials (Nalwa 1999). Nanoparticles in recent studies have been reported as "particles of twenty-first century" due to its rare patterns and special properties in comparison to standard or ordinary materials (Camargo et al. 2009). These are also composed of various shapes including oval, round, spherical, rod, cubic, cylindrical, platelets, prism and tubes.

Nanoparticle demonstrates very unique properties that rely upon their dimensions, pattern and arrangement which permits them to collaborate with plants, animals and microorganisms (Husen and Siddiqi 2014). Generally, nanoparticles are manufactured by two processes, i.e. bottom-up and top-bottom process. In bottom-up, the materials are constructed from the ground, i.e. the small atoms constitute together to form molecules and then those molecules aggregate together to form cluster of nanoparticles, whereas in top-bottom, the heavy materials are shredded to obtain nanoparticle (Husen and Siddiqi 2014). The bottom to top approach is considered to be more preferable for the formation of nanoparticles by homogenous system (reducing and capping agents), whereas in top to bottom approach, the reduction of nanoparticle is carried out by special ablation, cutting, sputtering and mechanical grinding (Zhang et al. 2011). Various methods are accessible for the biosynthesis of nanoparticle especially chemical method, physical and biological method (Siddiqi and Husen 2016a). The synthesis of nanomaterials from chemical method is done by chemical reduction, chemical irradiation, electrochemical and pyrolysis (Zhang et al. 2007) since the physical method is implemented for speed transformation life and also does not require any harmful pathogenic chemicals. These methods are known for high consumption of energy constituting arc-discharge, energy ball milling, vapour condensation and sputtering (Asanithi et al. 2012). Therefore, biological method is regarded as much better one rather than physical and chemical method. The nanoparticles synthesized by this method carried out with the usage of herbal extracts and microorganisms. These methods are very eco-friendly in nature, biodegradable, simple, cost-effective, do not require high energy (Babu et al. 2010, 2013b; Singh et al. 2021), free from toxic chemicals as well as scaled up for high production (Siddiqi and Husen 2016b).

In the field of nanotechnology, the synthesis of metal nanoparticles and metal oxide nanoparticles by using green synthesis methods including plant extract, bacteria, fungi, algal extract and yeast has earned high reputation (Siddiqi and Husen 2016a). The plant extracts containing carbohydrates, proteins, nucleic acids, pigments and secondary metabolites mainly act as reducing agent for the nanoparticle synthesis while in microorganism, enzymes, proteins, biomolecules and biosurfactants present in them performed as reducing or capping agents which produces nanoparticles without any toxic or harmful effects (Siddiqi et al. 2018).

Nanomaterials are divided into two classes: inorganic and organic nanoparticles. Inorganic nanoparticles integrate semiconductor nanomaterials (ZnO, ZnS, CdS), metallic nanomaterials (Au, Ag, Cu, Al) and magnetic nanoparticles (Co, Ni, Fe) although organic nanoparticles integrate only carbon nanomaterials such as fullerene, nanotubes, nanowires and quantum dots (Vadlapudi et al. 2014). Inorganic nanoparticles represent its specialty in chemical imaging, cellular and drug delivery for its proper unity and productive utility.

Nowadays, nanotechnology tends to show its interest in sustainable bioenergy production whose main aim is to obtain bioethanol through enzyme immobilization and nanocatalysts and also to provide eco-friendly bioenergy at low cost as well as toxic free.

19.3 Nanotechnology in Bioethanol Production

The requirement of bioenergy has been greatly expanded in view of the past century due to confine availability of fossil fuels (Rai et al. 2016b). Considering the increase in climatic change and its impact on global warming and also declining of fossil fuels production, there is always a need of alternative fuel such as biofuel in order to meet the present global demand (Rai et al. 2016b). Commercially, bioethanol was made by fermentation and transesterification process from vegetable waste, oils, fats and carbohydrates which undergo from huge cost of products and so many industrial or mechanical obstacles (Patumsawad 2011). So, seeing the natural and financial problem, nanotechnology is the best solution for the production of bioethanol from nanoparticles as it has various fascinating properties which can be recovered and reused to meet the needs (Verma et al. 2013).

Bioethanol obtained from feedstock like vegetables, oils, fats, sugarcanes and corns is considered to be first generation of bioethanol (Da Silva and Chandel 2014). The use of non-feedstock such as waste products from woods and agriculture for the bioethanol production is known as second generation of bioethanol (Eggert and Greaker 2014), whereas bioethanol manufactured with the help of microbial biomass is referred to as third generation of bioethanol (Ho et al. 2014).

Magnetic nanomaterials show wide application in the bioenergy field for the manufacture of bioethanol from immobilized enzymes such as cellulases and hemicellulases which can be magnetically retrieved and reclaimed for salvage. Many scientists have observed that carbon nanoparticles are used for the production of biofuels for their contrasting features like conductivity, porosity, 3D electro-active area, a rise in enzyme concentration (Holzinger et al. 2012). Metal or metal oxide nanoparticles follow different methods for the formation of bioethanol such as gas-phase deposition, oxidation reaction, hydrothermal reaction, sono-chemical method, biological method and nanoreactor synthesis (Rishton et al. 1997; Liu et al. 2007, and Moon et al. 2010).

Enzyme immobilization in nanoparticles enhances the reiterate ability, free surface region and eliminates the miserable constancy of enzymes as well as expensive large-scale industrial operations (Akia et al. 2014). The enzyme hemicellulase hydrolyses hemicellulose and converts it into pentose which further gets fermented by some microorganisms for the production of ethanol, namely *Candida tropicalis, Pachysolen tannophilus, Spathaspora passalidarum and Scheffersomyces stipitis* (Antunes et al. 2014). Recent studies also reveal that the enzyme lipases are also used for evolution of biodiesel through triglyceride transesterification of ethanol in the existence of biological catalyst (Da Rós et al. 2010) which exhibits better stability, reaction control and decrease in production cost (Xie and Wang 2012).

19.4 Interaction of Nanomaterials with Biomass

The conventional techniques of energy consumption cause many problems in our environment. Exhaustive utilization of fossil fuels and emission of greenhouse gases disturb our surrounding which resulted in change and fluctuation of climate. Many researchers studied to overcome the problem by replacing with alternative energy sources. The energy generated mostly from plants, animals and their by-products are called bioenergy. It accounts 80% of energy produced globally and exploited for heating, fuels and electricity (Strzalka et al. 2017). The energy from biomass can be released after the process of thermal conversion and pyrolysis. But, still there is wide gap in technical issues in production of energy. So, development of new technology to incorporate with biomass production and achieve high quality energy is required. The introduction of new technology "Nanotechnology" has become an important useful application in generation of energy from biomass. The advantage of nanomaterials and nanoparticles in other field has also been widely reported in the past research discipline. The conversion of biomass from plant is a hindrance due to the chemical composition like cellulose, hemicelluloses and lignin present in the plant (Singh 2019a). But with proper handling and treatment of lignocellulosic residues, it can be transformed into biofuels, bioethanol and biodiesel. Microalgae can also be used as another source for production of biofuels. It has same chemical properties when compared to fuel extract from fossil fuels. In digestion of lignocellulose biomass magnetic nanoparticles can be used as catalyst by hydrolysing the cellobiose. The use of acid functionalized magnetic nanoparticles with 6% of sulphur gave success result of 96.0% in conversion of cellobiose (Peña et al. 2014). The nano biocatalyst increases the hydrolysis process because of their high surface to volume ratio. Combine technology of microwave assisted with carbonaceous acid magnetic nanoparticles in pretreatment and hydrolysis of sugarcane bagasse, *Jatropha* hulls and *Plukenetia* hulls has shown potential yields of 58.3%,

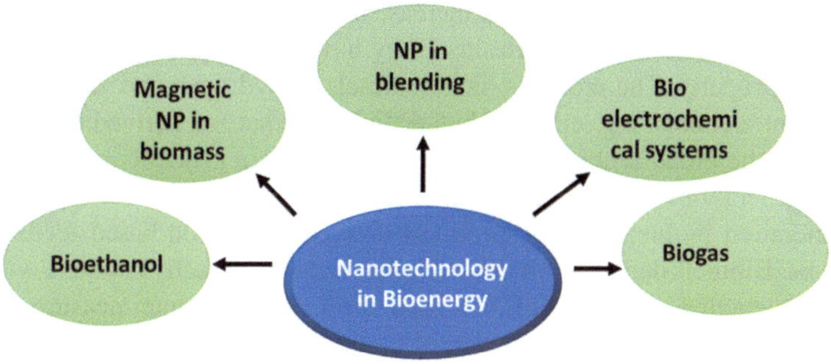

Fig. 19.1 Nanotechnology in bioenergy production

35.6% and 35.8%, respectively (Su et al. 2015). There is some limitation of magnetic nanoparticles in silica-based nanoparticles, nickel-based nanoparticles and carbon nanotubes. Silver based nanoparticles are used to immobilize lignocellulolytic enzyme such as cellulose by coating on surface of nanoparticles. Nickel-based nanoparticles are used to convert glucose into sorbitol molecule in the process of hydrogenation (Kobayashi et al. 2014) and carbon nanotubes are of single walled and multi-walled. They have different properties of electricity, thermal and mechanical strength (Ahmad and Khare 2018). Magnetic nanoparticles can also be used for release of lipid through hydrolysis process of microalgae cell. Although nanoparticles and nanomaterials are implied in production of biomass there is some unwanted factors which affect the release of energy. Nanoparticles have anti-bacterial properties of both gram positive and gram negative (Theivasanthi and Alagar 2011). The electrolysis method of copper nanoparticles synthesized has more anti-bacterial properties when compared with copper nanoparticles synthesized using chemical reduction method. Because of this anti-bacterial factor, the biochemical conversion of biomass is sometimes reduced and inhibits its activity. For example, use of nanomaterials like ZnO, CuO, MnO_2 and Al_2O_3 reduces the energy yield of biogas. Thus, proper analysis to find out the possible application of nanomaterials in biomass production is very important and required (Fig. 19.1).

19.4.1 Biofuel

Every organic substance that we derive directly or indirectly as a result from process of photosynthesis is considered to be known as biomass (Jacobsson and Johnson 2000). Due to its state being so diverse, the definition varies according to its origin and usage. More generally biomass is used for making fuel to meet the increasing demand for transportation and allied energy industries. The fuel derived from biomass to address these current trends is known as biofuel (McKendry 2002). Biofuels have been considered as viable alternatives for conventional petro-fuels. Being renewable sources of energy in addition to being nontoxic and biodegradable, biofuels offer a lot of advantages than fossil fuels (Singh 2019b). Furthermore, tests conducted around the world show no increase in carbon dioxide emission levels with the use of biofuels in their powertrains. They have low sulphur content and a good lubricating property. The products like biodiesel, biohydrogen and various industrially relevant alcohols, e.g. ethanol, butanediol, etc. that are derived from biomass come under the broad umbrella of biofuels. Biofuels are generally categorized into different generations based on the kind of feedstock source used for its production. First-generation biofuels include products derived from food-based feedstocks or food crops. Biofuels derived from these biomass sources directly compete with food, area of production and cost of the food crop. As a countermeasure, second-generation feedstock for the production of biofuels was looked after, mostly sources arising out of non-food crops. The agricultural and industrial residues, e.g. sugarcane molasses, cornflour, plant oils are regarded as first-generation feedstock sources (Cardona et al. 2010). The non-food part of the food crops like stem, husk was taken as a second-generation source which is termed as lignocellulosic biomass for the production of biofuel. Use of inedible oil from non-food sources such as *Jatropha curcas* and mahua for production of biodiesel has been used in industrial level (Qiul et al. 2011). Genetic engineering tools have also been employed to alleviate production of free fatty acids (FFA) and saturated fats from Jatropha plants. Brown grease is also used for production of biodiesel. It is composed of triglycerides, diglycerides, monoglycerides and free fatty acids. It was subsequently discontinued as the production costs went up due to the use of a large amount of base. Used cooking oil (UCO) has been transesterified using base catalyst to convert it to useful biofuel to be used as transportation fuel. It is a cheaper alternative for biodiesel as it also does not conflict with food crop land, avoid cost of waste disposal and treatment (Kawentar and Budiman 2013). Recent studies have shown the potential of microalgae to be an alternative source of biooils. Many classes of microalgae like Bacillariophyceae, Chlorophyceae, and Chrysophyceae that contain high percentage of lipids whose extraction is mainly based on alkaline pretreatment of the feedstock are emerging as new sources of biofuel (Khan et al. 2009). Impressed with initial experimental results workers are genetically engineering and modifying algal strains for enhanced production of lipids with wastewater as nutrient source for lipid synthesis for biofuel production. These photosynthetic microorganisms are the most primitive and are the biggest primary producers of the planet encompassing huge aquatic areas. The discovery of high lipid percentage in relation of their cellular compound fraction compounded with less operational costs pose it to be the next big thing as raw material for biofuel in future (Miao and Wu 2006).

19.5 Applications of Nanotechnology in Bioenergy

Nanotechnology defines the area where nanomaterials are employed to achieve desired objectives. The applicability of nanomaterials in various fields of research has grown explosively over the past years also covering interdisciplinary areas. The dimensional ambit of nanomaterials lies in the range of approximately up to 100 nm (Sekoai et al. 2019). This tiny size gives it an advantage of high active surface area required for efficient reaction outcome. Nanomaterials or nanoparticles also possess multiple morphological variations that give them a wider application in various fields. The existence of unique optical scale properties of nanomaterials finds its usefulness in the field of bioenergy, mechatronics, medicine, pharmaceutical field, ionic liquids and much more (Nizami and Rehan 2018).

Nanomaterial application in the field of bioenergy has been reported as both in direct and indirect ways. Nanoparticles possess high catalytic activity, crystallinity, durability, efficient storage, stability as well as adsorption capacity. They influence those metabolic reactions involved in biological processes responsible for biofuel production upgrading their output capability using nanofibers, metallic nanoparticles and nanotubes. Nanoparticles exert negative effect on inhibitory compounds and additive impact on electron transfer capacity by behaving as catalysts. Nanocrystals, nanomagnets, etc. are used as nano-additives to enhance blending capacity of biofuel with petro-fuels (Palaniappan 2017).

19.5.1 Magnetic Nanoparticles to Pretreat Lignocellulosic Biomass

The conversion of lignocellulosic biomass (LCB) to biofuel requires depolymerization of cellulose to its monomers along with delignification as the initial step (Singh 2017, 2019b). Acid functionalized magnetic nanoparticles (MNPs) are used to hydrolyse cellobiose (β-glucose) from LCB (Peña et al. 2014). The acid functionalized MNPs can adhere different enzymes and act as a nano biocatalyst. Its high active surface area to volume ratio enables higher hydrolysing rate compared to chemical pretreatments. Consequently, after hydrolysis the nanoparticles can be recovered with the help of an external magnetic field and reused again which is a cost-effective approach.

19.5.2 Magnetic Nanoparticles to Pretreat Microalgae Biomass

Immobilized enzymes in nanoparticles can improve biofuel recovery from feedstocks like microalgae. It has also been shown that nanoparticles can be used as nutritional additive in enhancing growth in photosynthetic microorganisms. Nanomaterials possessing optical properties can increase the rate of photosynthesis in photosynthetic microbes by transferring solar energy to the reaction centres of photosystems with minimal loss. However, there is a disadvantage associated with nanomaterials used as nutrient source as they induce oxidation stress endangering the cell viability for repeatable generation (Valko et al. 2005). Workers are trying to address the issue and exploring optimized ways to handle this using various life cycle analysis (LCA) methodologies. Hydrophobic MNPs with immobilized lipase enzyme were employed to extract lipids from microalgae (Lee et al. 2015) and in *Spirulina* cell lysis was carried out through a chitosan modified nanofiber membrane via negative surface charge coupling leading to extraction of fatty acids. The purification factor of 3.3-fold and a resource recovery of 66% indicated a beneficial potential for use of nanoparticle mediated lipid extraction and bioenergy generation (Ng et al. 2019).

19.5.3 Biodiesel Blended in Nano-additives

Nano-additives have been investigated to be used in internal combustion (IC) engine with biodiesel as it improves cetane number, flash point, viscosity of the fuel and many more. Metal oxide nanoparticles and carbon nanotubes have experimented on a trial basis whose initial results of fuel physicochemical properties have indicated it to be an oxygen buffer (Sajith et al. 2010). As an oxygen buffer it reduces emissions of nitrogen oxides resulting in a simultaneous oxidation process of hydrocarbons in IC engine with high efficiency. It also showed that nanoparticle additives in biodiesel are thermally stable, increases fluid layer resistance and viscosity thereby lubricating the fuel injection pump for higher calorific value. Despite its benefits, its high production costs have made it unfit for commercial application till date (Mehta et al. 2014).

19.5.4 Bio-Electrochemical Systems

Bio-electrochemical systems (BES) are defined as platforms where electrochemically active bacteria degrade organic matter from various sources such as wastewater, biomass and resulting in production of bioelectricity, hydrogen and other valuable products, e.g. ethanol, hydrogen peroxide and formic acid (Gul and Ahmad 2019). Microbial fuel cell (MFC), plant MFC, photosynthetic MFC, microbial electrosynthesis cell (MEC) and bio-photovoltaics are different forms of BES. BES setup typically consists of two reaction chambers separated by a cation or anion exchange membrane. Fuel reactions may be carried out in both chambers depending on the need of end products. The chamber where the fuel is burnt or electron is generated is normally termed as anode chamber and the section where this electron is recaptured completing the fuel cycle is termed as cathode chamber (Santoro et al. 2017). The scope of using nanomaterials lies in modifying the fuel rods used in both anodic and cathodic chambers, better known as anodes and cathodes. Use of nanomaterials like quantum dot, metal nanoparticles, carbon nanotubes, etc. on anode and cathode would reduce the shortcomings of BES and improve end product generation. Higher input costs related to nanomaterials, low growth of microbes, nanoparticle toxicity are some issues related with BES when used for biofuel generation (Kalathil and Pant 2016). However, workers are trying to mitigate these costs by employing cheaper, effective alternatives in generating power as well as biofuel. BES systems allow us to use non useful organic feed and turn them into electricity in optimal room temperature and pH neutral conditions. The process is not harsh as it does not require any chemical intervention. Pilot scale research is necessary to validate the use of nanomaterials in large scale proportion for biofuel production.

19.5.5 Nanotechnology in Biogas Production

Worldwide industrialization and population growth have led to a significant increase in energy demand over the past few decades. Currently 80% of total global energy derives from fossil fuels and mostly produced by Brazil and the USA (Hussein 2015). Fossil fuels resources are depleting at a rapid rate due to intensive consumption of fuel. Thus, researchers are more focused on exploring the renewable energy source to cater the world's demand of the fuel (Bazmi and Zahedi 2011). One such renewable energy source is biogas production by anaerobic digestion (AD). Conventional biogas production utilizes various wastes such as animal manure (Bidart et al. 2014), agricultural waste (Karellas et al. 2010) and organic wastes (Zhang et al. 2014) and can convert only 30 to 40% of the organic matter which is not very effective in comparison with energy sources (Lund 2007). Thus, there is a desperate need of modern approaches for the improvement of biogas production. One such modern approach is nanotechnology which has been widely used in various fields. The special physical and chemical properties attributed to the nanoparticles especially high surface area make them to use in diverse field. However, the use of nanoparticles has mixed effect in biogas production. For example, iron oxide nanoparticles (Fe_3O_4) (NPs) additive in biogas production can enhance the yield. Masi et al. described that methane production was significantly increased within 2 days when they added 5 mM Fe^{2+} (Ram et al. 1999). Fe_3O_4 NPs having 7 nm in size were used in anaerobic waste digester at a concentration of 100 ppm and results showed that there is an increment of 180% biogas production compared to the control (Casals et al. 2014). Another study showed that 11% increase in the biogas production upon addition of 10 mg/L cerium oxide (CeO_2), however, there is 30% decrease upon use of 640 mg/L of CeO_2 (Duc 2013). The use of nano-zero-valance iron (NZVI) has positive effect on aerobic digestion and showed the increase of 30.4 of biogas and 40% of methane after 17 days (Su et al. 2013). Xiu et al. also studied the effect of the addition of 1 mg/l NZVI which stimulates the methanogenesis process in which methane production was increased from 58 mmol to 275 mmol upon using NZVI (1 g/l) (Xiu et al. 2010). Al-Ahmad et al. achieved positive effects on biogas production using metal nanoparticles (Fe, Pd, Pt, Ni, Co, Ag and Cu) encapsulated porous silicon dioxide (SiO_2) at the concentration of 5 mg/L showed increase in methane production in case of Ni (70%), Co (48%), Fe (7%) and Pt (6%). Lo et al. studied the effect of micro/nano-fly ash (MNFA) and micro/nano-bottom ash (MNBA) at the concentrations of 0.12, 3, 6, 18 and 30 g/g and 0.6, 12, 36, 60 and 120 g/g, respectively. The results showed significant increase in the biogas production upon addition of 36 g/g MNBA in comparison with other samples as well as control sample.

In contrast to the various nanoparticles, metal oxide nanoparticles showed negative impact on the production of the biogas and biofuel. Otero-González et al. investigated the long-term effect of the copper oxide (CuO) nanoparticles having 37 nm size at the concentration of 1.4 mg/l of 37 nm. This study revealed the significant decrease in methane production in the anaerobic digestive reactor having the (CuO) nanoparticles nanoparticle after 126 days (Otero-González et al. 2014).

Luna-delRisco et al. studied the nano- and micro-particles of copper oxide (CuO; 5 mm–30 nm) and zinc oxide (ZnO: 15 mm, 50–70 nm). The results showed that the nano-sized CuO particles show more negative impact on production of biogas compared to the nano-sized ZnO particles at the same concentration of 15 mg/l (Luna-delRisco et al. 2011). In the same line, Mu and Chen studied the long-term effect of the ZnO nanoparticles at the concentration of 1, 35 and150 mg/g for 105 days of anaerobic digestion. The addition of 35 and 150 mg/g resulted in the decreased production of methane gas up to 81.7% and 24.9% compared to control sample value, respectively (Mu and Chen 2011). The effect of various nanoparticles, namely TiO_2, SiO_2, Al_2O_3 and ZnO NPs on the biogas production was studied by Mu et al. at the concentrations of 6, 30 and 150 mg/g. It was observed that the production of biogas was not affected by the addition of TiO_2, SiO_2 and Al_2O_3 at all given concentrations including the ZnO nanoparticle at 6 mg/g concentrations. However, this study also reveals that the ZnO nanoparticle at the higher concentrations 30 and 150 mg/g decreases the biogas production at the rate of 77.2% and 18.9%, respectively. Estrella et al. studied the effect of different metal oxide NPs such as Al_2O_3, CeO_2, CuO, Fe_2O_3, Mn_2O_3, TiO_2, SiO_2 and ZnO at the concentrations of 1500 mg/l each, revealed that Mn_2O_3, ZnO, CeO_2, CuO decreased the methanogenesis activities to 52, 53, 80 and 87%, respectively, whereas Al_2O_3, Fe_2O_3, TiO_2 and SiO_2 had no significant effect on methanogenesis activities (Gonzalez-Estrella et al. 2013). In contrast to these studies, Garcia et al. studied the effect of TiO_2 NPs at two different concentrations (1120 mg/l,7.5 nm) on the waste sludge for 50 days showed the biogas production in 10% with respect to the control (García et al. 2012).

In conclusion, different nanomaterials have reported to be additive and negative during anaerobic digestion for the biogas production. The metal oxide nanoparticles showed ZnO, CuO, Mn_2O_3 and Al_2O_3 showed negative effect, whereas nano-zero-valance iron (NZVI), porous SiO_2 nanoparticles encapsulated with Ni, Co, Fe and Pt showed increased methanogenesis. Among all nanoparticles, Fe_3O_4 enhanced the methane production by 234% in aerobic digestion and 180% in biogas. The negative impact of these nanoparticles is due to the metal toxicity associated with respective metal ions. However, some metal nanoparticles such as TiO_2 and CeO_2 showed mixed effects depending on their concentration and the digestion time in the bioreactor. These investigations suggest that the various metal oxide nanoparticles show different activity on the production biogas and biofuel at different concentrations and sizes.

19.6 Safety Issues

The incorporation of nanoparticles and nanomaterials in bioenergy production has brought some positive effects but, when is released to the environment, directly or indirectly it may cause health hazards to some extent in living beings. Since nanoparticles are ultra-fine particle, they can be absorbed immediately in living systems and enter our body. The inhalation of nanoparticles and nanomaterials causes toxicity in our health and prolonged exposure can have adverse effect in environment (Jamuna and Ravishankar 2014). This tiny particle can enter our body by process of inhalation, injection and penetration through fraction dermis layer (Gupta et al. 2012). The intake of this nanoparticles may deposit in our respiratory system and may cause lung inflammation and heart problems. Upadhyay et al. (2015) reported that the effects of inhaled nanoparticles can cause asthma, bronchitis and other abnormalities in lungs tissue. After nanoparticles are deposited in lungs, they have the ability to move into other organs like brain, liver and foetus in pregnant women. They also have the potential to enter from mucous membrane of nose and go directly to our brain through olfactory nerve. The nanoparticles can bring damage to mitochondria and break the normal metabolism of cell (Chen et al. 2008). Sighting the negative effects of nanoparticles in our biological system, deep evaluation and extensive studies to reduce or remove toxicity for our safety are utmost essential.

19.7 Conclusion

The global warming due causes by accumulation of toxic gases in our atmosphere has posed the human health. The problems arise out of petrol and coal gases released in the atmosphere have necessitated to search for sustainable alternate renewable energy. Increase in demand for alternate fuels such as biofuel and bioenergy is also due to rapid decline and ever-increasing crude oil price in the market. The use of nanotechnology in biofuel production to escalate the productivity seems to be a reality. Encouragingly, various types of nanomaterials such as carbon nanotubes, nanofibers, magnetic nanoparticles and other metal oxides are found to show positive effects in biofuel production. Magnetic nanoparticles have been extensively tested in biofuel production due to their easy recovery system from the bioreactor. Although the use of nanotechnology in biofuel production has been beneficial and recommended in large-scale production, there are certain safety issues that need to be addressed meticulously for long-term studies.

References

Ahmad R, Khare SK (2018) Immobilization of *Aspergillus niger* cellulase on multiwall carbon nanotubes for cellulose hydrolysis. Bioresour Technol 252:72–75

Akia M, Yazdani F, Motaee E, Han D, Arandiyan H et al (2014) A review on conversion of biomass to biofuel by nanocatalysts. Biofuel Res J 1(1):16–25

Antunes FAF, Chandel AK, Milessi TSS, Santos JC, Rosa CA, Da Silva SS et al (2014) Bioethanol production from sugarcane bagasse by a novel Brazilian pentose fermenting yeast Scheffersomyces shehatae UFMG-HM 52.2: evaluation of fermentation medium. Int J Chem Eng. https://doi.org/10.1155/2014/180681

Asanithi P, Chaiyakun S, Limsuwan P et al (2012) Growth of silver nanoparticles by DC magnetron sputtering. J Nanomater. https://doi.org/10.1155/2012/963609

Babu PJ, Sharma P, Borthakur BB, Das RK, Nahar P, Bora U (2010) Synthesis of gold nanoparticles using Mentha arvensis leaf extract. Int J Green Nanotechnol Phys Chem 2(2): P62–P68

Babu PJ, Sharma P, Saranya S, Bora U (2013a) Synthesis of gold nanoparticles using ethonolic leaf extract of Bacopa monnieri and UV irradiation. Mater Lett 93:431–434

Babu PJ, Sharma P, Saranya S, Tamuli R, Bora U (2013b) Green synthesis and characterization of biocompatible gold nanoparticles using Solanum indicum fruits. Nanomaterials Nanotechnol 3: 4

Baranwal A, Srivastava A, Kumar P, Bajpai VK, Maurya PK, Chandra P et al (2018) Prospects of nanostructure materials and their composites as antimicrobial agents. Front Microbiol 9:422

Bazmi AA, Zahedi G (2011) Sustainable energy systems: role of optimization modeling techniques in power generation and supply—a review. Renew Sust Energ Rev 15:3480–3500

Bidart C, Fröhling M, Schultmann F et al (2014) Livestock manure and crop residue for energy generation: macro-assessment at a national scale. Renew Sust Energ Rev 38:537–550

Camargo PHC, Satyanarayana KG, Wypych F et al (2009) Nanocomposites: synthesis, structure, properties and new application opportunities. Mater Res 12(1):1–39

Cardona CA, Quintero JA, Paz IC et al (2010) Production of bioethanol from sugarcane bagasse: status and perspectives. Bioresour Technol 101(13):4754–4766

Casals E, Barrena R, García A, González E, Delgado L, Busquets-Fité M et al (2014) Programmed iron oxide nanoparticles disintegration in anaerobic digesters boosts biogas production. Small 10(14):2801–2808

Chen L, Yokel RA, Hennig B, Toborek M et al (2008) Manufactured aluminum oxide nanoparticles decrease expression of tight junction proteins in brain vasculature. J Neuroimmune Pharmacol 3(4):286–295

Cherubini F (2010) The bio refinery concept: using biomass instead of oil for producing energy and chemicals. Energy Convers Manag 51(7):1412–1421

Da Rós PC, Silva GA, Mendes AA, Santos JC, De Castro HF et al (2010) Evaluation of the catalytic properties of Burkholderia cepacia lipase immobilized on non-commercial matrices to be used in biodiesel synthesis from different feedstocks. Bio Resoure Technol 101(14):5508–5516

Da Silva SS, Chandel AK (2014) Biofuels in Brazil. Springer, Switzerland

Duc NM (2013) Effects of CeO2 and ZnO nanoparticles on anaerobic digestion and toxicity of digested sludge. Asian Institute of Technology, Thailand

Durán N, Marcato PD, De Souza GI, Alves OL, Esposito E et al (2007) Antibacterial effect of silver nanoparticles produced by fungal process on textile fabrics and their effluent treatment. J Biomed Nanotechnol 3(2):203–208

Eggert H, Greaker M (2014) Promoting second generation biofuels: does the first generation pave the road? Energies 7(7):4430–4445

García A, Delgado L, Ja T, Casals E, González E, Puntes V et al (2012) Effect of cerium dioxide, titanium dioxide, silver, and gold nanoparticles on the activity of microbial communities intended in wastewater treatment. J Hazard Mater 199–200:64–72

Gogoi N, Babu PJ, Mahanta C, Bora U (2015) Green synthesis and characterization of silver nanoparticles using alcoholic flower extract of Nyctanthes arbortristis and in vitro investigation of their antibacterial and cytotoxic activities. Mater Sci Eng C 46:463–469

Gonzalez-Estrella J, Sierra-Alvarez R, Field JA et al (2013) Toxicity assessment of inorganic nanoparticles to acetoclastic and hydrogenotrophic methanogenic activity in anaerobic granular sludge. J Hazard Mater 260:278–285

Gul MM, Ahmad KS (2019) Bioelectrochemical systems: sustainable bio-energy powerhouses. Biosens Bioelectron 142:111576

Gupta I, Duran N, Rai M et al (2012) Nano-silver toxicity: emerging concerns and consequences in human health. In: Nano-Antimicrobials. Springer, Berlin, pp 525–548

Ho SH, Ye X, Hasunuma T, Chang JS, Kondo A et al (2014) Perspectives on engineering strategies for improving biofuel production from microalgae—a critical review. Biotechnol Adv 32(8):1448–1459

Holzinger M, Le Goff A, Cosnier S et al (2012) Carbon nanotube/enzyme biofuel cells. Electrochim Acta 82:179–190

Huber GW, Iborra S, Corma A et al (2006) Synthesis of transportation fuels from biomass: chemistry, catalysts, and engineering. Chem Rev 106(9):4044–4098

Husen A, Siddiqi KS (2014) Phytosynthesis of nanoparticles: concept, controversy and application. Nanoscale Res Lett 9(1):229

Hussein AK (2015) Applications of nanotechnology in renewable energies—a comprehensive overview and understanding. Renew Sust Energ Rev 42:460–476

Jacobsson S, Johnson A (2000) The diffusion of renewable energy technology: an analytical framework and key issues for research. Energy Policy 28(9):625–640

Jahn W (1999) Chemical aspects of the use of gold clusters in structural biology. J Struct Biol 127(2):106–112

Jamuna Bai A, Ravishankar Rai V (2014) Environmental risk, human health, and toxic effects of nanoparticles. Nanomater Environ Prot:523

Kalathil S, Pant D (2016) Nanotechnology to rescue bacterial bidirectional extracellular electron transfer in bio electrochemical systems. RSC Adv 6:30582–30597

Karellas S, Boukis I, Kontopoulos G et al (2010) Development of an investment decision tool for biogas production from agricultural waste. Renew Sust Energ Rev 14:1273–1282

Kawentar WA, Budiman A (2013) Synthesis of biodiesel from second-used cooking oil. Energy Procedia 32:190–199

Khan SA, Hussain MZ, Prasad S, Banerjee UC et al (2009) Prospects of biodiesel production from microalgae in India. Renew Sust Energ Rev 13(9):2361–2372

Kobayashi H, Hosaka Y, Hara K, Feng B, Hirosaki Y, Fukuoka A et al (2014) Control of selectivity, activity and durability of simple supported nickel catalysts for hydrolytic hydrogenation of cellulose. Green Chem 16(2):637–644

Lee YC, Lee K, Oh YK et al (2015) Recent nanoparticle engineering advances in microalgal cultivation and harvesting processes of biodiesel production: a review. Bioresour Technol 184: 63–72

Liu JF, Lu MF, Chai P, Fu L, Wang ZL, Cao XQ, Meng J et al (2007) The magnetic and structural properties of hydrothermal-synthesized single-crystal Sn1− xFexO2 nanograins. J Magn Magn Mater 317(1–2):1–7

Luna-delRisco M, Orupõld K, Dubourguier H-C et al (2011) Particle-size effect of CuO and ZnO on biogas and methane production during anaerobic digestion. J Hazard Mater 189(1–2):603–608

Lund H (2007) Renewable energy strategies for sustainable development. Energy 32:912–919

McKendry P (2002) Energy production from biomass (part 2): conversion technologies. Bioresour Technol 83(1):47–54

Mehta RN, Chakraborty M, Parikh PA et al (2014) Nanofuels: combustion, engine performance and emissions. Fuel 120:91–97

Miao X, Wu Q (2006) Biodiesel production from heterotrophic microalgal oil. Bioresour Technol 97(6):841–846

Momin JK, Jayakumar C, Prajapati JB et al (2013) Potential of nanotechnology in functional foods. Emirates J Food Agricult (EJFA) 25(1):10–19

Moon JW, Rawn CJ, Rondinone AJ, Love LJ, Roh Y, Everett SM, Lauf RJ, Phelps TJ et al (2010) Large-scale production of magnetic nanoparticles using bacterial fermentation. J Ind Microbiol Biotechnol 37(10):1023–1031

Mu H, Chen Y (2011) Long-term effect of ZnO nanoparticles on waste activated sludge anaerobic digestion. Water Res 45(17):5612–5620

Nalwa HS (ed) (1999) Handbook of nanostructured materials and nanotechnology, five-volume set. Academic Press

Ng IS, Tang MS, Show PL, Chiou ZM, Tsai JC, Chang YK et al (2019) Enhancement of C-phycocyanin purity using negative chromatography with chitosan-modified nanofiber membrane. Int J Biol Macromol 132:615–628

Nizami AS, Rehan M (2018) Towards nanotechnology-based biofuel industry. Biof Res J 5(2):798–799

Otero-González L, Ja F, Sierra-Alvarez R et al (2014) Inhibition of anaerobic waste water treatment after long-term exposure to low levels of CuO nanoparticles. Water Res 58:160–168

Panda MK, Panda SK, Singh YD, Jit BP, Behara RK, Dhal NK (2020a) Role of nanoparticles and nanomaterials in drug delivery: an overview. Adv Pharm Biotechnol:247–265

Panda MK, Singh YD, Behera RK, Dhal NK (2020b) Biosynthesis of nanoparticles and their potential application in food and agricultural sector. In: Green nanoparticles. Springer, Cham, pp 213–225

Palaniappan K (2017) An overview of applications of nanotechnology in biofuel production. World Appl Sci J 35:1305–1311

Patumsawad S (2011) 2nd generation biofuels: technical challenge and R&D opportunity in Thailand. J Sustain Energy Environ (Special Issue) 1:47–50

Peña L, Hohn KL, Li J, Sun XS, Wang D et al (2014) Synthesis of propyl-sulfonic acid-functionalized nanoparticles as catalysts for cellobiose hydrolysis. J Biomater Nanobiotechnol 5(4):241

Qiul J, Fan X, Zou H et al (2011) Development of biodiesel from inedible feedstock through various production processes. Review. Chem Technol Fuels Oils 47(2):102

Qu X, Alvarez PJ, Li Q et al (2013) Applications of nanotechnology in water and wastewater treatment. Water Res 47(12):3931–3946

Rai M, Dos Santos JC, Soler MF, Marcelino PRF, Brumano LP, Ingle AP, Gaikwad S, Gade A, Da Silva SS et al (2016a) Strategic role of nanotechnology for production of bioethanol and biodiesel. Nanotechnol Rev 5(2):231–250

Rai M, Ingle AP, Gaikwad S, Gupta I, Gade A, Silvério da Silva S et al (2016b) Nanotechnology based anti-infective to fight microbial intrusions. J Appl Microbiol 120(3):527–542

Ram MSAI, Singh L, Suryanarayana MVS, Alam SI et al (1999) Effect of iron, nickel and cobalt on bacterial activity and dynamics during anaerobic oxidation of organic matter. Water Air Soil Pollut 117(1–4):305–312

Ramsurn H, Gupta RB (2013) Nanotechnology in solar and biofuels. ACS Sustain Chem Eng 1(7):779–797

Rishton SA, Lu Y, Altman RA, Marley AC, Bian XP, Jahnes C, Viswanathan R, Xiao G, Gallagher WJ, Parkin SSP et al (1997) Magnetic tunnel junctions fabricated at tenth-micron dimensions by electron beam lithography. Microelectron Eng 35(1–4):249–252

Sajith V, Sobhan C, Peterson G et al (2010) Experimental investigations on the effects of cerium oxide nanoparticle fuel additives on biodiesel. Adv Mech Eng 2:581407

Santoro C, Arbizzani C, Erable B, Ieropoulos I et al (2017) Microbial fuel cells: from fundamentals to applications. A review. J Power Sources 356:225–244

Sarsar V, Selwal KK, Selwal MK et al (2013) Green synthesis of silver nanoparticles using leaf extract of Mangifera indica and evaluation of their antimicrobial activity. J Microbial Biotechnol Res 3(5):27–32

Sekoai PT, Ouma CNM, Du Preez SP, Modisha P, Engelbrecht N, Bessarabov DG, Ghimire A et al (2019) Application of nanoparticles in biofuels: an overview. Fuel 237:380–397

Serrano E, Rus G, Garcia-Martinez J et al (2009) Nanotechnology for sustainable energy. Renew Sust Energ Rev 13(9):2373–2384

Siddiqi KS, Husen A (2016a) Fabrication of metal nanoparticles from fungi and metal salts: scope and application. Nanoscale Res Lett 11(1):98

Siddiqi KS, Husen A (2016b) Green synthesis, characterization and uses of palladium/platinum nanoparticles. Nanoscale Res Lett 11(1):482

Siddiqi KS, Husen A, Rao RA et al (2018) A review on biosynthesis of silver nanoparticles and their biocidal properties. J Nanobiotechnol 16(1):14

Singh YD (2017) Profiling of selected grass species for bioethanol production (Doctoral dissertation)

Singh YD (2019a) Comprehensive characterization of indigenous lignocellulosic biomass from Northeast India for biofuel production. SN Appl Sci 1(5):458

Singh YD (2019b) Cellulosic bioethanol production from *Eragrostis airoides* Nees grass collected from Northeast India. SN Appl Sci 1(8):889

Singh YD, Satapathy KB (2018) Conversion of lignocellulosic biomass to bioethanol: an overview with a focus on pretreatment. Int J Eng Technol 15:17–43

Singh YD, Mahanta P, Bora U (2017) Comprehensive characterization of lignocellulosic biomass through proximate, ultimate and compositional analysis for bioenergy production. Renew Energy 1(103):490–500

Singh YD, Ningthoujam R, Panda MK, Jena B, Babu PJ, Mishra AK (2021) Insight from nanomaterials and nanotechnology towards covid-19. Sens Int:100099

Slawson RM, Trevors JT, Lee H et al (1992) Silver accumulation and resistance in Pseudomonas stutzeri. Arch Microbiol 158(6):398–404

Stoimenov PK, Klinger RL, Marchin GL, Klabunde KJ et al (2002) Metal oxide nanoparticles as bactericidal agents. Langmuir 18(17):6679–6686

Strzalka R, Schneider D, Eicker U et al (2017) Current status of bioenergy technologies in Germany. Renew Sust Energ Rev 72:801–820

Su L, Shi X, Guo G, Zhao A, Zhao Y et al (2013) Stabilization of sewage sludge in the presence of nanoscale zero-valent iron (nZVI): abatement of odor and improvement of biogas production. J Mater Cycles Waste Manage 15(4):461–468

Su TC, Fang Z, Zhang F, Luo J, Li XK et al (2015) Hydrolysis of selected tropical plant wastes catalyzed by a magnetic carbonaceous acid with microwave. Sci Rep 5:17538

Theivasanthi T, Alagar M (2011) Studies of copper nanoparticles effects on micro-organisms. arXiv preprint arXiv:1110-1372

Upadhyay S, Ganguly K, Palmberg L et al (2015) Wonders of nanotechnology in the treatment for chronic lung diseases. J Nanomed Nanotechnol 6(6):1

Vadlapudi V, Kaladhar DSVGK, Behara M, Sujatha B, Naidu GK et al (2014) Synthesis of green metallic nanoparticles (NPs) and applications. Orient J Chem 29(4):1589–1595

Valko M, Morris H, Cronin MT et al (2005) Metals, toxicity and oxidative stress. Curr Med Chem 12:1161–1208

Verma ML, Barrow CJ, Puri M et al (2013) Nanobiotechnology as a novel paradigm for enzyme immobilisation and stabilisation with potential applications in biodiesel production. Appl Microbiol Biotechnol 97(1):23–39

Xie W, Wang J (2012) Immobilized lipase on magnetic chitosan microspheres for transesterification of soybean oil. Biomass Bioenergy 36:373–380

Xiu ZM, Jin ZH, Li TL, Mahendra S, Lowry GV, Alvarez PJJ et al (2010) Effects of nanoscale zero-valent iron particles on a mixed culture dechlorinating trichloroethylene. Bioresour Technol 101(4):1141–1146

Zhang C, Su H, Baeyens J, Tan T et al (2014) Reviewing the anaerobic digestion of food waste for biogas production. Renew Sustain Energy Rev 38:383–392

Zhang Q, Li N, Goebl J, Lu Z, Yin Y et al (2011) A systematic study of the synthesis of silver nanoplates: is citrate a "magic" reagent? J Am Chem Soc 133(46):18931–18939

Zhang W, Qiao X, Chen J et al (2007) Synthesis of silver nanoparticles—effects of concerned parameters in water/oil micro emulsion. Mater Sci Eng B 142(1):1–15

FSC
www.fsc.org
MIX
Papier aus verantwortungsvollen Quellen
Paper from responsible sources
FSC® C105338

If you have any concerns about our products, you can contact us on
ProductSafety@springernature.com

In case Publisher is established outside the EU, the EU authorized representative is:
Springer Nature Customer Service Center GmbH
Europaplatz 3, 69115 Heidelberg, Germany

Printed by Libri Plureos GmbH
in Hamburg, Germany